LONDON MATHEMATICAL SOCIETY

Managing Editor: Professor M. Reid, Mathematics Institute, University of Warwick, Coventry CV4 7AL,
United Kingdom

The titles below are available from booksellers, or from Cambridge University Press at
www.cambridge.org/mathematics

London Mathematical Society Lecture Note Series: 372

Moonshine: The First Quarter Century and Beyond

Proceedings of a Workshop on the Moonshine Conjectures and Vertex Algebras

Edited by

JAMES LEPOWSKY

Rutgers University, New Jersey

JOHN MCKAY

Concordia University, Montréal

MICHAEL P. TUITE

National University of Ireland, Galway

CAMBRIDGE
UNIVERSITY PRESS

CAMBRIDGE UNIVERSITY PRESS
Cambridge, New York, Melbourne, Madrid, Cape Town, Singapore,
São Paulo, Delhi, Dubai, Tokyo

Cambridge University Press
The Edinburgh Building, Cambridge CB2 8RU, UK

Published in the United States of America by Cambridge University Press, New York

www.cambridge.org
Information on this title: www.cambridge.org/9780521106641

First published 2010

Printed in the United Kingdom at the University Press, Cambridge

A catalogue record for this publication is available from the British Library

Library of Congress Cataloguing in Publication data
Moonshine : the first quarter century and beyond : proceedings of a workshop
on the moonshine conjectures and vertex algebras / [edited by]
James Lepowsky, John McKay, Michael P. Tuite.
p.   cm. – (London mathematical society lecture note series ; 372)
ISBN 978-0-521-10664-1 (pbk.)
1. Modular functions – Congresses.   2. Group theory – Congresses.
I. Lepowsky, J. (James)   II. McKay, John, 1939–
III. Tuite, Michael P.   IV. Title.   V. Series.
QA343.M73   2010
515′.982–dc22
2010000320

ISBN 978-0-521-10664-1 Paperback

Contents

Preface

In 1979 John Conway and Simon Norton published their famous paper entitled "Monstrous Moonshine." This paper greatly expanded on earlier observations and ideas of John McKay and John Thompson and on an observation of Andrew Ogg stimulated by a lecture of Jacques Tits on the conjectured Fischer-Griess Monster sporadic finite simple group. The paper presented a number of conjectures relating the conjugacy classes of the Monster to certain meromorphic modular invariant functions, called Hauptmoduln (= principal moduli), for a particular set of genus zero modular groups. The search for an explanation of this remarkable connection between finite group theory and number theory involved the development and application of many diverse areas of mathematics including vertex (operator) algebras, Borcherds algebras, or generalized Kac-Moody algebras, automorphic forms and elliptic cohomology, together with string theory and conformal field theory in theoretical physics. Robert Griess constructed the Monster; Igor Frenkel, James Lepowsky and Arne Meurman constructed a "Moonshine Module" for the Monster by means of vertex operator theory, proving the McKay-Thompson conjecture; and Richard Borcherds proved the remaining Conway-Norton conjectures for the Moonshine Module, which carries the structure of a vertex operator algebra. Many new problems remain — problems that could not even have been formulated in 1979.

To mark the 25th anniversary of the publication of the Monstrous Moonshine paper, a workshop entitled "Moonshine – the First Quarter Century and Beyond, a Workshop on the Moonshine Conjectures and Vertex Algebras" was hosted by the International Centre for Mathematical Sciences at Heriot-Watt University, Edinburgh from 5th July to 13th July in 2004 (www.icms.org.uk/archive/meetings/2004/moonshine). The aim of this workshop was to review the impact of Monstrous Moonshine on mathematics and theoretical physics and to highlight possible new directions. As part of the workshop, the London Mathematical Society also sponsored a Spitalfield day wherein talks for a more general audience were presented by Robert L. Wilson, Geoffrey Mason and John Conway. The workshop and Spitalfields Day were a tremendous success with many outstanding talks and a high level of interaction and discussion among over fifty researchers who attended the workshop from around the globe. This volume consists of seventeen papers based on most of the talks presented at the meeting. They contain a mixture of expository and current research material (or both) and represent a very good snapshot of the current range of research activity that has stemmed from the Moonshine Conjectures.

The following is a brief overview of the papers in this volume. P. Bantay's paper is concerned with the association of a premodular category to any finite crossed module. In the paper of J. Bruinier and J. Funke, various relationships between the Kudla-Millson and Borcherds lifts from elliptic modular forms to automorphic forms are discussed. G. Buhl's paper is concerned with how C_2-cofiniteness implies the existence of finite generating sets and Poincaré-Birkhoff-Witt-like spanning sets for a vertex operator algebra and its modules. In the paper of A. Degeratu and K. Wendland, a new conjecture (due to John McKay) is examined whereby conjugacy classes of the Monster group (and its centralisers) are related to the Picard groups of bases in certain elliptically fibered Calabi-Yau threefolds. C. Dong and Z. Zhao's paper is a study of the modular properties of trace functions in orbifold theory for $\mathbb{Z}$-graded vertex operator superalgebras. B. Doyon's paper is concerned with sufficient conditions and explicit constructions of twisted modules for vertex operator algebras.

In the paper of J. Duncan, vertex operator algebras are discussed whose automorphism group is a sporadic simple group. T. Gannon's paper reviews the meaning of the hauptmodul property in Monstrous Moonshine and speculates on a new proof of the Moonshine conjectures. In the paper of E. Jurisich, Borcherds' proof of the Conway-Norton conjectures is outlined based on the homology of a certain subalgebra of the monster Lie algebra and the Euler-Poincaré identity. H. Li's paper is a survey on the connection of certain infinite-dimensional Lie algebras, including twisted and untwisted affine Lie algebras, toroidal Lie algebras and quantum torus Lie algebras, with vertex algebras. G. Mason's first paper is based on his Spitalfields Day talk and reviews the relationship between vertex operator algebras and elliptic modular functions and on how this may be generalized to higher genus Riemann surfaces. His second paper discusses orbifold theory for rational vertex operator algebras and its use in understanding aspects of generalized Moonshine.

In the paper of A. Matsuo, K. Nagatomo and A. Tsuchiya, module categories over quasi-finite algebras are described and applied to the representation theory of C_2-cofinite vertex operator algebras. A. Milas's paper is concerned with the Wronskian of the characters of a rational vertex operator algebra. The paper of C. Thomas discusses the relationship between generalized Moonshine for a Monster centraliser group and the elliptic cohomology of the centraliser group's classifying space. M. Tuite's paper concerns permutation orbifolds and their possible application in understanding the genus zero property in Monstrous and generalized Moonshine. Finally, R. A. Wilson's paper is a survey of recent computational results involving the Monster group.

We would like to thank the International Centre for Mathematical Sciences (ICMS) and Heriot-Watt University for hosting the workshop. In particular we thank the director of the ICMS John Toland for his support and Tracey Dart for her outstanding and expert help in running the workshop. We also thank the UK Engineering and Physical Science Research Council (EPSRC) who funded the workshop and the London Mathematical Society (LMS) who sponsored the Spitalfields Day. We would like to thank the other members of the Workshop Organising Committee: Andy Baker, Sasha Ivanov and Viacheslav Nikulin. We are particularly grateful to Andy Baker for all his hard work in organizing the workshop and in leading the application processes with the ICMS, the EPSRC and the LMS. We also thank Chris Eilbeck for his help in running the workshop and for his photographic record (which can be viewed at http://www.ma.hw.ac.uk/chris/icms/moonshine). We are very grateful to the editors and staff at Cambridge University Press for their wonderful and expert help at every stage of the publication process. Finally, we pay tribute to the late Charles Thomas who sadly passed away since the workshop. His presented talk and paper published here are a testament to the originality and beauty of his research.

James Lepowsky, John McKay and Michael Tuite

Schedule of Talks

Monday 5 July

John Conway (Princeton University): Failing to Construct IM!

Robert Griess (University of Michigan): Construction of the Monster

Arne Meurman (Lund University): FLM construction and the McKay-Thompson conjecture (I)

Geoff Mason (University of California at Santa Cruz): Rational orbifold models: past, present and future

Tuesday 6 July

Arne Meurman (Lund University): FLM construction and the McKay-Thompson conjecture (II)

Elizabeth Jurisich (College of Charleston): Borcherds' proof of the Conway-Norton conjecture (I)

Charles Thomas (University of Cambridge): Is there a pattern behind the cohomology of the 26 sporadic simple groups?

Haisheng Li (Rutgers University): Constructions of vertex operator algebras and their modules

Chongying Dong (University of California at Santa Cruz): Orbifolds and generalised Moonshine

Wednesday 7 July

Robert Griess (University of Michigan): Relations between finite groups and vertex operator algebras

Elizabeth Jurisich (College of Charleston): Borcherds' proof of the Conway-Norton conjecture (II)

Terry Gannon (University of Alberta): The algebraic meaning of being a Hauptmodul

Gerald Höhn (Heidelberg University/Kansas State University): Generalized Moonshine for the baby monster

John Duncan (Yale University): Lattice free Moonshine for Co_1

Thursday 8 July

Jan Bruinier (University of Köln): Borcherds products and Hilbert modular surfaces

Chongying Dong (University of California at Santa Cruz): Permutation orbifolds and Moonshine

Antun Milas (Rensselaer Polytechnic Institute/University of Albany-SUNY): Virasoro vertex operator algebras

Friday 9 July

Masahiko Miyamoto (University of Tsukuba): Involutions and the structure of the Moonshine module

Atsushi Matsuo (University of Tokyo): On generalizations of Zhu's algebra and the zero mode algebra

Katrin Wendland (University of Warwick): Friendly giant meets point-like instantons? On a new conjecture by John McKay

Simon Norton (University of Cambridge): Irrational Moonshine

Benjamin Doyon (Rutgers University): Twisted modules for vertex operator algebras

Saturday 10 July
London Mathematical Society Spitalfields Day, held at the William Robertson Building, George Square, Edinburgh

Robert L. Wilson (Rutgers University): Explaining the Moonshine recursions: vertex algebras and free Lie algebras

Geoff Mason (University of California at Santa Cruz): Vertex operators and arithmetic: how a single photon illuminates number theory

John Conway (Princeton University): Symmetry in space

Monday 12 July

Yi-Zhi Huang (Rutgers University): Tensor product theory (I)

Jeff Harvey (University of Chicago): D-brane spectrum in a string theory with monster symmetry

Alex Ryba (City University of New York): Modular Moonshine

Michael Tuite (National University of Ireland Galway): Monstrous Moonshine from orbifolds

Peter Bantay (Eötvös Loránd University): 2D group theory and Moonshine

Nils Scheithauer (University of Heidelberg): Automorphic forms related to Conway's group

Tuesday 13 July

Masahiko Miyamoto (University of Tsukuba): Automorphisms of lattices and VOAs

Yi-Zhi Huang (Rutgers University): Tensor product theory (II)

Chris Cummins (Concordia University): Congruence subgroups of small genus

Robert A. Wilson (University of Birmingham): My three constructions of the Monster

Geoffrey Buhl (Rutgers University): Spanning sets and C_2-cofiniteness of vertex operator algebras

John McKay (Concordia University): A new perspective – and outstanding problems

Characters of Crossed Modules and Premodular Categories

P. Bantay

Institute for Theoretical Physics,
Eötvös University, Budapest

Abstract

A general procedure is presented, associating a premodular category to a finite crossed module, generalizing the representation category of the double of a finite group, and the extent to which the resulting premodular category fails to be modular is explained.

1. Introduction

Modular Tensor Categories (MTCs for short) [1, 16] have attracted much attention in recent years, which is due to the recognition of their importance in both pure mathematics – 3-dimensional topology, representations of Vertex Operator Algebras (VOAs for short) – and theoretical physics (Rational Conformal Field Theory, Topological Field Theories). They are also closely related to Moonshine [4, 7, 10]: a most interesting (and mysterious) example of a Modular Tensor Category, which is responsible for some of the deeper aspects of Moonshine, is the MTC associated to the Moonshine orbifold, i.e. the fixed point VOA of the Moonshine module under the action of the Monster: note that this MTC is yet to be rigorously constructed.

As in every branch of science, a deeper understanding of Modular Tensor Categories requires a suitable supply of examples. Since the work of Huang [12], we know that the module category of any rational VOA (satisfying some technical conditions) is modular, but this important result doesn't help us that much, because VOAs are pretty complicated objects usually hard to deal with. This leads to the desire of associating MTCs to simpler and more accessible algebraic objects. There are several such constructions, a most notable case being the one that associates to a finite group the module category of its

1

(Drinfeld) double [2, 8]. The aim of the present note is to sketch a generalization of this last construction, associating to any (finite) crossed module a premodular category, i.e. a braided tensor category that falls short of being modular. The idea behind is to use 'higher dimensional groups', whose simplest instance are crossed modules [5, 18], for constructing Modular Tensor Categories. In the sequel we will examine to which extent this idea may be put to work.

The plan of the paper is the following. In the next section we'll recall some basic definitions and results about crossed modules. In Section 3 we introduce our basic object of study, the tensor category associated to the crossed module, and discuss some of its properties. Section 4 describes the notion of characters of crossed modules, the main technical tool in our study. Section 5 discusses the premodular structure of the category, and the extent to which it fails to be modular. We conclude by some remarks on the possible applications of the results presented.

We have decided to present only an outline of the theory, without going into detailed proofs, since we felt that their inclusion would not help to clarify the arguments, but could hide the main line of thought. Detailed proofs of all the results to be presented could be supplied by exploiting the close analogy with the character theory of finite groups.

2. Crossed modules

To begin with, let's recall that an action of the group G on the group M is a homomorphism $G \to \mathrm{Aut}\,(M)$ or, what is the same, a map $\mu : M \times G \to M$ such that

(1) $\mu\,(m_1 m_2, g) = \mu\,(m_1, g)\,\mu\,(m_2, g)$ for all $m_1, m_2 \in M$ and $g \in G$;
(2) $\mu\,(m, g_1 g_2) = \mu\,(\mu\,(m, g_1)\,, g_2)$ for all $m \in M$ and $g_1, g_2 \in G$.

As is customary, we'll use the exponential notation $\mu\,(m, g) = m^g$ in the sequel.

A crossed module [5, 11, 18] is nothing but a 4-tuple $\mathcal{X} = (\mathcal{X}_1, \mathcal{X}_2, \mu, \partial)$, where $\mathcal{X}_1, \mathcal{X}_2$ are groups, μ is an action of $\mathcal{X}_1$ on $\mathcal{X}_2$, and $\partial : \mathcal{X}_2 \to \mathcal{X}_1$ is a homomorphism, called the boundary map, that satisfies

XMod1: $\partial\,(m^g) = g^{-1}\,(\partial m)\,g$ for all $m \in \mathcal{X}_2$ and $g \in \mathcal{X}_1$;
XMod2: $m^{\partial n} = n^{-1} m n$ for all $m, n \in \mathcal{X}_2$.

A crossed module is finite if both $\mathcal{X}_1$ and $\mathcal{X}_2$ are finite groups. Examples of crossed modules abound in algebra and topology, let's just cite two, coming from group theory, that will guide our investigations later.

Example 1. For a group G, we'll denote by $\mathcal{R}G$ the crossed module $(G, \mathbf{1}, \mu, \partial)$, where $\mathbf{1}$ denotes the trivial subgroup of G, i.e. $\mathbf{1} = \{1\}$, and both the action μ and the boundary map ∂ are trivial.

Example 2. If G is a group, $\mathcal{D}G$ is the crossed module $(G, G, \mu, \mathbf{id})$, where μ is the conjugation action, i.e. $\mu(m, g) = g^{-1}mg$, and $\mathbf{id} : g \mapsto g$ is the trivial map.

A standard consequence of the defining properties of a crossed module is that $K = \ker \partial$ is a central subgroup of $\mathcal{X}_2$, $I = \operatorname{im} \partial$ is a normal subgroup of $\mathcal{X}_1$, and one has an exact sequence

$$\mathbf{1} \to K \to \mathcal{X}_2 \to \mathcal{X}_1 \to C \to \mathbf{1} \tag{1}$$

where $C = \mathcal{X}_1/I$ is the cokernel of ∂ [5]. In particular, $|\mathcal{X}_2| \, |C| = |K| \, |\mathcal{X}_1|$ for a finite crossed module.

Finally, a morphism $\phi : \mathcal{X} \to \mathcal{Y}$ between the crossed modules $\mathcal{X} = (\mathcal{X}_1, \mathcal{X}_2, \mu_{\mathcal{X}}, \partial_{\mathcal{X}})$ and $\mathcal{Y} = (\mathcal{Y}_1, \mathcal{Y}_2, \mu_{\mathcal{Y}}, \partial_{\mathcal{Y}})$ is a pair (ϕ_1, ϕ_2), where $\phi_i : \mathcal{X}_i \to \mathcal{Y}_i$ are group homomorphisms for $i = 1, 2$, and the following relations hold:

$$\partial_{\mathcal{Y}} \circ \phi_2 = \phi_1 \circ \partial_{\mathcal{X}}$$
$$\mu_{\mathcal{Y}} \circ (\phi_2 \times \phi_1) = \phi_2 \circ \mu_{\mathcal{X}} \,,$$

which simply express the commutativity of the diagrams

$$
\begin{array}{ccc}
\mathcal{X}_2 & \xrightarrow{\partial_{\mathcal{X}}} & \mathcal{X}_1 \\
{\scriptstyle \phi_2}\downarrow & & \downarrow{\scriptstyle \phi_1} \\
\mathcal{Y}_2 & \xrightarrow{\partial_{\mathcal{Y}}} & \mathcal{Y}_1
\end{array}
\qquad
\begin{array}{ccc}
\mathcal{X}_2 \times \mathcal{X}_1 & \xrightarrow{\mu_{\mathcal{X}}} & \mathcal{X}_2 \\
{\scriptstyle \phi_2 \times \phi_1}\downarrow & & \downarrow{\scriptstyle \phi_2} \\
\mathcal{Y}_2 \times \mathcal{Y}_1 & \xrightarrow{\mu_{\mathcal{Y}}} & \mathcal{Y}_2
\end{array}
$$

3. The category

To any finite crossed module $\mathcal{X} = (\mathcal{X}_1, \mathcal{X}_2, \mu, \partial)$ we'll associate a braided tensor category $\mathscr{M}(\mathcal{X})$, which falls short of being modular. Let's begin by describing the objects and morphisms of $\mathscr{M}(\mathcal{X})$. Here and in the sequel, we use the notation

$$\delta(x, y) = \begin{cases} 1 & \text{if } x = y, \\ 0 & \text{otherwise.} \end{cases}$$

An object of $\mathscr{M}(\mathcal{X})$ is a triple (V, P, Q), where V is a complex linear space, while P and Q are maps $P : \mathcal{X}_2 \to \operatorname{End}(V)$ and $Q : \mathcal{X}_1 \to \operatorname{GL}(V)$ such that for all $g, h \in \mathcal{X}_1$ and $m, n \in \mathcal{X}_2$

$$P(m)\,P(n) = \delta(m,n)\,P(m) \tag{2}$$

$$\sum_{m \in \mathcal{X}_2} P(m) = \mathbf{id}_V \tag{3}$$

$$Q(g)\,Q(h) = Q(gh) \tag{4}$$

$$P(m)\,Q(g) = Q(g)\,P\!\left(m^g\right) \tag{5}$$

By the dimension of an object (V, P, Q) we'll mean the dimension of the linear space V. A morphism $\phi : (V_1, P_1, Q_1) \to (V_2, P_2, Q_2)$ between two objects of $\mathcal{M}(\mathcal{X})$ is a linear map $\phi : V_1 \to V_2$ such that $\phi \circ P_1(m) = P_2(m) \circ \phi$ for all $m \in \mathcal{X}_2$ and, $\phi \circ Q_1(g) = Q_2(g) \circ \phi$ for all $g \in \mathcal{X}_1$. In general, we won't distinguish isomorphic objects of $\mathcal{M}(\mathcal{X})$.

Let's look at a couple of illustrating examples of objects of $\mathcal{M}(\mathcal{X})$ for a finite crossed module $\mathcal{X} = (\mathcal{X}_1, \mathcal{X}_2, \mu, \partial)$.

Example 3. The triple $\mathbf{1} = (V, P, Q)$, with $V = \mathbb{C}$, $P(m) = \delta(m, 1)\,\mathbf{id}_V$ and $Q(g) = \mathbf{id}_V$, is a one dimensional object of $\mathcal{M}(\mathcal{X})$, that we'll call the trivial object.

Example 4. The triple $\mathbf{R} = (V, P, Q)$, with $V = \mathbb{C}(\mathcal{X}_1 \times \mathcal{X}_2)$ and $P(m)\,\phi :$ $(x, y) \mapsto \delta(m, y^x)\,\phi(x, y)$, $Q(g)\,\phi : (x, y) \mapsto \phi(xg, y)$ for $\phi \in V$ and $(x, y) \in \mathcal{X}_1 \times \mathcal{X}_2$, is an object of $\mathcal{M}(\mathcal{X})$, that we'll call the regular object. Clearly, $\dim \mathbf{R} = |\mathcal{X}_1|\,|\mathcal{X}_2|$.

Example 5. The triple $\mathbf{0} = (V, P, Q)$, with $V = \mathbb{C}(K \times C)$ (remember the notations $K = \ker \partial$, $I = \operatorname{im} \partial$ and $C = \operatorname{coker} \partial = \mathcal{X}_1/I$ from Eq.1) and $P(m)\,\phi : (x, Iy) \mapsto \delta(m, x^y)\,\phi(x, Iy)$, $Q(g)\,\phi : (x, Iy) \mapsto \phi(x, Iyg)$ for $\phi \in V$, is an object of $\mathcal{M}(\mathcal{X})$, that we'll call the vacuum object.

Note that the above objects, which exist for any finite crossed module $\mathcal{X}$, need not be distinct. For example, in the category $\mathcal{M}(\mathcal{R}G)$ (see Example 1) one has $\mathbf{0} = \mathbf{R}$, while in $\mathcal{M}(\mathcal{D}G)$ one has $\mathbf{0} = \mathbf{1}$.

Given an object (V, P, Q) of $\mathcal{M}(\mathcal{X})$, a linear subspace $W < V$ is invariant if $P(m)W \subset W$ and $Q(g)W \subset W$ for all $m \in \mathcal{X}_2$ and $g \in \mathcal{X}_1$. An object (V, P, Q) is reducible if it has a nontrivial invariant subspace, otherwise it is irreducible. For a finite crossed module $\mathcal{X}$ there are only finitely many isomorphism classes of irreducible objects in $\mathcal{M}(\mathcal{X})$, which follows from the following generalization of Burnside's classical theorem [13, 15]:

$$\sum_{p \in \operatorname{Irr}(\mathcal{X})} d_p^2 = |\mathcal{X}_1|\,|\mathcal{X}_2|, \tag{6}$$

where we denote by $\operatorname{Irr}(\mathcal{X})$ the set of (isomorphism classes of) irreducible objects of $\mathcal{M}(\mathcal{X})$, and d_p denotes the dimension of the irreducible $p \in \operatorname{Irr}(\mathcal{X})$.

The notion of direct sum of objects of $\mathcal{M}(\mathcal{X})$ is the obvious one:

$$(V_1, P_1, Q_1) \oplus (V_2, P_2, Q_2) = (V_1 \oplus V_2, P_1 \oplus P_2, Q_1 \oplus Q_2). \quad (7)$$

The analogue of Maschke's theorem states that, for a finite crossed module $\mathcal{X}$, any object of $\mathcal{M}(\mathcal{X})$ decomposes uniquely (up to ordering) into a direct sum of irreducible objects.

The tensor product of the objects (V_1, P_1, Q_1) and (V_2, P_2, Q_2) is the triple $(V_1 \otimes V_2, P_{12}, Q_{12})$, where $P_{12} : m \mapsto \sum_{n \in \mathcal{X}_2} P_1(n) \otimes P_2\left(n^{-1}m\right)$ and $Q_{12} : g \mapsto Q_1(g) \otimes Q_2(g)$. The category $\mathcal{M}(\mathcal{X})$ may be shown to be a monoidal tensor category, which in general fails to be symmetric, but it is always braided, the braiding being provided by the map

$$R_{12} : V_1 \otimes V_2 \rightarrow V_2 \otimes V_1$$
$$v_1 \otimes v_2 \mapsto \sum_{m \in \mathcal{X}_2} Q_2(\partial m) v_2 \otimes P_1(m) v_1$$

At this point it is worthwhile to take a look the category $\mathcal{M}(\mathcal{X})$ for the two canonical examples of crossed modules considered in Section 2, namely $\mathcal{R}G$ and $\mathcal{D}G$ for a finite group G. In the first case, since $\mathcal{X}_2 = \mathbf{1}$, the map $P : \mathcal{X}_2 \rightarrow \mathrm{End}(V)$ is trivial: $P(m) = \delta(m, 1)\,\mathbf{id}$, while the map $Q : \mathcal{X}_1 \rightarrow \mathrm{Aut}(V)$ provides a representation of the finite group $\mathcal{X}_1 = G$. Thus, for $\mathcal{X} = \mathcal{R}G$ the category $\mathcal{M}(\mathcal{X})$ is nothing but the category of representations of the finite group G. On the other hand, for $\mathcal{X} = \mathcal{D}G$ the map P is no longer trivial, and a little thought reveals that in this case $\mathcal{M}(\mathcal{X})$ is just the module category of the (Drinfeld) double of the finite group G [2, 3, 8]. It is known that this last tensor category is modular, and describes the properties of the so-called holomorphic G-orbifold models [9]. So, from this point of view, the category $\mathcal{M}(\mathcal{X})$ may be viewed as a common generalization of the module categories of a finite group and of its double.

4. Characters

The notion of group characters is an extremely powerful tool in the study of group representations [13]. Not only do characters distinguish inequivalent representations, but they prove invaluable in actual computations, e.g. the decomposition into irreducibles, the computation of tensor products, etc. As it turns out, a close analogue of group characters exists for the (isomorphism classes of) objects of $\mathcal{M}(\mathcal{X})$. Namely, the character of an object (V, P, Q) of $\mathcal{M}(\mathcal{X})$ is the complex valued function $\psi : \mathcal{X}_2 \times \mathcal{X}_1 \rightarrow \mathbb{C}$ given by

$$\psi(m, g) = \mathrm{Tr}_V(P(m)\,Q(g)). \quad (8)$$

Clearly, characters of isomorphic objects are equal, and it follows from the orthogonality relations to be presented a bit later that characters distinguish inequivalent objects of $\mathcal{M}(\mathcal{X})$. The character ψ of an object of $\mathcal{M}(\mathcal{X})$ is a class function of the crossed module $\mathcal{X}$, i.e. a complex valued function ψ : $\mathcal{X}_2 \times \mathcal{X}_1 \to \mathbb{C}$ that satisfies

(1) $\psi(m, g) = 0$ unless $m^g = m$, for $m \in \mathcal{X}_2$ and $g \in \mathcal{X}_1$;
(2) $\psi\left(m^h, h^{-1}gh\right) = \psi(m, g)$ for all $m \in \mathcal{X}_2$ and $g, h \in \mathcal{X}_1$.

The set of class functions of a finite crossed module $\mathcal{X}$ form a finite dimensional linear space $\mathscr{Cl}(\mathcal{X})$, which carries the natural scalar product

$$\langle \psi_1, \psi_2 \rangle = \frac{1}{|\mathcal{X}_1|} \sum_{m \in \mathcal{X}_2, g \in \mathcal{X}_1} \overline{\psi_1(m, g)} \psi_2(m, g), \tag{9}$$

where $\psi_1, \psi_2 \in \mathscr{Cl}(\mathcal{X})$, and the bar denotes complex conjugation[1].

Characters behave well under direct sums and tensor products: the character of a direct sum is just the (pointwise) sum of the characters of the summands, while the character of a tensor product is given by the formula

$$\psi_{A \otimes B}(m, g) = \sum_{n \in \mathcal{X}_2} \psi_A(n, g) \psi_B\left(n^{-1}m, g\right), \tag{10}$$

if ψ_A, ψ_B are the characters of the factors.

Irreducible characters, i.e. the characters of the irreducible objects of $\mathcal{M}(\mathcal{X})$, play a distinguished role, since any character may be written (uniquely) as a linear combination of irreducible ones with non-negative integer coefficients. The basic result about irreducible characters is the following analogue of the generalized orthogonality relations for group characters [13, 15]:

$$\frac{1}{|\mathcal{X}_1|} \sum_{h \in \mathcal{X}_1} \psi_p(m, h) \psi_q\left(m, h^{-1}g\right) = \frac{1}{d_p} \delta_{pq} \psi_p(m, g) \tag{11}$$

for $p, q \in \mathrm{Irr}(\mathcal{X})$, where

$$d_p = \sum_{m \in \mathcal{X}_2} \psi_p(m, 1) \tag{12}$$

denotes the dimension of the irreducible p. From this one can deduce at once that the characters of the irreducible representations form an orthonormal basis in the space $\mathscr{Cl}(\mathcal{X})$ of class functions, and that they also satisfy the second orthogonality relations

1 Note that for the character ψ of an object of $\mathcal{M}(\mathcal{X})$ one has $\overline{\psi(m, g)} = \psi\left(m, g^{-1}\right)$.

$$\sum_{p\in\mathrm{Irr}(\mathcal{X})} \psi_p\,(m,g)\,\psi_p\,(n,h) = \sum_{z\in\mathcal{X}_1} \delta\left(n,m^z\right)\delta\left(h^{-1},z^{-1}gz\right). \tag{13}$$

Note that the irreducible characters ψ_p may be computed explicitly for any finite crossed module $\mathcal{X}$, e.g. one has $\psi_{\underline{1}}\,(m,g) = \delta\,(m,1)$ for the identity object $\underline{1}$ of $\mathscr{M}(\mathcal{X})$ (cf. Example 3).

Using the orthogonality relations, one may express the fusion rule coefficient N_{pq}^r, i.e. the multiplicity of the irreducible $r \in \mathrm{Irr}\,(\mathcal{X})$ in the tensor product of the irreducibles p and q, through the formula

$$N_{pq}^r = \frac{1}{|\mathcal{X}_1|}\sum_{m,n\in\mathcal{X}_2}\sum_{g\in\mathcal{X}_1} \psi_p\,(m,g)\,\psi_q\,(n,g)\,\overline{\psi_r\,(mn,g)}. \tag{14}$$

To each irreducible $p \in \mathrm{Irr}\,(\mathcal{X})$ one may associate the complex number

$$\omega_p = \frac{1}{d_p}\sum_{m\in\mathcal{X}_2}\psi_p\,(m,\partial m), \tag{15}$$

(remember that d_p denotes the dimension of the irreducible p), which turns out to be a root of unity (of order dividing the exponent of $I = \mathrm{im}\,\partial$), and one may show that[2]

$$\psi_p\,(m,g\partial m) = \omega_p\psi_p\,(m,g), \tag{16}$$

for all $m \in \mathcal{X}_2, g \in \mathcal{X}_1$. Combined with the orthogonality relations Eq.(11), this leads to (remember that $K = \ker\partial$)

$$\sum_{p\in\mathrm{Irr}(\mathcal{X})} d_p^2\omega_p^{-1} = |\mathcal{X}_1|\,|K|, \tag{17}$$

to be compared with Eq.(6).

To conclude, let's just note that the close analogy with ordinary group characters goes much further, e.g. one may introduce the Frobenius-Schur indicator

$$\nu_p = \frac{1}{|\mathcal{X}_1|}\sum_{m\in\mathcal{X}_2,g\in\mathcal{X}_1} \delta\left(m^g,m^{-1}\right)\psi_p\left(m,g^2\right). \tag{18}$$

of the irreducible character ψ_p, and show that ν_p may take only the values 0 and ± 1, in perfect parallel with the classical case [13]. Of course, this is related to the fact that ordinary characters of the finite group G are nothing but the characters of the crossed module $\mathcal{R}G$ of Example 1: from this perspective, ordinary character theory of groups is just a special case of the more general theory presented in this note.

---

2 This follows from Schur's lemma, upon noting that $\sum_{m\in\mathcal{X}_2} P\,(m)\,Q\,(\partial m)$ commutes with $P\,(n)\,Q\,(g)$ for all $(n,g) \in \mathcal{X}_2 \times \mathcal{X}_1$.

5. The S matrix and the structure of the vacuum

Up to now, we have seen the close parallel between the structure of the category $\mathcal{M}(\mathcal{X})$ and the representation category of a finite group. We now turn to describe the premodular structure, related to the existence of the so-called S matrix. This is a square matrix, with rows and columns labeled by the irreducibles of $\mathcal{M}(\mathcal{X})$, and with matrix elements

$$S_{pq} = \frac{1}{|\mathcal{X}|} \sum_{m,n \in \mathcal{X}_2} \overline{\psi_p(m, \partial n)\, \psi_q(n, \partial m)} \tag{19}$$

for $p, q \in \mathrm{Irr}(\mathcal{X})$, where $|\mathcal{X}| = |\mathcal{X}_2|\,|C| = |K|\,|\mathcal{X}_1|$ (remember Eq.(1)). This matrix is obviously symmetric, and a simple computation shows that

$$S_{\mathbf{1}p} = \frac{d_p}{|\mathcal{X}|} > 0, \tag{20}$$

where $\mathbf{1}$ denotes the identity object of $\mathcal{M}(\mathcal{X})$ (cf. Example 3). The definition of S is motivated by the case of group doubles $\mathcal{D}G$, when it describes the modular properties of the corresponding holomorphic G-orbifolds.

A most important feature of the above S matrix is its relation to the fusion rule coefficients N_{pq}^r appearing in Eq.(14), for one may show that

$$\sum_{r \in \mathrm{Irr}(\mathcal{X})} N_{pq}^r S_{rs} = \frac{S_{ps} S_{qs}}{S_{\mathbf{1}s}} \tag{21}$$

holds, which is an avatar of Verlinde's celebrated formula [17]. A closely related result states that

$$\sum_{r \in \mathrm{Irr}(\mathcal{X})} N_{pq}^r \omega_r^{-1} S_{\mathbf{1}r} = \omega_p^{-1} \omega_q^{-1} S_{pq}, \tag{22}$$

where the roots of unity ω_p are given by Eq.(15). But this is not the end of the story since, upon introducing the diagonal matrix $T_{pq} = \omega_p \delta_{pq}$, one may show that

$$STS = T^{-1}ST^{-1}. \tag{23}$$

Should S satisfy the relation $S^4 = 1$, Eq.(23) would mean that the matrices S and T give a finite dimensional representation of the modular group $\mathrm{SL}_2(\mathbb{Z})$, which conforms with Verlinde's theorem [14, 17], i.e.

(1) T is diagonal and of finite order;
(2) S is symmetric;
(3) Verlinde's formula Eq.(21) holds.

Should this be the case, $\mathcal{M}(\mathcal{X})$ would be a Modular Tensor Category. As it turns out, in general this is not the case, because the matrix S of Eq.(19) does only satisfy the weaker property

$$S^8 = S^4. \tag{24}$$

This means that S is not necessarily invertible: it might have a nontrivial kernel. This is the extent to which $\mathcal{M}(\mathcal{X})$ fails to be modular in general.

The lack of invertibility of S is related to the reducibility of the vacuum object $\underline{0}$ (cf. Example 5). Denoting by μ_p the multiplicity of the irreducible p in $\underline{0}$, and by $D = |C| |K|$ the dimension of $\underline{0}$, one may show that

$$\mu_p = D \left[S^2 \right]_{\underline{1}p,} \tag{25}$$

and that $\mu_p > 0$ if and only if there exists an α such that

$$S_{pq} = \alpha S_{\underline{1}q} \quad \text{for all} \quad q \in \text{Irr} (\mathcal{X}) , \tag{26}$$

in which case $\alpha = \mu_p = d_p$ and $\omega_p = 1$. In other words, the irreducible objects of $\mathcal{M}(\mathcal{X})$ that satisfy Eq.(26) for some constant α are precisely the irreducible constituents of the vacuum $\underline{0}$. The invertibility of S requires that the only such object is the identity $\underline{1}$, and this condition may be shown to be equivalent to the bijectivity of the boundary map ∂, which in turn is equivalent to $\mathcal{X}$ being isomorphic to $\mathcal{D}G$ for some finite group G. Note also that for $\mathcal{X} = \mathcal{R}G$ every irreducible of $\mathcal{M}(\mathcal{X})$ satisfies Eq.(26), since in this case $\underline{0} = \mathbf{R}$.

Finally, we note that while $\mathcal{M}(\mathcal{X})$ fails to be modular in case ∂ is not bijective, it can nevertheless be turned into an MTC! Indeed, according to the modularizability criterion of Bruguieres [6], one can associate a well-defined MTC (unique up to isomorphism) to any premodular category in which Eq.(26) implies $\omega_p = 1$ and $\alpha = d_p$. But we won't pursue this line any further in the present note, and leave the construction of the corresponding MTC to some future work.

6. Discussion

As we have sketched in the previous sections, to any finite crossed module $\mathcal{X}$ one may associate a premodular category $\mathcal{M}(\mathcal{X})$. In special instances this construction gives back the module category of a finite group or that of its (Drinfeld) double, but in general one gets new premodular categories, which are very close to being modular: they satisfy the modularizability criterion of [6], i.e. they can be turned into a Modular Tensor Category. This opens the way to the construction of a huge number of Modular Tensor Categories starting from (relatively) simple algebraic structures.

As stressed before, the category $\mathcal{M}(\mathcal{X})$ may be viewed as a generalization of the module category of the double of a finite group G, which describes the properties of holomorphic G-orbifolds [2, 3, 8]. This leads to the speculation that for a general crossed module $\mathcal{X}$ the category $\mathcal{M}(\mathcal{X})$, or more precisely its modularisation, should describe the properties of some 'generalized' holomorphic orbifold related to $\mathcal{X}$. To find out whether this vague idea may be made to work seems to be a rewarding task.

Acknowledgments: This work was supported by research grants OTKA T047041, T037674, T043582, TS044839, the János Bolyai Research Scholarship of the Hungarian Academy of Sciences, and EC Marie Curie RTN, MRTN-CT-2004-512194.

References

[1] B. Bakalov and A. A. Kirillov, *Lectures on Tensor Categories and Modular Functors*, University Lecture Series, vol. 21, AMS, Providence, 2001.

[2] P. Bantay, *Orbifolds and Hopf algebras*, Phys. Lett. **B245** (1990), 477–479.

[3] _____, *Orbifolds, Hopf algebras and the Moonshine*, Lett. Math. Phys. **22** (1991), 187–194.

[4] R. E. Borcherds, *Monstrous Moonshine and monstrous Lie superalgebras*, Invent. Math. **109** (1992), 405.

[5] R. Brown, *Higher dimensional group theory*, Low-dimensional topology (R. Brown and T. L. Thickstun, eds.), London Math. Soc. Lecture Note Series, no. 48.

[6] A. Bruguieres, *Categories premodulaires, modularisations et invariants des varietes de dimension 3*, Math. Ann. **316** (2000), 215.

[7] J. H. Conway and S. P. Norton, *Monstrous Moonshine*, Bull. London Math. Soc. **11** (1979), 308.

[8] R. Dijkgraaf, V. Pasquier, and P. Roche, Nucl. Phys. Proc. Suppl. **18B** (1990), 60.

[9] R. Dijkgraaf, C. Vafa, E. Verlinde, and H. Verlinde, *The operator algebra of orbifold models*, Commun. Math. Phys. **123** (1989), 485.

[10] T. Gannon, *Monstrous Moonshine: the first twenty-five years*, Bull. London Math. Soc. **38** (2005), 1.

[11] N. D. Gilbert, *Derivations, automorphisms and crossed modules*, Comm. in Algebra **18** (1990), 2703.

[12] Y.-Z. Huang, *Vertex Operator Algebras, the Verlinde conjecture, and Modular Tensor Categories*, Proc.Natl.Acad.Sci. **102** (2005), 5352.

[13] I. M. Isaacs, *Character theory of finite groups*, Pure and Applied Mathematics, vol. 22, Academic Press, New York, 1977.

[14] G. Moore and N. Seiberg, *Classical and quantum Conformal Field Theory*, Commun. Math. Phys. **123** (1989), 177.

[15] J. P. Serre, *Linear representations of finite groups*, GTM, vol. 42, Springer, Berlin-New York, 1977.

[16] V. G. Turaev, *Quantum invariants of knots and 3-manifolds*, Studies in Mathematics, vol. 18, de Gruyter, Berlin, 1994.

[17] E. Verlinde, *Fusion rules and modular transformations in 2d-conformal field theory*, Nucl. Phys. **B300** (1988), 360.

[18] J. H. C. Whitehead, *On operators in relative homotopy groups*, Ann. of Math. (1948), no. 49, 610.

On the Injectivity of the Kudla-Millson Lift and Surjectivity of the Borcherds Lift

Jan Hendrik Bruinier

Mathematisches Institut, Universität zu Köln,
Weyertal 86–90, D-50931 Köln, Germany

Jens Funke*

Department of Mathematical Sciences, New Mexico State University,
P.O.Box 30001, 3MB, Las Cruces, NM 88003, USA

Abstract

We consider the Kudla-Millson lift from elliptic modular forms of weight $(p + q)/2$ to closed q-forms on locally symmetric spaces corresponding to the orthogonal group $O(p, q)$. We study the L^2-norm of the lift following the Rallis inner product formula. We compute the contribution at the Archimedian place. For locally symmetric spaces associated to even unimodular lattices, we obtain an explicit formula for the L^2-norm of the lift, which often implies that the lift is injective. For $O(p, 2)$ we discuss how such injectivity results imply the surjectivity of the Borcherds lift.

1. Introduction

In previous work [8], we studied the Kudla-Millson theta lift (see e.g. [19]) and Borcherds' singular theta lift (e.g. [3, 6]) and established a duality statement between these two lifts. Both of these lifts have played a significant role in the study of certain cycles in locally symmetric spaces and Shimura varieties of orthogonal type. In this paper, we study the injectivity of the Kudla-Millson theta lift, and revisit part of the material of [6] from the viewpoint of [8], to obtain surjectivity results for the Borcherds lift. Moreover, we provide evidence for the following principle: The vanishing of the standard L-function of a cusp form of weight $1 + p/2$ at $s_0 = p/2$ corresponds to the existence of a certain "exceptional automorphic product" on $O(p, 2)$ (see Theorem 1.8).

We now describe the content of this paper in more detail. We begin by recalling the Kudla-Millson lift in a setting which is convenient for the application to the Borcherds lift. Let (V, Q) be a non-degenerate rational quadratic space of signature (p, q). We write $(\cdot, \cdot)$ for the bilinear form corresponding to the

* Partially supported by NSF grant DMS-0305448.

quadratic form Q. We write r for the Witt index of V, i.e., the dimension of a rational maximal isotropic subspace. Throughout we assume for simplicity that the dimension $m = p + q$ of V is even. We realize the symmetric space D associated to V as the Grassmannian of oriented negative q-planes in $V(\mathbb{R})$.

Let $L \subset V$ be an even lattice of level N, and write $L^{\#}$ for the dual lattice. The quadratic form on L induces a non-degenerate $\mathbb{Q}/\mathbb{Z}$-valued quadratic form on the discriminant group $L^{\#}/L$. Recall that the Weil representation ρ_L of the quadratic module $(L^{\#}/L, Q)$ is a unitary representation of $SL_2(\mathbb{Z})$ on the group ring $\mathbb{C}[L^{\#}/L]$, which can be defined as follows [3], [6]. If $(\mathfrak{e}_\gamma)_{\gamma \in L^{\#}/L}$ denotes the standard basis of $\mathbb{C}[L^{\#}/L]$, then ρ_L is given by the action of the generators $T = \left(\begin{smallmatrix} 1 & 1 \\ 0 & 1 \end{smallmatrix}\right)$ and $S = \left(\begin{smallmatrix} 0 & -1 \\ 1 & 0 \end{smallmatrix}\right)$ of $SL_2(\mathbb{Z})$ by

$$\rho_L(T)(\mathfrak{e}_\gamma) = e(\gamma^2/2)\mathfrak{e}_\gamma,$$

$$\rho_L(S)(\mathfrak{e}_\gamma) = \frac{e(-(p-q)/8)}{\sqrt{|L^{\#}/L|}} \sum_{\delta \in L^{\#}/L} e(-(\gamma, \delta))\mathfrak{e}_\delta,$$

where $e(w) := e^{2\pi i w}$. This representation factors through the group $SL_2(\mathbb{Z}/N\mathbb{Z})$.

Let $\Gamma \subset O(L)$ be a torsion-free subgroup of finite index which acts trivially on $L^{\#}/L$. Then

$$X = \Gamma \backslash D$$

is a real analytic manifold. For $x \in L^{\#}$ with $Q(x) > 0$, we let

$$D_x = \{z \in D; \ z \perp x\}.$$

Note that D_x is a subsymmetric space attached to the orthogonal group H_x, the stabilizer of x in H. Put $\Gamma_x = \Gamma \cap H_x$. The quotient

$$Z(x) = \Gamma_x \backslash D_x \longrightarrow X$$

defines a (in general relative) cycle in X. For $h \in L^{\#}/L$ and $n \in \mathbb{Q}$, the group Γ acts on $L_{h,n} = \{x \in L + h; \ Q(x) = n\}$ with finitely many orbits, and we define the composite cycle

$$Z(h, n) = \sum_{x \in \Gamma \backslash L_{h,n}} Z(x).$$

Kudla and Millson constructed Poincaré dual forms for such cycles by means of the Weil representation, see e.g. [19]. They constructed a Schwartz form $\varphi_{KM} \in [\mathcal{S}(V(\mathbb{R})) \otimes \mathcal{Z}^q(D)]^{O(V)(\mathbb{R})}$ on $V(\mathbb{R})$ taking values in $\mathcal{Z}^q(D)$, the closed differential q-forms on D. Let ω_∞ be the Schrödinger model of the Weil representation of $SL_2(\mathbb{R})$ acting on the space of Schwartz functions $\mathcal{S}(V(\mathbb{R}))$, associated to the standard additive character. We obtain a $\mathbb{C}[L^{\#}/L]$-valued theta function on the upper half plane $\mathbb{H}$ by putting

$$\Theta(\tau, z, \varphi_{KM}) = v^{-m/4} \sum_{h \in L^{\#}/L} \sum_{x \in L+h} (\omega_{\infty}(g_{\tau})\varphi_{KM})(x, z)\mathfrak{e}_h.$$

Here $\tau = u + iv \in \mathbb{H}$ and $g_{\tau} = \begin{pmatrix} 1 & u \\ 0 & 1 \end{pmatrix} \begin{pmatrix} \sqrt{v} & 0 \\ 0 & \sqrt{v}^{-1} \end{pmatrix} \in SL_2(\mathbb{R})$ is the standard element moving the base point $i \in \mathbb{H}$ to τ. In the variable τ, this theta function transforms as a (non-holomorphic) modular form of weight $\kappa = m/2$ for $SL_2(\mathbb{Z})$ of type ρ_L. In the variable z, it defines a closed q-form on X. Kudla and Millson showed that the Fourier coefficient at $e^{2\pi i n \tau}\mathfrak{e}_h$ is a Poincaré dual form for the cycle $Z(h, n)$.

Let $S_{\kappa,L}$ denote the space of $\mathbb{C}[L^{\#}/L]$-valued cusp forms of weight κ and type ρ_L for the group $SL_2(\mathbb{Z})$. We define a lifting $\Lambda : S_{\kappa,L} \to \mathcal{Z}^q(X)$ by the theta integral

$$f \mapsto \Lambda(f) = \int_{SL_2(\mathbb{Z}) \backslash \mathbb{H}} \langle f(\tau), \Theta(\tau, z, \varphi_{KM}) \rangle \frac{du\, dv}{v^2}, \qquad (1.1)$$

where $\langle \cdot, \cdot \rangle$ denotes the standard scalar product on $\mathbb{C}[L^{\#}/L]$.

In the present paper, we consider the question whether Λ is injective. We compute the L^2-norm of the differential form $\Lambda(f)$ in the sense of Riemann geometry by means of the Rallis inner product formula [27]. First, using the see-saw

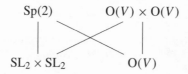

and the Siegel-Weil formula (see e.g. [20], [21], [23], [30]), the inner product can be expressed as a convolution integral of f against the restriction of a genus 2 Eisenstein series to the diagonal (see Proposition 4.7).

Such convolution integrals can be evaluated by means of the doubling method, see e.g. [5], [13], [25], [27]. If f is a Hecke eigenform of level N, one obtains a special value of the partial standard L-function of f (where the Euler factors corresponding to the primes dividing the level N and ∞ are omitted) times a product of "bad" local factors corresponding to the primes dividing N and ∞. If $m > 4$, then, by the Euler product expansion, the special value of the partial standard L-function is non-zero. Therefore the lift $\Lambda(f)$ vanishes precisely if at least one of the "bad" local factors vanishes. By the analysis of the present paper we determine the local factor at infinity.

In the special case where L is even and unimodular, the level of L is $N = 1$, so that ∞ is the only "bad" place. The space $S_{\kappa,L}$ is equal to the space $S_\kappa(\Gamma(1))$ of scalar valued cusp forms of weight κ for $\Gamma(1) = \mathrm{SL}_2(\mathbb{Z})$. We obtain the following explicit formula for the L^2-norm of the lift (see Theorem 4.9):

Theorem 1.1. *Assume that $m > 3 + r$, where r is the Witt index of V. Let $f \in S_\kappa(\Gamma(1))$ be a Hecke eigenform, and write $\|f\|_2^2$ for its Petersson norm, and $D_f(s)$ for its standard L-function. Then $\Lambda(f)$ is square integrable and*

$$\frac{\|\Lambda(f)\|_2^2}{\|f\|_2^2} = C \cdot \frac{D_f(m/2 - 1)}{\zeta(m/2)\zeta(m - 2)},$$

where $C = C(p, q)$ is an explicit real constant, which does not depend on f. The constant C vanishes if and only if $p = 1$.

Corollary 1.2. *Assume that $m > \max(4, 3 + r)$ and that L is even unimodular. When $p \neq 1$, the theta lift Λ is injective. When $p = 1$, the lift vanishes identically.*

It would be interesting to compute the bad local factors at finite primes (or at least to show their non-vanishing) as well. However, in our setting, this requires first a Hecke theory for vector valued modular forms in $S_{\kappa,L}$. Its foundations are developed in [9], but a newform theory is not yet available. It seems conceivable that one could prove more general injectivity results along these lines. For the relationship between the vector-valued modular forms in $S_{\kappa,L}$ and the adelic language, see [17].

Note that in this context, J.-S. Li [15] has used the theta correspondence and the doubling method for automorphic representations in great generality to obtain non-vanishing results for cohomology when passing to a sufficiently large level.

In the body of the paper, we actually consider the generalization of the Kudla-Millson lift due to Funke and Millson [12]. It maps cusp forms in $S_{\kappa,L}$ to closed differential q-forms with values in certain local coefficient systems. Moreover, we use an adelic set-up for the theta and Eisenstein series in question.

1.1. Surjectivity of the Borcherds lift

We briefly discuss how the injectivity results on the Kudla-Millson lift imply surjectivity results for the Borcherds lift. We revisit part of the material of [6] in the light of the adjointness result of [8] between the regularized theta lift and the Kudla Millson lift. We restrict ourselves to the Hermitean case of signature $(p, 2)$ where X is a p-dimensional complex algebraic manifold. The special cycles $Z(h, n)$ are algebraic divisors on X, also called Heegner divisors or rational quadratic divisors.

We say that a meromorphic modular form for Γ has a Heegner divisor, if its divisor on X is a linear combination of the $Z(h, n)$. A large supply of modular forms with Heegner divisor is provided by the Borcherds lift, see [2], [3]. We briefly recall its construction.

A meromorphic modular form for a congruence subgroup of $SL_2(\mathbb{Z})$ is called *weakly holomorphic*, if its poles are supported on the cusps. If $k \in \mathbb{Z}$, we write $M_{k,L}^!$ for the space of weakly holomorphic modular forms of weight k for $SL_2(\mathbb{Z})$ of type ρ_L. Any $f \in M_{k,L}^!$ has a Fourier expansion of the form

$$f(\tau) = \sum_{h \in L^\#/L} \sum_{n \in \mathbb{Z}+Q(h)} c(h, n)e(n\tau)\mathfrak{e}_h,$$

where only finitely many coefficients $c(h, n)$ with $n < 0$ are non-zero. We write V^- for the quadratic space $(V, -Q)$ of signature $(2, p)$ and L^- for the lattice $(L, -Q)$ in V^-.

Theorem 1.3 (Borcherds [3], Theorem 13.3). *Let $f \in M_{1-p/2,L^-}^!$ be a weakly holomorphic modular form with Fourier coefficients $c(h, n)$. Assume that $c(h, n) \in \mathbb{Z}$ for $n < 0$. Then there exists a meromorphic modular form $\Psi(z, f)$ for Γ (with some multiplier system of finite order) such that:*

  (i) *The weight of Ψ is equal to $c(0, 0)/2$.*
  (ii) *The divisor $Z(f)$ of Ψ is determined by the principal part of f at the cusp ∞. It equals*

$$Z(f) = \sum_{h \in L^\#/L} \sum_{n < 0} c(h, n)Z(h, n).$$

  (iii) *In a neighborhood of a cusp of Γ the function Ψ has an infinite product expansion analogous to the Dedekind eta function, see [3], Theorem 13.3 (5.).*

The proof of this result uses a regularized theta lift. Let $\varphi_0^{p,2} \in \mathcal{S}(V(\mathbb{R}))$ be the Gaussian for signature $(p, 2)$. The corresponding Siegel theta function

$$\Theta(\tau, z, \varphi_0) = v \sum_{h \in L^\#/L} \sum_{x \in L+h} (\omega_\infty(g_\tau)\varphi_0)(x, z)\mathfrak{e}_h$$

transforms like a non-holomorphic modular form of weight $p/2 - 1$ of type ρ_L in the variable τ. Hence the theta integral

$$\Phi(z, f) = \int_{\Gamma(1)\backslash\mathbb{H}} \langle f(\tau), \overline{\Theta(\tau, z, \varphi_0^{p,2})} \rangle \, d\mu \qquad (1.2)$$

formally defines a Γ-invariant function on D. Because of the singularities of f at the cusps, the integral diverges. However, Harvey and Moore discovered that it can be regularized essentially by viewing it as the limit $T \to \infty$ of the integral over the standard fundamental domain truncated at $\Im(\tau) = T$, see [3], [14]. It turns out that $\Phi(z, f)$ defines a smooth function on $X \setminus Z(f)$ which has a logarithmic singularity along $Z(f)$. Moreover,

$$\Phi(z, f) = -2 \log \|\Psi(z, f)\|_{\mathrm{Pet}} + \text{constant},$$

where $\| \cdot \|_{\mathrm{Pet}}$ denotes the Petersson metric on the line bundle of modular forms of weight $c(0, 0)/2$ over X. From this identity, the claimed properties of $\Psi(z, f)$ can be derived.

Modular forms for the group $\Gamma \subset O(L)$ arising via this lift are called automorphic products or Borcherds products. By (ii) they have a Heegner divisor.

Here we consider the question whether the Borcherds lift is surjective. More precisely we ask whether every meromorphic modular form for Γ with Heegner divisor is the lift $\Psi(z, f)$ of a weakly holomorphic form $f \in M_{1-p/2,L^-}^!$?

An affirmative answer to this question was given in [6] in the special case that the lattice L splits two hyperbolic planes over $\mathbb{Z}$. In the (more restrictive) case that L is unimodular, a different proof was given in [7] using local Borcherds products and a theorem of Waldspurger on theta series with harmonic polynomials [29].

The approach of [6] was to first simplify the problem and to consider the regularized theta lift for a larger space of "input" modular forms. Namely, we let $H_{k,L}$ be the space of *weak Maass forms* of weight k and type ρ_L. This space consists of the smooth functions $f : \mathbb{H} \to \mathbb{C}[L^{\#}/L]$ that transform with ρ_L in weight k under $\mathrm{SL}_2(\mathbb{Z})$, are annihilated by the weight k Laplacian, and satisfy $f(\tau) = O(e^{Cv})$ as $\tau = u + iv \to i\infty$ for some constant $C > 0$ (see [8] Section 3).

Any $f \in H_{k,L}$ has a Fourier expansion of the form

$$f(\tau) = \sum_{h \in L^{\#}/L} \sum_{n \in \mathbb{Q}} c^+(h, n) e(n\tau) \mathfrak{e}_h$$

$$+ \sum_{h \in L^{\#}/L} c^-(h, 0) v^{1-k} \mathfrak{e}_h + \sum_{\substack{n \in \mathbb{Q} \\ n \neq 0}} c^-(h, n) H(2\pi n v) e(nu) \mathfrak{e}_h, \qquad (1.3)$$

where only finitely many of the coefficients $c^+(h, n)$ (respectively $c^-(h, n)$) with negative (respectively positive) index n are non-zero. The function $H(w)$ is a Whittaker type function.

For $f \in H_{k,L}$, put $\xi_k(f) = R_{-k}(v^k \bar{f})$, where R_{-k} is the standard raising operator for modular forms of weight $-k$. This defines an antilinear map $\xi_k : H_{k,L} \to M^!_{2-k,L^-}$ to the space of weakly holomorphic modular forms in weight $2 - k$. It is easily checked that $M^!_{k,L}$ is the kernel of ξ_k. According to [8], Corollary 3.8, the sequence

$$0 \longrightarrow M^!_{k,L} \longrightarrow H_{k,L} \overset{\xi_k}{\longrightarrow} M^!_{2-k,L^-} \longrightarrow 0$$

is exact. We let $H^+_{k,L}$ be the preimage under ξ_k of the space of cusp forms S_{2-k,L^-} of weight $2 - k$ with type ρ_{L^-}. Hence we have the exact sequence

$$0 \longrightarrow M^!_{k,L} \longrightarrow H^+_{k,L} \overset{\xi_k}{\longrightarrow} S_{2-k,L^-} \longrightarrow 0 \,.$$

The space $H^+_{k,L}$ can also be characterized as the subspace of those $f \in H_{k,L}$ whose Fourier coefficients $c^-(h, n)$ with non-negative index n vanish. This implies that

$$f(\tau) = \sum_{h \in L^\#/L} \sum_{n<0} c^+(h, n) e(n\tau) \mathfrak{e}_h + O(1), \qquad \Im(\tau) \to \infty,$$

i.e., the singularity at ∞, called the *principal part* of f, looks like the singularity of a weakly holomorphic form.

For $f \in H_{1-p/2,L^-}$, we can define the regularized theta lift $\Phi(z, f)$ as in (1.2), see [6], [8]. This generalized lift is related to the Kudla-Millson lift Λ defined in (1.1) in the following way (see [8], Theorem 6.1).

Theorem 1.4. *Let $f \in H^+_{1-p/2,L^-}$ and denote its Fourier expansion as in (1.3). The $(1, 1)$-form $dd^c \Phi(z, f)$ can be continued to a smooth form on X. It satisfies*

$$dd^c \Phi(z, f) = \Lambda(\xi_{1-p/2}(f))(z) + c^+(0, 0)\Omega.$$

Here Ω denotes the invariant Kähler form on D normalized as in [8].

On the other hand, the following "weak converse theorem" is proved in [6], Theorem 4.23.

Theorem 1.5. *Assume that $p > r$. Let F be a meromorphic modular form for the group Γ with Heegner divisor*

$$\operatorname{div}(F) = \sum_h \sum_{n<0} c^+(h,n) Z(h,n)$$

(where $c^+(h,n) = c^+(-h,n)$ without loss of generality). Then there is a weak Maass form $f \in H^+_{1-p/2,L^-}$ with principal part $\sum_h \sum_{n<0} c^+(h,n) e(n\tau) \mathfrak{e}_h$ whose regularized theta lift satisfies

$$\Phi(z,f) = -2\log \|F\|_{\mathrm{Pet}} + \text{constant}. \tag{1.4}$$

Note that the proof in [6] is only given in the case that $p \geq 3$ (where the assumption on the Witt index is automatically fulfilled). However, the argument extends to the low dimensional cases. It is likely that the hypothesis on the Witt index can be dropped as well, but we have not checked this.

Corollary 1.6. *Assume that $p > r$. Let F be a meromorphic modular form for the group Γ with Heegner divisor as in Theorem 1.5. Let $f \in H^+_{1-p/2,L^-}$ be a weak Maass form whose regularized theta lift satisfies (1.4). Then*

$$\Lambda(\xi_{1-p/2}(f)) = 0.$$

Proof. The assumption on f implies that

$$dd^c \Phi(z,f) = -2 dd^c \log \|F\|_{\mathrm{Pet}} = c^+(0,0)\Omega.$$

On the other hand, according to Theorem 1.4, we have

$$dd^c \Phi(z,f) = \Lambda(\xi_{1-p/2}(f))(z) + c^+(0,0)\Omega.$$

If we combine these identities, we obtain the claim. $\qquad\square$

Corollary 1.7. *Assume the hypotheses of Corollary 1.6. If Λ is injective, then f is weakly holomorphic, and F is a constant multiple of the Borcherds lift $\Psi(z,f)$ of f in the sense of Theorem 1.3.*

Proof. By Corollary 1.6 we have $\Lambda(\xi_{1-p/2}(f)) = 0$. Since Λ is injective, we find that $\xi_{1-p/2}(f) = 0$. But this means that f is weakly holomorphic. $\qquad\square$

When the lattice L splits two hyperbolic planes over $\mathbb{Z}$, it was proved in [6] that Λ is injective by considering the Fourier expansion of the lift. In Section 4 of the present paper we show (for even unimodular lattices) how such injectivity results can be obtained by the Rallis inner products formula.

We end this section by stating a converse of Corollary 1.6. If $r > 0$, we let $\ell \in L$ be a primitive isotropic vector, and let $\ell' \in L^\#$ be a vector with $(\ell,\ell') = 1$. We let L_0 be the singular lattice $L \cap \ell^\perp$ and let K be the Lorentzian lattice $L_0/\mathbb{Z}\ell$.

Theorem 1.8. *Assume that $p \geq 2$ and $p > r$. Let $f \in H^+_{k,L^-}$ and assume that the Fourier coefficients $c^+(h, n)$ $(n < 0)$ of the principal part of f are integral. If $\xi_{1-p/2}(f) \in \ker(\Lambda)$, then there exists a meromorphic modular form F for Γ (with some multiplier system of finite order) such that:*

(i) *The weight of F is equal to $c^+(0, 0)/2$.*

(ii) *The divisor of F is equal to*

$$Z(f) = \sum_{h \in L^\#/L} \sum_{n<0} c^+(h, n) Z(h, n).$$

(iii) *In a neighborhood of a cusp of Γ, given by a primitive isotropic vector $\ell \in L$, the function F has an automorphic product expansion*

$$F(z) = Ce((\rho, z)) \prod_{\substack{\lambda \in K' \\ (\lambda, W)>0}} \prod_{\substack{\delta \in L^\#/L \\ \delta|L_0=\lambda}} \left(1 - e((\lambda, z)) + (\delta, \ell')\right)^{c^+(\lambda, Q(\lambda))}.$$

Here C is a non-zero constant, and we have used the notation of [3].

Proof. Theorem 1.4 and the fact that $\Lambda\left(\xi_{1-p/2}(f)\right) = 0$ imply that

$$dd^c\Phi(z, f) = c^+(0, 0)\Omega.$$

(In particular, if $c^+(0, 0) = 0$, then f is pluriharmonic.) Now we can argue as in [6], Lemma 3.13 and Theorem 3.16 to prove the claim. ☐

We note that the assumption on r and p is needed to guarantee that the multiplier system of F has finite order. (When f is not weakly holomorphic, we cannot argue with the embedding trick as in [4], Correction).

If f is weakly holomorphic, then $\xi_{1-p/2}(f) = 0$ and the Theorem reduces to Theorem 1.3. However, if Λ is not injective, and f is a weak Maass form such that $\xi_{1-p/2}(f)$ is a non-trivial element of the kernel, then Theorem 1.8 leads to *exceptional automorphic products*. If there are any cases where Λ is not injective, it would be very interesting to construct examples of such exceptional automorphic products.

Remark 1.9. If $p \geq 4$, the existence of the meromorphic modular form F with divisor (ii) is related to the fact that $H^1(X, \mathcal{O}_X) = 0$ in this case, which can be proved following the argument of [10] §3.1. Therefore the Chern class map $\mathrm{Pic}(X) \to H^2(X, \mathbb{Z})$ is injective.

We thank S. Böcherer, E. Freitag, W. T. Gan, S. Kudla, and J. Millson for very helpful conversations on the content of this paper. The second named author also thanks the Max Planck Institut für Mathematik in Bonn/Germany for its hospitality during the summer 2005 where substantial work on this paper was done.

2. Theta functions and the Siegel-Weil formula

Let (V, Q) be a non-degenerate rational quadratic space of of dimension m. We write $(\cdot, \cdot)$ for the bilinear form corresponding to the quadratic form Q so that $Q(x) = \frac{1}{2}(x, x)$. For simplicity we assume that m is even. We let $G = \mathrm{Sp}(n)$ be the symplectic group acting on a symplectic space of dimension $2n$ over $\mathbb{Q}$. The embedding of $\mathrm{U}(n)$ into $G(\mathbb{R})$ given by $\mathbf{k} = A + iB \mapsto k = \left(\begin{smallmatrix} A & B \\ -B & A \end{smallmatrix} \right)$ gives rise to a maximal compact subgroup $K_\infty \subset G(\mathbb{R})$. At the finite places, we pick the open compact subgroup $K_p = \mathrm{Sp}(n, \mathbb{Z}_p)$. Then $K = K_\infty \times \prod_p K_p$ is the corresponding maximal compact subgroup of $G(\mathbb{A})$, the symplectic group over the ring of adeles of $\mathbb{Q}$. We let $\omega = \omega_n$ be the Schrödinger model of the Weil representation of $G_\mathbb{A}$ acting on $\mathcal{S}(V^n(\mathbb{A}))$, the space of Schwartz-Bruhat functions on $V^n(\mathbb{A})$, associated to the standard additive character of $\mathbb{A}/\mathbb{Q}$ (which on $\mathbb{R}$ is given by $t \mapsto e(t) = e^{2\pi i t}$). Note that since m is even we do not have to deal with metaplectic coverings. We form the theta series associated to $\varphi \in \mathcal{S}(V^n(\mathbb{A}))$ by

$$\theta(g, h, \varphi) = \sum_{\mathbf{x} \in V^n(\mathbb{Q})} (\omega(g)\varphi)(h^{-1}\mathbf{x}), \tag{2.1}$$

with $g \in G(\mathbb{A})$ and $h \in \mathrm{O}(V)(\mathbb{A})$. We assume $\varphi = \varphi_\infty \otimes \varphi_f$ with $\varphi_\infty \in \mathcal{S}(V^n(\mathbb{R}))$ and $\varphi_f \in \mathcal{S}(V^n(\mathbb{A}_f))$.

We now briefly review the Siegel-Weil formula, see e.g. [16]. We put

$$I(g, \varphi) = \int_{\mathrm{O}(V)(\mathbb{Q}) \backslash \mathrm{O}(V)(\mathbb{A})} \theta(g, h, \varphi) dh, \tag{2.2}$$

where dh is the invariant measure on $\mathrm{O}(V)(\mathbb{Q}) \backslash \mathrm{O}(V)(\mathbb{A})$ normalized to have total volume 1. By Weil's convergence criterion [30], $I(g, \varphi)$ is absolutely convergent if either V is anisotropic or if

$$m - r > n + 1. \tag{2.3}$$

Here r is the Witt index of V, i.e., the dimension of a maximal isotropic subspace of V over $\mathbb{Q}$.

We set $n(b) = \left(\begin{smallmatrix} 1 & b \\ 0 & 1 \end{smallmatrix} \right)$ for b a symmetric $n \times n$ matrix and $m(a) = \left(\begin{smallmatrix} a & 0 \\ 0 & {}^t a^{-1} \end{smallmatrix} \right)$ for $a \in \mathrm{GL}(n)$. Then the Siegel parabolic is given by $P(\mathbb{A}) = N(\mathbb{A})M(\mathbb{A})$ with $N = \{n(b); \ b \in \mathrm{Mat}_n, \ b = {}^t b\}$ and $M = \{m(a); \ a \in \mathrm{GL}(n)\}$. Then using the Iwasawa decomposition $G(\mathbb{A}) = P(\mathbb{A})K$ we define

$$\Phi(g, s) = (\omega(g)\varphi)(0) \cdot \det |a(g)|_\mathbb{A}^{s - s_0}, \tag{2.4}$$

where

$$s_0 = \frac{m}{2} - \frac{n+1}{2}. \qquad (2.5)$$

Thus Φ defines a section in a certain induced parabolic induction space (see [16] (I.3.6)). Note that Φ is determined by its values on K. Since Φ comes from $\varphi \in \mathcal{S}(V^n(\mathbb{A}))$, we also see that Φ is a standard section, i.e., its restriction to K does not depend on s, and we write $\Phi(k) = \Phi(k, s)$ for $k \in K$. Furthermore, Φ factors as $\Phi = \Phi_\infty \otimes \Phi_f$.

We then define the Eisenstein series associated to Φ by

$$E(g, s, \Phi) = \sum_{\gamma \in P(\mathbb{Q}) \backslash G(\mathbb{Q})} \Phi(\gamma g, s), \qquad (2.6)$$

which for $\mathrm{Re}(s) > \rho_n := (n+1)/2$ converges absolutely and has a meromorphic continuation to the whole complex plane. The extension of Weil's work [30] by Kudla and Rallis in the convergent range is:

Theorem 2.1. *([20], [21].) Assume Weil's convergence criterion holds.*

(i) *Then $E(g, s, \Phi)$ is holomorphic at $s = s_0$.*
(ii) *We have*

$$I(g, \varphi) = c_0 E(g, s_0, \Phi),$$

where $c_0 = 1$ if $m > n + 1$ and $c_0 = 2$ if $m \leq n + 1$.

We translate the adelic Eisenstein series into more classical language, see [16] section IV.2. We let $K_f(N) \subset \prod_p K_p$ be a subgroup of finite index of level N, i.e.,

$$\Gamma := G(\mathbb{Q}) \cap (G(\mathbb{R}) K_f(N))$$

contains the principal congruence subgroup $\Gamma(N) \subset \mathrm{Sp}(n, \mathbb{Z})$. We assume that Φ_f is $K_f(N)$-invariant. Furthermore, if φ_f corresponds to the characteristic function of a coset of an even lattice L of level N in V, then we have

$$\Phi_f(\gamma) = \prod_{p \mid N} \Phi_p(\gamma)$$

for $\gamma \in \Gamma$. Via $G(\mathbb{A}) = G(\mathbb{Q}) G(\mathbb{R}) K_f(N)$ we see that the Eisenstein series $E(g, s, \Phi)$ is determined by its restriction to $G(\mathbb{R})$. We assume that the restriction of $\Phi(g, s)$ to K_∞ is given by

$$\Phi_\infty^\kappa(k, s) := \det(\mathbf{k})^\kappa. \qquad (2.7)$$

We denote the unique section at the Archimedian prime with this property by Φ_∞^κ. Let $g_\tau = n(u)m(a)$ with ${}^t aa = v$ be an element moving the base point

$i1_n$ of the Siegel upper half plane $\mathbb{H}_n$ to $\tau = u + iv$. Then we obtain a classical Eisenstein series of weight κ (and level N):

$$E(g_\tau, s, \Phi) = \sum_{\gamma \in (P(\mathbb{Q}) \cap \Gamma) \backslash \Gamma} \Phi^\kappa_\infty(\gamma g_\tau) \Phi_f(\gamma)$$

$$= \det(v)^{\kappa/2} \sum_{\gamma \in (P(\mathbb{Q}) \cap \Gamma) \backslash \Gamma} \left(\frac{\det(v)}{|\det(c\tau + d)|^2} \right)^{(s + \rho_n - \kappa)/2} \det(c\tau + d)^{-\kappa} \Phi_f(\gamma),$$

with $\gamma = \left(\begin{smallmatrix} a & b \\ c & d \end{smallmatrix} \right)$. In particular, if $N = 1$ then

$$E(g_\tau, s, \Phi) = \det(v)^{\kappa/2} E_\kappa^{(n)} \left(\tau, (s + \rho_n - \kappa)/2 \right), \tag{2.8}$$

where

$$E_\kappa^{(n)}(\tau, s) = \sum_{\gamma \in \Gamma_\infty \backslash \mathrm{Sp}(n, \mathbb{Z})} \left(\det \Im(\gamma \tau) \right)^s \det(c\tau + d)^{-\kappa} \tag{2.9}$$

is the classical Siegel Eisenstein series for $\mathrm{Sp}(n, \mathbb{Z})$ of weight κ.

For later use, we introduce an embedding ι_0 of $\mathrm{Sp}(n) \times \mathrm{Sp}(n)$ into $\mathrm{Sp}(2n)$ by

$$\begin{pmatrix} a & b \\ c & d \end{pmatrix} \times \begin{pmatrix} a' & b' \\ c' & d' \end{pmatrix} \mapsto \begin{pmatrix} a & & b & \\ & a' & & b' \\ c & & d & \\ & c' & & d' \end{pmatrix}. \tag{2.10}$$

3. Special Schwartz forms

We change the setting in this section and consider the real place only. We assume that V is now a real quadratic space of signature (p, q) of dimension m. Since it does not make any extra work we do allow m odd in this section. We pick an oriented orthogonal basis $\{v_i\}$ of V such that $(v_\alpha, v_\alpha) = 1$ for $\alpha = 1, \ldots, p$ and $(v_\mu, v_\mu) = -1$ for $\mu = p + 1, \ldots, m$, and we denote the corresponding coordinate functions by x_α and x_μ. We let K^V be the maximal compact subgroup of $\mathrm{O}(V)$ stabilizing $\mathrm{span}\{v_{p+1}, \ldots, v_m\}$. Thus $K^V \simeq \mathrm{O}(p) \times \mathrm{O}(q)$. We realize the symmetric space D associated to V as the Grassmannian of oriented negative q-planes in V. Thus D has two components

$$D = D_+ \amalg D_-.$$

Picking for the base point z_0 the space $\mathrm{span}\{v_{p+1}, \ldots, v_m\}$ together with the induced orientation, we see

$$D_+ = \{z \subset V; \quad \dim z = q, (,)|_z < 0, z \text{ has the same orientation as } z_0\}.$$

Thus $D_+ \simeq SO_0(V)/K_0^V$, where the subscript indicates the connected component of the identity. We associate to $z \in D$ the standard majorant $(\ ,\ )_z$ given by

$$(x, x)_z = (x_{z^\perp}, x_{z^\perp}) - (x_z, x_z),$$

where $x = x_z + x_{z^\perp} \in V$ is given by the orthogonal decomposition $V = z \oplus z^\perp$. We write $(\ ,\ )_0 = (\ ,\ )_{z_0}$.

Let $\mathfrak{o}(V)$ be the Lie algebra of $O(V)$ and let $\mathfrak{o}(V) = \mathfrak{p}^V \oplus \mathfrak{k}^V$ with $\mathfrak{k}^V = \mathrm{Lie}(K^V)$ be the associated Cartan decomposition. Then $\mathfrak{p} = \mathfrak{p}^V$ is isomorphic to the tangent space at the base point z_0 of D. With respect to the above basis of V we have

$$\mathfrak{p} \simeq \left\{ \begin{pmatrix} 0 & X \\ {}^t X & 0 \end{pmatrix}; \ X \in \mathrm{Mat}_{p,q}(\mathbb{R}) \right\}.$$

We let $X_{\alpha\mu}$ ($1 \leq \alpha \leq p$, $p+1 \leq \mu \leq p+q$) denote the element of $\mathfrak{p}$ which interchanges v_α and v_μ and annihilates all the other basis elements of V. We write $\omega_{\alpha\mu}$ for the element of the dual basis corresponding to $X_{\alpha\mu}$.

We let $\omega = \omega_n$ be the Weil representation of the metaplectic cover $\mathrm{Mp}(n, \mathbb{R})$ of $\mathrm{Sp}(n, \mathbb{R})$ acting on the Schwartz functions $\mathcal{S}(V^n)$. We let $K = \widetilde{U}(n)$ be the maximal compact subgroup of $\mathrm{Mp}(n, \mathbb{R})$ given by the inverse image of the standard maximal compact subgroup $U(n)$ in $\mathrm{Sp}(n, \mathbb{R})$. Recall that K admits a character $\det^{1/2}$ whose square descends to the determinant character of $U(n)$. We also write ω for the associated Lie algebra action on the space of K-finite vectors in $\mathcal{S}(V^n)$. It is given by the so-called polynomial Fock space $S(V^n) \subset \mathcal{S}(V^n)$. It consists of those Schwartz functions on V^n of the form $p(\mathbf{x})\varphi_0(\mathbf{x})$, where $p(\mathbf{x})$ is a polynomial function on V^n. Here $\varphi_0(\mathbf{x})$ is the standard Gaussian on V^n. More precisely, for $\mathbf{x} = (x_1, \ldots, x_n) \in V^n$ and $z \in D$, we let

$$\varphi_0(\mathbf{x}, z) = \exp\left(-\pi \sum_{i=1}^n (x_i, x_i)_z\right),$$

and set $\varphi_0(\mathbf{x}) = \varphi_0(\mathbf{x}, z_0)$. We view

$$\varphi_0 \in [\mathcal{S}(V^n) \otimes C^\infty(D)]^{O(V)} \simeq [\mathcal{S}(V^n) \otimes {\textstyle\bigwedge^0}(\mathfrak{p}^*)]^{K^V},$$

where the isomorphism is given by evaluation at the base point z_0 of D. In the following we will identify corresponding objects under this isomorphism.

Kudla and Millson (see [18]) constructed (in much greater generality) Schwartz forms φ_{KM} on V taking values in $\mathcal{A}^q(D)$, the differential q-forms on D. More precisely,

$$\varphi_{KM} \in [\mathcal{S}(V) \otimes \mathcal{A}^q(D)]^{O(V)} \simeq [\mathcal{S}(V)_\alpha \otimes \bigwedge^q(\mathfrak{p}^*)]^{K^V}.$$

Here $\mathcal{S}(V)_\alpha$ is the Schwartz space for V twisted by the spinor norm character α on $O(V)$. On $K^V = O(p) \times O(q)$, α is given by $1 \otimes \det$. The Schwartz form φ_{KM} is given by

$$\varphi_{KM} = \frac{1}{2^{q/2}} \prod_{\mu=p+1}^{p+q} \left[\sum_{\alpha=1}^{p} \left(x_\alpha - \frac{1}{2\pi} \frac{\partial}{\partial x_\alpha} \right) \otimes A_{\alpha\mu} \right] \varphi_0.$$

Here $A_{\alpha\mu}$ denotes the left multiplication by $\omega_{\alpha\mu}$. More generally, we consider the Schwartz forms

$$\varphi_{q,\ell} \in [\mathcal{S}(V)_\alpha \otimes \bigwedge^q(\mathfrak{p}^*) \otimes \mathrm{Sym}^\ell(V)]^{K^V}$$

with values in the ℓ-th symmetric powers of V introduced by Funke and Millson [12]. The forms $\varphi_{q,\ell}$ are given by

$$\varphi_{q,\ell} = \left[\frac{1}{2} \sum_{\alpha=1}^{p} \left(x_\alpha - \frac{1}{2\pi} \frac{\partial}{\partial x_\alpha} \right) \otimes 1 \otimes A_{v_\alpha} \right]^\ell \varphi_{KM}$$

$$= \frac{1}{2^\ell} \sum_{\alpha_1,\ldots,\alpha_\ell=1}^{p} \left[\prod_{i=1}^{\ell} \left(x_{\alpha_i} - \frac{1}{2\pi} \frac{\partial}{\partial x_{\alpha_i}} \right) \otimes 1 \otimes \prod_{i=1}^{\ell} A_{v_{\alpha_i}} \right] \varphi_{KM}.$$

Here A_v denotes the multiplication with the vector v in the symmetric algebra of V. Note that $\mathrm{Sym}^\ell(V)$ is not an irreducible representation of $O(V)$, and we denote by $\varphi_{q,[\ell]}$ the projection of $\varphi_{q,\ell}$ onto $\mathcal{H}^\ell(V)$, the harmonic ℓ-tensors in V. It consists of those symmetric ℓ-tensors which are annihilated by the signature (p,q)-Laplacian $\Delta = \sum_{\alpha=1}^{p} \frac{\partial^2}{\partial v_\alpha^2} - \sum_{\mu=p+1}^{m} \frac{\partial^2}{\partial v_\mu^2}$. Here we view v_α and v_μ as independent variables. It can be also characterized as the space of symmetric ℓ-tensors in V which are orthogonal with respect to the induced inner product on $\mathrm{Sym}^\ell(V)$ to vectors of the form $r^2 w$. Here $w \in \mathrm{Sym}^{\ell-2}(V)$ and r^2 denotes the multiplication with $\sum_{\alpha=1}^{p} v_\alpha^2 - \sum_{\mu=p+1}^{m} v_\mu^2$. Recall that we have $\mathrm{Sym}^\ell(V) = \mathcal{H}^\ell(V) \oplus r^2 \mathrm{Sym}^{\ell-2}(V)$ as representations of $O(V)$.

The Schwartz form $\varphi_{q,\ell}$ (and also $\varphi_{q,[\ell]}$) is an eigenfunction of weight $m/2 + \ell$ under the action of $k \in K$, see [12, 18], i.e.,

$$\omega(k)\varphi_{q,\ell} = \det(\mathbf{k})^{m/2+\ell}\varphi_{q,\ell}. \tag{3.1}$$

Here $\mathbf{k}$ is the element in $\widetilde{U}(1)$ corresponding to $k \in \widetilde{SO}(2) \subset \mathrm{Mp}(1,\mathbb{R})$. Moreover, $\varphi_{q,\ell}(x)$ is a closed differential form on D.

We normalize the inner product on $\mathrm{Sym}^\ell(V)$ inductively by setting

$$(w_1 \cdots w_\ell, w_1' \cdots w_\ell') = \frac{1}{\ell} \sum_{j=1}^{\ell} (w_1, w_j')(w_2 \cdots w_\ell, w_1' \cdots \widehat{w_j'} \cdots w_\ell').$$

With this normalization we easily see that for the restriction of $(\ ,\ )$ to the positive definite subspace $\mathrm{span}\{v_\alpha;\ 1 \le \alpha \le p\}$ of V we have

$$\sum_{\substack{\alpha_1,\dots,\alpha_\ell=1 \\ \beta_1,\dots,\beta_\ell=1}}^{p} \left(\prod_{i=1}^{\ell} v_{\alpha_i}, \prod_{i=1}^{\ell} v_{\beta_i} \right) = p^\ell.$$

We let $\widetilde{\mathrm{Sym}}^\ell(V)$ be the local system on D associated to $\mathrm{Sym}^\ell(V)$. Then for the wedge product, we have $\wedge : \mathcal{A}^r(D, \widetilde{\mathrm{Sym}}^\ell(V)) \times \mathcal{A}^s(D, \widetilde{\mathrm{Sym}}^\ell(V)) \to \mathcal{A}^{r+s}(D)$ by taking the inner product on the fibers $\mathrm{Sym}^\ell(V)$. We are ultimately more interested in the form $\varphi_{q,[\ell]}$, but calculations with $\varphi_{q,\ell}$ are more convenient. In this context the following lemma will be important later.

Lemma 3.1. *Let* $\eta \in \mathcal{A}^{(p-1)q}(D, \widetilde{\mathrm{Sym}}^{\ell-2}(V))$. *Then*

$$\varphi_{q,\ell} \wedge r^2 \eta = -\frac{1}{2\pi} (\omega(R)\varphi_{q,\ell-2}) \wedge \eta.$$

Here $R = \frac{1}{2} \left(\begin{smallmatrix} 1 & i \\ i & -1 \end{smallmatrix} \right) \in \mathfrak{sl}(2,\mathbb{C})$ *is the standard* SL(2)*-raising operator.*

Proof. By the adjointness of $\frac{1}{\ell(\ell-1)}\Delta$ and r^2 with respect to the inner product in $\mathrm{Sym}^\bullet(V)$, we have $\varphi_{q,\ell} \wedge r^2\eta = \frac{1}{\ell(\ell-1)}(\Delta\varphi_{q,\ell}) \wedge \eta$. Note that Δ operates on the coefficient part of $\varphi_{q,\ell}$. Then switching to the Fock model of the Weil representation, see the proof of Lemma 3.5, and using (3.9) one easily sees $\Delta\varphi_{q,\ell} = -\frac{\ell(\ell-1)}{2\pi}\omega(R)\varphi_{q,\ell-2}$. We leave the details to the reader. $\qquad\square$

We let $*$ denote the Hodge $*$-operator on D. Then $\varphi_{q,\ell}(x_1) \wedge *\varphi_{q,\ell}(x_2)$ with $\mathbf{x} = (x_1, x_2) \in V^2$, being a top degree differential form, gives rise to a scalar-valued Schwartz function $\phi_{q,\ell}$ on V^2 defined by

$$\phi_{q,\ell}(\mathbf{x}, z)\mu = \varphi_{q,\ell}(x_1, z) \wedge *\varphi_{q,\ell}(x_2, z). \tag{3.2}$$

Here μ is the volume form on D induced by the Riemannian metric coming from the Killing form on $\mathfrak{g}$. For convenience we scale the metric such that the restriction of μ to the base point z_0 is given by

$$\mu = \omega_{1,p+1} \wedge \cdots \wedge \omega_{1,p+q} \wedge \omega_{2,p+1} \wedge \cdots \wedge \omega_{p,p+q}. \tag{3.3}$$

Note that

$$\phi_{q,\ell} \in [\mathcal{S}(V^2) \otimes C^\infty(D)]^{O(V)} \simeq [\mathcal{S}(V^2) \otimes \bigwedge\nolimits^0 (\mathfrak{p}^*)]^{K^V}.$$

Lemma 3.2. *We have*

$$\phi_{q,\ell}(\mathbf{x}) = \frac{p^\ell}{2^{q+2\ell}} \sum_{\alpha_1,\dots,\alpha_{q+\ell}=1}^{p} \prod_{i=1}^{q+\ell} \left(x_{\alpha_i 1} - \frac{1}{2\pi}\frac{\partial}{\partial x_{\alpha_i 1}} \right) \left(x_{\alpha_i 2} - \frac{1}{2\pi}\frac{\partial}{\partial x_{\alpha_i 2}} \right) \varphi_0(\mathbf{x})$$

$$= \frac{p^\ell}{2^{q+2\ell}} \left(\sum_{\alpha=1}^{p} \left(x_{\alpha 1} - \frac{1}{2\pi} \frac{\partial}{\partial x_{\alpha 1}} \right) \left(x_{\alpha 2} - \frac{1}{2\pi} \frac{\partial}{\partial x_{\alpha 2}} \right) \right)^{q+\ell} \varphi_0(\mathbf{x}).$$

Example 3.3. For signature $(p, 2)$, we have

$$\phi_{q,0}(\mathbf{x}) = \sum_{\alpha=1}^{p} \left(x_{\alpha 1}^2 - \frac{1}{4\pi} \right) \left(x_{\alpha 2}^2 - \frac{1}{4\pi} \right) \varphi_0(\mathbf{x}) + 4 \sum_{\substack{\alpha,\beta=1 \\ \alpha \neq \beta}}^{p} x_{\alpha 1} x_{\beta 1} x_{\alpha 2} x_{\beta 2} \varphi_0(\mathbf{x}).$$

Note that (3.1) immediately implies:

Lemma 3.4. *For $k_1, k_2 \in \widetilde{SO}(2) \subset \mathrm{Mp}(1, \mathbb{R})$, we have*

$$\omega(\iota_0(k_1, k_2)) \phi_{q,\ell} = \det(\mathbf{k_1 k_2})^{m/2+\ell} \phi_{q,\ell}.$$

The action of the full maximal compact $K \subset \mathrm{Mp}(2, \mathbb{R})$ on $\phi_{q,\ell}$ via the Weil representation is more complicated, as we now explain. We let

$$\mathfrak{g} = \mathfrak{k} \oplus \mathfrak{p}_+ \oplus \mathfrak{p}_- \tag{3.4}$$

be a Harish-Chandra decomposition of $\mathfrak{g} = \mathfrak{sp}(2, \mathbb{C})$, where $\mathfrak{k} = \mathrm{Lie}(K)_{\mathbb{C}}$,

$$\mathfrak{p}_+ = \left\{ p_+(X) = \frac{1}{2} \begin{pmatrix} X & iX \\ iX & -X \end{pmatrix}; \ X \in \mathrm{Mat}_2(\mathbb{C}), \, {}^t X = X \right\}, \tag{3.5}$$

and $\mathfrak{p}_- = \overline{\mathfrak{p}_+}$. Note that $\mathfrak{p}_+$ is the holomorphic tangent space of $\mathbb{H}_2$ at the base point $i1_2$ and is spanned by the raising operators

$$R_1 = R_{11} = p_+ \begin{pmatrix} 1 & 0 \\ 0 & 0 \end{pmatrix}, \qquad R_2 = R_{22} = p_+ \begin{pmatrix} 0 & 0 \\ 0 & 1 \end{pmatrix}, \tag{3.6}$$

$$R_{12} = \frac{1}{2} p_+ \begin{pmatrix} 0 & 1 \\ 1 & 0 \end{pmatrix}. \tag{3.7}$$

Note that $R_1 = \iota_0(R, 0)$ and $R_2 = \iota_0(0, R)$ are the images of the SL_2-raising operator R in $\mathfrak{sp}(2, \mathbb{C})$ under the two standard embeddings of $\mathfrak{sl}(2)$ into $\mathfrak{sp}(2)$.

Recall that the adjoint action of K on $\mathfrak{p}_+$ is isomorphic to the standard action of K on $\mathrm{Sym}^2(\mathbb{C}^2)$. Explicitly, the intertwiner is given by $R_{rs} \mapsto e_r e_s$, where e_1, e_2 denotes the standard basis of $\mathbb{C}^2$. We obtain an isomorphism of K-modules

$$\mathrm{Sym}^\bullet \mathrm{Sym}^2 \mathbb{C}^2 = \bigoplus_{j=0}^{\infty} \mathrm{Sym}^j \mathrm{Sym}^2 \mathbb{C}^2 \simeq U(\mathfrak{p}_+) \tag{3.8}$$

of the symmetric algebra on $\mathrm{Sym}^2 \mathbb{C}^2$ with the universal enveloping algebra of $\mathfrak{p}_+$.

Lemma 3.5. *We have*

$$\phi_{q,\ell} = \frac{p^\ell(-1)^{q+\ell}}{2^\ell \pi^{q+\ell}} \omega(R_{12})^{q+\ell}\varphi_0.$$

Proof. We indicate a quick proof using the Fock model of the Weil representation. For more details for what follows, see the appendix of [12]. There is an intertwining map $\iota : S(V^n) \to \mathcal{P}(\mathbb{C}^{n(p+q)})$ from the polynomial Fock space to the infinitesimal Fock model of the Weil representation acting on the space of complex polynomials $\mathcal{P}(\mathbb{C}^{n(p+q)})$ in $n(p+q)$ variables such that $\iota(\varphi_0) = 1$. We denote the variables in $\mathcal{P}(\mathbb{C}^{n(p+q)})$ by $z_{\alpha i}$ $(1 \le \alpha \le p)$ and $z_{\mu i}$ $(p+1 \le \mu \le p+q)$ with $i = 1, \ldots, n$. Moreover, the intertwining map ι satisfies

$$\iota\left(x_{\alpha i} - \frac{1}{2\pi}\frac{\partial}{\partial x_{\alpha i}}\right)\iota^{-1} = \frac{1}{2\pi i}z_{\alpha i}.$$

Hence in the Fock model, we have

$$\phi_{q,\ell} = \frac{p^\ell}{2^{q+2\ell}}\left(\frac{1}{2\pi i}\right)^{2(q+\ell)}\left[\sum_{\alpha=1}^{p} z_{\alpha 1}z_{\alpha 2}\right]^{q+\ell}.$$

On the other hand, for the action of the raising operators, we find

$$\omega(R_{rs}) = \frac{1}{8\pi}\sum_{\alpha=1}^{p} z_{\alpha r}z_{\alpha s} - 2\pi\sum_{\mu=p+1}^{m}\frac{\partial^2}{\partial z_{\mu r}\partial z_{\mu s}}. \tag{3.9}$$

In the Fock model, we therefore have $\omega(R_{12})^{q+\ell}\varphi_0 = \left[\frac{1}{8\pi}\sum_{\alpha=1}^{p} z_{\alpha 1}z_{\alpha 2}\right]^{q+\ell}$, and the lemma follows. $\qquad\square$

We obtain:

Proposition 3.6. *For $k \in K \simeq \widetilde{\mathrm{U}}(2)$, we have*

$$\omega(k)\phi_{q,\ell} = \frac{p^\ell(-1)^{q+\ell}}{2^\ell \pi^{q+\ell}}\det(\mathbf{k})^{(p-q)/2}(\mathrm{Ad}(k)R_{12})^{q+\ell}\varphi_0.$$

Proof. This follows immediately from Lemma 3.5 and the fact that the Gaussian φ_0 has weight $(p-q)/2$. $\qquad\square$

Remark 3.7. The Kudla-Millson forms φ_{KM} cannot be expressed in terms of elements in $\mathfrak{p}_+$.

Proposition 3.6 reduces the K-action on $\phi_{q,\ell}$ to the representation theory of the group $\mathrm{U}(2)(\mathbb{C}) = \mathrm{GL}_2(\mathbb{C})$ on $\mathrm{Sym}^\bullet \mathrm{Sym}^2 \mathbb{C}^2$, which is given as follows.

Lemma 3.8. *The* $\mathrm{GL}_2(\mathbb{C})$*-representation* $\mathrm{Sym}^j \, \mathrm{Sym}^2 \, \mathbb{C}^2$ *decomposes as*

$$\mathrm{Sym}^j \, \mathrm{Sym}^2 \, \mathbb{C}^2 \simeq \bigoplus_{i=0}^{[j/2]} \mathrm{Sym}^{2j-4i} \, \mathbb{C}^2 \otimes \det^{2i}$$

into its irreducible constituents. The summand for $i = [j/2]$ *is given by*

$$\mathrm{Sym}^{2j-4[j/2]} \, \mathbb{C}^2 \otimes \det^{2[j/2]} = \begin{cases} \det^j & \text{if } j \text{ is even,} \\ \mathrm{Sym}^2 \, \mathbb{C}^2 \otimes \det^{j-1} & \text{if } j \text{ is odd,} \end{cases} \quad (3.10)$$

and is generated by the vector

$$\alpha_j = \sum_{i=0}^{[j/2]} \binom{[j/2]}{i} (-1)^i (e_1^2)^i (e_2^2)^i (e_1 e_2)^{j-2i}$$

$$= \begin{cases} \left[(e_1 e_2)^2 - e_1^2 e_2^2 \right]^{j/2} & \text{if } j \text{ is even,} \\ (e_1 e_2) \left[(e_1 e_2)^2 - e_1^2 e_2^2 \right]^{[j/2]} & \text{if } j \text{ is odd.} \end{cases} \quad (3.11)$$

Proof. For the first statement, see e.g. [11], p.81/82. For (3.11), note that in

$$\mathrm{Sym}^2 \, \mathrm{Sym}^2 \, \mathbb{C}^2 = \mathrm{Sym}^4 \, \mathbb{C} \oplus \det^2,$$

the vector

$$\alpha_2 = (e_1 e_2)^2 - e_1^2 e_2^2$$

generates the one-dimensional sub-representation. Then, for j even, α_j is given by the image of $(\alpha_2)^{j/2} \in \mathrm{Sym}^{j/2} \, \mathrm{Sym}^2 \, \mathrm{Sym}^2 \, \mathbb{C}^2$ under the projection onto $\mathrm{Sym}^j \, \mathrm{Sym}^2 \, \mathbb{C}^2$. The argument for j odd is analogous. $\quad\square$

By slight abuse of notation, we also write α_j for the corresponding element in $U(\mathfrak{p}_+)$ and define another Schwartz function $\xi = \xi_{q,\ell} \in \mathcal{S}(V^2)$ by

$$\xi = \xi_{q,\ell} = \frac{p^\ell (-1)^{q+\ell}}{2^\ell \pi^{q+\ell}} \omega(\alpha_j) \varphi_0. \quad (3.12)$$

Proposition 3.9. *For the Schwartz function* $\phi_{q,\ell}$*, there exists a* $\psi \in \mathcal{S}(V^2)$ *such that*

$$\phi_{q,\ell} = \xi_{q,\ell} + \omega(R_1) \omega(R_2) \psi. \quad (3.13)$$

Proof. We have

$$(e_1 e_2)^{q+\ell} - \alpha_{q+\ell} = e_1^2 e_2^2 \sum_{i=1}^{[(q+\ell)/2]} \binom{[(q+\ell)/2]}{i} (-1)^i (e_1^2)^{i-1} (e_2^2)^{i-1} (e_1 e_2)^{q+\ell-2i}.$$

Using the intertwiner with $U(\mathfrak{p}_+)$, we recall that e_i^2 corresponds to R_i. Thus ψ is given by

$$\psi = \frac{p^\ell(-1)^{q+\ell}}{2^\ell \pi^{q+\ell}} \sum_{i=1}^{[(q+\ell)/2]} \binom{[(q+\ell)/2]}{i} (-1)^i \omega \left(R_1^{i-1} R_2^{i-1} R_{12}^{q+\ell-2i} \right) \varphi_0. \quad \Box$$

One easily sees using (3.9):

Lemma 3.10. *The Schwartz function ξ vanishes identically if and only if $p = 1$ and $q + \ell > 1$.*

Example 3.11. For $q = 2$, $p > 1$, and $\ell = 0$, we have

$$\phi_{2,0} \cdot \Omega^p = C \varphi_{KM} \wedge \varphi_{KM} \wedge \Omega^{p-2} + C' \omega(R_1) \omega(R_2) \varphi_0 \cdot \Omega^p$$

for some nonzero constants C and C'. Here Ω denotes the Kähler form on the Hermitian domain D. But we will not need this.

In view of Lemma 3.8 and Proposition 3.6, we see for $q + \ell$ even that

$$\omega(k)\xi = \det(\mathbf{k})^{m/2+\ell}\xi \tag{3.14}$$

for $k \in K$. We let $\Xi(g, s)$ be the section in the induced representation corresponding to the Schwartz function ξ via (2.4).

Proposition 3.12. *Let $q + \ell$ be even. Then Ξ is the standard section (2.7) at the infinite place of weight $m/2 + \ell$. More precisely,*

$$\Xi(s) = C(s)\Phi_\infty^{m/2+\ell}(s) \tag{3.15}$$

for a certain (explicit) polynomial $C(s)$. Moreover,

$$C(s_0) \neq 0$$

with $s_0 = (m - 3)/2$ as in (2.5) for $p > 1$, while $C(s) \equiv 0$ for $p = 1$.

Proof. The identity (3.15) follows from (3.14) and the uniqueness of $\Phi_\infty^{m/2+\ell}$. The precise statement follows from considerations in [22]. The element $\alpha_{q+\ell}$ is trivially a highest weight vector of weight $\mu = (q + \ell, q + \ell)$ of $GL_2(\mathbb{C})$. Therefore we can take $\alpha_{q+\ell}$ equal to the element $u_\mu^0 \in U(\mathfrak{p}_+)$ (or $u_\mu \in U(\mathfrak{g})$) in the notation of [22], p.31/32. Then by Corollary 1.4 of [22], we have $\Xi(s) = u_\mu \Phi_\infty^{(p-q)/2}(s) = c P_\mu^{(p-q)/2}(s)\Phi_\infty^{m/2+\ell}(s)$, for a certain polynomial $P_\mu^{(p-q)/2}$ and a nonzero constant c. One easily sees $P_\mu^{(p-q)/2}(s_0) \neq 0$ for $p > 1$. See also [22], p. 38. For $p = 1$, Ξ vanishes identically, since already $\xi = 0$ by Lemma 3.10. $\quad \Box$

Remark 3.13. For $q + \ell$ odd, we see in the same way

$$\Xi(s) = C(s) R_{12} \Phi_\infty^{m/2+\ell-1}(s)$$

for a certain polynomial $C(s)$. Note that $\alpha_{q+\ell}$ is *not* a highest weight vector for $\mathrm{Sym}^2 \mathbb{C}^2 \otimes \det^{q+\ell-1}$ (which has weight $(q + \ell + 1, q + \ell - 1)$).

4. The L^2-norm of the theta lift

We now return to the global situation and retain the notation of Section 2. Let V be a non-degenerate quadratic space over $\mathbb{Q}$ of signature (p, q) and even dimension $m = p + q$. We let $L \subset V$ be an even lattice and write $L^\#$ for the dual lattice. We let $H = \mathrm{GSpin}(V)$. For each prime p, we let $L_p = L \otimes \mathbb{Z}_p$ and let K_p^H be the subgroup of $H(\mathbb{Q}_p)$ which leaves L_p stable and acts trivially on $L_p^\#/L_p$. Then $K_f^H = \prod_p K_p^H$ is an open compact subgroup of $H(\mathbb{A}_f)$. We let K_∞^H be a maximal compact subgroup of $H(\mathbb{R})$. By strong approximation we write

$$H(\mathbb{A}) = \coprod_j H(\mathbb{Q}) H(\mathbb{R})_0 h_j K_f^H \tag{4.1}$$

with $h_j \in H(\mathbb{A}_f)$. Then we put

$$X = X_{K_f^H} = H(\mathbb{Q}) \backslash (D \times H(\mathbb{A}_f)) / K_f^H \tag{4.2}$$

such that

$$X \simeq \coprod_j X_j \tag{4.3}$$

with $X_j = \Gamma_j \backslash D_+$, where $\Gamma_j = H(\mathbb{Q}) \cap (H(\mathbb{R})_0 h_j K_f^H h_j^{-1})$. We let $\varphi_f \in \mathcal{S}(V(\mathbb{A}_f))^{K_f^H}$ be a K_f^H-invariant Schwartz function on the finite adeles. Then φ_f corresponds to a linear combination of characteristic functions on the discriminant group $L^\#/L$. Since $\varphi_{q,\ell}$ is an eigenfunction of weight

$$\kappa = m/2 + \ell$$

under the action of $U(1)$, we can form the classical theta function on $\mathbb{H}$, the upper half space, by setting

$$\theta(\tau, z, \varphi_{q,\ell} \otimes \varphi_f) = v^{-\kappa/2} \sum_{x \in V(\mathbb{Q})} \varphi_f(x) \omega_\infty(g_\tau) \varphi_{q,\ell}(x, z)$$

$$= v^{-\ell/2} \sum_{x \in V(\mathbb{Q})} \varphi_f(x) \varphi_{q,\ell}(\sqrt{v} x, z) e^{\pi i(x,x)u}.$$

Here $\tau = u + iv \in \mathbb{H}$, and $g_\tau = \begin{pmatrix} 1 & u \\ 0 & 1 \end{pmatrix} \begin{pmatrix} \sqrt{v} & 0 \\ 0 & \sqrt{v}^{-1} \end{pmatrix} \in SL_2(\mathbb{R}) \subset SL_2(\mathbb{A})$ is the standard element moving the base point $i \in \mathbb{H}$ to τ. Then $\theta(\tau, z, \varphi_{q,\ell} \otimes \varphi_f)$ transforms like a non-holomorphic modular form of weight κ for the principal congruence subgroup $\Gamma(N)$ of $SL_2(\mathbb{Z})$ taking values in the differential q-forms on X. Here N is the level of L, i.e., the smallest positive integer such that $\frac{1}{2}N(x, x) \in \mathbb{Z}$ for all $x \in L^\#$. In particular, if L is unimodular, $\theta(\tau, z, \varphi_{q,\ell} \otimes \varphi_f)$ is a form for the full modular group $SL_2(\mathbb{Z})$.

We write $S_\kappa(\Gamma(N))$ for the space of cusp forms of weight κ for $\Gamma(N)$. We normalize the Petersson scalar product be putting

$$(f, g) = \frac{1}{[\Gamma(1) : \Gamma(N)]} \int_{\Gamma(N) \backslash \mathbb{H}} f(\tau) \overline{g(\tau)} v^\kappa \, d\mu(\tau) \qquad (4.4)$$

for $f, g \in S_\kappa(\Gamma(N))$. Here $d\mu(\tau) = \frac{du\,dv}{v^2}$ is the invariant measure on $\mathbb{H}$. For $f \in S_\kappa(\Gamma(N))$, we consider the theta lift

$$\Lambda(f) = \left(f, \theta(\tau, \varphi_{q,\ell} \otimes \varphi_f) \right) = \int_{\Gamma(N) \backslash \mathbb{H}} f(\tau) \overline{\theta(\tau, \varphi_{q,\ell} \otimes \varphi_f)} v^\kappa \, d\mu(\tau).$$
$$(4.5)$$

It defines a linear map

$$\Lambda : S_\kappa(\Gamma(N)) \longrightarrow \mathcal{Z}^q(X, \widetilde{\operatorname{Sym}}^\ell(V)) \qquad (4.6)$$

into the $\widetilde{\operatorname{Sym}}^\ell(V)$-valued closed differential q-forms on X.

In order to show the injectivity of Λ, we study its L^2-norm given by

$$\|\Lambda(f)\|_2^2 = \int_X \Lambda(f) \wedge *\overline{\Lambda(f)}. \qquad (4.7)$$

We will use the *doubling method* to compute $\|\Lambda(f)\|_2^2$, see [5, 13, 25, 27].

Proposition 4.1. *Assume that* $m > 3 + r$ *so that Weil's convergence criterion* (2.3) *in genus 2 holds. Then* $\Lambda(f)$ *is square integrable, and*

$$\|\Lambda(f)\|_2^2 = \left(f(\tau_1) \otimes \overline{f(\tau_2)}, \tilde{I}(\tau_1, -\bar{\tau}_2, \phi_{q,\ell} \otimes \phi_f) \right), \qquad (4.8)$$

where (,) *denotes the Petersson scalar product on* $\Gamma(N) \times \Gamma(N)$ *and*

$$\tilde{I}(\tau_1, \tau_2, \phi_{q,\ell} \otimes \phi_f) = \int_X \theta(\tau_1, \tau_2, z, \phi_{q,\ell} \otimes \phi_f)\mu \qquad (4.9)$$

is the integral over the locally symmetric space of the theta series

$$\theta(\tau_1, \tau_2, z, \phi_{q,\ell} \otimes \phi_f) = (v_1 v_2)^{-\kappa/2} \sum_{\mathbf{x} \in V^2(\mathbb{Q})} \phi_f(\mathbf{x})(\omega_\infty(\iota_0(g_{\tau_1}, g_{\tau_2}))\phi_{q,\ell}(\mathbf{x}, z),$$
$$(4.10)$$

(which by (3.1) defines a modular form of weight κ on $\Gamma(N) \times \Gamma(N)$). Here
$\phi_f = \varphi_f \otimes \varphi_f \in \mathcal{S}(V^2(\mathbb{A}_f))$.

Proof. The formula (4.8) implies the square integrability since the right hand side of (4.8) is absolutely convergent by Weil's convergence criterion (2.3). We have

$$\|\Lambda(f)\|_2^2 = \int_X \left(\int_{\Gamma(N)\backslash\mathbb{H}} f(\tau_1)\overline{\theta(\tau_1, \varphi_{q,\ell} \otimes \varphi_f)} v_1^\kappa d\mu(\tau_1) \right)$$
$$\wedge \overline{\left(\int_{\Gamma(N)\backslash\mathbb{H}} f(\tau_2)\overline{\theta(\tau_2, *\varphi_{q,\ell} \otimes \varphi_f)} v_2^\kappa d\mu(\tau_2) \right)}.$$

Interchanging the integration, we obtain

$$\int\int f(\tau_1)\overline{f(\tau_2)}\left(\int_X \theta(\tau_1, \varphi_{q,\ell} \otimes \varphi_f) \wedge \overline{\theta(\tau_2, *\varphi_{q,\ell} \otimes \varphi_f)} \right)(v_1 v_2)^\kappa d\mu(\tau_1)d\mu(\tau_2).$$

Since $\varphi_{q,\ell}$ is real valued, we easily see by the explicit formulas of the Weil representation that

$$\overline{\theta(\tau_2, *\varphi_{q,\ell} \otimes \varphi_f)} = \theta(-\bar{\tau}_2, *\varphi_{q,\ell} \otimes \varphi_f)$$

and therefore

$$\theta(\tau_1, \varphi_{q,\ell} \otimes \varphi_f) \wedge \overline{\theta(\tau_2, *\varphi_{q,\ell} \otimes \varphi_f)} = \theta(\tau_1, -\bar{\tau}_2, z, \phi_{q,\ell} \otimes \phi_f)\mu$$

by (3.2). This implies the assertion. $\qquad\square$

Remark 4.2. For signature $(p, 2)$, the lift $\Lambda(f)$ is actually always square integrable, see [6, 8]. We expect this to be true for other signatures as well even if Weil's convergence criterion does not hold. In that case, one would need to regularize the theta integral $\tilde{I}$ as in [23].

Note that the Schwartz function ξ introduced by (3.12) is K^V-invariant. We can therefore consider $\xi \in [\mathcal{S}(V^2) \otimes C^\infty(D)]^{O(V)(\mathbb{R})}$ by setting

$$\xi(\mathbf{x}, z) = \xi(h_\infty^{-1}\mathbf{x})$$

with $h_\infty \in O(V)(\mathbb{R})$ such that $h_\infty z_0 = z$. In particular, $\xi(\mathbf{x}, z_0) = \xi(\mathbf{x})$.

Proposition 4.3. *Define* $\theta(\tau_1, \tau_2, z, \xi \otimes \phi_f)$ *and* $\tilde{I}(\tau_1, \tau_2, \xi \otimes \phi_f)$ *in the same way as for* $\phi_{q,\ell}$ *in* (4.10), (4.9). *Then*

$$\|\Lambda(f)\|_2^2 = \left(f(\tau_1) \otimes \overline{f(\tau_2)}, \tilde{I}(\tau_1, -\bar{\tau}_2, \xi \otimes \phi_f) \right).$$

Proof. By Proposition 3.9 and Proposition 4.1, we see (omitting ϕ_f from the notation)

$$\|\Lambda(f)\|_2^2 = \left(f(\tau_1) \otimes \overline{f(\tau_2)}, \tilde{I}(\tau_1, -\bar{\tau}_2, \phi_{q,\ell}) \right)$$

$$= \left(f(\tau_1) \otimes \overline{f(\tau_2)}, \tilde{I}(\tau_1, -\bar{\tau}_2, \xi) \right)$$
$$+ \left(f(\tau_1) \otimes \overline{f(\tau_2)}, \tilde{I}(\tau_1, -\bar{\tau}_2, \omega(R_1)\omega(R_2)\psi) \right)$$
$$= \left(f(\tau_1) \otimes \overline{f(\tau_2)}, \tilde{I}(\tau_1, -\bar{\tau}_2, \xi) \right)$$
$$+ \left(f(\tau_1) \otimes \overline{f(\tau_2)}, R_1 R_2 \tilde{I}(\tau_1, -\bar{\tau}_2, \psi) \right).$$

By the adjointness of the Maass lowering and raising operators with respect to the Petersson scalar product, the latter summand vanishes. □

Corollary 4.4. *Let* $p = 1$ *and* $q + \ell > 1$. *Then* Λ *vanishes identically.*

Proof. This is obvious from Proposition 4.3 and $\xi = 0$ (Lemma 3.10). □

Remark 4.5. We could have defined the lift Λ of f by using the Schwartz form $\varphi_{q,[\ell]}$ instead of the form $\varphi_{q,\ell}$. Using Lemma 3.1 we see by the argument of the proof of Proposition 4.3 that the L^2-norms $\|\Lambda(f)\|$ coincide.

We want to relate the integral $\tilde{I}(\tau_1, \tau_2, \xi \otimes \phi_f)$ to the pullback of a genus 2 Eisenstein series via the Siegel-Weil formula. We first need to relate the integral over the locally symmetric space X to an integral over $O(V)(\mathbb{Q}) \backslash O(V)(\mathbb{A})$. We do this following [17], pp. 332. First we define the theta series associated to ξ more generally for $g \in \mathrm{Sp}(2, \mathbb{A})$ and $h = (h_\infty h_f) \in O(V)(\mathbb{A})$ by

$$\theta(g, h, \xi \otimes \phi_f) = \sum_{\mathbf{x} \in V^2(\mathbb{Q})} \omega(g) \xi(h_\infty^{-1}\mathbf{x}, z_0) \phi_f(h_f^{-1}\mathbf{x}),$$

where z_0 is the base point of D. Note that

$$\theta(\tau_1, \tau_2, z, \xi \otimes \phi_f) = (v_1 v_2)^{-\kappa/2} \theta(\iota_0(g_{\tau_1}, g_{\tau_2}), h_\infty, \xi \otimes \phi_f)$$

with $h_\infty \in O(V)(\mathbb{R})$ such that $z = h_\infty z_0$.

Proposition 4.6. *We have*

$$\frac{1}{\mathrm{vol}(X, \mu)} \tilde{I}(\tau_1, \tau_2, \xi \otimes \phi_f) = (v_1 v_2)^{-\kappa/2} \int_{O(V)(\mathbb{Q}) \backslash O(V)(\mathbb{A})} \theta(\iota_0(g_{\tau_1}, g_{\tau_2}), h, \xi \otimes \phi_f) dh.$$

Proof. Arguing as in the proof of Proposition 4.17 of [17], we first obtain

$$\frac{1}{\mathrm{vol}(X, \mu)} \tilde{I}(\tau_1, \tau_2, \xi \otimes \phi_f) = (v_1 v_2)^{-\kappa/2} \frac{1}{2} \int_{SO(V)(\mathbb{Q}) \backslash SO(V)(\mathbb{A})} \theta(\iota_0(g_{\tau_1}, g_{\tau_2}), h, \xi \otimes \phi_f) dh,$$

where dh is the Tamagawa measure on $SO(V)(\mathbb{A})$. But now the sign representation of $O(V_v)/SO(V_v)$ does not occur in the local theta correspondence for the pair $(\mathrm{Sp}(2), \mathrm{O}(V))$ for any place v if $\dim V = m > 2$, see [26]. Then arguing as in [17], p. 326, we see

$$\frac{1}{2} \int_{\mathrm{SO}(V)(\mathbb{Q}) \backslash \mathrm{SO}(V)(\mathbb{A})} \theta(g, h, \xi \otimes \phi_f) dh = \int_{\mathrm{O}(V)(\mathbb{Q}) \backslash \mathrm{O}(V)(\mathbb{A})} \theta(g, h, \xi \otimes \phi_f) dh,$$

from which the proposition follows. $\qquad \square$

Proposition 4.7. *Let* $\Xi(s) \otimes \Phi_f(s)$ *be the section associated to* $\xi \otimes \phi_f$ *via* (2.4) *and let* $s_0 = (m-3)/2$. *Then*

$$\frac{1}{\mathrm{vol}(X, \mu)} \|\Lambda(f)\|_2^2 = (v_1 v_2)^{-\kappa/2} \left(f(\tau_1) \otimes \overline{f(\tau_2)}, E(\iota_0(g_{\tau_1}, g_{-\bar{\tau}_2}), s_0, \Xi \otimes \Phi_f) \right).$$

Proof. Using Proposition 4.6 and the Siegel-Weil formula, Theorem 2.1, we find

$$\frac{1}{\mathrm{vol}(X, \mu)} \tilde{I}(\tau_1, \tau_2, \xi \otimes \phi_f) = (v_1 v_2)^{-\kappa/2} E(\iota_0(g_{\tau_1}, g_{\tau_2}), s_0, \Xi \otimes \Phi_f).$$

Now the assertion follows from Proposition 4.3. $\qquad \square$

Corollary 4.8. *Assume that* $q + \ell$ *is even and* $p > 1$. *Let* $\Phi_\infty^\kappa(s)$ *be the standard section defined by* (2.7), *and let* $\Phi_f(s)$ *be the section associated to* ϕ_f *via* (2.4). *Then*

$$\frac{1}{\mathrm{vol}(X, \mu)} \|\Lambda(f)\|_2^2 = C(s_0)(v_1 v_2)^{-\kappa/2} \left(f(\tau_1) \otimes \overline{f(\tau_2)}, E(\iota_0(g_{\tau_1}, g_{-\bar{\tau}_2}), s_0, \Phi_\infty^\kappa \otimes \Phi_f) \right),$$

where $C(s_0)$ *is the nonzero constant in Proposition 3.12.*

Proof. We have $\Xi(g, s) = C(s)\Phi_\infty^\kappa(g, s)$ by Proposition 3.12. Hence the Corollary immediately follows from Proposition 4.7. $\qquad \square$

Suppose that f is an eigenform of level N and let S denote the set of primes dividing N together with ∞. Then the doubling method [5, 13, 25, 27] expresses a convolution integral as on the right hand side above as a product of the standard L-function $L^S(s_0 + \frac{1}{2}, f)$ with the Euler factors corresponding to $p \in S$ omitted times a product of "bad" local factors corresponding to the primes in S. If $m > 4$ then $s_0 + \frac{1}{2}$ lies in the region of convergence of the Euler product of $L^S(s, f)$. Hence the L-value does not vanish. Therefore the lift $\Lambda(f)$ vanishes precisely if at least one of the "bad" local factors vanishes. By the analysis of the present paper we determine the local factor at infinity.

We now specialize to the case when the lattice L is even and *unimodular*. Then φ_f corresponds to the characteristic function of L and $\Phi_f(s) = 1$. The level of L is $N = 1$, so that ∞ is the only "bad" place. By the above analysis we obtain a very explicit formula for $\|\Lambda(f)\|_2^2$ as we shall now explain.

In this case $\theta(\tau, z, \varphi_{q,\ell})$ is a modular form of weight $\kappa = m/2 + \ell$ for $SL_2(\mathbb{Z})$ and vanishes unless $q + \ell$ is even, which we assume from now on as well. Then κ is even, because $8 \mid p - q$. By Corollary 4.8 and (2.8) we have

$$\frac{1}{\text{vol}(X, \mu)} \|\Lambda(f)\|_2^2 = C(s_0) \left(f(\tau_1) \otimes \overline{f(\tau_2)}, E_\kappa^{(2)}(\tau_1, -\bar{\tau}_2, -\ell/2) \right),$$

(4.11)

where $E_\kappa^{(2)}(\tau_1, \tau_2, s)$ is the pullback of the classical genus 2 Siegel Eisenstein series $E_\kappa^{(2)}(\tau, s)$ (see (2.9)) to the diagonal.

We recall the definition of the standard L-function of a Hecke eigenform $f \in S_\kappa(\Gamma(1))$. We use the normalization of [1], [5], [24]. We denote the Fourier coefficients of f by $c(n)$ and assume that f is normalized, i.e., $c(1) = 1$. Let p be a prime. The Satake parameters $\alpha_{0,p}, \alpha_{1,p}$ of f at p are defined by the factorization of the Hecke polynomial

$$(1 - c(p)X + p^{\kappa-1}X^2) = (1 - \alpha_{0,p}X)(1 - \alpha_{0,p}\alpha_{1,p}X).$$

(4.12)

Hence

$$\alpha_{0,p}^2\alpha_{1,p} = p^{\kappa-1}, \qquad\qquad \alpha_{0,p}(1 + \alpha_{1,p}) = c(p).$$

According to Deligne's theorem, formerly the Ramanujan-Petersson conjecture, we have $|\alpha_{1,p}| = 1$. The standard L-function of f is defined by the Euler product

$$D_f(s) = \prod_p \left[(1 - p^{-s})(1 - \alpha_{1,p}^{-1}p^{-s})(1 - \alpha_{1,p}p^{-s}) \right]^{-1}.$$

(4.13)

It converges for $\Re(s) > 1$. The corresponding completed L-function

$$\Psi_f(s) = \pi^{-\frac{3s}{2}} \Gamma\left(\frac{s+1}{2}\right) \Gamma\left(\frac{s+\kappa-1}{2}\right) \Gamma\left(\frac{s+\kappa}{2}\right) D_f(s)$$

(4.14)

has a meromorphic continuation to $\mathbb{C}$ and satisfies the functional equation

$$\Psi_f(s) = \Psi_f(1 - s)$$

(4.15)

(see e.g. [5], [28]). It is well known (see [28], Introduction, [31]) that $D_f(s)$ can be interpreted as the Rankin L-series

$$D_f(s) = \zeta(2s) \sum_{n=1}^\infty c(n^2)n^{-s-\kappa+1} = \frac{\zeta(2s)}{\zeta(s)} \sum_{n=1}^\infty c(n)^2 n^{-s-\kappa+1}.$$

Theorem 4.9. *Assume that $m > 3 + r$ so that Weil's convergence criterion (2.3) in genus 2 holds. Furthermore, assume that $q + \ell$ is even and that*

L is even unimodular. Let $f \in S_\kappa(\Gamma(1))$ *be a Hecke eigenform, and write* $\|f\|_2^2 = (f, f)$ *for its Petersson norm normalized as in* (4.4). *We have*

$$\frac{1}{\text{vol}(X, \mu)} \cdot \frac{\|\Lambda(f)\|_2^2}{\|f\|_2^2} = C(s_0)\mu(1, \kappa, -\ell/2)\frac{D_f(m/2 - 1)}{\zeta(m/2)\zeta(m - 2)},$$

where

$$\mu(1, \kappa, -\ell/2) = 2^{3-m/2}(-1)^{\kappa/2}\pi\frac{\Gamma(m/2 + \ell/2 - 1)}{\Gamma(m/2 + \ell/2)}.$$

Proof. The statement follows from (4.11) by means of [5], identities (14) and (22). $\square$

Remark 4.10. By the same argument it is easily seen that $(\Lambda(f), \Lambda(g)) = 0$ for two different normalized Hecke eigenforms f and g.

Corollary 4.11. *Assume that* $m > \max(4, 3 + r)$, $p > 1$, $q + \ell$ *even, and that* *L is even unimodular. Then the theta lift* $\Lambda : S_\kappa(\Gamma(1)) \to \mathcal{Z}^q(X, \widetilde{\text{Sym}}^\ell(V))$ *is injective.*

Proof. This follows from Theorem 4.9, Proposition 3.12, and the convergence of the Euler-product for $D_f(m/2 - 1)$ in this case. $\square$

References

[1] A. N. Andrianov, *The multiplicative arithmetic of Siegel modular forms*, Russian Math. Surveys **34** (1979), 75–148.

[2] R. E. Borcherds, *Automorphic forms on* $O_{s+2,2}(\mathbb{R})$ *and infinite products*, Invent. Math. **120** (1995), 161–213.

[3] R. E. Borcherds, *Automorphic forms with singularities on Grassmannians*, Inv. Math. **132** (1998), 491–562.

[4] R. Borcherds, *The Gross-Kohnen-Zagier theorem in higher dimensions*, Duke Math. J. **97** (1999), 219–233. Correction in: Duke Math J. **105** No. 1 p.183–184.

[5] S. Böcherer, *Über die Funktionalgleichung automorpher L-Funktionen zur Siegelschen Modulgruppe*, J. Reine Angew. Math. 362 (1985), 146–168.

[6] J. Bruinier, *Borcherds products on* O(2, l) *and Chern classes of Heegner divisors*, Springer Lecture Notes in Mathematics **1780**, Springer-Verlag (2002).

[7] *J. H. Bruinier and E. Freitag*, Local Borcherds products, Annales de l'Institut Fourier **51.1** (2001), 1–26.

[8] J. Bruinier and J. Funke, *On two geometric theta lifts*, Duke Math J. **125** (2004), 45–90.

[9] J. Bruinier and O. Stein, *The Weil representation and Hecke operators for vector valued modular forms*, preprint (2007).

[10] E. Freitag, *Stabile Modulformen*, Math. Ann. **230** (1977), 197–211.

[11] W. Fulton and J. Harris, *Representation Theory, A First Course*, Graduate Texts in Mathematics **129**, Springer, 1991.

[12] J. Funke and J. Millson, *Cycles with local coefficients for orthogonal groups and vector-valued Siegel modular forms*, American J. Math. **128** (2006), 899–948.

[13] P. Garrett, *Pullbacks of Eisenstein series; Applications.* In: Automorphic forms of several variables, Taniguchi Symposium, Katata, 1983, Birhäuser (1984).

[14] *J. Harvey and G. Moore*, Algebras, BPS states, and strings, Nuclear Phys. B **463** (1996), no. 2–3, 315–368.

[15] *J.-S. Li, Nonvanishing theorems for the cohomology of certain arithmetic quotients*, J. Reine Angew. Math. **428** (1992), 177–217.

[16] S. Kudla, *Some extensions of the Siegel-Weil formula.* In: Eisenstein series and applications, 205–237, Progr. Math. **258**, Birkhäuser, Boston (2008).

[17] S. Kudla, *Integrals of Borcherds forms*, Compositio Math. **137** (2003), 293–349.

[18] S. Kudla and J. Millson, *The Theta Correspondence and Harmonic Forms I*, Math. Ann. **274** (1986), 353-378.

[19] S. Kudla and J. Millson, *Intersection numbers of cycles on locally symmetric spaces and Fourier coefficients of holomorphic modular forms in several complex variables*, IHES Pub. **71** (1990), 121–172.

[20] S. Kudla and S. Rallis, *On the Weil-Siegel formula*, J. Reine Angew. Math. **387** (1988), 1–68.

[21] S. Kudla and S. Rallis, *On the Weil-Siegel formula II*, J. Reine Angew. Math. **391** (1988), 65–84.

[22] S. Kudla and S. Rallis, *Degenerate principal series and invariant distributions*, Israel J. Math. **69** (1990), 25–45.

[23] S. Kudla and S. Rallis, *A regularized Weil-Siegel formula: the first term identity*, Annals of Math. **140** (1994), 1–80.

[24] R. P. Langlands, *Problems in the theory of automorphic forms*, Lecture Notes in Math. **170** (1970), 18–86, Springer-Verlag.

[25] I. Piatetski-Shapiro, S. Rallis, *L*-functions for classical groups. Lecture Notes in Mathematics **1254**, Springer-Verlag, Berlin (1987).

[26] S. Rallis, *On the Howe duality conjecture*, Compositio Math. **51** (1984), 333–399.

[27] S. Rallis, *Injectivity properties of liftings associated to Weil representations*, Compositio Math. **52** (1984), 139–169.

[28] G. Shimura, *On the holomorphicity of certain Dirichlet series*, Proc. London Math. Soc. **31** (1975), 79–98.

[29] J.-L. Waldspurger, *Engendrement par des séries têta de certains espaces de formes modulaires*, Invent. Math. **50** (1979), 135–168.

[30] A. Weil, *Sur la formule de Siegel dans la théorie des groupes classiques*, Acta Math. **113** (1965) 1–87.

[31] D. Zagier, *Modular forms whose Fourier coefficients involve zeta-functions of quadratic fields*. In: Modular Functions of One Variable VI, Lecture Notes in Math. **627**, Springer-Verlag (1977), 105–169.

Ordered Spanning Sets for Vertex Operator Algebras and their Modules[1]

Geoffrey Buhl[2]

Mathematics Department
Rutgers University
Piscataway, NJ 08854-8019
gbuhl@math.rutgers.edu

Abstract

Moonshine relates three fundamental mathematical objects: the Monster sporadic simple group, the modular function $j(\tau)$, and the moonshine module vertex operator algebra $V^\natural$. Examining the relationship between modular functions and the representation theory of vertex operator algebras reveals rich structure. In particular, C_2-cofiniteness (also called Zhu's finiteness condition) implies the existence of finite generating sets and Poincaré-Birkhoff-Witt-like spanning sets for vertex operator algebras and their modules. These spanning sets feature desirable ordering restrictions, e.g., a difference-one condition.

1. Introduction

The theory of vertex operator algebra blossomed from two major accomplishments: the proof of the McKay-Thompson conjecture by Frenkel, Lepowsky, and Meurman [FLM88] who constructed the Moonshine module $V^\natural$ and the proof of the Conway-Norton conjecture by Borcherds [Bor92] using the Moonshine module. These two conjectures make up what is commonly referred to as Monstrous Moonshine, relating the modular function $j(\tau)$ and the Monster group by way of a third fundamental mathematical object, the Moonshine module vertex operator algebra $V^\natural$. The study of vertex operator algebras continues to reveal relations within mathematics and with physics.

Representation theory is a particularly rich aspect of the theory of vertex operator algebras with fundamental connections to number theory, the theory of simple groups, and string and conformal field theories in physics. A core idea in the representation theory of vertex operator algebras and conformal

---

1 A contribution to the Moonshine Conference at ICMS, Edinburgh, July 2004.
2 Supported by a NSF Postdoctoral Fellowship for the Mathematical Sciences.

field theory is "rationality", a term used in a variety of ways to describe certain desirable properties of a vertex operator algebra and its modules. Complete reducibility of modules is one property that "rationality" invariably encompasses, but not always solely. In both mathematics and physics, "rationality" is a term that suffers from a variety of meanings. Compounding this difficulty is the variety of module definitions that appear in mathematics and physics literature. This combination makes the concept "rationality is complete reducibility of modules" murky at best.

For certain vertex operator algebras, we can achieve some clarity. An assumption on the "size" of the vertex operator algebra has important implications for its representation theory. This size condition is called C_2-cofiniteness, and it implies the existence of a finite generating set and Poincaré-Birkhoff-Witt-like ordered spanning sets for the algebra and modules. An assumption of C_2-cofiniteness on an algebra ensures that even the most basic notion of a module has "suitable" structure. In addition, the assumption of C_2-finiteness clarifies the concept of complete reducibility for modules of a vertex operator algebra. Understanding the implications of C_2-cofiniteness is especially important in light of the recent developments in the representation theory of "non-rational", C_2-cofinite theories [Abe] [CF06] [HLZ].

2. Vertex operator algebras and quotient spaces

For an introduction to the theory of vertex operator algebras, I refer the reader to "Introduction to Vertex Operator Algebras and Their Representations" by Lepowsky and Li [LL04]. Throughout this exposition, I will assume that the vertex operator algebras are of "CFT-type". That is, a vertex operator algebra V is of CFT-type if $V = \bigoplus_{n \geq 0} V_n$ and $V_0 = \mathbb{C}\mathbf{1}$. The weight of a homogenous vector is its $L(0)$-eigenvalue, $L(0)u = (\mathrm{wt}u)u$. The weight of an operator, or "mode", u_n is also given by the $L(0)$-action, $L(0)u_n v = \mathrm{wt}(u_n)u_n L(0)v = (\mathrm{wt}u - n - 1)u_n L(0)v$ for $n \in \mathbb{Z}$.

One of the powerful tools in the study of these infinite-dimesional objects, vertex operator algebras, has been to look at quotient spaces. This technique's most important example is Zhu's algebra $A(V)$ [Zhu96].

Definition 2.1. For V a vertex operator algebra, let

$$O(V) = \mathrm{span}\left\{\mathrm{Res}_x \frac{(1+x)^{\mathrm{wt}u}}{x^2} Y(u,x)v : u, v \in V\right\},$$

and let $A(V) = V/O(V)$.

Zhu's algebra $A(V)$ is an associative algebra with identity, and it acts on lowest weight vectors of modules. This concept has been expanded to act on larger "slices" of modules. The nth Zhu's algebra $A_n(V)$, also an associative algebra, acts on the bottom n levels of modules [DLM98c].

Definition 2.2. For V a vertex operator algebra and $n \in \mathbb{N}$, let

$$O_n(V) = \text{span} \left\{ \text{Res}_x \frac{(1+x)^{\text{wt}u+n}}{x^{2n+2}} Y(u, x)v : u, v \in V \right\},$$

and let $A_n(V) = V/O_n(V)$, where $A_0(V) = A(V)$.

Under certain assumptions, all of the nth Zhu's algebras are semisimple and hence finite-dimensional. The representation theories of a vertex operator algebra V and its Zhu's algebras $A_n(V)$ are intimately related [Zhu96] [DLM98a] [DLM98c].

Another family of subspaces spaces used to create interesting quotient spaces are the C_n spaces. The subspace $C_2(V) = \text{span}\{\text{Res}_x x^{-2} Y(u, x)v : u, v \in V\}$ was introduced in Zhu's modularity paper [Zhu96]. One of the crucial assumptions needed to prove the modularity properties of certain graded traces is finite-dimensionality of the quotient space $V/C_2(V)$. This property is known as C_2-cofiniteness or Zhu's finiteness condition.

This quotient space $V/C_2(V)$ has the structure of a Poisson algebra. A Poisson algebra has two operations: an associative product $\cdot$ and a Lie bracket $[,]$ with compatibility of these operations given by Liebniz's Law $[x, y \cdot z] = [x, y] \cdot z + y \cdot [x, z]$. For $V/C_2(V)$, the product is given by $u \cdot v = \text{Res}_x x^{-1} Y(u, x)v = u_{-1}v$ and the Poisson bracket is given by $[u, v] = \text{Res}_x Y(u, x)v = u_0 v$. We can expand the definition of $C_2(V)$ to obtain a family of subspaces.

Definition 2.3. For a vertex operator algebra V and for $n \geq 2$, let

$$C_n(V) = \text{span}\{\text{Res}_x x^{-n} Y(u, x)v : u, v \in V\}.$$

Then V is called C_n-*cofinite* if $V/C_n(V)$ is finite-dimensional.

The case where $n = 1$ is more nuanced and depending on an author's focus, is approached differently. Focusing on the algebra, the naive extension of the definition, $\text{span}\{\text{Res}_x x^{-1} Y(u, x)v : u, v \in V\}$, is not particularly interesting since the creation axiom for vertex operator algebras ensures that this subspace is all of V. A more interesting subspace is the following.

Definition 2.4. (cf. [KL99]) For a vertex operator algebra $V = \bigoplus_{n \geq 0} V_n$, let

$$C_1(V) = \text{span} \left\{ \text{Res}_x x^{-1} Y(u, x)v, L(-1)u : u, v \in \bigoplus_{n>0} V_n \right\}.$$

Then V is called C_1-*cofinite* if $V/C_1(V)$ is finite-dimensional.

The assumption of C_n-cofiniteness of a vertex operator algebra controls the size of other quotient spaces. For example, a simple calculation shows that if a vertex operator algebra V is C_2-cofinite, then $A(V)$ is finite-dimensional. The $L(-1)$ derivation property implies $C_n(V) \subseteq C_{n-1}(V)$, and thus C_n-cofiniteness implies C_{n-1}-cofiniteness for $n \geq 2$. In fact C_2-cofiniteness implies that a great deal of quotient spaces are finite-dimensional [GN03].

There are other interesting quotient spaces. For example, if we define $L(-1)V = \text{span}\{L(-1)v : v \in V\}$, we can consider the quotient space $V/L(-1)V$. This has the structure of a commutative algebra under the operation $u \cdot v = \text{Res}_x x^{-1} Y(u, x)v = u_{-1}v$.

3. Modules

There are a wide variety of definitions of modules for vertex operator algebras. This variety stems from the amount of grading assumed for a given module and finite-dimensionality of the graded pieces (or lack thereof). Some modules are ungraded and others admit a grading by $\mathbb{N}$, $\mathbb{Q}$, $\mathbb{R}$, or $\mathbb{C}$. A $\mathbb{N}$-grading emphasizes lower-truncation, while the other gradings are given by the $L(0)$-eigenvalues. With a grading in place we may impose a further restriction: the graded pieces must be finite-dimensional.

Not only are there a variety of definitions for modules, the situation is further muddled by different names for the same objects (e.g.,"$\mathbb{N}$-graded weak " and "admissible"). Other adjectives modifying "module" in the literature are: weak, strong, ordinary, lowest-weight, and generalized. Because of the variety in language and structure, an explicit description of some of the different modules is warranted. A natural starting point is modules for vertex algebras, which are naturally ungraded. Every vertex operator algebra is a vertex algebra if one ignores the Virasoro vector and related axioms.

Definition 3.1. For a vertex algebra $(V, Y, \mathbf{1})$, a *vertex algebra module* (M, Y_M) is a vector space M with a linear map

$$Y_M : \quad V \to \text{End}(M)[[x, x^{-1}]] \qquad (3.1)$$

$$v \mapsto Y_M(v, x) = \sum_{n \in \mathbb{Z}} v_n x^{-n-1}. \qquad (3.2)$$

In addition Y_M satisfies the following:

1) $v_n w = 0$ for $n >> 0$ where $v \in V$ and $w \in M$
2) $Y_M(\mathbf{1}, x) = Id_M$
3) For all $u, v \in V$,

$$x_0^{-1} \delta \left(\frac{x_1 - x_2}{x_0} \right) Y_M(u, x_1) Y_M(v, x_2)$$

$$- x_0^{-1} \delta \left(\frac{x_2 - x_1}{-x_0} \right) Y_M(v, x_2) Y_M(u, x_1)$$

$$= x_2^{-1} \delta \left(\frac{x_1 - x_0}{x_2} \right) Y_M(Y(u, x_0)v, x_2). \qquad (3.3)$$

For a vertex operator algebra $V = (V, Y, \omega, \mathbf{1})$, we can consider objects (M, Y_M) as defined above for the vertex algebra structure of V.

Definition 3.2. A *weak module* for a vertex operator algebra V is a vertex algebra module for the vertex algebra structure of V.

Weak modules for vertex operator algebras have additional structure that is a consequence of the vertex algebra module axioms. They admit a representation of the Virasoro algebra and modules for a vertex operator also obey the $L(-1)$-derivation property.

Proposition 3.3. *Let $V = (V, Y, \omega, \mathbf{1})$ be a vertex operator algebra and $M = (M, Y_M)$ a weak module for V.*

1) $Y_M(\omega, x) = \sum_{n \in \mathbb{Z}} L_M(n) x^{-n-2}$ where

$$[L_M(m), L_M(n)] = (m - n) L_M(m + n) + \frac{m^3 - m}{12} \delta_{m+n,0} c$$

2) $Y_M(L(-1)v, x) = \frac{d}{dx} Y_M(v, x)$ for all $v \in V$

Even with this additional structure known, weak modules of vertex operator algebras still lack suitable structure. Some grading is necessary, and in particular a lower-truncated grading is desirable. A lower-truncated grading guarantees the existence of "lowest weight" vectors.

Definition 3.4. A weak module M for a vertex operator algebra V is called $\mathbb{N}$-*gradable* if it admits an $\mathbb{N}$-grading, $M = \bigoplus_{n \in \mathbb{N}} M(n)$, such that if $v \in V_r$ then $v_m M(n) \subseteq M(n + r - m - 1)$.

The additional structure we have imposed on these modules is a lower-truncated grading, and we ensure that the grading is compatible with the vertex operator algebra action. These modules are also called "admissible" in the literature. The grading of these $\mathbb{N}$-gradable weak modules differs from the grading of vertex operator algebras in the following way. The grading of vertex operator algebra is given by the eigenvalues of $L(0)$, while this is not necessarily true for for $\mathbb{N}$-gradable modules. A third type of module is one where the grading is given by the $L(0)$-action.

Definition 3.5. A weak module M for a vertex operator algebra $V = (V, Y, \mathbf{1}, \omega)$ is a V-*module* if M is $\mathbb{C}$-graded with $M = \bigoplus_{\lambda \in \mathbb{C}} M_\lambda$, and

1) $\dim(M_\lambda) < \infty$,
2) $M_{\lambda+n=0}$ *for fixed* λ *and* $n << 0$,
3) $L(0)w = \lambda w = \mathrm{wt}(w)w$, *for* $w \in M_\lambda$.

The grading has been expanded to $\mathbb{C}$ to account for all possible $L(0)$-eigenvalues, and there is a lower truncation condition. In addition, each graded piece must be finite-dimensional. Such a finiteness condition is not imposed on $\mathbb{N}$-gradable weak modules. One result of this finiteness condition and lower-truncation condition for V-modules is that V-modules are $\mathbb{N}$-gradable weak modules. In practice, $\mathbb{N}$-gradable weak modules have enough structure to develop interesting theory. We will see that for C_2-cofinite vertex operator algebras, weak modules are $\mathbb{N}$-gradable as well. In his work on modularity, Zhu used what he called strong modules. The definition of a strong module is the same as the definition of an ordinary module except that the axiom "$\dim(M_\lambda) < \infty$" is omitted.

It is possible to extend the definition of C_n-cofiniteness to modules using $C_n(M) = \mathrm{span}\{\mathrm{Res}_x x^{-n} Y_M(u, x)w | u \in V, w \in M\}$ for $n \geq 2$. Because there is no creation axiom for modules, it can be interesting to extend the idea of C_1-cofiniteness to modules in the naive way.

Definition 3.6. For a vertex operator algebra $V = \bigoplus_{n \geq 0} V_n$ and a module M, let

$$c_1(M) = \mathrm{span}\left\{\mathrm{Res}_x x^{-1} Y_M(u, x)w : u \in \bigoplus_{n > 0} V_n, w \in M\right\}.$$

Then M is called c_1-*cofinite* if $M/c_1(M)$ is finite-dimensional.

This definition appears in the work of Nahm, who studied vertex operator algebras for which all irreducible $\mathbb{N}$-gradable weak modules are c_1-cofinite [Nah94]. He called such algebras *quasirational*. Quasirationality or c_1-cofiniteness of all irreducible modules is a important assumption in Huang's work on modular tensor categories and the Verlinde conjecture

[Hua03] [Hua05a] [Hua05b]. Huang's work also requires that the algebras be C_2-cofinite, which implies c_1-cofiniteness of the modules [ABD04].

4. Complete reducibility

One desirable property of vertex operator algebras that is featured in both mathematics and physics is complete reducibility of modules, the primary feature of "rationality". The definition differs from author to author, with each rendition of "rationality" encompassing some minimum amount of "goodness" needed for the author's theory to work. The "goodness" invariably includes some form of complete reducibility of modules and may also include some finiteness condition, i.e., finite number of irreducible modules, the graded pieces of irreducible modules are finite-dimensional, or even C_2-cofiniteness for some authors. Calling a vertex operator algebra or conformal field theory "rational" endows that object with some physical importance, but the cost can sometimes be misinterpretation. A common type of complete reducibility imposed on a vertex operator algebras is the following.

Every $\mathbb{N}$-gradable weak module is the direct sum of irreducible $\mathbb{N}$-gradable weak modules.

In mathematical literature, this property is sometimes called rationality, but certainly not consistently. A clearer naming would be complete reducibility of $\mathbb{N}$-gradable weak modules (in terms of irreducible $\mathbb{N}$-gradable weak modules). I will use "complete reducibility of $\mathbb{N}$-gradable weak modules" to convey this form of complete reducibility. The assumption of this form of complete reducibility is necessary to prove many important results in vertex operator algebra theory. As mentioned above, the concept of rationality sometimes includes some finiteness assumptions. Zhu's formulation of rationality included two additional conditions: there exists a finite number of irreducible $\mathbb{N}$-gradable weak modules, and each graded piece of an irreducible $\mathbb{N}$-gradable weak module is finite-dimensional. However Dong, Li, and Mason demonstrated that Zhu's additional conditions are consequences of complete reducibility of $\mathbb{N}$-gradable modules [DLM98b]. In other words, Zhu's seemingly stronger formulation of rationality is equivalent to complete reducibility of $\mathbb{N}$-gradable weak modules. In fact, the Dong-Li-Mason results imply that complete reducibility of $\mathbb{N}$-gradable weak modules is equivalent to: every $\mathbb{N}$-gradable weak module is the direct sum of irreducible V-modules. Some vertex operator algebras feature a stronger form of complete reducibility:

Every weak module is the direct sum of irreducible V-modules.

This property is called *regularity*, and examples of vertex operator algebras that satisfy this form of complete reducibility are the Moonshine module vertex operator algebra $V^\natural$, the Virasoro vertex operator algebras $L(c_{p,q}, 0)$, and vertex operator algebras associated to positive definite even lattices [DLM97]. We will see that many more vertex operator algebras are regular. Zhu conjectured that complete reducibility of $\mathbb{N}$-gradable modules implies C_2-cofiniteness. This remains an important open question. However for the stronger form of complete reducibility, Li proved that regular vertex operator algebras are C_2-cofinite [Li99].

5. Spanning sets for algebras and modules

One of the important consequences of C_1- or C_2-cofiniteness for a vertex operator algebra is that the algebra is finitely generated and has a Poincaré-Birkhoff-Witt-like spanning sets featuring desirable ordering restrictions.

Proposition 5.1. *(cf. [LL04]) For a subset S of a vertex operator algebra $V = (V, Y, \omega, \mathbf{1})$, the subalgebra of V generated by S is*

$$\langle S \rangle = \mathrm{span}\{u_{n_1}^{(1)} \cdots u_{n_r}^{(r)} \mathbf{1} \mid r \in \mathbb{N}, u^{(1)}, \ldots, u^{(r)} \in S \cup \{\omega\}, n_1, \ldots, n_r \in \mathbb{Z}\}.$$

Different types of spanning sets feature different restrictions on the basic form, $u_{n_1}^{(1)} \cdots u_{n_r}^{(r)} \mathbf{1}$, of spanning set elements. Some restrictions describe how often the index n_i of or the weight of a mode $u_{n_i}^{(i)}$ can appear in a spanning set element, while other restrictions limit the $u^{(i)}$'s to certain subsets of V. One way to think about the index restrictions on spanning set elements is in terms of a difference condition, similar to a difference condition on partitions. A difference-n condition on modes means that the indices of adjacent modes must differ by at least n. That is, for adjacent modes $u_{m_i}^{(i)}$ and $u_{m_{i+1}}^{(i+1)}$ in a spanning set element, $m_{i+1} - m_i \geq n$.

A natural question is: for a vertex operator algebra V, what sets S generate V? Certainly a minimal set S is desirable, and this is what Karel and Li have explored.

Proposition 5.2. *(cf. [KL99]) For a vertex operator algebra V, let X be a set of homogeneous representatives of a spanning set for the quotient space $V/C_1(V)$. Then V is spanned by the elements of the form*

$$u_{n_1}^{(1)} \cdots u_{n_r}^{(r)} \mathbf{1},$$

where $r \in \mathbb{N}$, $u^{(1)}, \ldots, u^{(r)} \in X$, $n_1, \ldots, n_r \in \mathbb{Z}$, and $\mathrm{wt}(u_{n_1}^1) \geq \cdots \geq \mathrm{wt}(u_{n_r}^r) > 0$.

In addition to showing that representatives of a basis for $V/C_1(V)$ generate V, Karel and Li also show that this set is a minimal generating set of V. So C_1-cofinite vertex operator algebras are finitely generated. Karel and Li also prove an analogous spanning set for $\mathbb{N}$-gradable weak modules.

Proposition 5.3. *(cf. [KL99]) For a vertex operator algebra V and an irreducible $\mathbb{N}$-gradable weak module $M = \bigoplus_{n \geq 0} M(n)$, let X be a set of homogeneous representatives of a spanning set for the quotient space $V/C_1(V)$. Then M is spanned by the elements of the form*

$$u^{(1)}_{n_1} \cdots u^{(r)}_{n_r} w,$$

where $r \in \mathbb{N}$, $u^{(1)}, \ldots, u^{(r)} \in X$, $n_1, \ldots, n_r \in \mathbb{Z}$, $w \in M(0)$, and $\mathrm{wt}(u^1_{n_1})$ $\geq \cdots \geq \mathrm{wt}(u^r_{n_r}) > 0$.

For both the algebra and module spanning sets, the ordering restriction on the modes $u^{(i)}_{n_i}$ is in terms of the weight of the mode, and there is no restriction on how often an index n_i of a mode $u^{(i)}_{n_i}$ can appear in a spanning set element. However, one can prove an alternate version of the algebra spanning set, where the ordering restriction on the operators is in terms of the indices of modes, i.e., the n_i's.

Proposition 5.4. *For a vertex operator algebra V, let X be a set of homogeneous representatives of a spanning set for the quotient space $V/C_1(V)$. Then V is spanned by the elements of the form*

$$u^{(1)}_{n_1} \cdots u^{(r)}_{n_r} \mathbf{1},$$

where $r \in \mathbb{N}$, $u^{(1)}, \ldots, u^{(r)} \in X$, $n_1, \ldots, n_r \in \mathbb{Z}$, and $n_1 \leq \cdots \leq n_r < 0$.

The proof of this algebra spanning set is the same as the proof of the Karel-Li spanning result, since the mechanism for reordering the modes is the same. This same mechanism is just used to impose a different ordering. It is possible to extend this spanning set to modules. Since there is no creation axiom for modules, modes u_n with $n \geq 0$ need to be limited in some way in the expression of spanning set elements.

Lemma 5.5. *Given an $\mathbb{N}$-gradable weak module $M = \bigoplus_{n \geq 0} M(n)$ and X a finite set of vectors in V, there exists $T \in \mathbb{N}$ such that $u_n w = 0$ for all $n \geq T$, $u \in X$, and $w \in M(0)$.*

Proof: We have $u_{\mathrm{wt}u + L} w = 0$ for all $v \in V$, $L \geq 0$, and $w \in M(0)$. Let $T = \max_{u \in X}\{\mathrm{wt}u\}$. $\qquad\square$

In particular, if X is a set of representatives of a basis of $V/C_1(V)$ for a C_1-cofinite vertex operator algebra, such a T exists.

Proposition 5.6. *For a C_1-cofinite vertex operator algebra V and an irreducible $\mathbb{N}$-gradable weak module $M = \bigoplus_{n \geq 0} M(n)$, let X be a set of homogeneous representatives of a spanning set for the quotient space $V/C_1(V)$. Then M is spanned by the elements of the form*

$$u_{n_1}^{(1)} \cdots u_{n_r}^{(r)} w,$$

where $r \in \mathbb{N}$, $u^{(1)}, \ldots, u^{(r)} \in X$, $n_1, \ldots, n_r \in \mathbb{Z}$, $w \in M(0)$, and $n_1 \leq \cdots \leq n_r < T$ (as above).

Gaberdiel and Neitzke developed another type of spanning set for a vertex operator algebra using a set of representatives of a basis of the quotient space $V/C_2(V)$. Though this generating set is not minimal, it does have stronger ordering restrictions than the spanning set of Karel and Li.

Proposition 5.7. *(cf. [GN03]) For a vertex operator algebra V, let X be a set of homogeneous representatives of a spanning set for the quotient space $V/C_2(V)$. Then V is spanned by the elements of the form*

$$u_{n_1}^{(1)} \cdots u_{n_r}^{(r)} \mathbf{1},$$

where $r \in \mathbb{N}$, $u^{(1)}, \ldots, u^{(r)} \in X$, $n_1, \ldots, n_r \in \mathbb{Z}$, and $n_1 < \cdots < n_r < 0$.

By enlarging the generating set, Gaberdiel and Neitzke were able to introduce a repetition restriction. Each index of a mode can only appear once in the expression of a spanning set element, or in other words this is a no-repetion restriction on the indices of modes. One corollary of Proposition 5.7 is that C_2-cofiniteness implies C_n-cofiniteness for $n \geq 2$. The converse, mentioned above, is also true, yielding the following result.

Corollary 5.8. *(cf. [GN03]) If a vertex operator algebra V is C_n-cofinite for some $n \geq 2$ then V is C_n-cofinite for all $n \geq 2$.*

C_1-cofiniteness of a vertex operator algebra is a strictly weaker condition since the vertex operator algebra constructed from a Heisenberg algebra is C_1-cofinite, but is not C_2-cofinite.

A more natural way to view the no-repetition restriction is in terms of a difference condition. The Gaberdiel and Neitzke algebra spanning set obeys a difference-one condition, and the reformulation of the Karel and Li algebra spanning set obeys a difference-zero condition. A natural extension of the Gaberdiel-Neitzke result would be a module spanning set satisfying a difference-one condition. This next result is a partial solution to this difference-one module spanning set question.

Proposition 5.9. *(cf. [Buh02]) For a C_2-cofinite vertex operator algebra V and an irreducible $\mathbb{N}$-gradable weak module $M = \bigoplus_{n \geq 0} M(n)$, let X be a set of homogeneous representatives of a spanning set for the quotient space $V/C_2(V)$. Then M is spanned by the elements of the form*

$$u^{(1)}_{n_1} \cdots u^{(r)}_{n_r} w,$$

where $r \in \mathbb{N}$, $u^{(1)}, \ldots, u^{(r)} \in X$, $n_1, \ldots, n_r \in \mathbb{Z}$, $w \in M(0)$, and $u^1_{n_1} \leq \cdots \leq u^r_{n_r} < T$ (with T as above) where $n_{j-1} < n_j$ if $n_j < 0$ and $n_j = n_{j+1}$ for at most Q indices j for $n_j \geq 0$, where $Q \in \mathbb{N}$, and Q is fixed for V.

In this module spanning set, the modes with negative indices obey a difference-one condition, but the non-negative modes do not. However, the non-negative modes may repeat only a globally finite number of times. This spanning set was useful in proving a number of results, yet it still is not a true difference-one condition module spanning set. Miyamoto provides a further refinement obtaining a full difference-one module spanning set.

Proposition 5.10. *(cf. [Miy04]) For a C_2-cofinite vertex operator algebra V and an irreducible $\mathbb{N}$-gradable weak module $M = \bigoplus_{n \geq 0} M(n)$, let X be a set of homogeneous representatives of a spanning set for the quotient space $V/C_2(V)$. Then M is spanned by the elements of the form*

$$u^{(1)}_{n_1} \cdots u^{(r)}_{n_r} w,$$

where $r \in \mathbb{N}$, $u^{(1)}, \ldots, u^{(r)} \in X$, $n_1, \ldots, n_r \in \mathbb{Z}$, $w \in M(0)$, and $u^1_{n_1} < \cdots < u^r_{n_r} < T$ (as above).

Viewed in terms of difference conditions, this means that C_1-cofiniteness implies a difference-zero condition on elements of a spanning set of a vertex operator algebra and its modules, and C_2-cofiniteness implies a difference-one condition on elements of a spanning set of a vertex operator algebra and its modules.

Orbifold theory and twisted modules are important aspects of the representation theory of vertex operator algebras. A paper by Yamauchi [Yam04] addresses twisted modules, and the full statement of his difference-one spanning set theorem applies in this larger generality.

Again an underlying assumption in this exposition is that the vertex operator algebras are of CFT-type. Miyamoto's result is true for vertex operator algebras that are not of CFT-type [Miy04]. In particular, he assumed that $V = \bigoplus_{n \geq 0} V_n$, but V_0 not necessarily one-dimensional.

6. Finiteness Results

As mentioned in previous sections, C_2-cofiniteness implies the finite-dimensionality of many quotient spaces of the algebra and implies the existence of a finite generating set for the algebra. The assumption of C_2-finiteness has implications for the representation theory of vertex operator algebras beyond the difference-one module spanning set.

Theorem 6.1. *Let V be a C_2-cofinite vertex operator algebra. Then:*

1. *V has finite number of irreducible V-modules up to isomorphism. [KL99]*
2. *Weak modules for V are $\mathbb{N}$-gradable weak modules.[ABD04]*
3. *Irreducible $\mathbb{N}$-gradable weak modules for V are irreducible V-modules. [KL99]*
4. *Irreducible weak modules for V are irreducible V-modules. [ABD04]*
5. *The associative algebra $A(V)$ is finite-dimensional.*

Practically, this means that under the assumption of C_2-cofiniteness, we do not need to be concerned about the myriad types of modules. The weakest definition of modules is sufficient, as weak modules are gradable and lower truncated. Further, any irreducible module has a grading given by the $L(0)$-action and each graded piece is finite-dimensional. Some of these results were extended by Miyamoto, in his extended generality described above.

Theorem 6.2. *[Miy04] For V a vertex operator algebra, the following are equivalent:*

1. *V is C_2-cofinite.*
2. *Every weak module is a direct sum of generalized eigenspaces of $L(0)$.*
3. *Every weak module is an $\mathbb{N}$-gradable weak module $M = \bigoplus_{n \geq 0} M(n)$ such that $M(n)$ is a direct sum of generalized eigenspaces of $L(0)$.*
4. *V is finitely generated and every weak module is an $\mathbb{N}$-gradable weak module.*

In light of this, we see that C_2-cofiniteness is equivalent to all modules having suitable properties for an interesting representation theory, with the lone exception of complete reducibility. However, the assumption of C_2-cofiniteness unifies notions of complete reducibility.

Theorem 6.3. *[ABD04] [Li99] For a C_2-cofinite vertex operator algebra V, the following are equivalent:*

1. *Every weak module for V is the direct sum of irreducible V-modules.*
2. *Every $\mathbb{N}$-gradable weak module is the direct sum of irreducible $\mathbb{N}$-gradable weak modules.*

In particular, this means that all known vertex operator algebras with complete reducibility of $\mathbb{N}$-gradable weak modules are regular. Theorem 6.1 should be compared with the following theorem for vertex operator algebras with complete reducibility of $\mathbb{N}$-gradable weak modules.

Theorem 6.4. *Let V be a vertex operator algebra for which every $\mathbb{N}$-gradable weak module is the direct sum of irreducible V-modules. Then:*

1. *V has a finite number of irreducible V modules up to isomorphism.* *[DLM98a]*
2. *Irreducible $\mathbb{N}$-gradable weak modules for V are irreducible V-modules.* *[DLM98a]*
3. *The associative algebra $A(V)$ is semisimple and finite-dimensional.* *[DLM98c]*

This is compelling evidence that complete reducibility of $\mathbb{N}$-gradable weak modules and C_2-cofiniteness are somehow related. Zhu conjectured that complete reducibility of $\mathbb{N}$-gradable weak modules implies C_2-cofiniteness [Zhu96]. The converse of this conjecture has been disproved. Building on the work of Kausch and Gaberdiel [GK96], Abe and Carqueville-Flohr construct examples of C_2-cofinite vertex operators for which there exist $\mathbb{N}$-gradable weak modules that are not completely reducible. Specifically, Abe constructs a family of C_2-cofinite vertex operator algebras with central charge $-2d$ for $d \in \mathbb{Z}_+$ with reducible indecomposible modules [Abe]. Carqueville and Flohr prove that the vertex operator algebras constructed from the triplet algebras $c_{p,1}$ are C_2-cofinite and also have reducible indecomposible modules [CF06]. However Zhu's conjecture as to whether complete reducibility of $\mathbb{N}$-gradable weak modules implies C_2-cofiniteness remains open.

References

[ABD04] Toshiyuki Abe, Geoffrey Buhl, and Chongying Dong. Rationality, regularity, and C_2-cofiniteness. *Trans. Amer. Math. Soc.*, 356(8):3391–3402, 2004.

[Abe] Toshiyuki Abe. A Z_2-orbifold model of the symplectic fermionic vertex operator superalgebra. arXiv:math.QA/0503472.

[Bor92] Richard E. Borcherds. Monstrous moonshine and monstrous Lie superalgebras. *Invent. Math.*, 109(2):405–444, 1992.

[Buh02] Geoffrey Buhl. A spanning set for VOA modules. *J. Algebra*, 254(1):125–151, 2002.

[CF06] Nils Carqueville and Michael Flohr. Nonmeromorphic operator product expansion and c_2-cofiniteness for a family of w-algebras. *J.PHYS.A*, 39:951, 2006.

[DLM97] Chongying Dong, Haisheng Li, and Geoffrey Mason. Regularity of rational vertex operator algebras. *Adv. Math.*, 132(1):148–166, 1997.

[DLM98a] Chongying Dong, Haisheng Li, and Geoffrey Mason. Twisted representations of vertex operator algebras. *Math. Ann.*, 310(3):571–600, 1998.

[DLM98b] Chongying Dong, Haisheng Li, and Geoffrey Mason. Twisted representations of vertex operator algebras and associative algebras. *Internat. Math. Res. Notices*, (8):389–397, 1998.

[DLM98c] Chongying Dong, Haisheng Li, and Geoffrey Mason. Vertex operator algebras and associative algebras. *J. Algebra*, 206(1):67–96, 1998.

[FLM88] Igor Frenkel, James Lepowsky, and Arne Meurman. *Vertex operator algebras and the Monster*, volume 134 of *Pure and Applied Mathematics*. Academic Press Inc., Boston, MA, 1988.

[GK96] Matthias R. Gaberdiel and Horst G. Kausch. A rational logarithmic conformal field theory. *Phys. Lett. B*, 386(1-4):131–137, 1996.

[GN03] Matthias R. Gaberdiel and Andrew Neitzke. Rationality, quasirationality and finite W-algebras. *Comm. Math. Phys.*, 238(1-2):305–331, 2003.

[HLZ] Yi-Zhi Huang, James Lepowsky, and Lin Zhang. A logarithmic generalization of tensor product theory for modules for a vertex operator algebra. *Internat. J. Math., to appear*, arXiv:math.QA/0311235.

[Hua03] Yi-Zhi Huang. Riemann surfaces with boundaries and the theory of vertex operator algebras. In *Vertex operator algebras in mathematics and physics (Toronto, ON, 2000)*, volume 39 of *Fields Inst. Commun.*, pages 109–125. Amer. Math. Soc., Providence, RI, 2003.

[Hua05a] Yi-Zhi Huang. Differential equations and intertwining operators. *Commun. Contemp. Math.*, 7(3):375–400, 2005.

[Hua05b] Yi-Zhi Huang. Vertex operator algebras, the Verlinde conjecture, and modular tensor categories. *Proc. Natl. Acad. Sci. USA*, 102(15):5352–5356 (electronic), 2005.

[KL99] Martin Karel and Haisheng Li. Certain generating subspaces for vertex operator algebras. *J. Algebra*, 217(2):393–421, 1999.

[Li99] Haisheng Li. Some finiteness properties of regular vertex operator algebras. *J. Algebra*, 212(2):495–514, 1999.

[LL04] James Lepowsky and Haisheng Li. *Introduction to vertex operator algebras and their representations*, volume 227 of *Progress in Mathematics*. Birkhäuser Boston Inc., Boston, MA, 2004.

[Miy04] Masahiko Miyamoto. Modular invariance of vertex operator algebras satisfying C_2-cofiniteness. *Duke Math. J.*, 122(1):51–91, 2004.

[Nah94] Werner Nahm. Quasi-rational fusion products. *Internat. J. Modern Phys. B*, 8(25-26):3693–3702, 1994.

[Yam04] Hiroshi Yamauchi. Modularity on vertex operator algebras arising from semisimple primary vectors. *Internat. J. Math.*, 15(1):87–109, 2004.

[Zhu96] Yongchang Zhu. Modular invariance of characters of vertex operator algebras. *J. Amer. Math. Soc.*, 9(1):237–302, 1996.

Friendly Giant Meets Pointlike Instantons?
On a New Conjecture by John McKay

Anda Degeratu

Dept. of Mathematics,
Duke University, Durham NC 27708, USA

Katrin Wendland

Dept. of Mathematics
UNC Chapel Hill, Chapel Hill, NC 25599, USA; and
Mathematics Institute, University of Warwick
CV4-7AL Coventry, UK

Abstract

A new conjecture due to John McKay claims that there exists a link between (1) the conjugacy classes of the Monster sporadic group and its offspring, and (2) the Picard groups of bases in certain elliptically fibered Calabi-Yau threefolds. These Calabi-Yau spaces arise as F-theory duals of point-like instantons on ADE type quotient singularities. We believe that this conjecture, may it be true or false, connects the Monster with a fascinating area of mathematical physics which is yet to be fully explored and exploited by mathematicians. This article aims to clarify the statement of McKay's conjecture and to embed it into the mathematical context of heterotic/F-theory string-string dualities.

1. Introduction

John McKay has observed a remarkable connection between the three sporadic groups: the Monster, the Baby Monster, the Fischer group, and the three affine Dynkin diagrams: E_8, E_7, E_6 [McK80]. Let us present this statement in more detail, following Borcherds [Bor02, Bor01] and Glauberman and Norton [GN01].

The Monster group $\mathbb{M}$ has a total number of 194 conjugacy classes, two of which contain elements of order 2; we denote them by $2A$ and $2B$. The class $2A$ is the conjugacy class of the Fischer involution in $\mathbb{M}$ so that its centralizer is a double cover of the Baby Monster $\mathbb{B}$. There are 9 conjugacy classes of $\mathbb{M}$ which can be written in the form $[tt_i]$, $i = 0, \ldots, 8$, with t, t_i of type $2A$. The orders of $tt_i \in \mathbb{M}$ are $1, 2, 3, 4, 5, 6, 2, 3, 4$, numbers which are familiar as the numbers which label[1] the affine Dynkin diagram E_8.

The Baby Monster $\mathbb{B}$ has 5 conjugacy classes of elements which can be written as a product of two elements of type $2A$. These have orders $2, 4, 3, 2, 1$, the numbers labeling the affine Dynkin diagram F_4. The diagram is related to the affine Dynkin diagram of E_8 as follows: Omitting the node labeled 2 on the left side of the diagram in the latter gives the E_7 Dynkin diagram, which we then extend to get its affine version. Folding this by its $\mathbb{Z}_2$ automorphism gives the affine Dynkin diagram of F_4. On the level of the associated sporadic groups the omission of the node labeled 2 corresponds to taking the centralizer of an element of type $2A$ in $\mathbb{M}$, which gives the Baby Monster $\mathbb{B}$.

The Fischer group Fi_{24} has 3 classes of elements that are products of two elements of type $2A$. The orders are $2, 3, 1$, the numbers labeling the affine Dynkin diagram of G_2. This diagram is obtained from the affine Dynkin diagram of E_6 via a folding under a symmetry of order 3.

John McKay has pointed out another mysterious appearance of the number 194 of conjugacy classes of the Monster, together with various E_8's. 194 occurs as the Picard number of the base in an elliptically fibered Calabi-Yau 3-fold with section, which was studied by Aspinwall, Morrison, and Katz [AM97, AKM00] in the context of the so–called heterotic – F-theory duality. The 3-fold is the F-theory dual of the $E_8 \times E_8$ heterotic data consisting of 24 pointlike instantons in an E_8 quotient singularity on a $K3$ surface. This is the most degenerate situation of a heterotic – F-theory pair: The maximal number of pointlike instantons is moved into the worst possible quotient singularity on $K3$. Moreover, on the F-theory side, our 3-fold has Euler characteristic 960, the current record among Calabi-Yau 3-folds. McKay's observation adds that on the F-theory side we also find the maximal number of conjugacy classes of a sporadic group ... Though the evidence may be scarce, if McKay's numerology is true, then this points to a very interesting connection: surface orbifold singularities for the exceptional simply-laced Dynkin diagrams should correspond to elliptically fibered Calabi-Yau 3-folds over rational surfaces with Picard number equal to the number of conjugacy classes in the three sporadic simple groups.

Motivated by these observations, we originally set out to prove or disprove McKay's conjecture. Using the results of [AM97] it is not too hard to see that

---

1 In terms of representation theory, if $\alpha_1, \ldots, \alpha_8$ denote the fundamental roots in a chosen root system for E_8, and if α_0 is the negative of the maximal root of the system, then there exist integers $c_0, c_1, \ldots, c_8$ with $c_0 = 1$, so that $\sum_{i=0}^{8} c_i \alpha_i = 0$. Here $(c_0, \ldots, c_8) = (1, 2, 3, 4, 5, 6, 2, 3, 4)$.

the conjecture at least needs refinement, since 24 pointlike instantons on other ADE type singularities under the heterotic – F-theory duality do not produce any convincing numerology. In fact, because F_4 and G_2 are non-simply laced but give the Dynkin data corresponding to the Baby Monster and the Fischer group, any naive attempt to collect further evidence for McKay's conjecture by placing pointlike instantons on other rational double points was bound to fail. We will briefly comment on possible remedies in the Conclusions. However, instead of ending this work here by announcing that McKay's conjecture as yet awaits confirmation by further data points, we prefer to report on the fascinating areas of mathematical physics which this very conjecture relates to: On our journey we quickly got entangled in the amazing features of string-string dualities. Though friendly giants have not yet been sighted, pointlike instantons will certainly make their appearance in this work. In summary, the reader should be warned that the title of this work can be misleading: It honestly states the outset of this project but does not reflect the fact that the pointlike instantons or rather the heterotic – F-theory duality which supposedly confronts them with the friendly giant is playing the main part in our study. We aim to give a mathematical account of some foundations of string-string duality and include some of the open questions which we plan to address in future work. Much of our discussion is collected from the vast literature on this topic. However, we attempt to carefully separate physics lore from mathematical derivations, to pinpoint the open questions, to explain computations in algorithmic form, and to illustrate them with examples where appropriate, in a form which is not yet available in the literature. A number of original observations are found in this work, but on the large this is a review article. We hope to convince the reader that this field of mathematical physics deserves more attention than it has so far received from mathematicians.

The paper is organized as follows:

In Section 2 we present aspects of heterotic $E_8 \times E_8$ and type IIA string theory which are relevant for our later discussion. The main focus is on the form of the massless spectrum; we describe first the massless spectrum of the $D = 10$ theories, and then its modification as we compactify to a lower dimensional theory. We treat the cases when the compactifying manifolds are $K3$ surfaces, T^4, $K3 \times T^2$, and smooth Calabi-Yau 3-folds, and in each of these cases we determine the numbers n_V, n_T, n_H of linearly independent vector-, tensor-, and hypermultiplets in the massless spectrum. More precisely, we study carefully the behavior of the Yang-Mills multiplet in 10-dimensional heterotic theories under compactification. We modify an Atiyah-Hitchin-Singer index theorem [AHS78, Thm. 6.1] to relate the multiplicities arising in this context to the dimensions of the moduli spaces of anti-selfdual connections on E_8-bundles over $K3$. These moduli spaces must have non-negative dimensions to yield consistent theories, imposing bounds on the instanton numbers of the bundles in play. Although these observations must be known to physicists, we have not been able to find explanations, along the lines we are giving, in the literature. We also discover a correction to the count of neutral versus charged massless hypermultiplets which seems to have escaped mention so far.

In Section 3 we discuss the anomaly cancellation condition, which in physics arises as a consistency condition for string theory. Our focus is the heterotic $E_8 \times E_8$ theory compactified on a $K3$ surface, where this condition takes the form

$$n_H - n_V + 29n_T = 273.$$

We give two ways to derive it: (1) directly on the space-time, and (2) on the $K3$ surface by employing a purely index theoretical argument. It is not clear from the literature whether the physics community is aware of the equivalence of (1) and (2). We end the section by discussing a number of examples. Similar (and more extensive) collections of examples have appeared previously, see particularly [BIK+96]; our list is representative in view of the later comparison to purely geometrical techniques which are shown to reproduce the numbers n_H, n_V, n_T in Section 5.3.

Section 4 presents first the conjectured duality between heterotic and type IIA theories. In the so–called F-theory limit a duality is induced between heterotic $E_8 \times E_8$ string theory compactified on a $K3$ surface and F-theory compactified on a Calabi-Yau 3-fold which is $K3$-fibered. The existence of such a limit with the desired properties requires the Calabi-Yau 3-fold to be elliptically fibered, with a section, over a rational surface. The duality predicts that the numbers of linearly independent vector-, tensor-, and hypermultiplets of two dual theories agree.

In Section 5 we analyze the implications of this conjecture from the perspective of the geometry of the elliptically fibered Calabi-Yau 3-fold, and we explain how the above mentioned anomaly cancellation condition (on the heterotic side) induces a classically unknown relation among the geometric invariants of the Calabi-Yau 3-fold (on the F-theory side). The significance of this interpretation of the duality was first observed in [GM03]. We give a detailed summary of this latter work. Particularly, incorporating the more general results of [Mir83], we describe an algorithm to calculate the Euler numbers of these Calabi-Yau 3-folds. This includes a brief discussion of "charged hypermultiplets", together with some new evidence for the geometric realization of these mysterious structures. We plan to give a more complete construction in the near future. We end this section by presenting examples where we check that in each case the results match the data on the heterotic side. We also include examples where the heterotic analysis is impossible, because the bundles degenerate in a way which still needs to be understood: the case of pointlike instantons on ADE type quotient singularities on a $K3$ surface which was pioneered in [AM97]. We give details of the geometric analysis on the F-theory side, most of which must have been known to the authors of [AM97], but which have not appeared elsewhere.

We conclude with a discussion in Section 6 where we take stock of our results, relate them back to McKay's conjecture, and suggest some further steps to take.

Two appendices summarize background material on Rarita-Schwinger fields and on characteristic classes, particularly for $K3$ surfaces.

Acknowledgements

John McKay deserves our gratitude for stating a conjecture which has led us into a fascinating area of mathematical physics, and for his steady support and interest. We are grateful to Miles Reid, who was the first to inform us of this conjecture. We thank Gavin Brown, Terry Gannon, Yang-Hui He, Vishnu Jejjala, and Ilarion Melnikov for discussions and encouragement, and particularly Paul Aspinwall, Antonella Grassi, Mark Gross, David Morrison, and Mark Stern for their comments and explanations.

For their hospitality we wish to thank the following institutes, where part of this work was performed: The Isaac Newton Institute in Cambridge, UK, during the programme on "Higher Dimensional Complex Geometry", the "Mathem. Forschungsinstitut Oberwolfach" during the mini workshop "Complex Geometry: Mirror Symmetry and Related Topics", and the Max Planck Institute for Mathematics in Sciences in Leipzig, Germany (AD). We gratefully acknowledge support from the AMS under an AWM travel grant for a visit of AD to the Mathematics Institute of the University of Warwick, UK, whose hospitality also deserves our thanks.

AD was partly supported by NSF grant number DMS-0505767. KW was partly supported by Nuffield Foundation grant number NAL/00755/G.

2. Vocabulary from heterotic and type IIA theories

Let us present some of the standard lore of superstring theory. Consistency conditions require one to work on a ten-dimensional real space-time with Minkowski signature, where there are five basic theories: two $\mathcal{N} = 2$ (type IIA and type IIB), and three $\mathcal{N} = 1$ theories (heterotic $E_8 \times E_8$, heterotic $SO(32)$, and type I). At first level the difference between these theories is given by the number $\mathcal{N}$ of supersymmetries, and by their "massless field content". In fact, when space-time is the ten dimensional Minkowski space, then this information is sufficient to tell the five theories apart, and since we are not capable to address string theory in general, we will focus on issues related to these massless particles.

Interesting structure arises when the space-time is of the form $M^{1,D-1} \times X^d$, where $d + D = 10$, $M^{1,D-1}$ is flat D-dimensional Minkowski space, and the "internal" X^d is a real d-dimensional manifold which admits Ricci-flat Kähler-Einstein metrics. On $M^{1,D-1}$ it is convenient to work with light-cone coordinates, i.e. with $x^{\pm} = \left(x^0 \pm x^{D-1}\right)/\sqrt{2}$ and the remaining $(D - 2)$ transverse directions, which are space-like. The massless particles are labeled by irreducible representations of $\mathrm{Spin}(D - 2)$, the double cover of the little group $\mathrm{SO}(D - 2)$, which acts on the transverse directions. Both have the same Lie algebra $\mathfrak{so}(D - 2)$ which we shall use for convenience from now on.

In this section we describe the massless field content of our main protagonists, namely the type IIA and heterotic $E_8 \times E_8$ string theories. We do this first on $M^{1,9}$ and then compactify to $M^{1,D-1} \times X^d$ with $D = 6$ and $D = 4$, respectively.

Much of this section consists of standard material [GSW87, Pol98]. Our exposition aims to present the mathematical details, some of which we have not been able to find explicitly in the literature.

2.1. Massless spectra in $\mathbf{D} = 10$

As stated above, massless particles in 10 dimensions are given by irreducible representations of $\mathfrak{so}(8)$. Let $\mathbf{8}_v$, $\mathbf{8}_+$, $\mathbf{8}_-$ denote the real vector, the positive, and the negative real spinor representation of $\mathfrak{so}(8)$. Spinors transforming in $\mathbf{8}_{\pm}$ are called Majorana-Weyl spinors (see e.g. [Pol98, II, Appendix B] for a useful account on spinors and supersymmetry in various dimensions). Recall that $\mathbf{8}_v$, $\mathbf{8}_+$, $\mathbf{8}_-$ are related to one another by the triality automorphism of $\mathfrak{so}(8)$ [Stu03]. For each of the superstring theories, the massless particles arise as tensor products of "left" and "right moving" representations, where the left hand side is always $\mathbf{8}_v \oplus \mathbf{8}_+$.

Type IIA for $D = 10$

In type IIA theories the massless spectrum is $(\mathbf{8}_v \oplus \mathbf{8}_+) \otimes (\mathbf{8}_v \oplus \mathbf{8}_-)$. Expanding into irreducible representations of $\mathfrak{so}(8)$, one has the explicit field content, organized into four sectors with the following standard notations:

$$
\begin{array}{llll}
\text{the NS-NS sector:} & \mathbf{8}_v {\otimes} \mathbf{8}_v & = \mathbf{35} \oplus \mathbf{28} \oplus \mathbf{1} & \to \quad g_{MN} \oplus B_{MN} \oplus \phi \\
\text{the R-NS sector:} & \mathbf{8}_+ {\otimes} \mathbf{8}_v & = \mathbf{56}_+ \oplus \mathbf{8}_- & \to \quad \Psi_M^+ \oplus \Psi^- \\
\text{the NS-R sector:} & \mathbf{8}_v {\otimes} \mathbf{8}_- & = \mathbf{56}_- \oplus \mathbf{8}_+ & \to \quad \Psi_M^- \oplus \Psi^+ \\
\text{the R-R sector:} & \mathbf{8}_+ {\otimes} \mathbf{8}_- & = \mathbf{56}_v \oplus \mathbf{8}_v & \to \quad C_{MNP} \oplus C_M
\end{array}
$$

$$(2.1.1)$$

Here, g_{MN} gives the metric on $M^{1,9}$ up to scaling and is called the graviton; the 2-form B_{MN} on $M^{1,9}$ is the B-field[2]; the scalar ϕ is the dilaton; the two fields $\Psi_M^{\pm}$ in the $\mathbf{56}_{\pm}$ are the Majorana-Weyl gravitinos of opposite chiralities; the $\Psi^{\mp}$ are Majorana-Weyl dilatinos; C_{MNP} is called RR 3-form, C_M is an RR 1-form. The graviton has spin 2, all p-forms have spin 1, while the gravitinos have spin $\frac{3}{2}$ and pure spinors have spin $\frac{1}{2}$. The fields $\Psi_M^{\pm}$ with spin $\frac{3}{2}$ carry both vector and spinor indices; in the physics literature such fields are called Rarita-Schwinger fields, see Appendix A for more details.

Remark. Consider the group Spin(n) (or Spin($1, n+1$)). If $n \equiv 0$(mod4) then the smallest real irreducible representation of the real Clifford algebra Cl(n) decomposes into two inequivalent representations. On the other hand if $n \equiv 0$(mod2) then the smallest irreducible complex representation of the complex Clifford algebra $\mathbb{C}$l(n) decomposes into two inequivalent complex irreducible representations. If these last two representations are self-dual, and if they allow a real structure such that their real part is a real spin representation, then spinors in those real representations are called Majorana-Weyl spinors (for Spin(n), these are normally called pseudoreal, but we will not make this distinction). From Bott periodicity and the classification of these representations for small n, it can be seen that the condition for the existence of Majorana-Weyl spinors is a dimensional one: $n \equiv 0$(mod8). For example, for Spin(8) let $\mathbf{8}_{\pm}$ denote the real spin irreducible representations of real dimension 8, and let $\mathbf{8}_{\pm}^{\mathbb{C}}$ denote the complex spin irreducible representations. Then the Majorana-Weyl condition gives that

$$\mathbf{8}_{\pm}^{\mathbb{C}} = \mathbf{8}_{\pm} \otimes_{\mathbb{R}} \mathbb{C}, \qquad \mathbf{8}_{\pm} = \mathrm{Re}(\mathbf{8}_{\pm}^{\mathbb{C}}). \tag{2.1.2}$$

Heterotic $E_8 \times E_8$ for $D = 10$

The massless spectrum of heterotic $G_{10} = E_8 \times E_8$ strings is $(\mathbf{8}_v \oplus \mathbf{8}_+) \otimes (\mathbf{8}_v \oplus \mathrm{Ad}(G_{10}))$, such that there is supersymmetry only on the left hand side, with the following two irreducible representations of the $\mathcal{N} = 1$ super Poincaré algebra:

$$
\begin{aligned}
R(10): \ (\mathbf{8}_v \oplus \mathbf{8}_+) \otimes \mathbf{8}_v \quad &= \ \mathbf{35} \oplus \mathbf{28} \oplus \mathbf{1} \oplus \mathbf{56}_+ \oplus \mathbf{8}_- \\
Y(10): \ (\mathbf{8}_v \oplus \mathbf{8}_+) \otimes \mathrm{Ad}(G_{10}) \ &= \ (\mathbf{8}_v \otimes \mathrm{Ad}(G_{10})) \oplus (\mathbf{8}_+ \otimes \mathrm{Ad}(G_{10})) \\
&\quad \to \ A_M \oplus \Lambda^+
\end{aligned}
$$
$$\tag{2.1.3}$$

The irreducible representations of the super Poincaré algebra along with their decomposition under $\mathfrak{so}(D-2)$ are commonly called supermultiplets. Above,

---

2 The B-field is only locally given by a 2-form, as shall be of importance in Section 3.2.1. For the time being, however, we can safely view B_{MN} as a 2-form on $M^{1,9}$.

$R(10)$ is the supergravity multiplet, which agrees with the NS-NS plus the R-NS sector of type IIA, containing the graviton g_{MN}, the B-field B_{MN}, the dilaton ϕ, the gravitino Ψ_M^+, and the dilatino Ψ^-. Second, $Y(10)$ is the super-Yang-Mills multiplet, where A_M is a gauge field, i.e. a connection 1-form on a principal G_{10}-bundle on $M^{1,9}$. For definiteness we always identify the spin connection with the gauge connection. The superpartner Λ^+ of A_M is called gaugino and is a positive spinor on $M^{1,9}$ with values in the adjoint representation of G_{10}. We will see later that under compactification to $K3$ anomaly cancellation forces the instanton number of the connection A_M to be 24, see (2.2.22) and Section 3.

2.2. Compactification

Let us now "compactify" d dimensions, that is consider the above-mentioned $M^{1,D-1} \times X^d$ as background for our strings. We will refer to $M^{1,D-1}$ as space-time, and here assume that the compact space X^d is smooth. Space-time indices are denoted by μ, ν, while coordinates on X^d are indexed by i, j. We summarize the discussion in [GSW87, II, p. 366 ff].

The massless fields of the ten-dimensional theory decompose under the action of $\mathfrak{so}(D-2) \oplus \mathfrak{so}(d)$, and we view the components as fields on the Minkowski space $M^{1,D-1}$. At the same time they give sections in certain fiber bundles on X^d, depending on the type of field under consideration, together with associated geometrical differential operators, which can be of Dirac-type, Laplacians, or Yang-Mills or Einstein linearizations. These operators are "mass operators", and massless fields are characterized as sections in their kernels.

As a representative case consider spinor fields and the corresponding Dirac operator. The above tells us that we need to calculate the kernel of a Dirac operator $D\!\!\!/$ on X. In fact, pairs of elements of opposite helicity in this kernel can conspire and acquire mass, and it is believed that they tend to do so in nature unless a large gauge group remains unbroken, see [GSW87, II, p. 368]. Hence generically to calculate the dimension of the space of massless spinors we determine the index of the appropriate Dirac operator on the compactifying manifold X: This index gives the number of independent solutions to the massless equations of motion which cannot become massive. In other words, a positive index of $D\!\!\!/$ implies that, when viewed as fields on X, we have ind $D\!\!\!/$ massless fields having positive chirality, while a negative index implies that we have $(-\text{ind } D\!\!\!/)$ massless fields of negative chirality.

Similar comments hold for the other fields: Rarita-Schwinger fields, p-forms, gauge fields, or gravitational fields, where p-forms give massless fields iff they are harmonic. Some massless fields satisfy linear equations and these are easy to figure out using the appropriate index theorem on the compactifying manifold X. On the other hand, there are fields like the gauge fields and the gravitational fields which satisfy non-linear equations [GSW87, II, p. 398]: the Ricci-flat Einstein equation for the metric, and the Yang-Mills equation for the gauge fields. To count the number of independent massless solutions, one considers the kernel of the linearizations of these operators around a solution.

2.2.1. Compactification to $D = 6$

Consider superstring theory on $M^{1,5} \times X^4$. We need to decompose $\mathbf{8}_v$, $\mathbf{8}_+$, and $\mathbf{8}_-$ under the action of $\mathfrak{so}(4) \oplus \mathfrak{so}(4) \subset \mathfrak{so}(8)$. The vector representation decomposes as

$$\mathbf{8}_v = (\mathbf{4}, \mathbf{1}) \oplus (\mathbf{1}, \mathbf{4})$$

where $\mathbf{4}$ is the natural 4-dimensional vector representation of $\mathfrak{so}(4)$. To figure out how the spinor representations decompose we need to look first at the decomposition of the complexifications. Under $\mathfrak{so}(4) \oplus \mathfrak{so}(4)$ we have

$$\mathbf{8}_+^{\mathbb{C}} = (\mathbf{2}_+^{\mathbb{C}}, \mathbf{2}_+^{\mathbb{C}}) \oplus (\mathbf{2}_-^{\mathbb{C}}, \mathbf{2}_-^{\mathbb{C}}), \qquad \mathbf{8}_-^{\mathbb{C}} = (\mathbf{2}_+^{\mathbb{C}}, \mathbf{2}_-^{\mathbb{C}}) \oplus (\mathbf{2}_-^{\mathbb{C}}, \mathbf{2}_+^{\mathbb{C}}),$$

where $\mathbf{2}_{\pm}^{\mathbb{C}}$ are the complex irreducible spin representations of $\mathfrak{so}(4)$, which have complex dimension 2 each. There are also two real irreducible spin representations $\mathbf{4}_{\pm}$ of $\mathfrak{so}(4)$ of dimension 4 each, which are related to the respective complex spin representations by

$$\mathbf{4}_{\pm} = \mathrm{Re}(\mathbf{2}_{\pm}^{\mathbb{C}} \otimes \mathbb{C}^2). \tag{2.2.1}$$

The above and formula (2.1.2) give

$$\mathbf{8}_+ = \mathbf{4}_{++} \oplus \mathbf{4}_{--}, \qquad \mathbf{8}_- = \mathbf{4}_{+-} \oplus \mathbf{4}_{-+},$$

where the double indices refer to the behavior of the real four dimensional representations under the respective $\mathfrak{so}(4)$ actions.

To summarize, under $\mathfrak{so}(4) \oplus \mathfrak{so}(4)$ we have

$$\mathbf{8}_v = (\mathbf{4}, \mathbf{1}) \oplus (\mathbf{1}, \mathbf{4}), \qquad \mathbf{8}_+ = (\mathbf{4}_{++}) \oplus (\mathbf{4}_{--}), \qquad \mathbf{8}_- = (\mathbf{4}_{+-}) \oplus (\mathbf{4}_{-+}).$$
$$\tag{2.2.2}$$

Below, we determine the massless field content arising from compactification of ten-dimensional string theories to six dimensions. These fields are conveniently grouped into supermultiplets of the respective superalgebras.

The following multiplets can arise in our setting [Sei88], where we give the contribution from each irreducible representation of the space-time-$\mathfrak{so}(4)$:

- in six-dimensional theories with $\mathcal{N} = (1, 1)$ supersymmetry,

$$
\begin{array}{llll}
\text{supergravity multiplet:} & \text{bosonic:} & (\mathbf{9}) \oplus 4(\mathbf{4}) \oplus (\mathbf{3}_+) \oplus (\mathbf{3}_-) \oplus (\mathbf{1}), \\
& \text{fermionic:} & (\mathbf{12}_+) \oplus (\mathbf{12}_-) \oplus (\mathbf{4}_+) \oplus (\mathbf{4}_-), \\
\text{matter multiplet:} & \text{bosonic:} & (\mathbf{4}) \oplus 4(\mathbf{1}), \\
& \text{fermionic:} & (\mathbf{4}_+) \oplus (\mathbf{4}_-);
\end{array}
$$

$$(2.2.3)$$

- in six-dimensional theories with $\mathcal{N} = 1$ supersymmetry (notations as in [Wal88, Sch96])

$$
\begin{array}{llll}
R(6) & \text{supergravity multiplet:} & (\mathbf{9}) \oplus (\mathbf{12}_+) \oplus (\mathbf{3}_+) & \rightarrow \quad g_{\mu\nu} \oplus \psi_\mu^+ \oplus B_{\mu\nu}^+, \\
T(6) & \text{tensormultiplet:} & (\mathbf{3}_-) \oplus (\mathbf{4}_-) \oplus (\mathbf{1}) & \rightarrow \quad B_{\mu\nu}^- \oplus \psi^- \oplus \phi, \\
H(6) & \text{hypermultiplet:} & 4(\mathbf{1}) \oplus (\mathbf{4}_-) & \rightarrow \quad 4\varphi^\alpha \oplus \chi^-, \\
V(6) & \text{vectormultiplet:} & (\mathbf{4}) \oplus (\mathbf{4}_+) & \rightarrow \quad A_\mu \oplus \lambda^+.
\end{array}
$$

$$(2.2.4)$$

We explain below how the first situation arises in the cases of type IIA theory compactified on $K3$ and heterotic $E_8 \times E_8$ compactified on T^4, while the second situation arises in the case of heterotic $E_8 \times E_8$ theory compactified on $K3$.

Type IIA on $K3$

Let us assume that X^4 is a $K3$ surface and consider type IIA strings on $M^{1,5} \times X^4$, which yield an $\mathcal{N} = (1, 1)$ supersymmetric theory. The NS-NS sector in (2.1.1) gives under $\mathfrak{so}(4) \oplus \mathfrak{so}(4)$:

$$
\begin{aligned}
\mathbf{8}_v \otimes \mathbf{8}_v &= ((\mathbf{4}, \mathbf{1}) \oplus (\mathbf{1}, \mathbf{4})) \otimes ((\mathbf{4}, \mathbf{1}) \oplus (\mathbf{1}, \mathbf{4})) \\
&= \underbrace{(\mathbf{9}, \mathbf{1}) \oplus (\mathbf{4}, \mathbf{4}) \oplus (\mathbf{1}, \mathbf{1} \oplus \mathbf{9})}_{=35} \\
&\oplus \underbrace{(\mathbf{3}_+ \oplus \mathbf{3}_-, \mathbf{1}) \oplus (\mathbf{4}, \mathbf{4}) \oplus (\mathbf{1}, \mathbf{3}_+ \oplus \mathbf{3}_-)}_{=28} \oplus \underbrace{(\mathbf{1}, \mathbf{1})}_{=1}.
\end{aligned}
$$

Here we use the decomposition of the 6-dimensional representation $\Lambda^2(\mathbb{R}^4)$ of $\mathfrak{so}(4)$ into the two irreducible representations $\Lambda_\pm^2$ of dimension 3.

At the level of fields this means that the graviton g_{MN}, which transforms in the $\mathbf{35}$ of $\mathfrak{so}(8)$, decomposes into

$$
g_{MN} \longrightarrow g_{\mu\nu} \oplus g_{\mu i} \oplus g_{ij}
$$

where the six-dimensional graviton $g_{\mu\nu}$ gives the metric on the transversal directions of $M^{1,5}$ up to scaling (a spin 2 field), $g_{\mu i}$ are 1-forms (spin 1 fields), and g_{ij} are scalar fields (spin 0) on $M^{1,5}$. To count the dimension of the spaces of corresponding massless fields we need to consider each component as a section of a bundle on $K3$. $g_{\mu\nu}$ yields a scalar on $K3$ which is automatically massless. Hence $g_{\mu\nu}$ gives one massless field in the **9** of $\mathfrak{so}(4)$. Since $K3$ has no closed one-forms, $g_{\mu i}$ generically can only contribute to the massive spectrum. To first order in perturbation theory the scalars g_{ij} are the components of a Ricci-flat metric on X. The kernel of the linearized Einstein equation around g_{ij} gives the corresponding massless fields. The dimension of this kernel is 58, the real dimension of the space of Einstein metrics on $K3$.

Analogously, the B-field B_{MN} decomposes into

$$B_{MN} \longrightarrow B_{\mu\nu}^{+} \oplus B_{\mu\nu}^{-} \oplus B_{\mu i} \oplus B_{ij}^{+} \oplus B_{ij}^{-},$$

where $B_{\mu\nu}^{+}$ and $B_{\mu\nu}^{-}$ are selfdual and respectively anti-selfdual 2-forms, $B_{\mu i}$ a 1-form and B_{ij} scalar fields on $M^{1,5}$. The $B_{\mu\nu}^{\pm}$ give massless fields in the $\mathbf{3}_{\pm}$ of $\mathfrak{so}(4)$, and the one-forms $B_{\mu i}$ become massive. The $B_{ij}^{\pm}$ give two-forms on $K3$. The space of closed two-forms on $K3$ is 22 dimensional, such that the $B_{ij}^{\pm}$ contribute a 22-dimensional space of massless scalars to the spectrum, called the B-field parameters.

The scalar dilaton in the **1** of $\mathfrak{so}(8)$ descends to a scalar dilaton ϕ in the **1** of $\mathfrak{so}(4)$.

In the R-NS sector, to decompose $\mathbf{56}_{+} \oplus \mathbf{8}_{-} = \Psi_{M}^{+} \oplus \Psi^{-}$ it is again convenient to look at the complexifications:

$$\mathbf{56}_{+}^{\mathbb{C}} \longrightarrow (\mathbf{6}_{+}^{\mathbb{C}}, \mathbf{2}_{+}^{\mathbb{C}}) \oplus (\mathbf{6}_{-}^{\mathbb{C}}, \mathbf{2}_{-}^{\mathbb{C}}) \oplus (\mathbf{2}_{+}^{\mathbb{C}}, \mathbf{2}_{+}^{\mathbb{C}} \otimes \mathbf{4}^{\mathbb{C}}) \oplus (\mathbf{2}_{-}^{\mathbb{C}}, \mathbf{2}_{-}^{\mathbb{C}} \otimes \mathbf{4}^{\mathbb{C}}),$$

$$\mathbf{8}_{-}^{\mathbb{C}} \longrightarrow (\mathbf{2}_{-}^{\mathbb{C}}, \mathbf{2}_{+}^{\mathbb{C}}) \oplus (\mathbf{2}_{+}^{\mathbb{C}}, \mathbf{2}_{-}^{\mathbb{C}}),$$

$$\Psi_{M}^{+} \longrightarrow \psi_{\mu}^{+} \otimes \eta^{+} \oplus \psi_{\mu}^{-} \otimes \eta^{-} \oplus \chi^{+} \otimes \psi_{i}^{+} \oplus \chi^{-} \otimes \psi_{i}^{-},$$

$$\Psi^{-} \longrightarrow \psi^{-} \otimes \eta^{+} \oplus \psi^{+} \otimes \eta^{-}.$$

where we use $\mathbf{2}_{\pm}^{\mathbb{C}} \otimes \mathbf{4}^{\mathbb{C}} = \mathbf{6}_{\pm}^{\mathbb{C}} \oplus \mathbf{2}_{\mp}^{\mathbb{C}}$ as representations of $\mathfrak{so}(4)$. In the real setting this corresponds to $\mathbf{4}_{\pm} \otimes \mathbf{4} = \mathbf{12}_{\pm} \oplus \mathbf{4}_{\mp}$, where $\mathbf{12}_{\pm} = \mathrm{Re}(\mathbf{6}_{\pm}^{\mathbb{C}} \otimes \mathbb{C}^{2})$.

In the NS-R sector we similarly have:

$$\mathbf{56}_{-}^{\mathbb{C}} \longrightarrow (\mathbf{6}_{-}^{\mathbb{C}}, \mathbf{2}_{+}^{\mathbb{C}}) \oplus (\mathbf{6}_{+}^{\mathbb{C}}, \mathbf{2}_{-}^{\mathbb{C}}) \oplus (\mathbf{2}_{-}^{\mathbb{C}}, \mathbf{2}_{+}^{\mathbb{C}} \otimes \mathbf{4}^{\mathbb{C}}) \oplus (\mathbf{2}_{+}^{\mathbb{C}}, \mathbf{2}_{-}^{\mathbb{C}} \otimes \mathbf{4}^{\mathbb{C}}),$$

$$\mathbf{8}_{+}^{\mathbb{C}} \longrightarrow (\mathbf{2}_{+}^{\mathbb{C}}, \mathbf{2}_{+}^{\mathbb{C}}) \oplus (\mathbf{2}_{-}^{\mathbb{C}}, \mathbf{2}_{-}^{\mathbb{C}}),$$

$$\Psi_{M}^{-} \longrightarrow \psi_{\mu}^{-} \otimes \eta^{+} \oplus \psi_{\mu}^{+} \otimes \eta^{-} \oplus \chi^{-} \otimes \psi_{i}^{+} \oplus \chi^{+} \otimes \psi_{i}^{-},$$

$$\Psi^{+} \longrightarrow \psi^{+} \otimes \eta^{+} \oplus \psi^{-} \otimes \eta^{-}.$$

The massless fields arise as spinors in the kernel of the respective Dirac operators on $K3$, and the dimension of the space of massless spinors is given by the index of that Dirac operator. To obtain the number of independent gravitinos $\psi_\mu^\pm$ we use

$$\text{ind } \not{D} = \int_X \widehat{A}(R_0) = 2$$

with R_0 the Riemann curvature tensor on X. This means that the space of solutions to the massless field equation $\not{D}\psi = 0$ which cannot become massive has complex dimension 2. From (2.2.1), a pair of two complex spinors in the kernel of $\not{D}$ gives one positive chirality Rarita-Schwinger gravitino. Hence we get one positive chirality Rarita-Schwinger gravitino ψ_μ^+ from Ψ_M^+ and one negative chirality Rarita-Schwinger gravitino ψ_μ^- from Ψ_M^-, respectively. The same index calculation also shows that the ten-dimensional dilatinos $\Psi^\mp$ contribute six-dimensional dilatinos $\psi^\mp$. The fermions $\chi^\pm$ coming from the ten-dimensional gravitinos $\Psi_\mu^\pm$ are governed by the index of the Rarita-Schwinger Dirac operator $\widetilde{\not{D}}_{RS}$ on sections of $S^\pm \otimes T^*X$, where $S^\pm$ denote the spinor bundles on X. By (B.9) we have ind $(\widetilde{\not{D}}_{RS}) = -40$, amounting to 20 negative chirality spinors χ^- coming from Ψ_M^+ and 20 positive chirality spinors χ^+ coming from Ψ_M^-.

In the R-R sector we finally have:

$$\mathbf{56}_v \longrightarrow (\mathbf{4},\mathbf{1}) \oplus (\mathbf{3}_+ \oplus \mathbf{3}_-, \mathbf{4}) \oplus (\mathbf{4},\mathbf{3}_+ \oplus \mathbf{3}_-) \oplus (\mathbf{1},\mathbf{4}), \qquad \mathbf{8}_v \longrightarrow (\mathbf{4},\mathbf{1}) \oplus (\mathbf{1},\mathbf{4}),$$

$$C_{MNP} \longrightarrow C_{\mu\nu\rho} \oplus \quad C_{\mu\nu i} \quad \oplus \quad C_{\mu ij} \quad \oplus C_{ijk}, \quad C_M \longrightarrow C_\mu \oplus C_i.$$

The space-time two-forms $C_{\mu\nu i}$ and the zero-forms C_{ijk}, C_i give one-forms and three-forms on $K3$ respectively; generically they are massive since there are no harmonic one-forms or three-forms on $K3$. Massless fields arise from the three-form $C_{\mu\nu\rho}$ and from the one-forms $C_{\mu ij}$, C_μ, which are zero-forms and two-forms on $K3$, respectively. Since the spaces of closed zero- and two-forms on $K3$ are one- and 22-dimensional, in total we get a 24-dimensional space of massless fields, all of which transform in the $\mathbf{4}$ of $\mathfrak{so}(4)$. In summary,

Proposition 2.1. *The massless spectrum of type IIA strings compactified to six dimensions on an internal $K3$ surface contains 1 graviton $g_{\mu\nu}$ in the ($\mathbf{9}$), 81 (linearly independent) scalars g_{ij}, B_{ij}, ϕ in the ($\mathbf{1}$), a selfdual and an antiselfdual $B_{\mu\nu}^\pm$ in the ($\mathbf{3}_\pm$), 2 gravitinos $\psi_\mu^\pm$ in the ($\mathbf{12}_\pm$), 2 dilatinos $\psi^\mp$ in the ($\mathbf{4}_\mp$), 40 fermions $\chi^\pm$ in the ($\mathbf{4}_\pm$), and 24 vectors $C_{\mu\nu\rho}$, $C_{\mu ij}$, C_μ in the ($\mathbf{4}$) of $\mathfrak{so}(4)$. They are grouped into supermultiplets according to (2.2.3), where* *for definiteness we set* $C := (C_{\mu\nu\rho}, C^{(2,0)}_{\mu ij}, C^{(0,2)}_{\mu ij}, C_\mu)$ *with the superscripts indicating the Dolbeault grading of the cohomology of K3:*

$1 \times$ (*supergravity multiplet*) :

$$(\mathbf{9}) \oplus (\mathbf{12_+}) \oplus (\mathbf{12_-}) \oplus 4(\mathbf{4}) \oplus (\mathbf{3_+}) \oplus (\mathbf{3_-}) \oplus (\mathbf{4_+}) \oplus (\mathbf{4_-}) \oplus (\mathbf{1})$$
$$(g_{\mu\nu}) \oplus \ \psi^+_\mu \ \oplus \ \psi^-_\mu \ \oplus \ C \ \oplus B^+_{\mu\nu} \oplus B^-_{\mu\nu} \oplus \psi^+ \oplus \psi^- \oplus \phi$$

$20 \times$ (*matter multiplet*) :

$$80(\mathbf{1}) \ \oplus 20(\mathbf{4}) \oplus 20((\mathbf{4_+}) \oplus (\mathbf{4_-}))$$
$$(g_{ij}, B_{ij}) \oplus C^{(1,1)}_{ij} \oplus \ \ \chi^+ \ \ \oplus \chi^- \quad .$$

Heterotic $E_8 \times E_8$ strings on $K3$

We now follow [Wal88, p. 379], changing the chiralities in ten dimensions to satisfy the conventions in [GSW85] and adding in further details.

We again assume that X^4 is a $K3$ surface and consider $G_{10} = E_8 \times E_8$ heterotic strings on $M^{1,5} \times X^4$. This means that $M^{1,5} \times X^4$ carries a gauge bundle $\mathcal{E}$ with holonomy G_{10}. The field A_M in the Yang-Mills hypermultiplet $Y(10)$ of (2.1.3) yields its connection. To compactify we also need to assume that the holonomy of $\mathcal{E}$ decomposes into a product $H \times K \subset G_{10}$ with H the holonomy of $\mathcal{E}$ viewed as a bundle on $M^{1,5}$ and K the holonomy of $\mathcal{E}$ viewed as a bundle on X^4. This is detailed and used below (2.2.6). Viewed as a bundle on X, for reasons of consistency the connection of $\mathcal{E}$ is Hermitian-Yang-Mills, where the Donaldson-Uhlenbeck-Yau theorem [Don85, UY86] states that equivalently $\mathcal{E}$ is semi-stable. Since X is a Kähler manifold of complex dimension 2, Hermitian-Yang-Mills connections are precisely the anti-selfdual (ASD) connections.

We have listed the massless ten-dimensional multiplets of heterotic strings in (2.2.4), so a general theory has massless spectrum

$$R(6) \oplus n_T \, T(6) \oplus n_H \, H(6) \oplus n_V \, V(6), \tag{2.2.5}$$

where we need to determine n_T, n_H, n_V.

If we are compactifying a theory which possesses a Lorentz-invariant action, then self-dual and anti-self dual two-forms $B^\pm_{\mu\nu}$ are paired up, hence there is only one tensormultiplet, $n_T = 1$ [DHVW85, DHVW86]. This is the case we focus on first. On the other hand, a general theory need not arise from compactification, and $n_T > 1$ is allowed.

The supergravity multiplet $R(10)$ of (2.1.3) agrees with the NS-NS plus the R-NS sector of type IIA. Hence to compactify to six dimensions we can use the previous results obtaining as massless fields: $g_{\mu\nu}$, the graviton in the (**9**), 81 scalars g_{ij}, B_{ij}, ϕ in the (**1**), selfdual and anti-selfdual two-forms $B^\pm_{\mu\nu}$

in the $(\mathbf{3}_\pm)$, a gravitino ψ_μ^+ in the $(\mathbf{12}_+)$, a dilatino ψ^- in the $(\mathbf{4}_-)$, and 20 fermions χ^- in the $(\mathbf{4}_-)$. A quick glance at (2.2.4) reveals that the ψ^-, χ^- can either belong to a tensor or a hypermultiplet. However, we know that we need to produce precisely one tensormultiplet, so we can safely assume that ψ^- resides in the tensormultiplet, while the χ^- live in hypermultiplets. Altogether we have listed the content of the following multiplets:

$$R(10) \longrightarrow R(6) \oplus T(6) \oplus 20H(6). \qquad (2.2.6)$$

To determine n_H, n_V in (2.2.5), note that the chirality of the fields χ^-, λ^+ respectively distinguishes hypermultiplets from vectormultiplets, the only two types of multiplets that are left for $Y(10)$ to contribute to. Hence it suffices to consider the reduction of the only fermionic field Λ^+ in the super-Yang-Mills multiplet $Y(10)$ of (2.1.3). The gaugino Λ^+ is a positive spinor on $M^{1,9}$ with values in the adjoint representation of the gauge group $G_{10} = E_8 \times E_8$ of the ten-dimensional theory. On reducing from ten to six dimensions the gauge group is broken to some subgroup $H = H^1 \times H^2 \subset E_8 \times E_8$ with $H^i \subset E_8$, while the components A_i of the gauge field A with $K3$ indices, the Higgs bosons, acquire expectation values in $K = K^1 \times K^2 \subset E_8 \times E_8$, with $K^i \subset E_8$ the maximal simple subgroup of the centralizer of H^i. This means that the gauge bundle $\mathcal{E}$, which we always view as a sum of two E_8 bundles $\mathcal{E}^i$, gives a K-principal bundle on $K3$ with curvature taking values in adj(K).

The adjoint representation of E_8 now decomposes under each $H^i \times K^i$ as

$$\mathrm{adj}(E_8) = \bigoplus_{a \in A^i} L_a^i \otimes Q_a^i = \left(\mathrm{adj}(H^i) \otimes \mathbf{1}\right) \oplus \bigoplus_{a \in A_{matter}^i} \left(L_a^i \otimes Q_a^i\right), \quad (2.2.7)$$

where L_a^i and Q_a^i are representations of H^i and K^i. In particular,

$$\sum_{a \in A_{matter}^i} \dim Q_a^i \cdot \dim L_a^i = \sum_{a \in A^i} \dim Q_a^i \cdot \dim L_a^i - \dim(H^i)$$

$$= \dim(\mathrm{adj}(E_8)) - \dim(H^i) = 248 - \dim(H^i). \qquad (2.2.8)$$

Representations with labels in A_{matter}^i comprise the so-called matter multiplets. Here $\mathbf{1} \otimes \mathrm{adj}(K^i)$ gives the neutral matter multiplets, and all other $L_a^i \otimes Q_a^i$ with $a \in A_{matter}^i$ give charged matter multiplets. We have

$$\Lambda^+ = \sum_{i, a \in A^i} \left(\lambda_{a,i}^+ \otimes \eta_{a,i}^+ + \lambda_{a,i}^- \otimes \eta_{a,i}^-\right), \qquad (2.2.9)$$

where $\lambda_{a,i}^\pm$ is a section of $S_M^\pm \otimes L_a^i$, and $\eta_{a,i}^\pm$ is a section of $S^\pm \otimes Q_a^i$, with $S_M^\pm$, $S^\pm$ the spinor bundles on M and X, respectively. To determine the net number of massless fields we need to calculate the index of a Dirac operator $\not{D}_Q : S^+ \otimes \mathcal{E}^i_Q \to S^- \otimes \mathcal{E}^i_Q$ on $K3$, twisted by a K^i-bundle $\mathcal{E}^i_Q$ corresponding to the representation Q of K^i. Since

$$\text{ind}\,(\not{D}_Q) = -\,\text{ind}\,(\not{D}_{Q^*}),$$

the index vanishes if Q is a real (or pseudoreal) representation. For complex representations Q we have

$$\text{ind}\,(\not{D}_Q) = \int_X \widehat{A}(R_0)\text{ch}(Q),$$

where R_0 is the Riemannian curvature of the $K3$ surface and $\text{ch}(Q)$ is a form in the induced curvature F_0 of the associated bundle $\mathcal{E}^i_Q$. See Appendix B for further details on characteristic classes.

In general for a simple Lie group G with Lie algebra $\mathfrak{g}$, for any G-bundle $\mathcal{E}^i_Q$ corresponding to a representation Q of G we introduce the first Pontrjagin class $p_1(\mathcal{E}^i_Q)$:

$$p_1(\mathcal{E}^i_Q) := -\frac{1}{8\pi^2} \int_X \text{Trace}\,_Q(F_0^2). \tag{2.2.10}$$

Then expansion of $\text{ch}(Q)$ yields

$$\text{ind}\,(\not{D}_Q) = \dim Q \int_X \widehat{A}(R_0) - \frac{1}{8\pi^2} \int_X \text{Trace}\,_Q(F_0^2) \overset{(B.8)}{=} 2\dim Q + p_1(\mathcal{E}^i_Q). \tag{2.2.11}$$

Next, following [AHS78] for any G-bundle on X the instanton number is given by

$$k := \frac{1}{8\pi^2}\frac{1}{2c_2(G)} \int_X \text{Trace}\,_\mathfrak{g}(F_0^2), \tag{2.2.12}$$

where $c_2(G)$ is the dual Coxeter number of G. It is important to note that the instanton number is a topological invariant of the bundle. If the holonomy is a proper subgroup of G, then the instanton number is also equal to the expression on the right hand side of (2.2.12) where G is replaced by the holonomy group and $\mathfrak{g}$ by its Lie algebra. Hence the expression (2.2.12) could be used as the definition of the dual Coxeter number $c_2(G)$. Now with $\mathfrak{k}^i$ the Lie algebra of K^i, the instanton number of the E_8-bundle $\mathcal{E}^i$ is

$$k^i = \frac{1}{8\pi^2}\frac{1}{2c_2(K^i)} \int_X \text{Trace}\,_{\mathfrak{k}^i}(F_0^2) = \frac{1}{8\pi^2}\frac{1}{2c_2(E_8)} \int_X \text{Trace}\,_{\mathfrak{e}_8}(F_0^2).$$

Since $c_2(E_8) = 30$ and by (2.2.10) we thus have

$$p_1(\mathcal{E}^i) = -60k^i. \tag{2.2.13}$$

Using the so–called index ind (Q) of the representation Q, which is defined by

$$\forall\, Y,\, Z \in \mathfrak{k}^i : \quad \text{Trace}\,_Q(Y \circ Z) = \frac{\text{ind}\,(Q)}{c_2(K^i)}\text{Trace}\,_{\mathfrak{k}^i}(Y \circ Z),$$

we moreover get

$$p_1(\mathcal{E}_Q^i) = -2k^i \text{ ind}\,(Q). \qquad (2.2.14)$$

Hence (2.2.11) yields

$$\text{ind}\,(\not{D}_Q) = 2\dim\,(Q) - 2k^i \text{ ind}\,(Q). \qquad (2.2.15)$$

This number is related[3] to the dimension of the moduli space of irreducible ASD connections on $\mathcal{E}_Q^i$: The moduli space is either empty or has the hyperkähler dimension

$$k^i \text{ ind}\,(Q) - \dim\,(Q). \qquad (2.2.16)$$

The proof is essentially identical to the proof for the analogous theorem [AHS78, Thm. 6.1], with the only modification that now we work with ASD connections on a vector bundle $\mathcal{E}_Q^i$.

In order to complete the counting of the various multiplets below, we note

$$p_1(\mathcal{E}^i) \overset{(2.2.10)}{=} -\frac{1}{8\pi^2} \int_{K3} \text{Trace}\,_{\mathfrak{e}_8}(F_0^2)$$

$$\overset{(2.2.7)}{=} -\sum_{a \in A^i} \dim L_a^i \cdot \frac{1}{8\pi^2} \int_{K3} \text{Trace}\,_{Q_a^i}(F_0^2)$$

$$\overset{(2.2.14)}{=} -2k^i \sum_{a \in A^i} \dim L_a^i \cdot \text{ind}\,(Q_a^i).$$

Since A_{matter}^i differs from A^i only by $L_a^i = \text{adj}(H^i)$, with $Q_a^i = \mathbf{1}$ and ind $(\mathbf{1}) = 0$ this gives

$$k^i \sum_{a \in A_{matter}^i} \dim L_a^i \cdot \text{ind}\,(Q_a^i) = k^i \sum_{a \in A^i} \dim L_a^i \cdot \text{ind}\,(Q_a^i)$$

$$= -\frac{p_1(\mathcal{E}^i)}{2} \overset{(2.2.13)}{=} 30k^i. \qquad (2.2.17)$$

Finally returning to the reduction of the ten-dimensional gaugino Λ^+ under compactification to six dimensions in (2.2.9), we set

$$N(L_a^i) := \frac{1}{2}\text{ind}\,(\not{D}_{Q_a^i}) \overset{(2.2.15)}{=} \dim Q_a^i - k^i \cdot \text{ind}\,(Q_a^i), \qquad (2.2.18)$$

---

3 This and the derivation of (2.2.19) following Proposition 2.2 is present between the lines in the physics literature, but we could not find it phrased out explicitly.

where we have taken into account that pairs of spinors have to be considered in order to count Weyl spinors. As discussed at the beginning of Section 2.2, if $N(L_a^i)$ is positive, then we have positive chirality spinors from the Minkowski point of view, i.e. using (2.2.4) there are $N(L_a^i)$ vectormultiplets in the representation L_a^i of the unbroken gauge group. If $N(L_a^i)$ is negative, then accordingly we have $(-N(L_a^i))$ hypermultiplets in the representation L_a^i of the unbroken gauge group. In particular, $N(\mathrm{adj}(H^i)) = 1$. For all other labels $a \in A_{matter}^i$, by (2.2.16) we know that $(-N(L_a^i))$ gives the dimension of the moduli space of ASD connections on $\mathcal{E}_{Q_a^i}^i$. This number must be non-negative in order for such a theory with unbroken gauge group $H = H^1 \times H^2$ to exist. Hence

$$Y(10) \longrightarrow \sum_i \dim(H^i) \cdot V(6) \oplus \sum_{i,a \in A_{matter}^i} (-N(L_a^i)) \dim L_a^i \cdot H(6),$$

(2.2.19)

where

$$\sum_{a \in A_{matter}^i} (-N(L_a^i)) \dim L_a^i \overset{(2.2.18)}{=} k^i \cdot \sum_{a \in A_{matter}^i} \mathrm{ind}(Q_a^i) \cdot \dim L_a^i$$

$$- \sum_{a \in A_{matter}^i} \dim Q_a^i \cdot \dim L_a^i$$

$$\overset{(2.2.17),(2.2.8)}{=} 30 k^i - 248 + \dim(H^i). \quad (2.2.20)$$

Remark. As already pointed out, for a theory as above to exist we must have

$$N(L_a^i) = \dim Q_a^i - k^i \cdot \mathrm{ind}(Q_a^i) \leq 0, \quad \text{for all } a \in A_{matter}^i.$$

These inequalities give lower bounds on the instanton numbers k^1, k^2.

So far, we have worked with smooth bundles $\mathcal{E}^i$ on smooth $K3$ surfaces. However, interesting new structures arise when these bundles degenerate to pointlike instantons, i.e. when the curvature of $\mathcal{E}^i$ acquires singularities in the form of Dirac delta distributions. In [DK90, Definition 4.4.1] the resulting connections on $\mathcal{E}^i$ are called ideal ASD connections. In this situation the Pontrjagin classes (2.2.10) and the instanton numbers (2.2.12) are defined in terms of the smooth part F_0 of the curvature. According to [SW96, Wit96b] non-perturbative strong coupling singularities occur when instantons become pointlike, associated to tensionless strings. The process is accompanied by the emergence of an additional tensormultiplet for each pointlike instanton along with an additional neutral hypermultiplet contributing scalars which account for the location of the instanton on the $K3$ surface. The scalar in the additional tensormultiplet is believed to give the parameter of a non-classical phase transition. Altogether the contributions from the Yang-Mills sector amount to

$$l\, T(6) \ \oplus\ \left(l - \sum_{i,a \in A^i_{matter}} N(L^i_a)\dim L^i_a\right) H(6) \ \oplus\ \sum_i \dim(H^i)\, V(6)$$

$$\overset{(2.2.20)}{=} l\, T(6) \oplus \left(30\left(k^1 + k^2\right) - 496 + \dim(H) + l\right) H(6)$$

$$\oplus \dim(H)\, V(6) \tag{2.2.21}$$

if there are l distinct pointlike instantons. According to what was said after (2.2.8), of the hypermultiplets, $(k^1 c_2(K^1) + k^2 c_2(K^2) - \dim K + l)$ are neutral.

One constraint to our theory coming from the Green-Schwarz mechanism in ten dimensions (see (3.2.1) with $1/\alpha = -c_2(E_8) = -30$ as argued below (3.2.2)) is

$$\int_{K3} \left(\mathrm{tr}(R_0^2) - \frac{1}{30}\mathrm{Trace}_{\,\mathfrak{e}_8 \oplus \mathfrak{e}_8}(\check{F}_0^2)\right) = 0,$$

where $\check{F}_0$ denotes the total curvature of $\mathcal{E}$, i.e. F_0 plus contributions of Dirac delta distributions from pointlike instantons. With the correct normalization of the Dirac delta distribution one gets

$$24 = \chi(K3) \quad = \quad \frac{1}{16\pi^2} \int_X \mathrm{tr}(R_0^2) = \frac{1}{60}\frac{1}{8\pi^2} \int_X \mathrm{Trace}_{\,\mathfrak{e}_8 \oplus \mathfrak{e}_8}(F_0^2) + l$$

$$\overset{(2.2.10)}{=} -\frac{1}{60}p_1(\mathcal{E}^1) - \frac{1}{60}p_1(\mathcal{E}^2) + l \overset{(2.2.13)}{=} k^1 + k^2 + l. \tag{2.2.22}$$

Hence (2.2.6) and (2.2.21) give

Proposition 2.2. *Consider a theory with unbroken gauge group H which is obtained by compactification to six dimensions on a smooth internal K3 surface from a ten-dimensional $E_8 \times E_8$ heterotic string theory. Moreover, assume that the bundle on K3 has degenerated to receive l distinct pointlike instantons. Then the massless spectrum is given by*

$$R(6) \oplus (l+1)\, T(6) \oplus (244 + \dim H - 29l)\, H(6) \oplus \dim(H)\, V(6).$$

The n_H hypermultiplets receive 20 contributions that give moduli of the K3 surface. The bundle $\mathcal{E}$ on K3 decomposes into two E_8-bundles $\mathcal{E}^1$, $\mathcal{E}^2$ with instanton numbers k^1, k^2 such that $k^1 + k^2 + l = 24$, and with holonomy groups K^1, K^2, where $K^1 \times K^2 \subset E_8 \times E_8$ is the maximal simple subgroup *of the centralizer of H. Then with $c_2(K^i)$ denoting the respective dual Coxeter numbers, further*

$$k^1 c_2(K^1) + k^2 c_2(K^2) - \dim K + l$$

of the hypermultiplets are neutral and give moduli of the bundle $\mathcal{E}$ on $K3$.

Remark. To clear notations, we introduce n_H^0 and n_H^{ch}, the numbers of neutral and charged hypermultiplets, respectively, so that generically,

$$n_H = n_H^0 + n_H^{ch}, \qquad n_H^0 = 20 + k^1 c_2(K^1) + k^2 c_2(K^2) - \dim K + l. \quad (2.2.23)$$

If K contains factors that are non-simply laced Lie groups then the decomposition $n_H = n_H^0 + n_H^{ch}$ into uncharged and charged matter is a little more subtle: In (2.2.7) the corresponding summands $L_a^i \otimes Q_a^i$ with non-trivial L_a^i and Q_a^i can have non-zero subspaces that are uncharged under the gauge group. This increases n_H^0 and accordingly decreases n_H^{ch} compared to the above formula. We will briefly discuss this effect in Section 5.2.2 below. See also Section 3.3.5 for an example.

Heterotic $E_8 \times E_8$ strings on T^4

Let us now consider heterotic $G_{10} = E_8 \times E_8$ strings compactified to $M^{1,5} \times X^4$ where X^4 is a real four-torus. As opposed to the case where X^4 is a $K3$ surface these theories enjoy enhanced supersymmetry since the holonomy of the torus is trivial, and hence all components of the supercharges in the ten-dimensional theory yield components of supercharges of the compactified theory, see e.g. [Asp97, p. 38]. In this situation massless particles of opposite chirality do not pair up to become massive. However, since all fields give flat sections of the relevant bundles on the torus X^4, starting from the massless spectrum (2.1.3) of the ten-dimensional theory we can decompose all representations with respect to the space-time $\mathfrak{so}(4)$ as before and take the results at face value without any index calculations. The gauge group is generically broken to $H_{gen} = U(1)^{16}$. Since $\dim H_{gen} = 16$, and using the decompositions obtained previously, we generically have

35:	$g_{MN} \rightarrow g_{\mu\nu} \oplus g_{\mu i} \oplus g_{ij}$	in $1(\mathbf{9}) \oplus 4(\mathbf{4}) \oplus 10(\mathbf{1})$,
28:	$B_{MN} \rightarrow B_{\mu\nu}^+ \oplus B_{\mu\nu}^- \oplus B_{\mu i} \oplus B_{ij}$	in $1(\mathbf{3}_+) \oplus 1(\mathbf{3}_-) \oplus 4(\mathbf{4}) \oplus 6(\mathbf{1})$,
1:	$\phi \rightarrow \phi$	in $1(\mathbf{1})$,
56$_+$:	$\Psi_M^+ \rightarrow \psi_\mu^+ \oplus \psi_\mu^- \oplus \chi^+ \oplus \chi^-$	in $(\mathbf{12}_+) \oplus (\mathbf{12}_-) \oplus 4((\mathbf{4}_+) \oplus (\mathbf{4}_-))$,
8$_-$:	$\Psi^- \rightarrow \psi^- \oplus \psi^+$	in $(\mathbf{4}_+) \oplus (\mathbf{4}_-)$,
8$_v \otimes \mathrm{Ad}(G_{10})$:	$A_M \rightarrow A_\mu \oplus A_i$	in $16((\mathbf{4}) \oplus 4(\mathbf{1}))$,
8$_+ \otimes \mathrm{Ad}(G_{10})$:	$\Lambda^+ \rightarrow \lambda^+ \oplus \lambda^-$	in $16((\mathbf{4}_+) \oplus (\mathbf{4}_-))$.

Altogether we find one graviton $g_{\mu\nu}$ in the (**9**), 24 (linearly independent) vectors $g_{\mu i}$, $B_{\mu i}$, A_μ in the (**4**), 81 scalars g_{ij}, B_{ij}, ϕ, A_i in the (**1**), a selfdual and an anti-selfdual two-form $B_{\mu\nu}^\pm$ in the (**3**$_\pm$), 2 gravitinos $\psi_\mu^\pm$ in the (**12**$_\pm$), 40 fermions $\chi^\pm$, $\lambda^\pm$ in the (**4**$_\pm$), and 2 dilatinos $\psi^\mp$ in the (**4**$_\mp$) of $\mathfrak{so}(4)$. Comparison with (2.2.3) and Proposition 2.1 shows

Proposition 2.3. *The massless spectrum of heterotic $E_8 \times E_8$ strings compactified to six dimensions on an internal four-torus generically consists of 1 supergravity and 20 matter multiplets in (2.2.3) and agrees with the massless spectrum of type IIA strings compactified to six dimensions on an internal $K3$ surface by Proposition 2.1.*

The agreement of the generic massless spectra of heterotic $E_8 \times E_8$ strings on T^4 and type IIA strings on $K3$ nowadays is believed not to be a coincidence. We will address this issue in Section 4.

2.2.2. Compactification to D = 4

Consider now superstring theories on $M^{1,3} \times X^6$. We will need to decompose our representations of $\mathfrak{so}(8)$ under $\mathfrak{so}(2) \oplus \mathfrak{so}(6)$, where for $\mathfrak{so}(2)$ we denote the two-dimensional tensor, vector, and spin $\frac{3}{2}$ representations by $\mathbf{2}_t$, $\mathbf{2}_v$, $\mathbf{2}_s$, respectively, all of which are equivalent. There also exists a "spinor" in two dimensions, which transforms trivially under $\mathfrak{so}(2)$, denoted $\mathbf{1}_s$. The possible massless supermultiplets for $\mathcal{N} = 2$ supersymmetry are then

$$
\begin{array}{lll}
R(4) & \text{supergravity multiplet:} & (\mathbf{2}_t) \oplus 2(\mathbf{2}_s) \oplus (\mathbf{2}_v), \\
H(4) & \text{hypermultiplet:} & 4(\mathbf{1}_s) \oplus 4(\mathbf{1}), \\
V(4) & \text{vectormultiplet:} & (\mathbf{2}_v) \oplus 4(\mathbf{1}_s) \oplus 2(\mathbf{1}).
\end{array}
\tag{2.2.24}
$$

For our discussion it hence suffices to consider the bosonic parts of the massless spectra: These already uniquely determine the numbers of independent supermultiplets of each type. In what follows we describe this analysis for type IIA compactified on a Calabi-Yau 3-fold, and for heterotic $E_8 \times E_8$ compactified on $T^2 \times K3$.

Type IIA on a Calabi-Yau 3-fold
Assume that X^6 is a Calabi-Yau 3-fold with full SU(3) holonomy, i.e. with Betti numbers

$$
b_0 = b_6 = 1, \quad b_1 = b_5 = 0, \quad b_2 = b_4 = h^{1,1}(X), \quad b_3 = 2\left(h^{1,2}(X) + 1\right).
$$

To obtain the massless spectrum of type IIA strings compactified to four dimensions with internal X we decompose the bosonic massless fields of the ten-dimensional theory in the NS-NS and the R-R sectors as given in (2.1.1) as before:

35:	g_{MN}	$\to$	$g_{\mu\nu} \oplus g_{\mu i} \oplus g_{ij}$	in	$1(\mathbf{2}_t) \oplus 6(\mathbf{2}_v) \oplus 21(\mathbf{1})$,
28:	B_{MN}	$\to$	$B_{\mu\nu} \oplus B_{\mu i} \oplus B_{ij}$	in	$1(\mathbf{1}) \oplus 6(\mathbf{2}_v) \oplus 15(\mathbf{1})$,
1:	ϕ	$\to$	ϕ	in	$1(\mathbf{1})$,
56$_v$:	C_{MNP}	$\to$	$C_{\mu\nu i} \oplus C_{\mu ij} \oplus C_{ijk}$	in	$6(\mathbf{1}) \oplus 15(\mathbf{2}_v) \oplus 20(\mathbf{1})$,
8$_v$:	C_M	$\to$	$C_\mu \oplus C_i$	in	$1(\mathbf{2}_v) \oplus 6(\mathbf{1})$.

Since X carries no closed one-forms, $g_{\mu i}$, $B_{\mu i}$, $C_{\mu\nu i}$ and C_i become massive. $g_{\mu\nu}$ contributes the graviton in the $\mathbf{2}_t$ to the massless spectrum. The g_{ij} contribute massless scalars, according to the Ricci-flat deformations of this metric on X. There are two types of deformations of this Ricci-flat metric: (1) those corresponding to deformations of a chosen complex structure and giving in total $2h^{1,2}(X)$ parameters; (2) those corresponding to deformations of a chosen Kähler structure and giving a total of $h^{1,1}(X)$ parameters. This analysis uses the fact that the moduli space of Calabi-Yau 3-folds locally splits into the product of Kähler and complex structure deformations. In total, one gets $h^{1,1}(X) + 2h^{1,2}(X)$ independent scalars coming from g_{ij}. Moreover, $a := B_{\mu\nu}$ contributes a massless scalar, known as the axion, while B_{ij} gives $b_2 = h^{1,1}(X)$ massless scalars. The dilaton ϕ is a scalar as always. $C_{\mu ij}$ gives $b_2 = h^{1,1}(X)$ massless fields in the $\mathbf{2}_v$, while C_{ijk} gives $b_3 = 2\left(h^{1,2}(X) + 1\right)$ additional massless scalars. Finally, C_μ contributes one massless vector in the $\mathbf{2}_v$, known as the graviphoton. Altogether comparison with (2.2.24) yields

Proposition 2.4. *The massless spectrum of type IIA strings compactified to four dimensions on a smooth internal Calabi-Yau 3-fold X with full $SU(3)$ holonomy consists of the following supermultiplets, where schematically we list the bosonic field content:*

$$1 \times R(4): \qquad\qquad (\mathbf{2}_t) \oplus (\mathbf{2}_v) \to g_{\mu\nu} \oplus C_\mu,$$
$$\left(h^{1,2}(X) + 1\right) \times H(4): \qquad 4\left(h^{1,2}(X) + 1\right)(\mathbf{1}) \to C_{ijk} \oplus \delta g_{ij} \oplus a \oplus \phi,$$
$$h^{1,1}(X) \times V(4): h^{1,1}(X)(\mathbf{2}_v) \oplus 2h^{1,1}(X)(\mathbf{1}) \to C_{\mu ij} \oplus \delta g_{i\bar{j}} \oplus B_{ij}.$$

Heterotic $E_8 \times E_8$ strings on $T^2 \times K3$

We now calculate the massless spectrum of heterotic $G_{10} = E_8 \times E_8$ string theories compactified to four dimensions on an internal X^6 which is the product of a real two-torus T^2 and a $K3$ surface, essentially following [Asp97]. It is easiest to use the results of Proposition 2.2 for heterotic $E_8 \times E_8$ strings compactified to six dimensions on an internal $K3$ surface, and to compactify two further space-like dimensions of $M^{1,5}$ to a two-torus T^2. As before let K

denote the holonomy of the gauge bundle viewed as bundle on $K3$. K is the maximal simple subgroup of the centralizer for the unbroken gauge group H in $E_8 \times E_8$ under compactification to six dimensions with internal $K3$. We can choose the torus T^2 together with a flat bundle with holonomy $U(1)^r$, and then the observed gauge group in the $D = 4$ uncompactified dimensions is the centralizer of $U(1)^r \times K$ in $E_8 \times E_8$. This implies $U(1)^r \subset H$, and in fact generically the observed gauge group is broken to the Abelian group $H = U(1)^r$.

As for type IIA strings compactified to four dimensions we can restrict to the discussion of bosonic fields, because the supermultiplets (2.2.24) are already distinguished by their bosonic field content. We decompose the relevant representations of $\mathfrak{so}(4)$ with respect to $\mathfrak{so}(2) \oplus \mathfrak{so}(2)$, keeping track of the space-time indices:

$$\mathbf{9} \longrightarrow (\mathbf{2}_t) \oplus 2(\mathbf{2}_v) \oplus 3(\mathbf{1}), \quad \mathbf{1} \longrightarrow (\mathbf{1}),$$

$$\mathbf{3}_\pm \longrightarrow 1(\mathbf{2}_v) \oplus 1(\mathbf{1}), \quad \mathbf{4} \longrightarrow 1(\mathbf{2}_v) \oplus 2(\mathbf{1}).$$

Using this, let us discuss the fate of each six-dimensional supermultiplet in (2.2.4): The bosonic fields $g_{\mu\nu}$, $B^+_{\mu\nu}$ of the six-dimensional supergravity multiplet contribute a four-dimensional graviton $g_{\mu\nu}$ in the $(\mathbf{2}_t)$, three vectors $g_{\mu a}$, $B^+_{\mu a}$ in the $(\mathbf{2}_v)$, and four scalars g_{ab}, B^+_{ab} in the $(\mathbf{1})$, where a, b denote the coordinates on T^2. The graviton and one of the vectors fill up the bosonic field content of the four-dimensional supergravity multiplet $R(4)$. The remaining two vectors and four scalars give two four-dimensional vectormultiplets $V(4)$. The bosonic fields $B^-_{\mu\nu}$, ϕ of each six-dimensional tensormultiplet descend to one vector $B^-_{\mu a}$ in the $(\mathbf{2}_v)$ and two scalars $a = B_{\mu\nu}$, ϕ in the $(\mathbf{1})$, yielding the bosonic field content of a four-dimensional vector multiplet $V(4)$. The only bosonic fields in a six-dimensional hypermultiplet are quadruplets of scalars, which descend to quadruplets of scalars in four dimensions, yielding the bosonic field content of a four-dimensional hypermultiplet $H(4)$. Finally, the only bosonic field in a six-dimensional vectormultiplet is the gauge field A_μ, yielding a vector in the $(\mathbf{2}_v)$ and two scalars in the $(\mathbf{1})$ in four dimensions, i.e. the bosonic field content of a four-dimensional vectormultiplet $V(4)$. Summarizing, we have

$$R(6) \longrightarrow R(4) \oplus 2\,V(4), \quad T(6) \longrightarrow V(4),$$

$$H(6) \longrightarrow H(4), \quad V(6) \longrightarrow V(4).$$

Now from Proposition 2.2 we obtain

Proposition 2.5. *Consider a theory which is obtained by compactification to four dimensions on an internal product of a two-torus T^2 and a smooth $K3$*

surface from a ten-dimensional $E_8 \times E_8$ heterotic string theory where the gauge group on compactification to $K3$ is H. Moreover, assume that the bundle on the $K3$ surface has degenerated to receive l distinct pointlike instantons. Then the massless spectrum is given by

$R(4) \oplus (244 + \dim H - 29l) \, H(4) \oplus (3 + l + \mathrm{rk}\, H) \, V(4) \oplus (\dim H - \mathrm{rk}\, H) \, V(4).$

Here, 20 hypermultiplets account for the moduli of the $K3$ surface, and with notations as in Proposition 2.2 at least

$$n_H^0 - 20 = k^1 c_2(K^1) + k^2 c_2(K^2) - \dim K + l$$

hypermultiplets give the moduli of the gauge bundle on $K3$, with additional contributions in special cases as remarked after Proposition 2.2. One of the vectormultiplets contains the axion-dilaton pair, two of the vectormultiplets give moduli of T^2, and $\mathrm{rk}\, H$ vectormultiplets give moduli of the heterotic bundle on T^2. Further l vectormultiplets correspond to l distinct pointlike instantons. If H is non-Abelian, i.e. $\dim H - \mathrm{rk}\, H \neq 0$, then one has enhanced gauge symmetry, and the additional $\dim H - \mathrm{rk}\, H$ vectormultiplets do not allow a perturbative interpretation in this heterotic theory.

3. Anomalies

To yield string theory as a promising approach towards describing nature, parity violation has to be incorporated in a consistent manner. However, parity violating superstring theories can suffer from anomalies. Let us give a brief summary, following [GSW87, II §10]. Helpful introductions to this topic can also be found in [AGG85, PS95, Wei05, Pol98, BM03, SS04, Adl, Har05].

Very roughly speaking, an anomaly arises when under quantization of a classical system in the resulting quantum field theory a classical symmetry is broken. In terms of the Feynman calculus there then are divergent radiative corrections which do not allow regularization. Such divergent Feynman diagrams are always one-loop diagrams [Adl69, Bar69], with a chiral fermion around the loop and a classically conserved current attached, the conservation of which is not compatible with regularization. For our purposes we need to work in $D = 2n$ space-time dimensions, and below we will see that anomalies of the type of interest here can occur only for odd n. We restrict the discussion to specific local anomalies which lead to fatal inconsistencies of the resulting quantum field theory, in contrast to global anomalies which can yield a welcome technique to break global symmetries for phenomenological reasons.

Mathematically, in order to calculate the relevant propagators in a field theory one in particular needs the determinants of those differential operators that give the equations of motion for the various fields. In the bosonic case the differential operator in question is a Laplacian, and zeta function regularization is a well-understood technique to define a determinant of it. For the fermions, however, one has to deal with families of Dirac operators, which depend on the metric of the underlying manifold and on a gauge connection. By the results of [Ati84, AS84] the existence of, say, gauge covariant propagators is obstructed by the first Chern class of the determinant of the index bundle associated to the relevant family of Dirac operators. In other words, if this bundle is non-trivial, then a gauge covariant propagator cannot be defined and the theory is anomalous. The family index theorem [AS71] hence allows to calculate all quantities that govern potential anomalies. As is explained in [ASZ84], the relevant first Chern class can be evaluated by restricting to families parametrized on two-dimensional spheres. Hence the anomaly of a theory on a D-dimensional spin manifold M is given by the $(D + 2)$ form

$$\widehat{I}_{D+2} = \left(\widehat{A}(Z)\text{ch}(V)\right)_{D+2}.$$

Here, Z is a fiber bundle with base $\mathbb{S}^2$ and fiber M, i.e. it is a $(D + 2)$ dimensional manifold. Hence $\widehat{A}(Z)\text{ch}(V)$ is the density which occurs in the family index theorem as the curvature of the determinant bundle of the index bundle associated to a two-dimensional family of Dirac operators on M, coupled to a vector bundle V on M. Indeed, $\int_M \widehat{A}(Z)\text{ch}(V)$ gives the Chern character of the index bundle on the base $\mathbb{S}^2$ of Z, which up to a constant agrees with the first Chern class of this bundle.

As can be seen from our discussion of massless spectra in Section 2, all chiral fields in our string theories that can contribute to the bundle V are sections of either a spin bundle, a Rarita-Schwinger bundle, or a bundle of two-forms. For all of them the Chern classes can be expressed as combinations of the Pontrjagin classes of M, and therefore they can only integrate over M to a non-vanishing class in degree 2 if $D = 4k + 2, k \in \mathbb{N}$, which we will assume from now on.

$\widehat{I}_{D+2}$ is related to the actual anomaly G by a process known as transgression: $\widehat{I}_{D+2}$ is closed and gauge invariant, $d\widehat{I}_{D+2} = 0 = \delta\widehat{I}_{D+2}$. Hence one locally has $\widehat{I}_{D+2} = dI_{D+1}$, where I_{D+1} can be viewed as closed one-form on $\mathcal{G}$ if the base $\mathbb{S}^2$ of Z is a two-sphere in $\mathcal{M}/\mathcal{G}$. Here $\mathcal{M}$ is the infinite dimensional space of parameters (be it the space of metrics, the space of gauge connections, or the product of these two), and $\mathcal{G}$ is the infinite Lie group of symmetries of $\mathcal{M}$. In other words, I_{D+1} is obtained precisely by Chern's transgression operation. Moreover, since $d\delta I_{D+1} = 0$ but in general I_{D+1} is not gauge invariant one finds $\delta I_{D+1} = dI_D$. Now I_D can be viewed as D-form on $\mathcal{M}$. It is ambiguous up to a closed form which is irrelevant for the actual anomaly $G = \int_M I_D$. The value of G gives the change of the effective action under gauge or coordinate transformations, indeed the "anomaly" of the effective theory in the sense of the breaking of classical symmetries.

In [ASZ84] one also finds the following argument: While the structure group of the spin frame bundle of Z in our formula for $\widehat{I}_{D+2}$ is $\mathfrak{so}(D + 2)$, it may be reduced to $\mathfrak{so}(D) \oplus \mathfrak{so}(2)$, and $\widehat{A}(Z)$ factors into $\mathfrak{so}(D)$ and $\mathfrak{so}(2)$ pieces, respectively. Only the $\mathfrak{so}(D)$ contributions I_{D+2} are relevant, since on the one hand the $\mathfrak{so}(2)$ pieces are universal, and on the other hand approximating them by 1 is sufficient for checking consistency by results of [WZ71, BZ84]. Note that this approximation, strictly speaking, does not restrict the relevant contributions to $\widehat{I}_{D+2}$ to forms on M. However, in all calculations this subtlety is irrelevant: $\widehat{I}_{D+2}$ is viewed as a formal object, given that $(D + 2)$ forms on M do not make sense, but solely the dependence on the coordinates of M is of interest. To implement this, let $\mathcal{R}$ be the Riemann curvature of the spin frame bundle of Z and $\overline{R}$ the Riemann curvature of M. According to the splitting principle, there exist two-forms $x_0, x_1, \ldots, x_{2k+1}$ such that

$$\frac{1}{2}\left(-\frac{1}{4}\right)^m \mathrm{tr}(\mathcal{R}^{2m}) = \frac{1}{4^m}\sum_{i=0}^{2k+1} x_i^{2m} \quad \longrightarrow \quad Y_{2m} := \frac{1}{4^m}\sum_{i=1}^{2k+1} x_i^{2m}$$

$$= \frac{1}{2}\left(-\frac{1}{4}\right)^m \mathrm{tr}(\overline{R}^{2m}),$$

where in restricting attention to Y_{2m} we assume that x_0 is the $\mathfrak{so}(2)$ piece of the curvature, while the x_i with $i > 0$ give the $\mathfrak{so}(D)$ pieces which are relevant for our discussion. We also follow [GSW87] and [Sch02] in suppressing the dependence on the base of Z.

3.1. Gravitational and gauge anomalies

By the above, potential anomalies of the type we are interested in can only occur for string theories with chiral fields in an external space-time of dimension $D = 4k + 2$, $k \in \mathbb{N}$. Of the theories considered in Section 2 this amounts to the cases $D = 10$ and $D = 6$. Contributions to the anomaly can come from space-time spinors $\Psi^{\mp}, \Lambda^+, \psi^-, \chi^-, \lambda^+$, i.e. from spin $\frac{1}{2}$ fields, or from Rarita-Schwinger fields $\Psi_M^{\pm}, \psi_\mu^{\pm}$, i.e. from spin $\frac{3}{2}$ fields, or from chiral two-forms $B_{\mu\nu}^{\pm}$. The indices of the respective Dirac operators contribute with a sign according to their chirality. This implies that theories of type IIA can never suffer from anomalies of this type, since spinors occur only in pairs of opposite chiralities in these theories, with their contributions to the anomaly cancelling.

Among the theories discussed in Section 2 we therefore only need to consider the heterotic $E_8 \times E_8$ theory in $D = 10$ space-time dimensions, see
(2.1.3), and the one in $D = 6$ space-time dimensions with internal $K3$ surface,
see Proposition 2.2. The heterotic $E_8 \times E_8$ theories with internal real four-
torus are anomaly free, since these theories enjoy enhanced supersymmetry,
and their spectrum is non-chiral, as can be seen from Proposition 2.3.

Before discussing anomaly cancellation, let us determine the form of each
possible contribution to the anomaly. This is very nicely explained in [ASZ84].
As a warm-up, consider a chiral spin $\frac{1}{2}$ field, i.e. a section ψ^+ of the spinor
bundle S^+ which obeys $\displaystyle{\not{D}}\psi^+ = 0$ for the ordinary untwisted Dirac operator
$\not{D}\colon S^+ \longrightarrow S^-$. We need the index bundle of the family $\not{D}$ over $\mathcal{M}$, which
depends on the curvature $\overline{R}$,

$$I_{1/2}(\overline{R}) = \widehat{A}(\overline{R})_{4k+4} \quad \text{where} \quad \widehat{A}(\overline{R}) = \prod_{i=1}^{2k+1} \frac{\frac{1}{2}x_i}{\sinh \frac{1}{2}x_i}.$$

More generally a spinor with values in some gauge bundle of curvature $\overline{F}$
contributes an anomaly given by the index of the corresponding twisted Dirac
operator,

$$I_{1/2}(\overline{R}, F) = \left((\mathrm{tr}(e^{i\mathrm{F}})\widehat{A}(\overline{R})) \right)_{4k+4}, \quad \text{where} \quad I_{1/2}(\overline{R}) = I_{1/2}(\overline{R}, 0).$$

For Rarita-Schwinger fields one needs the index of the Rarita-Schwinger complex (A.6). The contributions $\ominus S^+ \ominus S^-$ which equally occur in the domain and
in the image of the Rarita-Schwinger operator $\not{D}_{RS}$ cancel out, leaving us with
the Dirac operator on $\left(S^+ \otimes T^*M\right) \ominus S^+ = S^+ \otimes (T^*M \ominus \mathbf{1})$, see [ASZ84,
§IV.V]. In the physics literature, the virtual subtraction of $\mathbf{1}$ is referred to as
subtracting ghost contributions. The relevant index is hence given by

$$I_{3/2}(\overline{R}) = \left(\widehat{A}(\overline{R})(2 \sum_{i=1}^{2k+1} \cosh x_i - 1) \right)_{4k+4}.$$

Finally, for chiral two-forms one needs the density of the twisted Dirac
operator $\not{D}_A\colon S^+ \otimes S^- \longrightarrow S^- \otimes S^-$, i.e.

$$I_A(\overline{R}) = -\frac{1}{8}L(\overline{R})_{4k+4}, \quad \text{where} \quad L(\overline{R}) = \prod_{i=1}^{2k+1} \frac{x_i}{\tanh x_i},$$

the Hirzebruch L-genus.

Altogether every contribution to the anomaly can be expressed in terms of
known universal polynomials in the Y_{2m}. For later convenience let us give the
result for the various anomalies: For chiral spinors governed by a twisted Dirac

operator let $g := \mathrm{tr}(1)$, the dimension of the representation to which our chiral spinor belongs, and $g = 1$ for $F = 0$. For $D = 10$ one finds [GSW85, II, pp. 351-352],

$$I_{3/2}(\overline{R}) = \frac{1}{45360}(7920Y_6 - 9450Y_2Y_4 + 2205Y_2^3),$$

$$I_A(\overline{R}) = \frac{1}{45360}(-7936Y_6 + 9408Y_2Y_4 - 2240Y_2^3), \qquad (3.1.1)$$

$$I_{1/2}(\overline{R}, \overline{F}) = -\frac{1}{720}\mathrm{tr}F^6 - \frac{1}{144}\mathrm{tr}F^4Y_2 - \frac{1}{8}\mathrm{tr}F^2\left(\frac{1}{45}Y_4 + \frac{1}{18}Y_2^2\right)$$
$$- g\left(\frac{1}{2835}Y_6 + \frac{1}{1080}Y_2Y_4 + \frac{1}{1296}Y_2^3\right),$$

while for $D = 6$ we have [GSW87, II, pp. 349-351]

$$I_{3/2}(\overline{R}) = \frac{1}{72}(-43Y_2^2 + 98Y_4),$$

$$I_A(\overline{R}) = -\frac{1}{45}(5Y_2^2 - 7Y_4), \qquad (3.1.2)$$

$$I_{1/2}(\overline{R}, \overline{F}) = \frac{1}{24}\mathrm{tr}F^4 + \frac{1}{12}\mathrm{tr}F^2Y_2 + g\left(\frac{1}{180}Y_4 + \frac{1}{72}Y_2^2\right).$$

3.2. Anomaly cancellation

Before turning to the discussion of anomaly cancellation for the heterotic $E_8 \times E_8$ theories, let us mention that for $D = 10$ from (3.1.1) one obtains

$$I_{3/2}(\overline{R}) + I_A(\overline{R}) - I_{1/2}(\overline{R}, 0) = 0.$$

This means that a ten-dimensional theory is free of anomalies if it has as its chiral field content m complex negative chirality spin $\frac{1}{2}$ field, m complex positive chirality spin $\frac{3}{2}$ field, and m real self-dual antisymmetric tensor. In fact, with $m = 2$ this is precisely the chiral field content of type IIB string theory. This surprising fact was first observed in [AGW84].

3.2.1. Anomaly free heterotic $E_8 \times E_8$ theories in $D = 10$ dimensions

Let us now turn to heterotic $G_{10} = E_8 \times E_8$ theories in ten dimensions. According to (2.1.3) we obtain contributions to the anomaly from one gravitino Ψ_M^+, one dilatino Ψ^-, and one gaugino Λ^+, such that the anomaly is

$$\widehat{I}_{12} = I_{3/2}(\overline{R}) - I_{1/2}(\overline{R}, 0) + I_{1/2}(\overline{R}, \overline{F}).$$

Using (3.1.1) one determines the coefficient of Y_6 in $\widehat{I}_{12}$ to be a multiple of $(496 - g)$, which vanishes precisely because $g = \dim G_{10} = 496$.

Although this is an encouraging start, in pure supergravity in $D = 4k + 2$ dimensions the total anomaly $\widehat{I}_{4k+4}$ turns out to never vanish. However, according to the seminal papers [GS84, GS85b, GS85a], in heterotic theories a further anomalous diagram occurs which we have not yet discussed. It is a tree diagram in which the massless 2-form B of the supergravity multiplet is exchanged between two gauge bosons and either two gluons, or two gravitinos, or four gravitons. In more mathematical terms, as already mentioned in the footnote on page 61, the B-field is only locally given by a closed 2-form. It is a closed differential cochain and in particular transforms non-trivially under gauge transformations. Accordingly, it does contribute to gauge anomalies. In fact, the non-trivial Yang-Mills gauge transformation of the B-field gives a potential for a gauge-invariant three-form field strength H. With ω_L and ω_Y denoting the Lorentz and Yang-Mills Chern-Simons forms, the latter obeys

$$H = dB + \omega_L + \alpha\omega_Y \quad \Longrightarrow \quad dH = \mathrm{tr}(\overline{\mathrm{R}}^2) + \alpha\mathrm{Tr}(\overline{\mathrm{F}}^2). \qquad (3.2.1)$$

The anomalous tree diagram therefore contributes a term $(\alpha\mathrm{Tr}(\overline{\mathrm{F}}^2) + \mathrm{tr}(\overline{\mathrm{R}}^2))\mathrm{X}_{4k}$ with a $4k$-form X_{4k}. Hence to cancel the anomaly $\widehat{I}_{4k+4}$, we must have

$$\widehat{I}_{4k+4} \sim (\alpha\mathrm{Tr}(\overline{\mathrm{F}}^2) + \mathrm{tr}(\overline{\mathrm{R}}^2))\mathrm{X}_{4k}. \qquad (3.2.2)$$

One can show (see e.g. [GSW85]) that in ten dimensions this factorization holds if $G_{10} = E_8 \times E_8$ or $G_{10} = \mathrm{SO}(32)/\mathbb{Z}_2$ and with $\alpha = -1/30$ (note that also dim $\mathrm{SO}(32)/\mathbb{Z}_2 = 496$), so that anomalies are indeed cancelled in these theories. If X_{4k} can be integrated to $dX_{4k-1} = X_{4k}$, then the factorization (3.2.2) together with the known properties of the Chern-Simons forms allow to solve the descent equations and to calculate the actual anomaly coming from the tree diagram for the exchange of B described above.

The process of anomaly cancellation described here is known as the Green-Schwarz mechanism [GS84, GS85b, GS85a]. It remains one of the mysteries of string theory to understand why it works, and its discovery meant a breakthrough that triggered what is now known as the First String Revolution.

3.2.2. Anomaly free heterotic $E_8 \times E_8$ theories in $D = 6$ dimensions

Let us determine when a heterotic $E_8 \times E_8$ string theory in $D = 6$ dimensions with an internal $K3$ surface is anomaly free. We know that its massless spectrum is of the form (2.2.5). The requirement that the theory be anomaly-free poses a restriction on the numbers n_T, n_V, n_H. The Green-Schwarz mechanism described above applies just as in the ten-dimensional situation. It implies, by integration of (3.2.1) over the internal $K3$ surface, that the instanton number (2.2.12) of the heterotic bundle viewed as a bundle on $K3$ is 24 (see (2.2.22)). The Green-Schwarz mechanism cancels all contributions to the anomaly by terms of the form (3.2.2), apart from those proportional to $Y_4 \sim \mathrm{tr}(\overline{R}^4)$. Hence in order to check anomaly cancellation we need to collect the contributions to the coefficient of Y_4 in the total anomaly, carefully taking into account the various fields in our multiplets and their chiralities.

The fields from the supergravity multiplet $R(6)$ of (2.2.4) which contribute to the anomaly are the gravitino ψ_μ^+ and the self-dual tensor field $B_{\mu\nu}^+$, both of positive chirality. From (3.1.2) we read the coefficient in front of Y_4,

$$\frac{98}{72} + \frac{7}{45} = \frac{1}{180}(245 + 28) = \frac{273}{180}.$$

The fields from a tensormultiplet $T(6)$ of (2.2.4) which contribute to the anomaly are the anti-self-dual tensor field $B_{\mu\nu}^-$ and the spinor ψ^-, both of negative chirality, yielding the following coefficient in front of Y_4,

$$-\frac{7}{45} - \frac{1}{180} = -\frac{1}{180}(28 + 1) = -\frac{29}{180}.$$

The only field from each vectormultiplet $V(6)$ of (2.2.4) contributing to the anomaly is the spinor λ^+ of positive chirality, giving a coefficient

$$\frac{1}{180},$$

while from each hypermultiplet we get a contribution from the spinor χ^- of negative chirality, thus the coefficient

$$-\frac{1}{180}$$

in front of Y_4. Adding everything up we find

Proposition 3.1. *A heterotic $E_8 \times E_8$ string theory in $D = 6$ dimensions with internal $K3$ surface and n_T, n_V, n_H tensor-, vector- and hypermultiplets, respectively, is anomaly free iff*

$$n_H - n_V = 273 - 29n_T, \tag{3.2.3}$$

and the instanton number is 24.

In Proposition 2.2 we have determined the massless spectrum for examples of theories of the type addressed in the above proposition. Namely, consider a theory which is obtained by compactification from a ten-dimensional $E_8 \times E_8$ heterotic theory to six dimensions on a smooth internal $K3$ surface. Denote the unbroken gauge group by H and assume furthermore that the bundle on

$K3$ has degenerated to receive l distinct pointlike instantons. Then according to Proposition 2.2 the massless spectrum obeys

$$n_T = l + 1, \qquad n_V = \dim(H), \qquad n_H = 244 + \dim(H) - 29l.$$

We immediately see that the anomaly cancellation condition of Proposition 3.1 holds. Note that we have thus given two independent derivations for the formula (3.2.3) for this case. In other words, heterotic string theories arising from compactification of anomaly free theories in ten dimensions are automatically anomaly free, and this remains true when the gauge bundle acquires pointlike instantons.

3.3. Examples

This section is devoted to the presentation of a number of examples, where we carry out the calculations of the numbers n_T, n_V, n_H^0, and n_H^{ch} explained above. In view of the heterotic – F-theory duality these examples are representative, as we shall see in Section 5.3. In each case, anomaly cancellation (3.2.3) holds.

Assume that the heterotic bundle data specify two E_8 bundles with instanton numbers k^1, k^2 and l^1, l^2 distinct pointlike instantons, respectively, where the holonomy is given by $K^i \subset E_8$ and by (2.2.22) we have $k^1 + k^2 + l^1 + l^2 = 24$. In this case the unbroken gauge group of the heterotic theory is $H^1 \times H^2$ with $H^i \subset E_8$ the centralizer of K^i in E_8. We use the unbroken gauge group $H^1 \times H^2$ to label these examples as in [BIK+96], where a similar analysis is performed.

3.3.1. Completely broken gauge group

If the gauge group is completely broken, the $K3$ bundle has holonomy $K = E_8 \times E_8$. Assuming no pointlike instantons, by Proposition 2.2 and (2.2.23)

$$n_V = 0, \quad n_T = 1, \quad n_H^0 = 20 + 30(k^1 + k^2) - 496 = 244, \quad n_H^{ch} = 0.$$
$$(3.3.1)$$

3.3.2. Unbroken E_8 gauge group

With gauge group $H = E_8 \times \{id\}$ we have holonomy $K = \{id\} \times E_8$. An E_8 bundle can have trivial holonomy only if all its curvature is concentrated in pointlike instantons. So let us assume that there are $l = l^1$ distinct pointlike instantons, $k^1 = 0$, while the second bundle has full E_8 holonomy with instanton number k^2 and we assume $l^2 = 0$. From (2.2.7) we see that there is no charged matter, such that with Proposition 2.2 and (2.2.23)

$$n_V = 248, \quad n_T = l+1, \quad n_H^0 = 20+30k^2-248+l = 492-29l, \quad n_H^{ch} = 0.$$
$$(3.3.2)$$

3.3.3. Unbroken E$_7$ gauge group

With gauge group $H = E_7 \times \{id\}$ we have holonomy $K = SU(2) \times E_8$. Assume that there are l distinct pointlike instantons on the bundle with SU(2) holonomy and effective instanton number k^1, while the second bundle has full E_8 holonomy with instanton number k^2. From (2.2.7) we see that there is a contribution to the charged matter, $\mathbf{248} = (\mathbf{133}, \mathbf{1}) \oplus (\mathbf{56}, \mathbf{2}) \oplus (\mathbf{1}, \mathbf{3})$, in which $(\mathbf{56}, \mathbf{2})$ gives charged hypermultiplets. The multiplicity is computed from (2.2.18) and (2.2.19) and amounts to

$$N(\mathbf{56}) = k^1 \operatorname{ind}(\mathbf{2}) - \dim(\mathbf{2}) = \frac{1}{2}k^1 - 2.$$

In total from Proposition 2.2 and (2.2.23) we have

$$n_V = 133, \quad n_T = l + 1,$$
$$n_H^0 = 20 + 2k^1 + 30k^2 - 251 + l = -231 + 2k^1 + 30k^2 + l,$$
$$n_H^{ch} = 28k^1 - 112. \tag{3.3.3}$$

3.3.4. Unbroken E$_6$ gauge group

With gauge group $H = E_6 \times \{id\}$ we have holonomy $K = SU(3) \times E_8$. Assume that there are l distinct pointlike instantons on the bundle with SU(3) holonomy and effective instanton number k^1, while the second bundle has full E_8 holonomy with instanton number k^2. From (2.2.7) we see that there is a contribution to the charged matter, $\mathbf{248} = (\mathbf{78}, \mathbf{1}) \oplus (\mathbf{27}, \mathbf{3}) \oplus (\overline{\mathbf{27}}, \overline{\mathbf{3}}) \oplus (\mathbf{1}, \mathbf{8})$, in which $(\mathbf{27}, \mathbf{3}) \oplus (\overline{\mathbf{27}}, \overline{\mathbf{3}})$ gives charged hypermultiplets. The multiplicity is computed from (2.2.18) and (2.2.19) and amounts to

$$N(\mathbf{27}) = N(\overline{\mathbf{27}}) = k^1 \operatorname{ind}(\mathbf{3}) - \dim(\mathbf{3}) = \frac{1}{2}k^1 - 3.$$

To ease computations like this one, the book [MP81] is recommended, where indices of representations like the ones that occur here are tabulated. In total from Proposition 2.2 and (2.2.23) we have

$$n_V = 78, \quad n_T = l + 1,$$
$$n_H^0 = 20 + 3k^1 + 30k^2 - 256 + l = -236 + 3k^1 + 30k^2 + l,$$
$$n_H^{ch} = 27k^1 - 162. \tag{3.3.4}$$

3.3.5. Unbroken F_4 gauge group

With gauge group $H = F_4 \times \{id\}$ we have holonomy $K = G_2 \times E_8$. From (2.2.7) we see that there is a contribution to the charged matter, $\mathbf{248} = (\mathbf{52}, \mathbf{1}) \oplus (\mathbf{26}, \mathbf{7}) \oplus (\mathbf{1}, \mathbf{14})$, in which $(\mathbf{26}, \mathbf{7})$ contributes charged hypermultiplets. However, in this case a two-dimensional subspace of the $\mathbf{26}$ is in fact uncharged under the gauge group, i.e. the kernel of the representation when restricted to the Cartan torus $\mathfrak{t}$ of H is two-dimensional. The corresponding multiplets contribute to n_H^0 rather than n_H^{ch} (see the remark after Proposition 2.2). The multiplicity of $\mathbf{26}$ obtained from (2.2.18) and (2.2.19) is

$$N(\mathbf{26}) = k^1 \operatorname{ind}(\mathbf{7}) - \dim(\mathbf{7}) = k^1 - 7.$$

In total from Proposition 2.2 and (2.2.23) we have

$$n_V = 52, \quad n_T = 1,$$
$$n_H^0 = 20 + 4k^1 + 30k^2 - 262 + 2(k^1 - 7) = -256 + 6k^1 + 30k^2,$$
$$n_H^{ch} = 24k^1 - 168. \tag{3.3.5}$$

3.3.6. Unbroken Spin(10) gauge group

With gauge group $H = \mathrm{Spin}(10) \times \{id\}$ we have holonomy $K = \mathrm{SU}(4) \times E_8$. From (2.2.7) we see that there is a contribution to the charged matter, $\mathbf{248} = (\mathbf{45}, \mathbf{1}) \oplus (\mathbf{10}, \mathbf{6}) \oplus (\mathbf{16}, \mathbf{4} \oplus \mathbf{4}) \oplus (\mathbf{1}, \mathbf{15})$, in which $(\mathbf{10}, \mathbf{6}) \oplus (\mathbf{16}, \mathbf{4} \oplus \mathbf{4})$ contributes charged hypermultiplets. The multiplicities obtained from (2.2.18) and (2.2.19) are

$$N(\mathbf{10}) = k^1 \operatorname{ind}(\mathbf{6}) - \dim(\mathbf{6}) = k^1 - 6,$$
$$N(\mathbf{16}) = 2\left(k^1 \operatorname{ind}(\mathbf{4}) - \dim(\mathbf{4})\right) = k^1 - 8.$$

In total from Proposition 2.2 and (2.2.23) we have

$$n_V = 45, \quad n_T = 1, \quad n_H^0 = 20 + 4k^1 + 30k^2 - 263 = -243 + 4k^1 + 30k^2,$$
$$n_H^{ch} = 10(k^1 - 6) + 16(k^1 - 8) = 26k^1 - 188. \tag{3.3.6}$$

4. Heterotic – type IIA and F-theory duality

While the five basic string theories in ten-dimensional Minkowski space can be distinguished by their numbers of supersymmetries and their massless field content, after compactification to D dimensions with $D < 10$ this is not true anymore. As a consequence, so-called string-string dualities between various string theories were conjectured. In fact it is claimed that all string theories are connected by a web of dualities. Here we concentrate on the heterotic – type IIA duality and the heterotic – F-theory duality. For our purposes, the latter is best viewed as a certain limit of the former.

4.1. The heterotic – type IIA duality

We have already encountered an example of the phenomenon known as a string-string duality. Namely, in Proposition 2.3 we observed that in compact-ifications to six dimensions the massless spectra agree for heterotic $E_8 \times E_8$ strings with internal real four-torus on the one hand and for type IIA strings with internal $K3$ surface on the other hand. In fact, more can be said since the (classical) moduli spaces of the respective theories are known explicitly. The scalars in the massless supermultiplets give real coordinates of these moduli spaces, and for both theories one finds the moduli space

$$\mathbb{R} \times \quad O^+(4, 20; \mathbb{Z}) \backslash O^+(4, 20; \mathbb{R}) / (SO(4) \times O(20)), \qquad (4.1.1)$$

where the factor $\mathbb{R}$ accounts for the dilaton, and the remaining 80-dimensional quaternionic Kähler space corresponds to the 20×4 scalars from the 20 hypermultiplets.

On the basis of these stunning agreements physicists have made the daring conjecture [HT95, Wit95] that these string theories in fact are equivalent. One says that there is a string-string duality between heterotic and type IIA theo-ries, respectively. Much evidence in favor of this conjectured duality has been collected, including the fact that the low energy effective actions agree [Wit95].

It is important to note that this string-string duality cannot be seen purely perturbatively, since it maps the heterotic dilaton to the negative of the type IIA dilaton [Wit95], and thus the string coupling constant is inverted under the duality. Nevertheless, we have geometric interpretations of all scalars in the hypermultiplets in terms of an Einstein metric, a B-field, and a connection one-form of a flat bundle on a real four-torus on the one hand, and in terms of an Einstein metric and a B-field on a $K3$ surface on the other hand. Hence the conjectured duality induces a map between these geometric structures. This can be made very explicit as follows: For the $K3$ surface on the type IIA side we can always choose a complex structure such that this surface is elliptically fibered with section. In standard (singular) Weierstraß form it is given by an equation

$$y^2 = x^3 + xf(z) + g(z), \quad \text{with} \quad f(z) = \sum_{m=0}^{8} f^{(m)} z^m, \quad g(z) = \sum_{n=0}^{12} g^{(n)} z^n,$$

$$(4.1.2)$$

where (x, y) are affine coordinates in $\mathbb{CP}^2$ for the fiber, while z is an affine coordinate in the base $\mathbb{CP}^1$ of the fibration and $f^{(j)}$, $g^{(k)}$ are complex constants. We explicitly allow degenerations of our $K3$ surface where it obtains singularities of ADE type: For example, if $f(z) = \alpha z^4$ and $g(z) = z^5 + \beta z^6 + z^7$ with α, $\beta \in \mathbb{C}$, then near $z = 0$ (and similarly near $z = \infty$), the equation (4.1.2) gives

$$y^2 = x^3 + z^5 + \ldots,$$

with the familiar exponents $(2, 3, 5)$ of an E_8 quotient singularity. The actual $K3$ surface is obtained by minimally resolving all such singularities. This is why the above model is called the *singular* Weierstraß model. The possible singular fibers have been classified by Kodaira [Kod64].

More generally note that the Grassmannian $O^+(4, 20; \mathbb{R})/SO(4) \times O(20)$ of (4.1.1) is modelled on the cohomology of $K3$, $H^*(K3, \mathbb{R}) \cong \mathbb{R}^{4,20}$, and its points are given by positive definite oriented four-planes in $H^*(K3, \mathbb{R})$ which encode the geometric data of a real Einstein metric and a B-field on $K3$ [AM94]. If the $K3$ surfaces under inspection are restricted to have specific singularities, then this amounts to restricting these four-planes to $E^\perp$, where $E \subset H^{even}(K3, \mathbb{Z})$ is the lattice associated to the exceptional divisor in the resolution of that singularity. Hence with $m = \mathrm{rk}\, E$ the Grassmannian factor in (4.1.1) reduces to $O^+(4, 20-m; \mathbb{R})/SO(4) \times O(20-m)$. In the interpretation of (4.1.1) as moduli space of real four-tori equipped with semi-stable $E_8 \times E_8$ bundles, the restriction of four-planes to $E^\perp$ amounts to restricting to bundles on the four-torus with some unbroken gauge symmetry, i.e. with restricted holonomy. Specifically, to the lattice E one can associated a semi-simple Lie algebra $\mathfrak{g} \subset \mathfrak{e}_8 \oplus \mathfrak{e}_8$ because E comes from a collection of singularities of ADE type on $K3$. The restricted holonomy of the gauge bundle then is the centralizer K of $G \subset E_8 \times E_8$, where G has Lie algebra $\mathfrak{g}$. In particular, in the above example the toroidal bundle has trivial holonomy and full $E_8 \times E_8$ gauge symmetry with $\mathfrak{g} = \mathfrak{e}_8 \oplus \mathfrak{e}_8$.

One finds [MV96b] that the coefficients $f^{(j)}$, $g^{(k)}$ in (4.1.2) with $j \leq 3$, $k \leq 5$ give the data of one of the heterotic E_8 bundles, while the data of the second heterotic E_8 bundle are encoded in the coefficients with $j \geq 5$, $k \geq 7$. The remaining two parameters $f^{(4)}$ and $g^{(6)}$ on the heterotic side are interpreted as specifying complex structure and complexified Kähler structure of an elliptic curve. This may be surprising, since we expect to find the geometric data of a real four-torus. However, for such a torus one can always choose a complex structure such that it is elliptically fibered. Then $f^{(4)}$ and $g^{(6)}$ only give the data of the base of such a fibration. Correspondingly on the type IIA side we have not specified the Kähler class of our $K3$ surface.

Following [HT95, FHSV95, KV95, Sen96a, Vaf96, DMW96], the conjectured heterotic – type IIA duality in six dimensions has been generalized to a
conjectured heterotic – type IIA duality in four dimensions. Surprisingly, the
following naive idea seems to work: Consider heterotic $E_8 \times E_8$ strings in six
dimensions with internal space the product of a real two-torus and a $K3$ surface. Using an elliptic fibration of the $K3$ surface, the complex structure of
the internal space can always be chosen such that it is a fibration with section
over $\mathbb{CP}^1$ and with generic fiber a complex two-torus. A fiberwise application
of the six-dimensional heterotic – type IIA duality yields this theory dual to a
type IIA theory in four dimensions with internal Calabi-Yau 3-fold which is
$K3$-fibered over the "same" $\mathbb{CP}^1$ we used on the heterotic side. We will see
below that this idea carries tremendously far.

Note that any heterotic – type IIA duality in four dimensions requires a
matching of the massless spectra found in Propositions 2.4 and 2.5, respectively. Particularly for heterotic theories on a product of a real two-torus and
a $K3$ surface, the form of the moduli space associated to the scalars which
give geometric moduli in the vectormultiplets is known, at least at small string
coupling, where perturbative techniques hold: With $m = \text{rk } H$ the rank of the
unbroken gauge group H on compactification to $K3$ as in Proposition 2.5, one
has a space of the form

$$O^+(2, 2 + m; \mathbb{Z})\backslash O^+(2, 2 + m; \mathbb{R})/ (SO(2) \times O(2 + m)) \times SU(2)/U(1),$$
$$(4.1.3)$$

where the second factor accounts for the axion-dilaton pair. If a dual type IIA
theory exists, then the moduli space formed by the scalars of $(m + 3)$ of its
vectormultiplets must take the same form as above in a regime where some
parameter corresponding to the heterotic dilaton becomes small. In [AL96] it
was shown that this implies that the respective type IIA theory has an internal
Calabi-Yau 3-fold which is $K3$-fibered. If this fibration has a section, then the
size of the section corresponds to the value of the heterotic dilaton.

To match the hypermultiplet spectrum recall that on the heterotic side the
hypermultiplets appear in two disguises, neutral and charged (2.2.23). On the
type IIA side all $(h^{1,2}(X) + 1)$ hypermultiplets of Proposition 2.4 are neutral.
The charged hypermultiplets (as well as $(\dim H - \text{rk } H)$ of the vectormultiplets
in Proposition 2.5) arise from non-perturbative phenomena which we have not
yet accounted for on the type IIA side, since we have always assumed the
Calabi-Yau 3-fold to be smooth. On the heterotic side this restriction amounts
to assuming Abelian gauge groups H where no enhanced symmetry and no
charged matter occurs. Using Propositions 2.4 and 2.5 we altogether have

Conjecture 4.1. *There exists a duality between four-dimensional string theories which maps a heterotic $E_8 \times E_8$ string theory on a product of an elliptic curve and a $K3$ surface to a type IIA theory with an internal Calabi-Yau 3-fold X which is $K3$-fibered, if there exists a regime where both the respective perturbation theories converge. Moreover, if the heterotic theory arises from compactification from a ten-dimensional theory such that the gauge group on compactification to $K3$ is H and the gauge bundle viewed as a bundle on $K3$ acquires l distinct pointlike instantons, then X has Hodge numbers*

$$h^{1,2}(X) = n_H^0 - 1 = 19 + k^1 c_2(K^1) + k^2 c_2(K^2) - \dim K + l,$$
$$h^{1,1}(X) = 3 + \operatorname{rk} H + l$$

with notations as in Proposition 2.2 and with the comment concerning a possible enhancement of n_H^0 as stated after that proposition.

If in addition the Calabi-Yau 3-fold X on the type IIA side of the duality is elliptically fibered, then to capture its complex structure data it is often convenient to use a singular Weierstraß form (4.1.2), where now the coefficients of the polynomials f and g depend on a second affine parameter z_2 of the base $\mathbb{CP}^1$ of the $K3$-fibration of X:

$$y^2 = x^3 + x \sum_{m=0}^{8} z_1^m f^{(m)}(z_2) + \sum_{n=0}^{12} z_1^n g^{(n)}(z_2) \qquad (4.1.4)$$

with $\deg f^{(i)} = 8$ and $\deg g^{(j)} = 12$. Generalizing the discussion in the case of an elliptically fibered $K3$ surface and according to physics lore [MV96b, BIK$^+$96] the various parameters in (4.1.4) are assigned an interpretation in the dual heterotic theory. Namely, the parameters governing the polynomials $f^{(0)}, \ldots, f^{(3)}$, $g^{(0)}, \ldots, g^{(5)}$ correspond to the bundle parameters of one E_8 bundle, those governing the polynomials $f^{(5)}, \ldots, f^{(8)}$, $g^{(7)}, \ldots, g^{(12)}$ correspond to the bundle parameters of the second E_8 bundle, while $f^{(4)}$ and $g^{(6)}$ give the complex structure data of the $K3$ surface in the heterotic product of an elliptic curve with $K3$. This description can be very useful, for instance because in many examples a parameter count in the polynomials already leads to a correct determination of $h^{1,2}(X)$. Moreover, this setting is tailor made for the application of a spectral cover description of the heterotic bundles [FMW97, BJPS97, BCG$^+$98]. However, the ansatz has to be handled with care. First, the parameter count only works when all complex structure deformations of X respect the algebraic form of (4.1.4). It is not hard to construct examples where not all contributions to $h^{1,2}(X)$ are visible in terms of a parameter count in (4.1.4). Second, the assumption that a global form (4.1.4) of the equation for X exists does not always hold, because globally one need not have coordinate transforms that yield all fibers of an elliptic fibration in Weierstraß form. Finally, in (4.1.4) one has tacitly assumed that all fibrations have at least one global section.

4.2. The F-theory limit

For the heterotic – type IIA duality in four dimensions our presentation of the matching of multiplets in Conjecture 4.1 so far is exact only for Abelian gauge groups H. This is the generic case, but in our Proposition 2.5 we have already accounted for the possibility of enhanced gauge symmetry on the heterotic side, which goes along with the appearance of charged hyper-multiplets. The natural setup for considering non-Abelian gauge groups is a decompactification limit of the real two-torus in the heterotic theory. To perform such an operation, in (4.1.3) one has to choose a subspace of the form $O^+(2, 2; \mathbb{Z})\backslash O^+(2, 2; \mathbb{R})/SO(2) \times O(2)$, singling out the parameters of T^2 to make its volume large. In this limit, all parameters of T^2 are lost, and the theory becomes effectively six-dimensional. In particular, the gauge group is the centralizer in $E_8 \times E_8$ of the holonomy group of the $K3$-bundle. In general this can be a non-Abelian group, and the charged hypermultiplets capture the respective decomposition of the holonomy representation as in (2.2.7).

On the type IIA side, this decompactification process corresponds to taking a so–called F-theory limit. The physics literature on this theme is vast, see for example [Vaf96, Sen96b, Wit95, MV96a, MV96b, BIK+96, AM97]. The singling out of heterotic T^2 parameters in terms of $O^+(2, 2; \mathbb{Z})\backslash O^+(2, 2; \mathbb{R})/SO(2) \times O(2)$ corresponds to imposing the structure of an elliptic fibration with section on the Calabi-Yau 3-fold X of the type IIA side [MV96a]. Since X is also $K3$-fibered, altogether we obtain an elliptic fibration of X over a rational surface $Z \to \mathbb{CP}^1$. In the process of passing to the F-theory limit, the parameter for the size of the fiber of $Z \to \mathbb{CP}^1$ is lost. Moreover, the size of the elliptic fiber in $X \to Z$ stabilizes to a constant value, which a priori can be taken to be zero. This means that X becomes singular: The Kähler class of X belongs to a face of a Kähler cone.

Since Z is a rational surface, it possesses two special sections, the section at infinity and the zero section. According to [MV96a, MV96b, BIK+96] the Kodaira type of the generic fiber over these special sections gives the two factors H^1, H^2 of the unbroken gauge group corresponding to the two heterotic E_8-bundles in the dual theory. This is compatible with the discussion of (4.1.4): If for $Z \to \mathbb{CP}^1$ we view z_1 as affine coordinate on the fiber, while z_2 gives a coordinate on the base, then the zero section is located at $z_1 = 0$, where the behavior of the elliptic fibration $X \to Z$ is encoded in the polynomials $f^{(0)}, \ldots, f^{(3)}$, $g^{(0)}, \ldots, g^{(5)}$, and analogously for the section at infinity located at $(z_1)^{-1} = 0$ with $f^{(5)}, \ldots, f^{(8)}, g^{(7)}, \ldots, g^{(12)}$ – the bundle parameters as claimed. Under the assumption that the heterotic $K3$ surface is smooth one can check that no other families of degenerate fibers of non-trivial ADE type occur in the fibration. In particular,

$$h^{1,1}(X) = 1 + \mathrm{rk}(H^1) + \mathrm{rk}(H^2) + \rho(Z)$$

with $\rho(Z)$ the Picard number of the base Z. Note that if the base Z of the fibration $X \to Z$ is minimal, then Z is a Hirzebruch surface $\mathbb{F}_n$, and in this case its Picard number is $\rho(Z) = 2$. In general, introduce $l \in \mathbb{N}$ such that

$$\rho(Z) = 2 + l$$

and observe from Conjecture 4.1 that l corresponds to the number of pointlike instantons on the heterotic side.

Altogether we have:

Conjecture 4.2. *Consider the decompactification limit onto $K3$ of a heterotic $E_8 \times E_8$ theory on $T^2 \times K3$ with smooth $K3$ surface, unbroken gauge group H, and l distinct pointlike instantons. This is a six-dimensional theory with massless spectrum according to Proposition 2.2. Under heterotic – F-theory duality this theory is mapped to the F-theory limit of type IIA strings on a Calabi-Yau 3-fold X with the following properties:*

X is elliptically fibered with a section over a rational surface Z with Picard number $\rho(Z)$ and $K3$-fibered over the base of $Z \to \mathbb{CP}^1$. The fibers of $X \to Z$ are shrunken to zero volume, corresponding to the size of the heterotic T^2 becoming infinite. Similarly, the moduli of the fiber of $Z \to \mathbb{CP}^1$ have dropped out, matching the loss of the complex structure parameters for the heterotic T^2. The Kähler parameter giving the size of the base of $Z \to \mathbb{CP}^1$ corresponds to the scalar dilaton in a tensormultiplet of the heterotic theory. Additional l tensormultiplets accounting for the l pointlike instantons on the heterotic side match the l remaining Kähler parameters of Z. Of the heterotic vectormultiplets, $\mathrm{rk}(H)$ account for bundle parameters of the gauge bundle on T^2 in the original four-dimensional heterotic theory. These are recovered as Kähler moduli of X in terms of degenerate fibers of the elliptic fibration over the two special sections of Z. Namely, the generic fibers over these sections are ADE type Kodaira fibers which match the ADE types of the summands of the gauge algebra $\mathfrak{h}$. Altogether we have

$$\rho(Z) = l + 2, \qquad h^{1,1}(X) = 3 + \mathrm{rk}(H) + l.$$

Finally, the n_H^0 neutral hypermultiplets of the heterotic theory match the hyper-multiplets that are carried unaltered from the type IIA theory into the F-theory limit:

$$n_H^0 = h^{1,2}(X) + 1.$$

So far, we have deliberately omitted a discussion of charged hypermultiplets on the F-theory side, although the conjectured heterotic – F-theory duality predicts their existence. Indeed, since delicate and interesting issues arise from their investigation we devote much of the Section 5.2 to their study.

The decompactification limit of our heterotic theory on $T^2 \times K3$ gives a six-dimensional theory. Hence anomaly cancellation (3.2.3) is an issue. In fact, since the numbers of the various supermultiplets in this theory are related to the geometric invariants of the dual Calabi-Yau 3-fold by Conjecture 4.2, the duality predicts a classically unknown relation between these invariants:

Conjecture 4.3. *Let $X \to Z$ denote an elliptically fibered Calabi-Yau 3-fold with section, where Z is a rational surface. Assume that X gives the background of an F-theory limit of type IIA string theory which is dual to a consistent, that is an anomaly free decompactification limit of heterotic strings on some $T^2 \times K3$ to $K3$. Then there exists an associated gauge group H which arises from families of ADE-type Kodaira fibers in the fibration $X \to Z$, and a number n_H^{ch} of "charged hypermultiplets" which is a "charged dimension"*

$$\dim_{ch}(\varrho) = \dim(\varrho) - \dim(\ker \varrho_{|Cartan\ torus(H)})$$

of a representation ϱ of H. Moreover, the Picard number $\rho(Z)$ and the Hodge number $h^{1,2}(X)$ obey

$$h^{1,2}(X) + 29\rho(Z) - \dim(H) + n_H^{ch} = 301.$$

In the above conjecture we have made no assumptions to the effect that X is smooth or that the respective four-dimensional string theories arise from compactification. Indeed, the conjecture is supposed to hold in great generality. In the setting of Conjecture 4.2 and Proposition 2.2 we can prove the conjecture with little difficulty: We know that the invariants $\rho(Z)$, $h^{1,2}(X)$, n_H^{ch} and H are related to the numbers n_T, n_H, n_V of tensor-, hyper- and vectormultiplets as follows:

$$n_T = l + 1 = \rho(Z) - 1, \quad n_H = n_H^0 + n_H^{ch} = h^{1,2}(X) + 1 + n_H^{ch}, \quad n_V = \dim(H).$$

Hence the anomaly cancellation condition (3.2.3) implies

$$301 = 28 + n_H - n_V + 29n_T = h^{1,2}(X) + n_H^{ch} - \dim(H) + 29\rho(Z),$$

as claimed.

Although the four-dimensional heterotic – type IIA duality as well as the heterotic – F-theory duality remain highly non-trivial conjectures, Conjecture 4.3 has valuable predictive power: On the one hand it serves as an important test for the duality. On the other hand, we so far had to assume that the heterotic $K3$ surface is smooth, since techniques for a direct investigation of the degeneration phenomena which occur when pointlike instantons coalesce at singularities of the $K3$ surface are not known. However, using the F-theory dual of such a degeneration much more can be said, in particular the parameters entering Conjecture 4.3 can be calculated directly in the F-theory picture [AM97], see also Section 5.3.7.

5. F-theory on elliptically fibered Calabi-Yau 3-folds

In this section we explore the geometric setting of the F-theory side in Conjecture 4.2 intrinsically. We discuss how the invariants $h^{1,2}(X)$, $\rho(Z)$ as well as H and n_H^{ch}, which are related to one another in Conjecture 4.3, should be encoded in the very geometry of an elliptically fibered Calabi-Yau 3-fold $X \to Z$ with section. The duality predicts the form of the elliptic fibration. For simplicity in this section we assume that the fibration has precisely one section, because this is the situation assumed in several steps of [GM03], which we use severely. It should not be too hard to include cases where the Mordell-Weil group has non-zero rank, but we have not found a full account in the literature, and we have not yet completed the relevant calculations.

While $h^{1,2}(X)$, $\rho(Z)$ as well as H can indeed be obtained by a classical analysis, the invariant n_H^{ch} remains rather mysterious. We discuss these quantities separately, where we deal with the "classical" analysis in Section 5.1 and with "charged matter" in Section 5.2.

5.1. Some invariants of elliptically fibered Calabi-Yau 3-folds

As mentioned above, in many examples a parameter count in (4.1.4) yields a prediction for the value of $h^{1,2}(X)$ in the F-theory picture. This however does not always give the right answer, because deformations of the complex structure of our Calabi-Yau 3-fold need not all be given in terms of polynomial deformations of the singular Weierstraß form. On the other hand, there are classical geometric methods to calculate $h^{1,2}(X)$ from the data predicted by the duality.

Of the invariants involved in Conjecture 4.3, from the fibration $X \to Z$ one can directly read an associated "gauge group" H; this is a semisimple Lie group whose simple factors are in $1:1$ correspondence to families of degenerate fibers in $X \to Z$ of ADE Kodaira type and such that the ADE types match [Wit96a, AKM00]. Note that indeed according to [Mir83, §7] near smooth points of the reduced discriminant locus of our fibration such families are locally trivializable. Particular care has to be taken with fibers of types I_k, IV, I_k^*, and IV^*, because generically a family of such fibers in $\pi: X \to Z$, i.e. $\pi^{-1}(\Delta_i)$ for some divisor $\Delta_i \subset Z$, is not globally trivializable, because the fibration has monodromy [AG96]. While without monodromy one would associate factors of the gauge algebra according to

$$I_k \mapsto \mathfrak{su}(k) = \mathfrak{a}_{k-1}, \quad IV \mapsto \mathfrak{su}(3) = \mathfrak{a}_2,$$
$$I_k^* \mapsto \mathfrak{so}(2k+8) = \mathfrak{d}_{k+4}, \quad IV^* \mapsto \mathfrak{e}_6$$

to such families, monodromy can reduce the associated Lie algebra to the corresponding non-simply-laced algebra that is read from the respective Dynkin diagram under modding out of an outer automorphism:

$$I_{2k} \mapsto \mathfrak{sp}(k) = \mathfrak{c}_k, \quad I_{2k+1} \mapsto \mathfrak{sp}(k) = \mathfrak{c}_k, \quad IV \mapsto \mathfrak{sp}(1) = \mathfrak{a}_1,$$
$$I_k^* \mapsto \mathfrak{so}(2k+7) = \mathfrak{b}_{k+3} \quad \text{or, if } k = 0, \ \mathfrak{g}_2, \quad IV^* \mapsto \mathfrak{f}_4. \tag{5.1.1}$$

In calculations, the difference between families with and without monodromy cannot be seen in the "short" Weierstraß form (4.1.4). Rather, one needs a "long" Weierstraß form

$$y^2 + a_1(z_1, z_2)xy + a_3(z_1, z_2)y = x^3 + a_2(z_1, z_2)x^2 + a_4(z_1, z_2)x + a_6(z_1, z_2), \tag{5.1.2}$$

where as before x, y are affine coordinates of $\mathbb{CP}^2$ for the fiber and z_1, z_2 are appropriate local coordinates of the base Z of $X \to Z$. Here it is convenient to assume that the divisor Δ_i over which we want to study a family of degenerate fibers is given by $z_1 = $ const. Then the vanishing orders γ_1, γ_2, γ_3, γ_4, γ_6 of a_1, a_2, a_3, a_4, a_6 with respect to z_2 encode the type of generic fiber over Δ_i, including information about the monodromy. For example, vanishing orders $(\gamma_1, \gamma_2, \gamma_3, \gamma_4, \gamma_6) = (1, 2, 2, 3, 5)$ give fibers of type IV^* without monodromy, while $(\gamma_1, \gamma_2, \gamma_3, \gamma_4, \gamma_6) = (1, 2, 2, 3, 4)$ gives type IV^* fibers with monodromy. All relevant data are tabulated in [GM03, Table 1].

With H the total gauge group obtained from families of degenerate fibers one has

$$h^{1,1}(X) = 1 + \mathrm{rk}(H) + \rho(Z),$$

and all these data can be read off from the geometry as predicted by the duality. Therefore, to determine the remaining classical invariant $h^{1,2}(X)$ of Conjecture 4.3 we can equivalently compute the Euler characteristic of X,

$$\chi(X) = 2\left(h^{1,1}(X) - h^{1,2}(X)\right).$$

The rest of this section is devoted to describing an algorithm for the computation of $\chi(X)$.

If X is a Calabi-Yau 3-fold which is elliptically fibered over a rational surface Z and $K3$-fibered with a section over the base of $Z \to \mathbb{CP}^1$, then it arises from a singular Weierstraß fibration $\widetilde{X} \to \widetilde{Z}$ over a Hirzebruch surface $\widetilde{Z} = \mathbb{F}_n$ by a sequence of blowups in the base. Let us first collect some properties of this singular fibration. As explained above, the fibration over the section C_0 of $\mathbb{F}_n$ at infinity and over the zero section $C_\infty{}^4$ is governed by the bundle data of the dual heterotic theory. Particularly, $n = -C_0^2 = C_\infty^2$ is related to the topological data k^1, k^2, l^1, l^2 of the two heterotic E_8 bundles by

$$k^1 + l^1 = 12 - n, \qquad k^2 + l^2 = 12 + n \qquad (5.1.3)$$

if the ith bundle acquires l^i pointlike instantons. These identities arise from a number of conjectures in the physics literature [SW96, Wit96c, MV96a], and we view their validity as part of the conjectured duality. Denote by

$$y^2 = x^3 + a(z_1, z_2)x + b(z_1, z_2) \qquad (5.1.4)$$

the singular Weierstraß fibration $\widetilde{X} \to \widetilde{Z}$ with $\widetilde{Z} = \mathbb{F}_n$, with z_1, z_2 affine coordinates on the zero section C_∞ and respectively the fiber F of $\mathbb{F}_n$. Moreover $a(z_1, z_2)$ and $b(z_1, z_2)$ are polynomials as in (4.1.4) which define divisors $\widetilde{A}$ and $\widetilde{B}$ in $\widetilde{Z}$. The fibers of (5.1.4) degenerate over the discriminant $\widetilde{\Delta} \subset \widetilde{Z}$ which hence captures all the interesting topology of $\widetilde{X}$ and eventually of X. The discriminant is the zero locus of $\delta = 4a^3 + 27b^2$. Since $\widetilde{X} \to \widetilde{Z}$ is assumed to be a singular Calabi-Yau 3-fold, $\widetilde{X}$ must in particular have trivial canonical class. With $L = -K_{\widetilde{Z}}$ denoting the anticanonical class of $\widetilde{Z} = \mathbb{F}_n$ this implies (see e.g. [Asp97, §6.2])

$$\widetilde{\Delta} = 3\widetilde{A} = 2\widetilde{B} \quad \text{with } \widetilde{A} = 4L, \quad \widetilde{B} = 6L.$$

Recall also

$$L = 2C_0 + (2+n)F, \quad C_0^2 = -n, \quad F^2 = 0,$$
$$C_0 \cdot F = 1, \quad C_\infty = C_0 + nF, \quad C_\infty^2 = n. \qquad (5.1.5)$$

In the context of our duality the discriminant $\widetilde{\Delta}$ in general decomposes into several irreducible components,

$$\widetilde{\Delta} = \widetilde{\Delta}_{het} + \widetilde{\Delta}',$$

---

4 We apologize for this seemingly confusing notation, which however is compatible with [AM97] and thus facilitates a comparison to that work. See [GH78, p. 518] for the standard mathematical notations.

where the generic fibers over the irreducible $\widetilde{\Delta}'$ are of type I_1, while $\widetilde{\Delta}_{het}$ is in general reducible and in particular accounts for more exotic families of degenerate fibers dictated by the heterotic dual. $\widetilde{\Delta}_{het}$ in the present context always consists of a collection of smooth rational curves in $\widetilde{Z} = \mathbb{F}_n$.

As mentioned above, the Calabi-Yau 3-fold $X \to Z$ is obtained by a sequence of blowups of $\widetilde{Z}$ from (5.1.4) to yield Z. We always assume that sufficiently many blowups have been performed such that all fibers of $X \to Z$ are minimal according to Kodaira's list of degenerate fibers [Kod64]. As we shall see when we present some examples in Section 5.3, the blowups can be a delicate issue which needs to be dealt with by a detailed analysis of the singularities in (5.1.2). For the proper transforms of the various divisors we write

$$A = A_{het} + A', \quad B = B_{het} + B', \quad \text{and then} \quad \Delta = \Delta_{het} + \Delta', \quad (5.1.6)$$

where in particular Δ' is the proper transform of $\widetilde{\Delta}'$. For later bookkeeping let us denote by b_j the number of blowups of points on $\widetilde{\Delta}'$ of multiplicity α_j. In particular,

$$\Delta'(\Delta' + K_Z) = \widetilde{\Delta}'_{\widetilde{Z}}(\widetilde{\Delta}'_{\widetilde{Z}} + K_{\widetilde{Z}}) - \sum_j \alpha_j(\alpha_j - 1)b_j, \quad (5.1.7)$$

see e.g. [GM03, Corollary 6.3].

All contributions to the Euler characteristic $\chi(X)$ are captured by the discriminant $\Delta \subset Z$ of $X \to Z$ and the singular fibers over it. To keep track of all of them, we consider the decomposition of Δ into irreducible components, $\Delta = \bigcup_{i=1}^{r} \Delta_i \bigcup \Delta'$, where from the above all Δ_i are smooth rational curves and the generic fibers over Δ' are of Kodaira type I_1. Contributions to $\chi(X)$ can come from the generic fibers over each component of Δ, from intersection points between any two of these components, from singular points (cusps) on Δ', and from the Euler characteristic of Δ itself. See [Mir83, (3.1)] for confirmation that we may indeed assume that no other singularities but cusps occur in the residual discriminant Δ'. To tabulate all this information, we need to introduce some notation.

First, let $\chi_i \in \mathbb{N}$ denote the Euler characteristic of the generic fibers over Δ_i. The values of χ_i in each case are tabulated in [GM03, Table 3] in the column marked "m". Note that the corresponding number for Δ' is 1. By $P^1, \ldots, P^l$ we denote all intersection points of irreducible components of Δ. Particularly let I' denote the number of points P^i on Δ'. When counting intersection points of two given components of Δ care has to be taken since the intersection number of the respective divisors counts points with multiplicities, while our $P^1, \ldots, P^l$ are understood to be pairwise distinct. Since all Δ_i are assumed to be smooth rational curves, this issue is only relevant in interpreting $\Delta_i \cdot \Delta'$.

The necessary local case by case analysis has been carried out in [GM03], and the respective multiplicities are found in [GM03, Table 2] as exponents of t in the columns marked "l.e. at P_k", and where "transversal" amounts to multiplicity 1. Note that in several cases intersections of Δ_i and Δ' come in two different "types P_1, P_2", meaning that for $P \in \Delta_i \cap \Delta'$ the local geometry near $\pi^{-1}(P)$ depends on whether P is of type P_1 or P_2. A case by case analysis of the defining polynomials shows that there are never more than two types. One can use [GM03, Table E] to determine the numbers B_1, B_2 of each type of intersection and thus to disentangle the value of $\Delta_i \cdot \Delta'$ in these cases. Moreover,

$$- \varepsilon_1 B_1 - \varepsilon_2 B_2 \tag{5.1.8}$$

with ε_i taken from [GM03, Table 4] gives the contribution to I' from the collision $\Delta_i \cap \Delta'$. The Euler characteristic of each fiber $\pi^{-1}(P^i)$ can be found in [GM03, Table 4] if $P^i \in \Delta'$, and otherwise one uses [Mir83, Table (14.1)]. Even though in the latter work, Miranda does not take monodromy into account, these calculations are still valid for our purposes, since the computation of the Euler characteristic only depends on the geometry of the fiber over P_i which he describes in great detail.

Cusps of Δ' are denoted by $Q^1, \ldots, Q^C$, and each of them carries a special fiber of type II, contributing $\chi(\pi^{-1}(Q^i)) = 2$ to the Euler characteristic. Note that cusps in (5.1.4) are characterized by the simultaneous vanishing of a and b, such that their total number in the resolved Calabi-Yau 3-fold X is given by $A' \cdot B'$, as long as these cusps do not coalesce with an intersection point of Δ' with one of the other components of Δ. From [GM03, Proposition 8] one obtains the general formula for the number C of cusps, where an overcounting in $A' \cdot B'$ is observed when Δ' intersects Δ_i which carry generic fibers of types I_k, I_0^*, I_1^*, or I_2^*. One uses [GM03, Table 4] to read off the invariants entering here: If $\Delta' \cap \Delta_i$ has B_1 intersection points of "type P_1" and B_2 intersection points of "type P_2", then

$$C = A' \cdot B' - \mu_1 B_1 - \mu_2 B_2, \tag{5.1.9}$$

with μ_1, μ_2 taken from [GM03, Table 4].

Altogether we have

$$\chi(X) = \chi\left(\pi^{-1}\left(\Delta' - \bigcup_{j=1}^{I}\{P^j\} - \bigcup_{j=1}^{C}\{Q^j\}\right)\right) + \sum_{i=1}^{r} \chi\left(\Delta_i - \bigcup_{j=1}^{I}\{P^j\}\right)\chi_i$$

$$+ \sum_{j=1}^{I} \chi(\pi^{-1}(P^j)) + 2C.$$

From the above discussion we can determine all contributions, where the first one simplifies to

$$
\chi\left(\pi^{-1}(\Delta' - \bigcup_{j=1}^{I}\{P^j\} - \bigcup_{j=1}^{C}\{Q^j\})\right) \ = \ \chi(\pi^{-1}(\Delta')) - I' - C
$$

$$
= \ -\Delta'(\Delta' + K_Z) + 2C - I' - C
$$

$$
\overset{(5.1.7)}{=} \ -\widetilde{\Delta}'(\widetilde{\Delta}' + K_{\widetilde{Z}})
$$

$$
+ \sum_j \alpha_j(\alpha_j - 1)b_j - I' + C
$$

In summary,

Proposition 5.1. *Let* $X \to Z$ *be a Calabi-Yau 3-fold which arises via a sequence of blowups in the base of the singular Weierstraß fibration* $\widetilde{X} \to \widetilde{Z}$ *over a Hirzebruch surface* $\widetilde{Z} = \mathbb{F}_n$, *associated to the data coming from a heterotic* $E_8 \times E_8$ *theory compactified on a K3 surface. Assume that* $X \to Z$ *has precisely one section and that all fibers are minimal. Then the Euler characteristic of* X *is given by*

$$
\chi(X) = -\widetilde{\Delta}'(\widetilde{\Delta}' + K_{\widetilde{Z}}) + \sum_j \alpha_j(\alpha_j - 1)b_j
$$

$$
+ \sum_{i=1}^{r} \chi\left(\Delta_i - \bigcup_{j=1}^{I}\{P^j\}\right)\chi_i + \sum_{j=1}^{I}\chi(\pi^{-1}(P^j)) - I' + 3C,
$$

where generically $C = A' \cdot B'$, *but in general* C *is obtained from* (5.1.9) *and* I' *is obtained according to the discussion around* (5.1.8).

5.2. Charged matter

Above we have explained that the heterotic – F-theory duality yields sufficiently detailed predictions about the elliptically fibered Calabi-Yau 3-fold $X \to Z$ on the F-theory side such that one can recover geometric invariants like $h^{1,2}(X)$ and $h^{1,1}(X)$ along with $\rho(Z)$ and an associated "gauge group" H. In light of Conjecture 4.3 one could then simply define

$$
n_H^{ch} := 301 + \dim(H) - h^{1,2}(X) - 29\rho(Z)
$$

and accept it as a new invariant of elliptically fibered Calabi-Yau 3-folds with section that occur as F-theory duals of well-defined heterotic string theories. However, from the derivation on the heterotic side for this invariant around Proposition 2.2, we know that n_H^{ch} is a purely gauge theoretic quantity obtained through the representation theory of H. Hence rather than using Conjecture 4.3 as a definition of n_H^{ch} one should expect an intrinsic geometric interpretation of this quantity and view the anomaly cancellation condition as a classically unknown relation between geometric invariants attached to elliptically fibered Calabi-Yau 3-folds with section. This is the viewpoint we are going to take here.

In fact, if the heterotic – F-theory duality holds on the level of string theory, then we are forced into this viewpoint: According to Proposition 2.2 the quantity n_H^{ch} accounts for part of the hypermultiplet spectrum of the dual heterotic theory, namely the charged part, and hence must also occur on the F-theory side. Similarly, for non-Abelian H we have $\dim H - \mathrm{rk}\, H$ additional vector-multiplets to account for which so far have escaped our explanations on the F-theory side.

Since in the F-theory limit the fibers of the elliptic fibration shrink to zero size, i.e. the Calabi-Yau 3-fold X is taken to the boundary of the Kähler cone, one can expect the additional multiplets to arise due to this degeneration of X. Indeed, explanations along these lines can be found in the literature, see in particular [KV97, AKM00]. This work connects the appearance of additional multiplets to two phenomena which occur in this limit, distinguished by the way the fibers of the elliptic fibration degenerate: (1) in families, or (2) isolated. Let us attempt to summarize and comment on these explanations.

5.2.1. Families of degenerate fibers yielding new vectormultiplets

From Proposition 2.4 we know that on a smooth Calabi-Yau 3-fold X a type IIA string theory possesses $h^{1,1}(X)$ vectormultiplets. These in particular include an RR three-form $C_{\mu ij}$ in each multiplet which can be integrated over two-cycles in X to produce one-forms. The latter can be interpreted as analogs of Yang-Mills connection one-forms of a gauge theory which can become non-Abelian when (-2) curves in X shrink to zero size [AKM00].

More precisely, consider a family of degenerate elliptic fibers in an elliptically fibered Calabi-Yau 3-fold such that the degenerate fiber is a bouquet of rational curves. As explained above, to this type of family one associates a simple factor G_i of the "gauge group" H, an ADE type Lie group. Each rational curve in the bouquet has normal bundle $\mathcal{O} \oplus \mathcal{O}(-2)$ in X, and in the family of degenerate fibers it sweeps out a four-cycle S_i in X with normal bundle $\mathcal{O}(-2)$. These four-cycles are in $1:1$ correspondence with the generators of a Cartan torus of G_i. Indeed, following [Wit96a, KV97, AKM00] one defines $U(1)$ charges with respect to each S_i on any (-2) curve in the degenerate fiber via the intersection form. Here one views the (-2) curve as the cycle which a two-brane wraps to produce a new massless particle with "electric charge" given by its $U(1)$ charge if the size of the relevant (-2) curve shrinks to zero. Since the (-2) curves in the degenerate fiber are in $1:1$ correspondence with the roots of G_i, and because the rational curves underlying S_i have intersection form given by the Cartan matrix of G_i, one obtains the adjoint representation of G_i. In particular, the "gauge representation" associated to the S_i is enhanced from a rk G_i to a dim G_i dimensional representation, accounting for the missing vectormultiplets: One has to add one "charged" vectormultiplet corresponding to each root of G_i.

In fact, for definiteness recall from Section 5.1 that

$$\Delta = \bigcup_{i=1}^{r} \Delta_i \cup \Delta',$$

with smooth irreducible curves Δ_i, and Δ' the residual discriminant of the fibration $\pi: X \to Z$, an irreducible curve over which the generic fibers are of Kodaira type I_1. For each Δ_i the fiber over it gives rise to a simply-laced Lie algebra, which may be trivial if the fibers are of type I_1 or II. If there is no monodromy within $\pi^{-1}(\Delta_i)$, then this determines the associated factor G_i of H. If within $\pi^{-1}(\Delta_i)$ there is monodromy, then G_i is the corresponding non-simply-laced Lie group which descends from the simply laced one as listed on the algebra level in (5.1.1). To summarize, the enhanced gauge group is

$$H = \prod_{i=1}^{r} G_i \,,$$

and the total number of vector multiplets is dim $H = \sum_{i=1}^{r}$ dim G_i.

5.2.2. Colliding degenerate fibers yielding charged hypermultiplets

We have now accounted for all effects of the degeneration of X in the F-theory limit, apart from the special degenerate fibers which occur over collision points of irreducible components of the discriminant $\Delta \subset Z$ of the fibration $\pi: X \to Z$. It is important to note that the type of fiber over such a collision point is not simply obtained by adding the vanishing orders of the polynomials a_1, a_2, a_3, a_4, a_6 in (5.1.2). Rather, according to [Mir83] a generic curve $C \subset Z$ through the collision point yields a surface $\pi^{-1}(C)$ with a singularity at the collision point. Minimally resolving this singularity gives a degenerate fiber Y of the type predicted by simply adding the vanishing orders of $a_1, \ldots, a_6$. The isolated degenerate fiber of $\pi: X \to Z$ is hence obtained by contracting some of the irreducible curves in Y. The explanations in [KV97] amount to translating this result into the language of the "gauge theory" associated to the group H as described above. The geometric setting hints towards various interesting phenomena that are related to these collision points. Particularly in view of the gauge theory attached to families of degenerate fibers according to Section 5.2.1, one expects representations of $\mathfrak{g}_i \oplus \mathfrak{g}_j$ attached to the fibers over collision points which are non-trivial with respect to both summands, where $\mathfrak{g}_i$, $\mathfrak{g}_j$ are the gauge algebras associated to the two colliding families of degenerate fibers. This phenomenon is the expected origin of the charged hypermultiplets [Wit96a, KV97, AKM00].

For the geometry of elliptically fibered Calabi-Yau 3-folds $X \to Z$ which occur in the heterotic – F-theory duality this means that at least formally it should be possible to associate to each degenerate fiber over an intersection of irreducible components of the discriminant, (a) a representation of the total gauge group H, and (b) a prescription to calculate its charged dimension, such that n_H^{ch} is the sum of all these charged dimensions. For the situation where $\Delta = \cup_{i=1}^{r} \Delta_i \cup \Delta'$ obeys $\Delta_i \cap \Delta_j = \emptyset$ for all $i \neq j$ this idea has been carried out in [GM03]. This work hence covers all F-theory duals of heterotic theories compactified on smooth $K3$ surfaces and with smooth bundles. To calculate n_H^{ch} from the results of [GM03] one proceeds as follows:

We have already explained in Section 5.1 how to determine the number of intersection points $P^1, \ldots, P^I$ ($I = I'$ such that one of the colliding divisors is always Δ' under our assumptions), along with their multiplicities and the information whether or not monodromy is involved in one of the colliding families. While [GM03, Table 4] gives the type of fiber over each collision point, the associated representations are listed in [GM03, Table A]. Here, ρ_1, ρ_2 denote representations attached to collision points of "type P_1, P_2" respectively (see Section 5.1 for this terminology), and ρ_0 is a representation which is "non-isolated" in the sense that monodromy prevents a localization over the collision points.

While representations of type ρ_1, ρ_2 contribute to n_H^{ch} according to their dimensions at each collision point, greater care has to be taken when determining the contributions for representations of type ρ_0. First, if at a collision point $P^k \in \Delta_i \cap \Delta'$ monodromy occurs, then one needs to work with a branched cover Δ_i' of Δ_i which parametrizes the exceptional curves in one homology class [GM03, Corollary 1.3]. The difference $g_i' - g_i$ of genera between Δ_i' and Δ_i is obtained from [GM03, Tabel E] and replaces the number of collision points in $\Delta_i \cap \Delta'$ in the contribution to n_H^{ch} from ρ_0 [GM03, Theorem 8.2]. This is an effect of having "non-isolated" representations: Some of the points in $\Delta_i \cap \Delta'$ belong to the same orbit under this representation. There is a second crucial effect of monodromy. Recall that we distinguish between charged and uncharged hypermultiplets, where uncharged matter is characterized by the fact that it transforms trivially under the gauge group. Generically, on the heterotic side the representations $L_a^i \otimes Q_a^i$ in (2.2.7) with non-trivial L_a^i, Q_a^i yield trivial kernel when restricted to the Cartan torus t of the gauge group H. However, with monodromy this need not be the case such that one can obtain additional contributions to the uncharged matter, and accordingly a smaller contribution to the charged matter from ρ_0 as above. In [GM03] the relevant contribution of ρ_0 to n_H^{ch} is called the charged dimension

$$\dim (\rho_0)_{ch} := \dim (\rho_0) - \dim \ker(\rho_0)_{|t}$$

and can be obtained from [GM03, Table B].

The invariant n_H^{ch} altogether receives a contribution $(g_i' - g_i)\dim (\rho_0)_{ch}$ from ρ_0. If $\dim (\rho_0)_{ch} \neq \dim (\rho_0)$ then n_H^0 receives an additional contribution $(g_i' - g_i)(\dim (\rho_0) - \dim (\rho_0)_{ch})$ which must be added to our formula (2.2.23) in order for $h^{1,2}(X) = n_H^0 + 1$ to hold in Conjecture 4.2. To our knowledge, this latter correction has escaped mention in the literature, so far.

After having established the prescription for the calculation n_H^{ch} from [GM03] we naturally ask whether the invariants $h^{1,2}(X)$, $\rho(Z)$, $\dim (H)$, and n_H^{ch} associated to an elliptic fibration $X \to Z$ obey the anomaly cancellation condition of Conjecture 4.3. Indeed, one of the main results of [GM03] is the confirmation of this classically unknown identity in the cases they treat. Moreover, a verification of anomaly cancellation in the context of the Green-Schwarz mechanism is given. These results of [GM03] yield yet another striking and highly non-trivial piece of evidence in favor of the conjectured heterotic – F-theory duality. However, they arise from a local case-by-case analysis and do not give an entirely intrinsic explanation for the origin of the representations attached formally to collisions of families of degenerate fibers in $X \to Z$. Although the familiar branching rules from the heterotic dual are invoked to determine the relevant representations, as is also suggested in [KV97], the actual mathematical origin of the "gauge theory" associated to degenerate fibers remains mysterious.

Indeed, in light of the gauge theory attached to each family of degenerate fibers according to Section 5.2.1, there seems to be a natural explanation in terms of the local geometry of the isolated degenerate fibers which also gives a lead on how to calculate n_H^{ch} in general [AKM00]. As mentioned above, one expects that both factors G_i, G_j of the gauge group associated to two colliding families should act non-trivially on the representation associated to the collision. While generic rational curves in this isolated fiber will have vanishing charge with respect to one of the two groups, in some cases such a curve can have non-trivial charge with respect to both G_i and G_j. As before, this charge is encoded in the respective normal bundle of the rational curve in X. Generically, such a curve will have normal bundle $\mathcal{O} \oplus \mathcal{O}(-2)$ or $\mathcal{O}(-2) \oplus \mathcal{O}$, while in some cases curves with normal bundle $\mathcal{O}(-1) \oplus \mathcal{O}(-1)$ can occur. The latter are expected to be the sources of the charged hypermultiplets. Note however that the explanation definitely needs adjustment. In the setting which the authors of [GM03] restrict to, either G_i or G_j is trivial, so the above approach would predict absence of charged hypermultiplets. This prediction is false. We believe that in [AKM00] one needs to assume that neither of G_i, G_j is trivial.

The above idea is shown to work for colliding families of type I_n and I_m, respectively, in [AKM00], but it seems not to have been pushed further in the physics literature. However, we have found encouraging confirmation beyond this case in [Mir83], where the local geometry of fibers over such collision points is studied in detail. In particular, a list of "fundamental collisions" is given, to which all other collisions can be reduced by appropriate blowups of the base. The topology of the isolated fiber for each fundamental collision is worked out, along with the normal bundles of the irreducible components of these fibers. The only fundamental collisions with at least one irreducible component of the isolated fiber having normal bundle $\mathcal{O}(-1) \oplus \mathcal{O}(-1)$ are collisions of types

$$I_n + I_m, \qquad I_n + I_m^*, \qquad IV + I_0^*, \qquad \text{and} \qquad III + I_0^*. \qquad (5.2.1)$$

We view it as a striking confirmation of the ideas of [AKM00] that in a wealth of examples where n_H^{ch} can be calculated on the heterotic side and hence a prediction for its value is available, its geometric derivation involves only collisions of type (5.2.1), or non-fundamental collisions, or collisions which generically suffer from monodromy – and no counter example is known to us. In other words, for all cases that are covered by [Mir83] the ideas of [AKM00] can be confirmed. Nevertheless – and surprisingly – Miranda's list has not yet been extended to all relevant cases to provide a complete intrinsic explanation for charged hypermultiplets in the geometry of elliptically fibered Calabi-Yau 3-folds. In particular, monodromy is not taken into account in [Mir83] – and in fact is also not addressed in [KV97]. Of course, several further examples are discussed in [AKM00], including some with monodromy, and various methods that apply in special cases are known [KMP96, Wit96a, AG96, BIK+96, Sad96, KV97, IMS97, CPR98, Int98, DE99], but a general intrinsic understanding apparently has not yet been reached. We are currently working on filling this gap.

5.3. Examples

In this section we present a number of examples to illustrate the algorithms explained above. They are chosen representatively to show all the special features that to our knowledge can occur in these algorithms.

As a first step, using the heterotic input data we need to specify the respective singular Weierstraß fibration (5.1.2) over $\widetilde{Z} = \mathbb{F}_n$ in each case. We assume that the heterotic data k^i, l^i, K^i, H^i are given as in Section 3.3. Recall that n is then determined by (5.1.3). Moreover, the fibration $\widetilde{X} \to \widetilde{Z}$ degenerates over C_0 and C_∞, where H^1, H^2 determine the Kodaira type of the generic singular elliptic fiber. Using [AM97, Table 1] this gives the vanishing orders α_i, β_i, d_i of a, b, and δ in the short Weierstraß form (5.1.4) along C_0 and C_∞. Hence we can specify the components $\widetilde{A}'$, $\widetilde{B}'$, $\widetilde{\Delta}'$ of the discriminant yielding A', B', Δ' of (5.1.6) after blowup:

$$\widetilde{A}' = 4L - \alpha_0 C_0 - \alpha_\infty C_\infty, \quad \widetilde{B}' = 6L - \beta_0 C_0 - \beta_\infty C_\infty,$$
$$\widetilde{\Delta}' = 12L - d_0 C_0 - d_\infty C_\infty, \tag{5.3.1}$$

with the notation of (5.1). In fact, in most of the cases discussed below no further blowup will be necessary, such that $\widetilde{A}' = A'$, $\widetilde{B}' = B'$, $\widetilde{\Delta}' = \Delta'$ and the collision points of components of Δ which take center stage in the analysis are counted by the intersection numbers $\Delta' \cdot C_0$ and $\Delta' \cdot C_\infty$. In some cases, though, collisions turn out to be non-minimal, such that blowups are needed, as we shall describe in more detail where necessary.

As mentioned above, to correctly incorporate monodromies instead of the short Weierstraß form (5.1.4) one needs to use the long version (5.1.2) from which the former is obtained via

$$c_2 := a_1^2 + 4a_2, \quad c_4 := a_1 a_3 + 2a_4, \quad c_6 := a_3^2 + 4a_6,$$
$$a = -\frac{1}{48}\left(c_2^2 - 24c_4\right), \quad b = -\frac{1}{864}\left(-c_2^3 + 36c_2 c_4 - 216c_6\right).$$

For later convenience let us also introduce the following notation: We write $Y_{\alpha, \beta; \gamma_1, \gamma_2, \gamma_3, \gamma_4, \gamma_6}(\mathfrak{g})$ to denote a Y-type fiber, where α, β, γ_i are the vanishing orders of a, b in (5.1.4) and a_i in (5.1.2), respectively, and $\mathfrak{g}$ is the Lie algebra of the associated gauge group. We sometimes simply call such a fiber "of type $(\alpha, \beta; \gamma_1, \gamma_2, \gamma_3, \gamma_4, \gamma_6)$".

5.3.1. Completely broken gauge group

If the gauge group is trivial, $H = \{\text{id}\}$, this means that our $K3$ bundle has holonomy $K = E_8 \times E_8$. Since no additional singular fibers are imposed, we have

$$A' = 4L = 8C_0 + (8 + 4n)F,$$
$$B' = 6L = 12C_0 + (12 + 6n)F,$$
$$\Delta' = 12L = 24C_0 + (24 + 12n)F.$$

Because there are no collisions, $\widetilde{Z} = Z$ needs no further blowups and hence

$$\rho(Z) = 2, \qquad h^{1,1}(X) = 3.$$

Furthermore we directly obtain all contributions to the formula for $\chi(X)$ in Proposition 5.1:

$$\Delta' \cdot (\Delta' + K_{\widetilde{Z}}) = 1056, \quad \sum_j \alpha_j(\alpha_j - 1)b_j = 0, \quad \sum_{i=1}^r \chi\left(\Delta_i - \bigcup_{j=1}^l \{P^j\}\right)\chi_i = 0,$$

$$\sum_{j=1}^l \chi(\pi^{-1}(P^j)) = 0, \qquad I' = 0, \qquad C = A' \cdot B' = 192.$$

Hence

$$\chi(X) = -480, \quad h^{1,2}(X) = 243.$$

Moreover since there are no collisions, no charged hypermultiplets occur, $n_H^{ch} = 0$. First, anomaly cancellation according to Conjecture 4.3 is seen to hold (as of course follows from the results of [GM03]). Second, comparing these data to the ones obtained for the heterotic dual (3.3.1), we see that Conjecture 4.2 is met.

5.3.2. *Unbroken E_8 gauge group*

If the gauge group is $H = E_8 \times \{id\}$, this means that our heterotic $K3$ bundle has holonomy $K = \{id\} \times E_8$. So we are imposing II^* fibers on C_0 and

$$\widetilde{A}' = 4L - 4C_0 = 4C_0 + (8 + 4n)F,$$
$$\widetilde{B}' = 6L - 5C_0 = 7C_0 + (12 + 6n)F,$$
$$\widetilde{\Delta}' = 12L - 10C_0 = 14C_0 + (24 + 12n)F.$$

We have $\widetilde{\Delta}' \cdot C_0 = 2(12 - n)$ and find that each intersection has multiplicity $\alpha_j = 2$. These intersections are non-minimal, so $b_j = 12 - n$ blowups are necessary, yielding

$$\rho(Z) = 14 - n, \qquad h^{1,1}(X) = 23 - n.$$

After blowup, no collisions are left. The contributions to the formula for $\chi(X)$ in Proposition 5.1 are:

$$\Delta' \cdot (\Delta' + K_{\widetilde{Z}}) = 130n + 596, \quad \sum_j \alpha_j(\alpha_j - 1)b_j = 24 - 2n,$$

$$\sum_{i=1}^r \chi\left(\Delta_i - \bigcup_{j=1}^l \{P^j\}\right)\chi_i = 20, \qquad \sum_{j=1}^l \chi(\pi^{-1}(P^j)) = 0,$$

$$I' = 0, \qquad C = \widetilde{A}' \cdot \widetilde{B}' = 104 + 24n.$$

Hence

$$\chi(X) = -60n - 240, \quad h^{1,2}(X) = 143 + 29n.$$

Since there are no collisions, no charged hypermultiplets occur, $n_H^{ch} = 0$. Anomaly cancellation according to Conjecture 4.3 holds (in accord with [GM03]). Moreover, comparing these data to the ones obtained for the heterotic dual (3.3.2) with $l = -n + 12$ according to (5.1.3), we see that Conjecture 4.2 is met.

5.3.3. Unbroken E_7 gauge group

If the gauge group is $H = E_7 \times \{\mathrm{id}\}$, this means that our $K3$ bundle has holonomy $K = SU(2) \times E_8$. So we are imposing III^* fibers on C_0 and

$$\widetilde{A}' = 4L - 3C_0 = 5C_0 + (8 + 4n)F,$$
$$\widetilde{B}' = 6L - 5C_0 = 7C_0 + (12 + 6n)F,$$
$$\widetilde{\Delta}' = 12L - 9C_0 = 15C_0 + (24 + 12n)F.$$

We have $\widetilde{\Delta}' \cdot C_0 = 3(8 - n)$ and from [GM03, Table 2] find that each intersection has multiplicity $\alpha_j = 3$. These intersections are minimal, but we can choose to blow up $b_j = l$ of them, yielding

$$\rho(Z) = 2 + l, \quad h^{1,1}(X) = 10 + l.$$

After blowup, $(8 - n - l)$ collisions are left. From [GM03, Table 4] the Euler characteristic of the isolated fibers over such collision points is 9. The contributions to the formula for $\chi(X)$ in Proposition 5.1 hence are:

$$\Delta' \cdot (\Delta' + K_{\widetilde{Z}}) = 126n + 642, \qquad \sum_j \alpha_j(\alpha_j - 1)b_j = 6l,$$

$$\sum_{i=1}^{r} \chi\left(\Delta_i - \bigcup_{j=1}^{l}\{P^j\}\right)\chi_i = -54 + 9n + 9l, \quad \sum_{j=1}^{l} \chi(\pi^{-1}(P^j)) = 72 - 9n - 9l,$$

$$l' = 8 - n - l, \qquad C = \widetilde{A}' \cdot \widetilde{B}' - l = 23n + 116 - l.$$

The formula for the number of cusps C takes into account that each collision of C_0 with $\widetilde{\Delta}'$ izzs also a cusp of $\widetilde{\Delta}'$ which is resolved if we blow up. Altogether

$$\chi(X) = -56n - 284 + 4l, \quad h^{1,2}(X) = 28n + 152 - l.$$

By [GM03, Table A] each of the $(8 - n - l)$ collisions carries charged matter, contributing $\frac{1}{2}\mathbf{56}$ to n_H^{ch}. One hence has $n_H^{ch} = 224 - 28n - 28l$. Anomaly cancellation according to Conjecture 4.3 holds, and comparing these data to the ones obtained for the heterotic dual (3.3.3) with $k^1 + l = -n + 12$ and $k^2 = n + 12$ according to (5.1.3), we see that Conjecture 4.2 is met.

5.3.4. Unbroken E_6 gauge group

If the gauge group is $H = E_6 \times \{\text{id}\}$, this means that our $K3$ bundle has holonomy $K = \text{SU}(3) \times E_8$. So we are imposing IV^* fibers on C_0, but with trivial monodromy, which is the non-generic case. We have

$$\widetilde{A}' = 4L - 3C_0 = 5C_0 + (8 + 4n)F,$$
$$\widetilde{B}' = 6L - 4C_0 = 8C_0 + (12 + 6n)F,$$
$$\widetilde{\Delta}' = 12L - 8C_0 = 16C_0 + (24 + 12n)F, \quad \widetilde{\Delta}' \cdot C_0 = 2(12 - 2n) = 4(6 - n).$$

From [GM03, Table 2] each intersection between C_0 and $\widetilde{\Delta}'$ has multiplicity $\alpha_j = 4$. These intersections are minimal, but we can choose to blow up $b_j = l$ of them, yielding

$$\rho(Z) = 2 + l, \qquad h^{1,1}(X) = 9 + l.$$

After blowup, $(6 - n - l)$ collisions are left. From [GM03, Table 4] the Euler characteristic of the isolated fibers over such collision points is 9. The contributions to the formula for $\chi(X)$ in Proposition 5.1 hence are:

$$\Delta' \cdot (\Delta' + K_{\widetilde{Z}}) = 120n + 688, \qquad \sum_j \alpha_j(\alpha_j - 1)b_j = 12l,$$

$$\sum_{i=1}^{r} \chi\left(\Delta_i - \bigcup_{j=1}^{l}\{P^j\}\right)\chi_i = -32 + 8n + 8l, \qquad \sum_{j=1}^{l} \chi(\pi^{-1}(P^j)) = 54 - 9n - 9l,$$

$$I' = 6 - n - l, \qquad C = \widetilde{A}' \cdot \widetilde{B}' - 2l = 22n + 124 - 2l.$$

The formula for the number of cusps C takes into account that each collision of C_0 with $\widetilde{\Delta}'$ is also a cusp of $\widetilde{\Delta}'$ which is resolved if we blow up; in fact, the collision with $\widetilde{\Delta}'$ has multiplicity 4 and the number of cusps is reduced by 2 by each blowup. Altogether

$$\chi(X) = -54n - 300 + 6l, \qquad h^{1,2}(X) = 27n + 159 - 2l.$$

By [GM03, Table A] each of the $(6 - n - l)$ collisions carries charged matter, contributing **27** to n_H^{ch}. One hence has $n_H^{ch} = 162 - 27n - 27l$. Anomaly cancellation according to Conjecture 4.3 holds, and comparing these data to the ones obtained for the heterotic dual (3.3.4) with $k^1 + l = -n + 12$ and $k^2 = n + 12$ according to (5.1.3), we see that Conjecture 4.2 is met.

5.3.5. Unbroken F_4 gauge group

If the gauge group is $H = F_4 \times \{\text{id}\}$, this means that our heterotic $K3$ bundle has holonomy $K = G_2 \times E_8$. So we are imposing IV^* fibers on C_0 as in the previous subsection, but with nontrivial monodromy, which is the generic case. The results for $\widetilde{A}'$, $\widetilde{B}'$, $\widetilde{\Delta}'$, $\widetilde{\Delta}' \cdot C_0$ can be taken from the previous subsection, but now according to [GM03, Table 2] each intersection between C_0 and $\widetilde{\Delta}'$ has multiplicity $\alpha_j = 2$. These intersections are minimal, but the collisions cannot be removed by blowups. Hence we do not blow up at all, $b_j = 0$, and

$$\rho(Z) = 2, \qquad h^{1,1}(X) = 7.$$

We have $(12 - 2n)$ collisions between C_0 and Δ'. From [GM03, Table 4] the Euler characteristic of the fibers associated to such collision points is 6. Note that in this case we have monodromy, which will affect the calculation of the charged hypermultiplets. The contributions to the formula for $\chi(X)$ in Proposition 5.1 are:

$$\Delta' \cdot (\Delta' + K_{\widetilde{Z}}) = 120n + 688, \qquad \sum_j \alpha_j (\alpha_j - 1) b_j = 0,$$

$$\sum_{i=1}^{r} \chi \left(\Delta_i - \bigcup_{j=1}^{l} \{P^j\} \right) \chi_i = -80 + 16n, \qquad \sum_{j=1}^{l} \chi(\pi^{-1}(P^j)) = 72 - 12n,$$

$$I' = 12 - 2n, \qquad C = \widetilde{A}' \cdot \widetilde{B}' = 22n + 124.$$

Altogether

$$\chi(X) = -48n - 336, \qquad h^{1,2}(X) = 24n + 175.$$

By [GM03, Table A] each of the $(12 - 2n)$ collisions carries charged matter, with associated representation **26**. However, we have monodromy, and the charged dimension of this representation is only $\dim (\mathbf{26})_{ch} = 24$ according to [GM03, Table B]. Moreover, its multiplicity in n_H^{ch} is not $(12 - 2n)$ but rather $(g' - g) = 5 - n$ as can be obtained from [GM03, Table E]. One hence has $n_H^{ch} = 120 - 24n$. Anomaly cancellation according to Conjecture 4.3 holds, and comparing these data to the ones obtained for the heterotic dual (3.3.5) with $k^1 = -n + 12$ and $k^2 = n + 12$ according to (5.1.3), we see that Conjecture 4.2 is met.

5.3.6. *Unbroken Spin(10) gauge group*

If the gauge group is $H = \text{Spin}(10) \times \{\text{id}\}$, this means that our $K3$ bundle has holonomy $K = \text{SU}(4) \times E_8$. So we are imposing I_1^* fibers on C_0, and we have

$$\widetilde{A}' = 4L - 2C_0 = 6C_0 + (8 + 4n)F,$$
$$\widetilde{B}' = 6L - 3C_0 = 9C_0 + (12 + 6n)F,$$
$$\widetilde{\Delta}' = 12L - 7C_0 = 17C_0 + (24 + 12n)F, \qquad \widetilde{\Delta}' \cdot C_0 = 24 - 5n.$$

In this case determining the multiplicities of intersection points is a bit trickier than before, because there are two different "types P_1, P_2" of intersections. Using [GM03, Table E] one finds that there are $B_1 = 6 - n$ intersections of "type P_1", and $B_2 = 4 - n$ intersections of "type P_2". Since when counted with multiplicities the total number of intersections is $24 - 5n$, one finds that intersections of "type P_1" have multiplicity 2, while intersections of "type P_2" have multiplicity 3. We choose not to blow up any of these intersections, $b_j = 0$, yielding

$$\rho(Z) = 2, \qquad h^{1,1}(X) = 8.$$

In total, we have $B_1 + B_2 = 10 - 2n$ collisions, and from [GM03, Table 4] the Euler characteristic of the isolated fibers over collision points of both "types P_1, P_2" is 8. The contributions to the formula for $\chi(X)$ in Proposition 5.1 hence are:

$$\Delta' \cdot (\Delta' + K_{\widetilde{Z}}) = 112n + 734, \qquad \sum_j \alpha_j(\alpha_j - 1)b_j = 0,$$

$$\sum_{i=1}^{r} \chi\left(\Delta_i - \bigcup_{j=1}^{l}\{P^j\}\right)\chi_i = -56 + 14n, \qquad \sum_{j=1}^{l} \chi(\pi^{-1}(P^j)) = 80 - 16n,$$

$$I' = 10 - 2n, \qquad C = \widetilde{A}' \cdot \widetilde{B}' - \mu_1 B_1 - \mu_2 B_2 = 20n + 136.$$

For the number of cusps in this case we have to apply the general formula (5.1.9) with $\mu_1 = 0$ and $\mu_2 = 2$ according to [GM03, Table 4]. Altogether

$$\chi(X) = -52n - 312, \qquad h^{1,2}(X) = 26n + 164.$$

By [GM03, Tables A, B] each of the $(10 - 2n)$ collisions carries charged matter, where the $(6 - n)$ points of "type P_1" contribute **10**, and the $(4 - n)$ points of "type P_2" contribute **16**, each to n_H^{ch}. One hence has $n_H^{ch} = 124 - 26n$. Anomaly cancellation according to Conjecture 4.3 holds, and comparing these data to the ones obtained for the heterotic dual (3.3.6) with $k^1 = -n + 12$ and $k^2 = n + 12$ according to (5.1.3), we see that Conjecture 4.2 is met.

5.3.7. *24 pointlike instantons on singularities of type* E_8, E_7, *or* E_6

Recall that in our Proposition 2.2 we restricted the internal $K3$ surfaces of our heterotic string theories to be smooth, because the analysis which lead to the formulas for n_H, n_V, n_T cannot be performed as stated if pointlike instantons collide with singularities on $K3$. In fact, no direct technique to tackle that situation is known. However, the heterotic – F-theory duality comes to aid and allows to predict the massless spectrum even in such highly degenerate cases [AM97]. Let us describe three of the most degenerate situations, which where brought to our attention by McKay's conjecture as mentioned in the Introduction.

On the heterotic side, we assume total degeneration of the bundle data to 24 pointlike instantons. This amounts to a primordial gauge group

$E_8 \times E_8$ and thus to imposing II^* fibers on C_0 and C_∞ in $\widetilde{Z} = \mathbb{F}_n$, yielding

$$\widetilde{A}'_0 = 4L - 4C_0 - 4C_\infty = 8F,$$
$$\widetilde{B}'_0 = 6L - 5C_0 - 5C_\infty = 2C_0 + (12 + n)F,$$
$$\widetilde{\Delta}'_0 = 12L - 10C_0 - 10C_\infty = 4C_0 + (24 + 2n)F.$$

Furthermore, we impose singularities of type E_8, E_7, E_6 on the heterotic $K3$ surface. On the F-theory side, according to [AM97], this corresponds to imposing additional degenerations of the Weierstraß form over one fiber F of $\widetilde{Z} = \mathbb{F}_n$ of type II^*, III^*, IV^*, respectively. The most degenerate situation arises when all pointlike instantons coalesce on the singularity of the heterotic $K3$ surface. Since in F-theory each pointlike instanton corresponds to a (multiple) intersection of $\widetilde{\Delta}'_0$ with C_0 or C_∞, respectively, this amounts to degenerating $\widetilde{X}$ such that all intersections of $\widetilde{\Delta}'_0$ with C_0 and C_∞ are situated also on F. To desingularize the highly degenerate variety $\widetilde{X}$ one performs a chain of blowups of $\widetilde{Z}$, as we shall now describe, following [AM97].

To explain the general procedure let us blow up the singularity coming from the intersection of two rational curves Δ_1, Δ_2 which carry exceptional fibers of type $(\alpha, \beta; \gamma_1, \gamma_2, \gamma_3, \gamma_4, \gamma_6)$ and $(\alpha', \beta'; \gamma'_1, \gamma'_2, \gamma'_3, \gamma'_4, \gamma'_6)$, respectively, such that over the intersection of Δ_1 and Δ_2 we have a non-minimal fiber. In other words, we assume $\alpha + \alpha' \geq 4$, $\beta + \beta' \geq 6$. We take the coordinate t on Δ_1 and s on Δ_2, such that the fibration is locally given by

$$y^2 + s^{\gamma_1} t^{\gamma'_1} xy + s^{\gamma_3} t^{\gamma'_3} y = x^3 + s^{\gamma_2} t^{\gamma'_2} x^2 + s^{\gamma_4} t^{\gamma'_4} x + s^{\gamma_6} t^{\gamma'_6}.$$

To blow up, we set

$$s = s_1 t_1, \quad t = t_1, \quad x = t_1^2 x_1, \quad y = t_1^3 y_1,$$

i.e. we blow the base up in $s = t = 0$, as well as the fiber in $x = y = 0$. The equation becomes

$$y_1^2 + s_1^{\gamma_1} t_1^{\gamma_1 + \gamma'_1 - 1} x_1 y_1 + s_1^{\gamma_3} t_1^{\gamma_3 + \gamma'_3 - 3} y_1 = x_1^3 + s_1^{\gamma_2} t_1^{\gamma_2 + \gamma'_2 - 2} x_1^2 + s_1^{\gamma_4} t_1^{\gamma_4 + \gamma'_4 - 4} x_1$$
$$+ s_1^{\gamma_6} t_1^{\gamma_6 + \gamma'_6 - 6},$$

which along the new divisor $\{t_1 = 0\}$ has a Kodaira fiber of type

$$(\alpha'', \beta''; \gamma_1 + \gamma'_1 - 1, \gamma_2 + \gamma'_2 - 2, \gamma_3 + \gamma'_3 - 3, \gamma_4 + \gamma'_4 - 4, \gamma_6 + \gamma'_6 - 6). \quad (5.3.2)$$

Additionally moving instantons into the collision point the Weierstraß model for the fibration becomes

$$y^2 + s^{\gamma_1} t^{\gamma'_1} xy + s^{\gamma_3} t^{\gamma'_3} y = x^3 + s^{\gamma_2} t^{\gamma'_2} x^2 + s^{\gamma_4} t^{\gamma'_4} x + s^{\gamma_6} t^{\gamma'_6} (s + \lambda t^k) \quad (5.3.3)$$

for some constant λ, and $k \in \mathbb{N}$ accounting for the number of instantons. Blowing up as before we obtain

$$y_1^2 + s_1^{\gamma_1} t_1^{\gamma_1 + \gamma'_1 - 1} x_1 y_1 + s_1^{\gamma_3} t_1^{\gamma_3 + \gamma'_3 - 3} y_1 = x_1^3 + s_1^{\gamma_2} t_1^{\gamma_2 + \gamma'_2 - 2} x_1^2 + s_1^{\gamma_4} t_1^{\gamma_4 + \gamma'_4 - 4} x_1$$
$$+ s_1^{\gamma_6} t_1^{\gamma_6 + \gamma'_6 - 5} (s_1 + \lambda t^{k-1}). \quad (5.3.4)$$

We have introduced a new $\mathbb{CP}^1$ with fibers of type

$$(\alpha'', \beta''; \gamma_1 + \gamma'_1 - 1, \gamma_2 + \gamma'_2 - 2, \gamma_3 + \gamma'_3 - 3, \gamma_4 + \gamma'_4 - 4, \gamma_6 + \gamma'_6 - 5). \quad (5.3.5)$$

E_8 instantons on an E_8 singularity

An E_8 type singularity in the $K3$ surface on the heterotic side implies that $\widetilde{X}$ has II^* fibers not only over C_0 and C_∞ but also over a fiber F. Forcing these, the divisors $\widetilde{A}'_0$, $\widetilde{B}'_0$, and $\widetilde{\Delta}'_0$ become

$$\widetilde{A}' = \widetilde{A}'_0 - 4F = 4F,$$
$$\widetilde{B}' = \widetilde{B}'_0 - 5F = 2C_0 + (7+n)F,$$
$$\widetilde{\Delta}' = \widetilde{\Delta}'_0 - 10F = 4C_0 + (14 + 2n)F.$$

Hence

$$\widetilde{\Delta}' \cdot C_0 = 2(7-n), \quad \widetilde{\Delta}' \cdot C_\infty = 2(7+n), \quad \widetilde{\Delta}' \cdot F = 2 \cdot 2.$$

The first two formulas show that there are in total 14 instantons that can be moved into the E_8 singularity. There is an apparent difference between this number and the total instanton number 24 on the heterotic side; the interpretation of this mismatch in [AM97] says that the singularity on $K3$ "eats" 10 pointlike instantons. The last formula shows that there are two further collision points of multiplicity 2 of $\widetilde{\Delta}'$ with F, which will eventually be blown up.

We now move all $7 \pm n$ pointlike instantons into each collision point of two $II^*_{4,5;1,2,3,4,5}(\mathfrak{e}_8)$ fibers. This amounts to $k = 7 \pm n$ in (5.3.3). Performing one blow up (5.3.5) shows that we obtain a new divisor with $(4, 5; 1, 2, 3, 4, 5)$ type fibers, i.e. $II^*_{4,5;1,2,3,4,5}(\mathfrak{e}_8)$. According to (5.3.4), in the collision of this new divisor with C_0 or C_∞ one still has $k - 1$ pointlike instantons. Taking the two collisions of F with $\widetilde{\Delta}'$ into account, altogether we need to perform $b_j = 14 + 2$ blowups of points of multiplicity $\alpha_j = 2$ of the discriminant. Moreover, we have produced a chain of 15 $\mathbb{CP}^1$'s in Z with II^* curves in the fiber.

To smoothen the 3-fold where non-minimal fibers appear over collisions of any two curves with II^* fibers, another chain of blowups is necessary. Iterative application of (5.3.2) gives

$$II^*_{4,5;1,2,3,4,5}(\mathfrak{e}_8) + II^*_{4,5;1,2,3,4,5}(\mathfrak{e}_8)$$
$$\longrightarrow II^*_{4,5;1,2,3,4,5}(\mathfrak{e}_8) + I_{0|4,0;1,2,3,4,0} + II_{4,1;1,2,3,4,1}$$
$$+ IV_{4,2;1,2,3,4,2}(\mathfrak{su}(2)) + I^*_{0|4,3;1,2,3,4,3}(\mathfrak{g}_2) + II_{4,1;1,2,3,4,1}$$
$$+ IV^*_{4,4;1,2,3,4,4}(\mathfrak{f}_4) + II_{4,1;1,2,3,4,1} + I^*_{0|4,3;1,2,3,4,3}(\mathfrak{g}_2)$$
$$+ IV_{4,2;1,2,3,4,2}(\mathfrak{su}(2)) + II_{4,1;1,2,3,4,1} + I_{0|4,0;1,2,3,4,0}$$
$$+ II^*_{4,5;1,2,3,4,5}(\mathfrak{e}_8)$$

Here we have performed 11 blowups. We have contributions to the Picard number from: the Hirzebruch surface (2); blowups from residual intersections of $\widetilde{\Delta}'$ with F (2); the chain of $\mathbb{CP}^1$'s with II^* fibers over them (14); the additional contributions from the blowups of their collisions ($176 = 16 \times 11$). In total, we get

$$\rho(Z) = 2 + 2 + 14 + 176 = 194.$$

John McKay remarks that $\rho(Z) = 194$ is precisely the number of conjugacy classes in the Monster sporadic group $\mathbb{M}$.

As to the gauge group, each of the 15 curves in the fiber of Z which carry II^* singularities contributes an $\mathfrak{e}_8$. Furthermore, by the above each of the 16 blowups of a $II^* + II^*$ collision contributes $\mathfrak{f}_4 \oplus \mathfrak{g}_2^{\oplus 2} \oplus \mathfrak{su}(2)^{\oplus 2}$. Together with the primordial $\mathfrak{e}_8^{\oplus 2}$ over C_0 and C_∞, the total gauge algebra is

$$\mathfrak{e}_8^{\oplus 17} \oplus \mathfrak{f}_4^{\oplus 16} \oplus \mathfrak{g}_2^{\oplus 32} \oplus \mathfrak{su}(2)^{\oplus 32}.$$

The total gauge group H has dimension and rank

$$\dim H = 5592 \quad \text{and} \quad \text{rk } H = 296.$$

Next we need calculate $h^{1,2}(X)$. Using the algorithm described in Section 5.1 this amounts to calculating $\chi(X)$. To apply the formula given in Proposition 5.1, let us determine its various contributions:

$$\widetilde{\Delta}' \cdot (\widetilde{\Delta}' + K_{\widetilde{Z}}) = 76, \qquad \sum_j \alpha_j(\alpha_j - 1)b_j = 2 \times 16 = 32,$$

$$\sum_{i=1}^r \chi\left(\Delta_i - \bigcup_{j=1}^l \{P^j\}\right)\chi_i = 20, \qquad \sum_{j=1}^l \chi(\pi^{-1}(P^j)) = 960,$$

$$I' = 0, \qquad\qquad C = \widetilde{A}' \cdot \widetilde{B}' = 8.$$

From here we get

$$\chi(X) = 960, \quad h^{1,1}(X) = 1 + 194 + 296 = 491, \quad h^{1,2}(X) = 11.$$

Note that $h^{1,2}(X) + 1 = n_H^0$, the number of $K3$ parameters on the heterotic side, where the complex structure of $K3$ is constrained to having an E_8 singularity on $K3$, as predicted by Conjecture 4.2. Indeed, it is conjectured that the neutral hypermultiplets are not affected by all the degenerations due to pointlike instantons coalescing with $K3$ singularities.

Finally let us calculate n_H^{ch}. According to the explanations in Section 5.2.2 we need to consider collisions of families of degenerate curves of type (5.2.1). From the above, we have two collisions of type $IV(\mathfrak{su}(2)) + I_0^*(\mathfrak{g}_2)$ for each of the 16 chains obtained from blowing up $II^* + II^*$. This gives 32 contributions to the charged hypermultiplets. The resulting matter representation of $\mathfrak{g}_2 \oplus \mathfrak{su}(2)$ is $\frac{1}{2}((\mathbf{2}, \mathbf{1}) + (\mathbf{2}, \mathbf{7}))$ according to [Int98], where the prefactor indicates that these representations are quaternionic and contribute with half their dimension in n_H^{ch}. Altogether we have

$$n_H^{ch} = 256$$

and one checks that anomaly cancellation according to Conjecture 4.3 holds. It should be emphasized that the calculation of the charged matter representations in [Int98] uses the anomaly cancellation condition rather than giving a direct derivation. In fact, since $I_0^*(\mathfrak{g}_2)$ suffers monodromy, in this case not even a conjecture is known to us which describes such a direct derivation. We are in the process of filling this gap in the literature. That representations of the type exist which yield anomaly cancellation is already remarkable.

E_8 instantons on an E_7 singularity
By the same procedure as for an E_8 singularity on $K3$, we obtain:

$$\widetilde{A}' = \widetilde{A}'_0 - 3F = 5F,$$
$$\widetilde{B}' = \widetilde{B}'_0 - 5F = 2C_0 + (7+n)F,$$
$$\widetilde{\Delta}' = \widetilde{\Delta}'_0 - 9F = 4C_0 + (15 + 2n)F.$$

Hence

$$\widetilde{\Delta}' \cdot C_0 = 1 + 2(7-n), \quad \widetilde{\Delta}' \cdot C_\infty = 1 + 2(7+n), \quad \widetilde{\Delta}' \cdot F = 2 \cdot 2.$$

As before, 14 instantons can be moved into the E_7 singularity. In the first two formulas the summand 1 takes into account that colliding II^* and III^* fibers forces an additional intersection with $\widetilde{\Delta}'$, which has multiplicity 2 on F.

We now move all $7 \pm n$ pointlike instantons into each collision point of $II^*_{4,5;1,2,3,4,5}(\mathfrak{e}_8)$ and $III^*_{3,5;1,2,3,3,5}(\mathfrak{e}_7)$ fibers. As before we repeatedly apply (5.3.3) with $k = 7 \pm n$ obtaining 14 additional divisors with $III^*_{3,5;1,2,3,3,5}(\mathfrak{e}_7)$

fibers. Altogether we need to perform $b_j = 14$ blowups of points of multiplicity $\alpha_j = 2$ of the discriminant.

The chains of blowups needed to smoothen the 3-fold are as follows

$$II^*_{4,5;1,2,3,4,5}(\mathfrak{e}_8) + III^*_{3,5;1,2,3,3,5}(\mathfrak{e}_7)$$

$$\longrightarrow II^*_{4,5;1,2,3,4,5}(\mathfrak{e}_8) + I_{0|3,0;1,2,2,3,0} + II_{3,1;1,2,2,3,1}$$
$$+ IV_{3,2;1,2,2,3,2}(\mathfrak{su}(2)) + I^*_{0|3,3;1,2,2,3,3}(\mathfrak{g}_2) + II_{2,1;1,2,1,2,1}$$
$$+ IV^*_{3,4;1,2,2,3,4}(\mathfrak{f}_4) + II_{1,1;1,2,1,1,1} + I^*_{0|2,3;1,2,2,2,3}(\mathfrak{g}_2)$$
$$+ III_{1,2;1,2,2,1,2}(\mathfrak{su}(2)) + I_{0|0,1;1,2,2,0,1} + III^*_{3,5;1,2,3,3,5}(\mathfrak{e}_7),$$
$$+ I_{0|0,1;1,2,2,0,1} + III^*_{3,5;1,2,3,3,5}(\mathfrak{e}_7),$$

$$III^*_{3,5;1,2,3,3,5}(\mathfrak{e}_7) + III^*_{3,5;1,2,3,3,5}(\mathfrak{e}_7)$$

$$\longrightarrow III^*_{3,5;1,2,3,3,5}(\mathfrak{e}_7) + I_{0|0,2;1,2,3,0,2} + III_{1,3;1,2,3,1,3}(\mathfrak{su}(2))$$
$$+ I^*_{0|2,4;1,2,3,2,4}(\mathfrak{so}(7)) + III_{1,3;1,2,3,1,3}(\mathfrak{su}(2)) + I_{0|0,2;1,2,3,0,2}$$
$$+ III^*_{3,5;1,2,3,3,5}(\mathfrak{e}_7).$$

Recall that the collision $II^* + III^*$ forced an additional intersection with $\widetilde{\Delta}'$. In the process of blowing up, this collision with F moves onto one of the divisors carrying $I^*_0(\mathfrak{g}_2)$ fibers and transversally intersects it.

We have contributions to the Picard number from: the Hirzebruch surface (2); the chain of $\mathbb{CP}^1$'s with III^* fibers over them (14); the additional contributions from the blowups of their collisions ($90 = 14 \times 5 + 2 \times 10$). In total, we get

$$\rho(Z) = 2 + 14 + 90 = 106.$$

The total gauge algebra is

$$\mathfrak{e}_8^{\oplus 2} \oplus \mathfrak{e}_6^{\oplus 17} \oplus \mathfrak{su}(2)^{\oplus 2} \oplus \mathfrak{su}(3)^{\oplus 18} \oplus \mathfrak{g}_2^{\oplus 2} \oplus \mathfrak{f}_4^{\oplus 2}.$$

The total gauge group H has dimension and rank

$$\dim H = 3041 \quad \text{and} \quad \text{rk}\, H = 211.$$

The contributions to $\chi(X)$ in Proposition 5.1 amount to

$$\widetilde{\Delta}' \cdot (\widetilde{\Delta}' + K_{\widetilde{Z}}) = 82, \quad \sum_j \alpha_j(\alpha_j - 1)b_j = 2 \times 14 = 28,$$

$$\sum_{i=1}^r \chi\left(\Delta_i - \bigcup_{j=1}^I \{P^j\}\right)\chi_i = 8, \quad \sum_{j=1}^I \chi(\pi^{-1}(P^j)) = 630,$$

$$I' = 2, \quad C = \widetilde{A}' \cdot \widetilde{B}' = 10.$$

From here we get

$$\chi(X) = 612, \quad h^{1,1}(X) = 318, \quad h^{1,2}(X) = 12.$$

Again $h^{1,2}(X) + 1 = n_H^0$, the number of $K3$ parameters on the heterotic side, where the complex structure of $K3$ is constrained to having an E_7 singularity on $K3$, as predicted by Conjecture 4.2.

For n_H^{ch} we list collisions of curves of type (5.2.1): We have 4 collisions $IV(\mathfrak{su}(2)) + I_0^*(\mathfrak{g}_2)$ and $III(\mathfrak{su}(2)) + I_0^*(\mathfrak{g}_2)$ each as well as 2×14 collisions $III(\mathfrak{su}(2)) + I_0^*(\mathfrak{so}(7))$. According to [Int98], the associated representations are, respectively, $\frac{1}{2}((\mathbf{2}, \mathbf{1}) + (\mathbf{2}, \mathbf{7}))$ of $\mathfrak{su}(2) \oplus \mathfrak{g}_2$ and $\frac{1}{2}(\mathbf{2}, \mathbf{8})$ of $\mathfrak{su}(2) \oplus \mathfrak{so}(7)$ amounting to a total of

$$n_H^{ch} = 256.$$

The same comment as above applies to the derivation of the charged representations in [Int98]. One checks that anomaly cancellation according to Conjecture 4.3 holds. We do not yet understand why the remaining collisions of Δ' with families of $I_0^*(\mathfrak{g}_2)$ fibers do not contribute to the charged hypermultiplets.

E_8 instantons on an E_6 singularity

This time we have

$$\widetilde{A}' = \widetilde{A}'_0 - 3F = 4F,$$
$$\widetilde{B}' = \widetilde{B}'_0 - 4F = 2C_0 + (8 + n)F,$$
$$\widetilde{\Delta}' = \widetilde{\Delta}'_0 - 8F = 4C_0 + (16 + 2n)F$$
$$\widetilde{\Delta}' \cdot C_0 = 2(8 - n), \quad \widetilde{\Delta}' \cdot C_\infty = 2(8 + n), \quad \widetilde{\Delta}' \cdot F = 4.$$

Hence 16 instantons can be moved into the E_6 singularity. In order to consistently impose families of II^* and IV^* fibers like this, the IV^* fibers cannot suffer from monodromy. The collision then forces an additional intersection with $\widetilde{\Delta}'$ in the intersection $II^* + IV^*$ with multiplicity 2. This is accounted for in the last formula above and also contributes 2 to each of the intersections of $\widetilde{\Delta}'$ with C_0 and C_∞.

We now move all $7 \pm n$ free pointlike instantons into each collision point of $II^*_{4,5;1,2,3,4,5}(\mathfrak{e}_8)$ and $IV^*_{3,4;1,2,2,3,5}(\mathfrak{e}_6)$ fibers. As before we repeatedly apply (5.3.3) with $k = 7 \pm n$. Altogether 14 blowups introduce additional divisors with $IV^*_{3,4;1,2,2,3,5}(\mathfrak{e}_6)$ fibers. In this situation, the collisions $II^* + IV^*$ still force an additional intersection with $\widetilde{\Delta}'$, which make one further blowup of the discriminant necessary each, yielding new divisors with $IV^*_{3,4;1,2,2,3,4}(\mathfrak{f}_4)$ fibers over them. The latter intersect $\widetilde{\Delta}'$ with multiplicity 2 away from other collisions. Altogether we perform $b_j = 16$ blowups of points of multiplicity $\alpha_j = 2$ of the discriminant.

The chains of blowups needed to smoothen the 3-fold are as follows

$$II^*_{4,5;1,2,3,4,5}(\mathfrak{e}_8) + IV^*_{3,4;1,2,2,3,4}(\mathfrak{f}_4)$$
$$\longrightarrow II^*_{4,5;1,2,3,4,5}(\mathfrak{e}_8) + I_{0|3,0;1,2,2,3,0} + II_{3,1;1,2,2,3,1}$$
$$+ IV_{3,2;1,2,2,3,2}(\mathfrak{su}(2)) + I^*_{0|3,3;1,2,2,3,3}(\mathfrak{g}_2) + II_{2,1;1,2,1,2,1}$$
$$+ IV^*_{3,4;1,2,2,3,4}(\mathfrak{f}_4),$$
$$IV^*_{3,4;1,2,2,3,4}(\mathfrak{f}_4) + IV^*_{3,4;1,2,2,3,5}(\mathfrak{e}_6)$$
$$\longrightarrow IV^*_{3,4;1,2,2,3,4}(\mathfrak{f}_4) + I_{0|1,0;1,2,0,1,1} + IV_{2,2;1,2,1,2,3}(\mathfrak{su}(3))$$
$$+ I_{0|1,0;1,2,0,1,2} + IV^*_{3,4;1,2,2,3,5}(\mathfrak{e}_6),$$
$$IV^*_{3,4;1,2,2,3,5}(\mathfrak{e}_6) + IV^*_{3,4;1,2,2,3,5}(\mathfrak{e}_6)$$
$$\longrightarrow IV^*_{3,4;1,2,2,3,5}(\mathfrak{e}_6) + I_{0|1,0;1,2,0,1,3} + IV_{2,2;1,2,1,2,4}(\mathfrak{su}(3))$$
$$+ I_{0|1,0;1,2,0,1,3} + IV^*_{3,4;1,2,2,3,5}(\mathfrak{e}_6).$$

We have contributions to the Picard number from: the Hirzebruch surface (2); the chain of $\mathbb{CP}^1$'s with IV^* fibers over them (16); the additional contributions from the blowups of their collisions ($58 = 2 \times 5 + 16 \times 3$). In total, we get

$$\rho(Z) = 2 + 16 + 58 = 76.$$

The total gauge algebra is

$$\mathfrak{e}_8^{\oplus 2} \oplus \mathfrak{e}_6^{\oplus 15} \oplus \mathfrak{f}_4^{\oplus 2} \oplus \mathfrak{su}(2)^{\oplus 2} \oplus \mathfrak{g}_2^{\oplus 2} \oplus \mathfrak{su}(3)^{\oplus 16}.$$

The total gauge group H has dimension and rank

$$\dim H = 1932 \quad \text{and} \quad \operatorname{rk} H = 152.$$

The contributions to $\chi(X)$ in Proposition 5.1 amount to

$$\widetilde{\Delta}' \cdot (\widetilde{\Delta}' + K_{\widetilde{Z}}) = 88, \quad \sum_j \alpha_j(\alpha_j - 1)b_j = 2 \times 16 = 32,$$

$$\sum_{i=1}^r \chi\left(\Delta_i - \bigcup_{j=1}^I \{P^j\}\right)\chi_i = 4, \quad \sum_{j=1}^I \chi(\pi^{-1}(P^j)) = 456,$$

$$I' = 2, \qquad C = \widetilde{A}' \cdot \widetilde{B}' = 10.$$

From here we get

$$\chi(X) = 432, \quad h^{1,1}(X) = 229, \quad h^{1,2}(X) = 13.$$

Once again $h^{1,2}(X) + 1 = n_H^0$, the number of $K3$ parameters on the heterotic side, where the complex structure of $K3$ is constrained to having an E_6 singularity on $K3$, as predicted by Conjecture 4.2.

For n_H^{ch} we list collisions of curves of type (5.2.1): We have 2 collisions $IV(\mathfrak{su}(2)) + I_0^*(\mathfrak{g}_2)$ which according to [Int98] have associated representations $\frac{1}{2}((\mathbf{2}, \mathbf{1}) + (\mathbf{2}, \mathbf{7}))$ of $\mathfrak{su}(2) \oplus \mathfrak{g}_2$ amounting to a total

$$n_H^{ch} = 16.$$

One checks that anomaly cancellation according to Conjecture 4.3 holds. Again we do not understand why the remaining collisions of $\widetilde{\Delta}'$ with families of $IV^*(\mathfrak{f}_4)$ fibers do not contribute to the charged hypermultiplets.

6. Conclusions

Having devoted the bulk of this work to the description of our understanding of aspects of the heterotic – F-theory duality, we would like to return to the original motivation of this project, namely a new conjecture by John McKay. His conjecture relates geometric data of Calabi-Yau three-folds on the F-theory side of this duality to the Monster sporadic group and its offspring. Namely, as before let $X \to Z$ denote the Calabi-Yau three-fold which arises as F-theory dual of the heterotic theory with 24 pointlike E_8 instantons localized at an E_8 type quotient singularity on $K3$. Then following [AM97, AKM00] for the Picard number of the base we have shown $\rho(Z) = 194$, which as John McKay has observed agrees with the number of conjugacy classes of the Monster sporadic group $\mathbb{M}$. He conjectures that this is not a coincidence.

As described in the Introduction, McKay supports his conjecture by a known relation between the conjugacy classes of $\mathbb{M}$ and the Dynkin data of E_8 [McK80, GN01]. Also note that 24 pointlike instantons in an E_8 type quotient singularity give the most degenerate case of the heterotic – F-theory duality with the maximal number of pointlike instantons in the worst possible singularity on $K3$. Relating the F-theory dual Calabi-Yau three-fold $X \to Z$ to the largest finite sporadic group may not be completely unexpected, in particular as the Euler characteristic of this three-fold is 960, see Section 5.3.7, the largest value among all known elliptically fibered Calabi-Yau three-folds.

Naturally one would like to support McKay's conjecture by further data points. A possible lead is the above-mentioned relation between $\mathbb{M}$ and the Dynkin data of E_8, which roughly extends to relating the Baby monster $\mathbb{B}$ to E_7 and the Fischer group Fi_{24} to E_6 [GN01]. However, see Section 5.3.7, the respective Picard numbers $\rho(Z)$ of the bases of the F-theory dual Calabi-Yau three-folds corresponding to 24 pointlike E_8 instantons in quotient singularities of type E_7 and E_6 do not agree with the numbers of conjugacy classes of $\mathbb{B}$ and Fi_{24}. Hence McKay's conjecture requires some refinement. This may be related to the details of the identification of group data and Dynkin data for

$\mathbb{B} \leftrightarrow \tilde{E}_7$ and $\mathrm{Fi}_{24} \leftrightarrow \tilde{E}_6$. As explained in the Introduction, the relevant Dynkin data are obtained via folding $\tilde{E}_7$ and $\tilde{E}_6$ to the non-simply laced diagrams $\tilde{F}_4$ and $\tilde{G}_2$. Therefore one would like to find a way of implementing this folding procedure geometrically on the F-theory side. One promising possibility could amount to making use of less standard orbifold techniques, like Slodowy's interpretation of non-simply laced Dynkin diagrams in the description of the geometry of certain quotient singularities [Slo80].

In conclusion, at this point McKay's new conjecture is definitely not settled. However, it directs towards innovative and beautiful mathematics. Even if the foundation of the conjecture[5] is rather weak, we hope that by now the reader appreciates the importance and depth of the duality and its geometric meaning, be the conjecture true or wrong.

I have that sneaking hope, a hope unsupported by any facts or any evidence, that sometime in the twenty-first century physicists will stumble upon the Monster group, built in some unsuspected way into the structure of the universe.
(F. J. Dyson, "Unfashionable Pursuits", Math. Intelligencer 5 (1983), no. 3, 47–54)

A. Rarita-Schwinger fields

In the physics literature, massless fields are called Rarita-Schwinger fields, if they transform in the highest irreducible component RS of $S \otimes V$ where S denotes the spinor representation, and V denotes the vector representation of $\mathfrak{so}(D-2)$. However, this terminology is not used completely consistently and great care has to be taken with it: For example, let $M = M^{1,D-1}$ and denote by $S_M^{\pm}$ the corresponding spinor bundles. Naively, a massless Rarita-Schwinger field is a section in $S_M^+ \otimes T^* M$ which vanishes under the associated Dirac operator $\not{D}$. However, reduction to $\mathfrak{so}(D-2)$ yields $T_x^* M \cong V \oplus \mathbf{1} \oplus \mathbf{1}$ for every $x \in M$, and moreover $S^+ \otimes V = S^- \oplus \mathrm{RS}^+$, where now $S^{\pm}$ are the components of the spin representation of $\mathfrak{so}(1, D-1)$ arising under reduction to $\mathfrak{so}(D-2)$. Hence to extract RS^+ we need to work on a virtual bundle and with the Dirac operator

$$\not{D}_{RS}: \left(S^+ \otimes T^* M\right) \ominus S^+ \ominus S^+ \ominus S^- \longrightarrow \left(S^- \otimes T^* M\right) \ominus S^- \ominus S^- \ominus S^+.$$
(A.6)

By stretching of terminology, even if M is Euclidean one calls the above the Rarita-Schwinger complex of M, where $S^{\pm}$ denote the spinor bundles on M [ASZ84, §IV.V].

Remark. For $D = 4$ the bundles $S^{\pm}$ have dimension two each, such that $S^+ \otimes S^- \cong T^* M$, and if $\Lambda^2 S^+ \cong \mathbf{1}$ then $S^+ \otimes T^* M \ominus S^+ \ominus S^+ \ominus S^- \cong$

5 E_8 vs. 194.

$\mathrm{Sym}^2(S^+){\otimes}S^- \ominus (S^+ \oplus S^+)$, where Sym^2 denotes the two-symmetric tensor product. This latter virtual bundle is given as the standard domain of Rarita-Schwinger operators in [EGH80] if M is a $K3$ surface.

To compute the index of $\not{D}_{RS}$, one introduces the Dirac operator $\widetilde{\not{D}}_{RS}$ on $S^+ \otimes T^*M$ and uses the additivity of the index to get

$$\mathrm{ind}\,(\not{D}_{RS}) = \mathrm{ind}\,(\widetilde{\not{D}}_{RS}) - \mathrm{ind}\,(\not{D}), \tag{A.7}$$

where $\not{D}\colon S^+ \to S^-$.

Note that the definition of the index requires us to work with complex vector bundles, so for real fields we simply complexify real representations.

B. Characteristic classes and properties of $K3$ surfaces

We collect a few properties of characteristic classes in particular for $K3$, carefully keeping track of all the prefactors.

Let $\mathcal{E}$ be a complex bundle over a manifold X, with connection A (which is a $\mathfrak{u}(n)$ valued 1-form) and associated curvature $F_{\mathcal{E}}$. To it one associates the total Chern form:

$$c(F_{\mathcal{E}}) = \det\left(I_n + \frac{i}{2\pi}F_{\mathcal{E}}\right),$$

where I_n is the $n \times n$ identity matrix. The integral of the component in each degree gives the corresponding Chern class of $\mathcal{E}$:

$$c_1(\mathcal{E}) = \tfrac{i}{2\pi}\int_X \mathrm{Trace}\,(F_{\mathcal{E}}), \quad c_2(\mathcal{E}) = \tfrac{1}{8\pi^2}\int_X \left(\mathrm{Trace}\,(F_{\mathcal{E}}^2) - \left(\mathrm{Trace}\,(F_{\mathcal{E}})\right)^2\right), \ \dots\,.$$

We also have the Chern character

$$\mathrm{ch}(\mathcal{E}) = \int_X \mathrm{Trace}\left[\exp\left(\frac{i}{2\pi}F_{\mathcal{E}}\right)\right] = d + c_1(\mathcal{E}) + \frac{1}{2}\left(c_1^2(\mathcal{E}) - 2c_2(\mathcal{E})\right) + \dots,$$

where d is the rank of the complex bundle $\mathcal{E}$.

Now for a real vector bundle E, with connection 1-form valued in $\mathfrak{so}(n)$ this time, and with curvature F_E, the corresponding form is the total Pontrjagin form

$$p(F_E) = \det\left(I_n + \frac{1}{2\pi}F_E\right),$$

which gives the Pontrjagin classes of E when the component in each degree is integrated over X. Note that $F_E^T = -F_E$ such that only the even powers of F_E contribute. One sets

$$p_k(E) := \int_X p(F_E)_{4k} \quad \text{such that} \quad p_1(E) = -\frac{1}{8\pi^2}\int_X \mathrm{Trace}\,(F_E^2), \ \dots\,.$$

We can analyze the Pontrjagin classes in terms of Chern classes for a real vector bundle E,

$$p_k(E) = (-1)^k c_{2k}(E \otimes \mathbb{C}).$$

Conversely, if we are given a complex vector bundle $\mathcal{E}$, we introduce a real vector bundle E such that $\mathcal{E} \oplus \overline{\mathcal{E}} = \mathbb{C} \otimes E$ and therefore

$$p_1(E) = (c_1^2 - 2c_2)(\mathcal{E}) = -\frac{1}{4\pi^2} \int_X \text{Trace}\,(F_{\mathcal{E}}^2) = -\frac{1}{8\pi^2} \int_X \text{Trace}\,(F_E^2),$$

since a diagonal matrix in $\mathfrak{u}(n)$ with the entry ix_j is mapped to $\begin{bmatrix} 0 & -x_j \\ x_j & 0 \end{bmatrix}$ in $\mathfrak{so}(2n)$. This also gives a cross-check for the formula for the Pontrjagin class.

On a spin manifold X with Riemannian curvature R, we have the A-hat form

$$\widehat{A}(R) = 1 - \frac{1}{24} p_1(R) + \frac{1}{5760}(7p_1^2 - 4p_2) + \dots,$$

which when integrated over X gives the index of the Dirac operator $\not{D}$: $S^+ \to S^-$,

$$\text{ind}\,(\not{D}) = \int_X \widehat{A}(R).$$

For a twisted Dirac operator $\not{D}_{\mathcal{E}}$, with S^+ twisted by a complex vector bundle $\mathcal{E}$ of rank d, we have $\not{D}_{\mathcal{E}}: S^+ \otimes \mathcal{E} \longrightarrow S^- \otimes \mathcal{E}$ and

$$\text{ind}\,(\not{D}_{\mathcal{E}}) = \int_X \widehat{A}(R)\text{ch}(\mathcal{E}).$$

If we assume that X is a 4-manifold then the above considerations give

$$\text{ind}\,(\not{D}) = -\frac{1}{24} \int_X p_1(R) = \frac{1}{24 \cdot 8\pi^2} \int_X \text{Trace}\,(R^2),$$

and its twisted version

$$\text{ind}\,(\not{D}_{\mathcal{E}}) = -\frac{d}{24} \int_X p_1(R) + \frac{1}{2} \int_X (c_1^2(F_{\mathcal{E}}) - 2c_2(F_{\mathcal{E}}))$$

$$= \frac{d}{24 \cdot 8\pi^2} \int_X \text{Trace}\,(R^2) - \frac{1}{8\pi^2} \int_X \text{Trace}\,(F_{\mathcal{E}}^2) \,.$$

Now for a $K3$ surface we have the signature

$$\tau(K3) = \frac{1}{3} \int_{K3} p_1(R) = -16,$$

the Euler characteristic

$$\chi(K3) = \int_{K3} c_2(R) = 24,$$

and the A-hat genus

$$\widehat{A}(K3) = -\frac{1}{8}\tau = 2. \tag{B.8}$$

Also, on a spin four-manifold X the index of the twisted Dirac operator $\widetilde{\not{D}}_{RS}$ on $S^+ \otimes T^*X$ is

$$\mathrm{ind}\,(\widetilde{\not{D}}_{RS})(X) = \frac{20}{24} \int_X p_1(R).$$

By (A.7) the Rarita-Schwinger operator then has index

$$\mathrm{ind}\,(\not{D}_{RS})(X) = \frac{21}{24} \int_X p_1(R).$$

In particular, for a $K3$ surface X we obtain

$$\mathrm{ind}\,(\widetilde{\not{D}}_{RS})(X) = -40. \tag{B.9}$$

References

[Adl]   S. L. ADLER, *Anomalies*; hep-th/0411038.

[Adl69] ——, *Axial vector vertex in spinor electrodynamics*, Phys. Rev. **177** (1969), 2426–2438.

[AG96] P. S. ASPINWALL AND M. GROSS, *The SO(32) Heterotic String on a K3 Surface*, Phys. Lett. **B387** (1996), 735–742; hep-th/9605131.

[AGG85] L. ALVAREZ-GAUMÉ AND P. H. GINSPARG, *The structure of gauge and gravitational anomalies*, Ann. Physics **161** (1985), 423–526, Erratum-ibid.171:233,1986.

[AGW84] L. ALVAREZ-GAUMÉ AND E. WITTEN, *Gravitational anomalies*, Nucl. Phys. **B234** (1984), 269–379.

[AHS78] M. F. ATIYAH, N. J. HITCHIN, AND I. M. SINGER, *Self-duality in four-dimensional Riemannian geometry*, prsla **362** (1978), no. 1711, 425–461.

[AKM00] P. S. ASPINWALL, S. KATZ, AND D. R. MORRISON, *Lie groups, Calabi-Yau threefolds, and F-theory*, Adv. Theor. Math. Phys. **4** (2000), 95–126; hep-th/0002012.

[AL96] P. S. ASPINWALL AND J. LOUIS, *On the Ubiquity of K3 Fibrations in String Duality*, Phys. Lett. **B369** (1996), 233–242; hep-th/9510234.

[AM94] P. S. ASPINWALL AND D. R. MORRISON, *String theory on K3 surfaces*, in: Mirror symmetry II, B. Greene and S. T. Yau, eds., 1994, pp. 703–716; hep-th/9404151.

[AM97] ——, *Point-like instantons on K3 orbifolds*, Nucl. Phys. **B503** (1997), 533–564; hep-th/9705104.

[AS71] M. F. ATIYAH AND I. M. SINGER, *The index of elliptic operators IV*, Adv. Math. **93** (1971), 119–138.

[AS84] ——, *Dirac operators coupled to vector potentials*, Proc. Nat. Acad. Sci. U. S. A. **81** (1984), no. 8, Phys. Sci., 2597–2600.

[Asp97] P. S. ASPINWALL, *K3 surfaces and string duality*, in: Fields, strings and duality (Boulder, CO, 1996), World Sci. Publishing, River Edge, NJ, 1997, pp. 421–540; hep-th/9611137.

[ASZ84] O. ALVAREZ, I. M. SINGER, AND B. ZUMINO, *Gravitational anomalies and the family's index theorem*, Commun. Math. Phys. **96** (1984), no. 3, 409–417.

[Ati84] M. ATIYAH, *Anomalies and index theory*, in: Supersymmetry and supergravity/nonperturbative QCD (Mahabaleshwar, 1984), vol. 208 of Lecture Notes in Phys., Springer, Berlin, 1984, pp. 313–322.

[Bar69] W. A. BARDEEN, *Anomalous Ward identities in spinor field theories*, Phys. Rev. **184** (1969), 1848–1857.

[BCG$^+$98] M. BERSHADSKY, T. M. CHIANG, B. R. GREENE, A. JOHANSEN, AND C. I. LAZAROIU, *F-theory and linear sigma models*, Nucl. Phys. **B527** (1998), 531–570; hep-th/9712023.

[BIK$^+$96] M. BERSHADSKY, K. INTRILIGATOR, S. KACHRU, D. R. MORRISON, V. SADOV, AND C. VAFA, *Geometric singularities and enhanced gauge symmetries*, Nucl. Phys. **B481** (1996), 215–252; hep-th/9605200.

[BJPS97] M. BERSHADSKY, A. JOHANSEN, T. PANTEV, AND V. SADOV, *On four-dimensional compactifications of F-theory*, Nucl. Phys. **B505** (1997), 165–201; hep-th/9701165.

[BM03] A. BILAL AND S. METZGER, *Anomaly cancellation in M-theory: A critical review*, Nucl. Phys. **B675** (2003), 416–446; hep-th/0307152.

[Bor01] R. E. BORCHERDS, *Problems in Moonshine*, in: First International Congress of Chinese Mathematicians (Beijing, 1998), vol. 20 of AMS/IP Stud. Adv. Math., Amer. Math. Soc., Providence, RI, 2001, pp. 3–10.

[Bor02] R. E. BORCHERDS, *What is the monster?*, Notices of the A. M. S **49** (2002), no. 9, 1076–1077; arXiv:math.GR/0209328.

[BZ84] W. A. BARDEEN AND B. ZUMINO, *Consistent and covariant anomalies in gauge and gravitational theories*, Nucl. Phys. **B244** (1984), 421.

[CPR98] P. CANDELAS, E. PEREVALOV, AND G. RAJESH, *Matter from toric geometry*, Nucl. Phys. **B519** (1998), 225–238; hep-th/9707049.

[DE99] D.-E. DIACONESCU AND R. ENTIN, *Calabi-Yau spaces and five-dimensional field theories with exceptional gauge symmetry*, Nucl. Phys. **B538** (1999), 451–484; hep-th/9807170.

[DHVW85] L. J. DIXON, J. HARVEY, C. VAFA, AND E. WITTEN, *Strings on orbifolds*, Nucl. Phys. **B261** (1985), 678–686.

[DHVW86] ———, *Strings on orbifolds II*, Nucl. Phys. **B274** (1986), 285–314.

[DK90] S. K. DONALDSON AND P. B. KRONHEIMER, *The geometry of four-manifolds*, Oxford Mathematical Monographs, Oxford University Press, New York, 1990.

[DMW96] M. J. DUFF, R. MINASIAN, AND E. WITTEN, *Evidence for Heterotic/Heterotic Duality*, Nucl. Phys. **B465** (1996), 413–438; hep-th/9601036.

[Don85] S. K. DONALDSON, *Anti–self–dual Yang–Mills connections on complex algebraic surfaces and stable vector bundles*, Proc. Lond. Math. Soc. **3** (1985), 1–26.

[EGH80] T. EGUCHI, P. B. GILKEY, AND A. J. HANSON, *Gravitation, gauge theories and differential geometry*, Phys. Rev. **66** (1980), 213.

[FHSV95] S. FERRARA, J. A. HARVEY, A. STROMINGER, AND C. VAFA, *Second quantized mirror symmetry*, Phys. Lett. **B361** (1995), 59–65; hep-th/9505162.

[FMW97] R. FRIEDMAN, J. MORGAN, AND E. WITTEN, *Vector bundles and F theory*, Commun. Math. Phys. **187** (1997), no. 3, 679–743; hep-th/9701162.

[GH78] P. GRIFFITHS AND J. HARRIS, *Principles of algebraic geometry*, John Wiley & Sons, New York, 1978.

[GM03] A. GRASSI AND D. R. MORRISON, *Group representations and the Euler characteristic of elliptically fibered Calabi-Yau three-folds*, J. Alg. Geom. **12** (2003), 321–356; math.ag/0005196.

[GN01] G. GLAUBERMAN AND S. P. NORTON, *On McKay's connection between the affine E_8 diagram and the Monster*, Proceedings on Moonshine and related topics (Montréal, QC, 1999) (Providence, RI), CRM Proc. Lecture Notes, vol. 30, Amer. Math. Soc., pp. 37–42.

[GS84] M. B. GREEN AND J. H. SCHWARZ, *Anomaly cancellation in supersymmetric $D = 10$ gauge theory and superstring theory*, Phys. Lett. **B149** (1984), 117–122.

[GS85a] ———, *The hexagon gauge anomaly in type* I *superstring theory*, Nucl. Phys. **B255** (1985), no. 1, 93–114.

[GS85b] ———, *Infinity cancellations in* SO(32) *superstring theory*, Phys. Lett. **B151** (1985), no. 1, 21–25.

[GSW85] M. B. GREEN, J. H. SCHWARZ, AND P. C. WEST, *Anomaly free chiral theories in six dimensions*, Nucl. Phys. **B254** (1985), 327–348.

[GSW87] M. GREEN, J. H. SCHWARZ, AND E. WITTEN, *Superstring theory I & II*, Cambridge University Press, 1987.

[Har05] J. A. HARVEY, *TASI 2003 lectures on anomalies*; hep-th/0509097.

[HT95] C. M. HULL AND P. K. TOWNSEND, *Unity of superstring dualities*, Nucl. Phys. **B438** (1995), 109–137; hep-th/9410167.

[IMS97] K. A. INTRILIGATOR, D. R. MORRISON, AND N. SEIBERG, *Five-dimensional supersymmetric gauge theories and degenerations of Calabi-Yau spaces*, Nucl. Phys. **B497** (1997), 56–100; hep-th/9702198.

[Int98] K. A. INTRILIGATOR, *New string theories in six dimensions via branes at orbifold singularities*, Adv. Theor. Math. Phys. **1** (1998), 271–282; hep-th/9708117.

[KMP96] S. KATZ, D. R. MORRISON, AND M. R. PLESSER, *Enhanced Gauge Symmetry in Type II String Theory*, Nucl. Phys. **B477** (1996), 105–140; hep-th/9601108.

[Kod64] K. KODAIRA, *On the structure of compact complex analytic surfaces*, I. Am. J. Math. **86** (1964), 751–798.

[KV95] S. KACHRU AND C. VAFA, *Exact results for* $N = 2$ *compactifications of heterotic strings*, Nucl. Phys. **B450** (1995), 69–89; hep-th/9505105.

[KV97] S. KATZ AND C. VAFA, *Matter from geometry*, Nucl. Phys. **B497** (1997), 146–154; hep-th/9606086.

[McK80] J. MCKAY, *Graphs, singularities, and finite groups*, in: The Santa Cruz Conference on Finite Groups (Univ. California, Santa Cruz, Calif., 1979), Amer. Math. Soc., Providence, R. I., 1980, pp. 183–186.

[Mir83] R. MIRANDA, *Smooth models for elliptic threefolds*, in: The birational geometry of degenerations (Cambridge, Mass., 1981), vol. 29 of Progr. Math., Birkhäuser Boston, Mass., 1983, pp. 85–133.

[MP81] W. G. MCKAY AND J. PATERA, *Tables of dimensions, indices, and branching rules for representations of simple Lie algebras*,

Lecture Notes in Pure and Applied Mathematics, vol. 69, Marcel Dekker Inc., New York, 1981.

[MV96a] D. R. MORRISON AND C. VAFA, *Compactifications of F-Theory on Calabi–Yau Threefolds – I*, Nucl. Phys. **B473** (1996), 74–92; hep-th/9602114.

[MV96b] ——, *Compactifications of F-Theory on Calabi–Yau Threefolds – II*, Nucl. Phys. **B476** (1996), 437–469; hep-th/9603161.

[Pol98] J. POLCHINSKI, *String theory. Vols. I & II*, Cambridge Monographs on Mathematical Physics, Cambridge University Press, Cambridge, 1998.

[PS95] M. E. PESKIN AND D. V. SCHROEDER, *An introduction to quantum field theory*, Addison-Wesley Publishing Company Advanced Book Program, Reading, MA, 1995, Edited and with a foreword by David Pines.

[Sad96] V. SADOV, *Generalized Green-Schwarz mechanism in F theory*, Phys. Lett. **B388** (1996), 45–50; hep-th/9606008.

[Sch96] J. H. SCHWARZ, *Anomaly-Free Supersymmetric Models in Six Dimensions*, Phys. Lett. **B371** (1996), 223–230; hep-th/9512053.

[Sch02] ——, *Anomaly cancellation: A retrospective from a modern perspective*, Int. J. Mod. Phys. **A17S1** (2002), 157–166; hep-th/0107059.

[Sei88] N. SEIBERG, *Observations on the moduli space of superconformal field theories*, Nucl. Phys. **B303** (1988), 286–304.

[Sen96a] A. SEN, *F-theory and Orientifolds*, Nucl. Phys. **B475** (1996), 562–578; hep-th/9605150.

[Sen96b] ——, *M-Theory on $(K3 \times S_1)/Z_2$*, Phys. Rev. **D53** (1996), 6725–6729; hep-th/9602010.

[Slo80] P. SLODOWY, *Simple singularities and simple algebraic groups*, Lecture Notes in Mathematics, vol. 815, Springer, Berlin, 1980.

[SS04] C. A. SCRUCCA AND M. SERONE, *Anomalies in field theories with extra dimensions*, Int. J. Mod. Phys. **A19** (2004), 2579–2642; hep-th/0403163.

[Stu03] E. STUDY, *Geometrie der Dynamen*, Leipzig: Teubner, 1903.

[SW96] N. SEIBERG AND E. WITTEN, *Comments on String Dynamics in Six Dimensions*, Nucl. Phys. **B471** (1996), 121–134; hep-th/9603003.

[UY86] K. K. UHLENBECK AND S.-T. YAU, *On the existence of Hermitian–Yang-Mills connections in stable vector bundles*, Commun. Pure Appl. Math. Suppl. **39(S)** (1986), 257–293.

[Vaf96] C. VAFA, *Evidence for F-Theory*, Nucl. Phys. **B469** (1996), 403–418; hep-th/9602022.

[Wal88] M. WALTON, *The Heterotic string on the simplest Calabi-Yau manifold and its orbifold limits*, Phys. Rev. **D 37** (1988), 377–390.

[Wei05] S. WEINBERG, *The quantum theory of fields. Vol. II*, Cambridge University Press, Cambridge, 2005, Modern applications.

[Wit95] E. WITTEN, *String theory dynamics in various dimensions*, Nucl. Phys. **B443** (1995), 85–126; hep-th/9503124.

[Wit96a] ———, *Phase Transitions In M-Theory And F-Theory*, Nucl. Phys. **B471** (1996), 195–216; hep-th/9603150.

[Wit96b] ———, *Physical interpretation of certain strong coupling singularities*, Mod. Phys. Lett. **A11** (1996), 2649–2654; hep-th/9609159.

[Wit96c] ———, *Strong Coupling Expansion Of Calabi-Yau Compactification*, Nucl. Phys. **B471** (1996), 135–158; hep-th/9602070.

[WZ71] J. WESS AND B. ZUMINO, *Consequences of anomalous Ward identities*, Phys. Lett. **B37** (1971), 95.

Modularity of Trace Functions in Orbifold Theory for $\mathbb{Z}$-Graded Vertex Operator Superalgebras

Chongying Dong[1] and Zhongping Zhao

Department of Mathematics
University of California, Santa Cruz, CA 95064

Abstract

We study the trace functions in orbifold theory for $\mathbb{Z}$-graded vertex operator superalgebras and obtain a modular invariance result. More precisely, let V be a C_2-cofinite $\mathbb{Z}$-graded vertex operator superalgebra and G a finite automorphism group of V. Then for any commuting pair $(g, h) \in G$, the $h\sigma$-trace function associated to a simple g-twisted V-modules is holomorphic in the upper half plane, where σ is the canonical involution on V coming from the superspace structure of V. If V is further g-rational for every $g \in G$, the trace functions afford a representation for the full modular group $SL(2, \mathbb{Z})$.

1. Introduction

This work is a continuation of our study of the modular invariance for trace functions in orbifold theory. Motivated by generalized moonshine [N] and orbifold theory in physics [DVVV], the modular invariance of trace functions in orbifold theory has been studied for an vertex operator algebra [DLM3], under suitable conditions. This work has been generalized to a $\frac{1}{2}\mathbb{Z}$-graded vertex operator superalgebra [DZ2] (also see [H]), under suitable assumptions. In this paper we investigate the modular invariance of trace functions in orbifold theory for a $\mathbb{Z}$-graded vertex operator superalgebra.

There is an essential difference between a $\mathbb{Z}$-graded vertex operator superalgebra considered in this paper and a $\frac{1}{2}\mathbb{Z}$-graded vertex operator superalgebra studied in [DZ1]-[DZ2]. For a $\frac{1}{2}\mathbb{Z}$-graded vertex operator superalgebra $V = \oplus_{n \in \frac{1}{2}\mathbb{Z}} V_n$ the even part of V is $\sum_{n \in \mathbb{Z}} V_n$ and the odd part is $\sum_{n \in \mathbb{Z}} V_{n+\frac{1}{2}}$. In other words, the weight of an even vector is always integer while the weight

---

1 Supported by NSF grants, China NSF grant 10328102 and a faculty research grant from the University of California at Santa Cruz.

of an nonzero odd vector is not an integer. But for a $\mathbb{Z}$-graded vertex operator algebra $V = \oplus_{n \in \mathbb{Z}} V_n$, each V_n is a direct sum of even part $V_{\bar{0},n}$ and odd part $V_{\bar{1},n}$. So the weight of a vector is always an integer.

It is true that many $\mathbb{Z}$-graded vertex operator superalgebras can be obtained from $\frac{1}{2}\mathbb{Z}$-graded vertex operator superalgebras by changing the Virasoro elements (cf. [DM2]). In this case we can apply the results from [DZ1] and [DZ2] to these $\mathbb{Z}$-graded vertex operator superalgebras without extra work. Unfortunately, there are many $\mathbb{Z}$-graded vertex operator superalgebras which cannot be obtained in this way. So an independent study of $\mathbb{Z}$-graded vertex operator superalgebra becomes necessary, although the main ideas and methods in this paper are similar to those used in [Z], [DLM3] and [DZ2].

There is a subtle difference among these modular invariance results. In order to explain this we fix a finite automorphism group G of the vertex operator superalgebra. We use g and h for two commuting elements in G. For vertex operator superalgebras, there is a special automorphism σ of order 2 coming from the structure of the superspace. The involution σ can be expressed as $(-1)^F$ in the physics literature (cf. [GSW], [P]) where F is the fermion number. Here is the difference: for a vertex operator algebra, the space of all h-traces on g-twisted sectors is modular invariant [DLM3], for a $\frac{1}{2}\mathbb{Z}$-graded vertex operator superalgebra, the space of all $h\sigma$-traces on $g\sigma$-twisted sectors is modular invariant [DZ2], and for a $\mathbb{Z}$-graded vertex operator superalgebra, the space of all $h\sigma$-traces on g-twisted sectors is modular invariant. It is worth pointing out that the $h\sigma$-trace in the physics literature is called the super trace.

A systematic study of the modular invariance of trace functions for a vertex operator algebra V with $G = 1$ was first carried out in [Z]. This work was extended in [DLM3] to an arbitrary finite automorphism group G. Since the setting, ideas and most results in this paper are similar to those in [Z], [DLM2], [DLM3], [DZ1], and [DZ2], we refer the reader in many places to these papers for details.

The organization of this paper is as follows: In section 2, we review the definition of $\mathbb{Z}$-graded vertex operator superalgebra (VOSA) and various notions of g-twisted modules. Section 3 is devoted to studying the representation theory for $\mathbb{Z}$-graded VOSA. We introduce the associative algebra $A_g(V)$, and investigate the relation between the g-twisted modules and and $A_g(V)$-modules. Section 4 is the heart of the paper. We fix a $\mathbb{Z}$-graded vertex operator superalgebra V and a finite automorphism group G. We define the space $C(g, h)$ of 1-point functions on the torus associated to any commuting pair $g, h \in G$ and establish the modular invariance property. We prove that for a simple g-twisted module M, two commuting elements g and h of G such that M is σ-stable and h-stable, the graded $h\sigma$-trace function on M is a 1-point function.

Moreover, when V is g-rational, the collection of the trace functions associated to the inequivalent simple $h\sigma, h$ stable g-twisted V modules forms a basis of $C(g, h)$. In Section 5 we discuss an example to show the modularity of trace functions.

2. $\mathbb{Z}$-graded vertex operator superalgebras

Let $V = V_{\bar{0}} \oplus V_{\bar{1}}$ be $\mathbb{Z}_2$-graded vector space. For any $v \in V_{\bar{i}}$ with $i = 0, 1$ we define $\tilde{v} = \bar{i}$. Moreover, let $\epsilon_{u,v} = (-1)^{\tilde{u}\tilde{v}}$ and $\epsilon_v = (-1)^{\tilde{v}}$ for homogeneous $u, v \in V$.

Let z, z_0, z_1, z_2 be the independent commuting formal variables. We shall use the formal δ-function $\delta(z) = \sum_{n \in \mathbb{Z}} z^n$ and we refer the reader to [FLM3] on the basic properties of the δ-functions. The following definition of $\mathbb{Z}$-graded vertex operator superalgebra is a generalization of the notion of vertex operator algebra formulated in [B] and [FLM3].

Definition 2.1. A $\mathbb{Z}$-graded vertex operator superalgebra *($\mathbb{Z}$-graded VOSA) is a $\mathbb{Z} \times \mathbb{Z}_2$-graded vector space*

$$V = \bigoplus_{n \in \mathbb{Z}} V_n = V_{\bar{0}} \oplus V_{\bar{1}} = \bigoplus_{n \in \mathbb{Z}} (V_{\bar{0},n} \oplus V_{\bar{1},n}) \quad (\text{wt}v = n \text{ if } v \in V_n)$$

where

$$V_{\bar{i}} = \bigoplus_{n \in \mathbb{Z}} V_{\bar{i},n},$$

together with two distinguished vectors $\mathbf{1} \in V_{\bar{0},0}$, $\omega \in V_{\bar{0},2}$ *and equipped with a linear map*

$$V \to (\text{End } V)[[z, z^{-1}]],$$
$$v \mapsto Y(v, z) = \sum_{n \in \mathbb{Z}} v(n) z^{-n-1} \quad (v(n) \in \text{End } V)$$

satisfying the following axioms for $u, v \in V$:

(i) $u(n)v = 0$ *for sufficiently large* n;
(ii) *If* $u \in V_{\bar{i}}$ *and* $v \in V_{\bar{j}}$, *then* $u(n)v \in V_{\bar{i}+\bar{j}}$ *for all* $n \in \mathbb{Z}$ *where* $i, j = 0, 1$;
(iii) $Y(\mathbf{1}, z) = Id_V$ *and* $Y(v, z)\mathbf{1} = v + \sum_{n \geq 2} v(-n)\mathbf{1}z^{n-1}$;
(iv) *Set* $Y(\omega, z) = \sum_{n \in \mathbb{Z}} L(n) z^{-n-2}$: *Then*

$$[L(m), L(n)] = (m - n)L(m + n) + \frac{1}{12}(m^3 - m)\delta_{m+n,0}c \quad (2.1)$$

where $c \in \mathbb{C}$ is called the central charge, *and*

$$L(0)|_{V_n} = n, \ n \in \mathbb{Z}, \tag{2.2}$$

$$\frac{d}{dz} Y(v, z) = Y(L(-1)v, z); \tag{2.3}$$

(v) For $\mathbb{Z}_2$-homogeneous $u, v \in V$,

$$z_0^{-1} \delta \left(\frac{z_1 - z_2}{z_0} \right) Y(u, z_1) Y(v, z_2) - \epsilon_{u,v} z_0^{-1} \delta \left(\frac{z_2 - z_1}{-z_0} \right)$$

$$\times Y(v, z_2) Y(u, z_1) = z_2^{-1} \delta \left(\frac{z_1 - z_0}{z_2} \right) Y(Y(u, z_0)v, z_2) \tag{2.4}$$

where $(z_i - z_j)^n$ is expanded in nonnegative powers of z_j.

Fix a $\mathbb{Z}$-graded vertex operator superalgebra V and we set

$$Y[v, z] = Y(v, e^z - 1)e^{z \mathrm{wt} v} = \sum_{n \in \mathbb{Z}} v[n] z^{-n-1}. \tag{2.5}$$

Following the proof of Theorem 4.21 of [Z] we have:

Theorem 2.2. *$(V, Y[\ ], \mathbf{1}, \tilde{\omega})$ is a $\mathbb{Z}$-graded vertex operator superalgebra, where $\tilde{\omega} = \omega - \frac{c}{24}$.*

Let $Y[\tilde{\omega}, z] = \sum_{n \in \mathbb{Z}} L[n] z^{-n-2}$. Then $V = \bigoplus_{n \in \mathbb{Z}} V_{[n]}$ is again $\mathbb{Z}$-graded and $L[0] = n$ on $V_{[n]}$. We will write $\mathrm{wt}[v] = n$ if $v \in V_{[n]}$.

Definition 2.3. *A linear bijection g from a $\mathbb{Z}$-graded VOSA V to itself is called an* automorphism *of V if g preserves $\mathbf{1}$, ω and each $V_{\bar{i}}$, and*

$$gY(v, z)g^{-1} = Y(gv, z)$$

for $v \in V$.

Note that if V is a $\frac{1}{2}\mathbb{Z}$-graded vertex operator superalgebra, the assumption that g preserves each $V_{\bar{i}}$ in the definition of automorphism is unnecessary (cf. [DZ2]).

We denote the full automorphism group of V by $\mathrm{Aut}(V)$. If we define an action, say σ, on V associated to the superspace structure of V via $\sigma|V_{\bar{i}} = (-1)^i$, then σ is a central element of $\mathrm{Aut}(V)$ and will play a special role, as in [DZ2].

Let g be an automorphism of V of finite period T. Then we have the following eigenspace decomposition:

$$V = \bigoplus_{r \in \mathbb{Z}/T\mathbb{Z}} V^r \tag{2.6}$$

where

$$V^r = \{v \in V | gv = e^{-2\pi i r/T} v\}.$$

We now give various notions of g−twisted V-module. The twisted sectors and the twisted vertex operators for finite automorphisms of even lattice vertex operator algebras were first constructed in [L1] and [FLM2]. In [L2] and [DL], the twisted Jacobi identity was formulated and shown to hold for these operators. These results led to the notion of g-twisted V-module for a vertex operator algebra V and an automorphism g of V [D2] and [FFR].

Definition 2.4. *A* weak *g-twisted V-module is a vector space M equipped with a linear map*

$$\begin{aligned} V &\rightarrow (\text{End}M)[[z^{1/T}, z^{-1/T}]] \\ v &\mapsto Y_M(v, z) = \sum_{n \in \mathbb{Q}} v(n) z^{-n-1} \end{aligned}$$

which satisfies:

(i) *$v(m)w = 0$ for $v \in V$, $w \in M$ and $m >> 0$;*
(ii) *$Y_M(\mathbf{1}, z) = Id_M$;*
(iii) *For $v \in V^r$ and $0 \leq r \leq T - 1$*

$$Y_M(v, z) = \sum_{n \in \frac{r}{T} + \mathbb{Z}} v(n) z^{-n-1};$$

(iv) *For $u \in V^r$,*

$$z_0^{-1} \delta \left(\frac{z_1 - z_2}{z_0} \right) Y_M(u, z_1) Y_M(v, z_2) - \epsilon_{u,v} z_0^{-1} \delta \left(\frac{z_2 - z_1}{-z_0} \right) Y_M(v, z_2)$$

$$\times Y_M(u, z_1) = z_2^{-1} \left(\frac{z_1 - z_0}{z_2} \right)^{-r/T} \delta \left(\frac{z_1 - z_0}{z_2} \right) Y_M(Y(u, z_0)v, z_2)$$

$$(2.7)$$

Set

$$Y_M(\omega, z) = \sum_{n \in \mathbb{Z}} L(n) z^{-n-2}.$$

Then we have $Y_M(L(-1)v, z) = \frac{d}{dz} Y_M(v, z)$ for $v \in V$, and the $L(n)$ also satisfy the Virasoro algebra relations with central charge c (see [DLM1]).

Definition 2.5. *A* weak *g-twisted V-module M is* admissible *if it is $\frac{1}{T}\mathbb{Z}_+$-gradable:*

$$M = \bigoplus_{0 \leq n \in \frac{1}{T}\mathbb{Z}} M(n) \qquad (2.8)$$

such that for homogeneous $v \in V$,

$$v(m)M(n) \subseteq M(n + \mathrm{wt}v - m - 1). \tag{2.9}$$

Note that a uniform degree shift of M gives an isomorphic admissible V-module.

Definition 2.6. *A weak g-twisted V-module M is called an ordinary g-twisted V-module if it is $\mathbb{C}$-graded with*

$$M = \coprod_{\lambda \in \mathbb{C}} M_\lambda \tag{2.10}$$

where $M_\lambda = \{w \in M | L(0)w = \lambda w\}$ such that $\dim M_\lambda$ is finite and for fixed λ, $M_{\frac{n}{T} + \lambda} = 0$ for all small enough integers n.

It is not hard to prove that any ordinary g-twisted V-module is admissible. If $g = 1$ we have the notions of of weak, admissible and ordinary V-modules. If M is a simple g-twisted V-module, then

$$M = \bigoplus_{n=0}^{\infty} M_{\lambda + n/T} \tag{2.11}$$

for some $\lambda \in \mathbb{C}$ such that $M_\lambda \neq 0$ (cf. [Z]). λ is defined to be the *conformal weight* of M.

Definition 2.7.

   (i) *A $\mathbb{Z}$-graded VOSA V is called g-rational for an automorphism g of finite order if the category of admissible modules is completely reducible. V is called rational if it is 1-rational.*
  (ii) *V is called holomorphic if V is rational and V is the only irreducible V-module up to isomorphism.*
 (iii) *V is called g-regular if any weak g-twisted V-module is a direct sum of irreducible ordinary g-twisted V-modules.*

3. The associative algebra $A_g(V)$

In this section we construct the associative algebra $A_g(V)$ and study the relation between admissible g-twisted V-modules and $A_g(V)$-modules. The result is similar to those obtained in [DLM2] (also see [Z], [KW], [X], [DZ1]).

We assume that the order of g is T. For $0 \leq r \leq T - 1$ we define $\delta_r = \delta_{r,0}$. Let $O_g(V)$ be the linear span of all $u \circ_g v$, where for homogeneous $u \in V^r$(cf. (2.6)) and $v \in V$,

$$u \circ_g v = \text{Res}_z \frac{(1+z)^{\text{Wt}u-1+\delta_r+\frac{r}{T}}}{z^{1+\delta_r}} Y(u,z)v. \tag{3.1}$$

Set $A_g(V) = V/O_g(V)$ and define a second bilinear product $*_g$ on V for the above u, v as follows:

$$u *_g v = \text{Res}_z Y(u,z) \frac{(1+z)^{\text{Wt}u}}{z} v \tag{3.2}$$

if $r = 0$ and $u *_g v = 0$ if $r > 0$. It is easy to see that $A_g(V)$ is, in fact, a quotient of V^0.

As in [DLM2], [X] and [DZ2] we have

Theorem 3.1. $A_g(V) = V/O_g(V)$ *is an associative algebra with identity* $1 + O_g(V)$ *under the product $*_g$. Moreover, $\omega + O_g(V)$ lies in the center of* $A_g(V)$.

For a weak g-twisted V-module M, we define the space of the lowest weight vectors as follows:

$$\Omega(M) = \{w \in M | u(m-1+n)w = 0, \ u \in V_m, \ n > 0, \ m \in \mathbb{Z}\}.$$

We have (see [DLM2]):

Theorem 3.2. *Let M be a weak g-twisted V-module. Then*

    (i) *$\Omega(M)$ is an $A_g(V)$-module such that $v + O_g(V)$ acts as $o(v)$.*

    (ii) *If $M = \sum_{n \geq 0} M(n/T)$ is an admissible g-twisted V-module such that $M(0) \neq 0$, then $M(0) \subset \Omega(M)$ is an $A_g(V)$-submodule. Moreover, M is irreducible if and only if $M(0) = \Omega(M)$ and $M(0)$ is a simple $A_g(V)$-module.*

    (iii) *The map $M \to M(0)$ gives a 1-1 correspondence between the irreducible admissible g-twisted V-modules and simple $A_g(V)$-modules.*

We also have (see [DLM2] and [DZ1]):

Theorem 3.3. *Suppose that V is a g-rational vertex operator superalgebra. Then the following hold:*

    (i) *$A_g(V)$ is a finite dimensional semisimple associative algebra.*

    (ii) *V has only finitely many irreducible admissible g-twisted modules up to isomorphism.*

    (iii) *Every irreducible admissible g-twisted V-module is ordinary.*

    (iv) *V is g^{-1}-rational.*

4. Modularity of trace functions

We are working in the setting of section 5 in [DLM3]. In particular, g, h are commuting elements in $Aut(V)$ with finite orders $o(g) = T, o(h) = T_1$, A is the subgroup of $Aut(V)$ generated by g and h, $N = lcm(T, T_1)$ is the exponent of A, $\Gamma(T, T_1)$ is the subgroup of matrices $\begin{pmatrix} a & b \\ c & d \end{pmatrix}$ in $SL(2, \mathbb{Z})$ satisfying $a \equiv d \equiv 1 \pmod{N}$, $b \equiv 0 \pmod{T}$, $c \equiv 0 \pmod{T_1}$ and $M(T, T_1)$ is the ring of holomorphic modular forms on $\Gamma(T, T_1)$ with natural gradation $M(T, T_1) = \oplus_{k \geq 0} M_k(T, T_1)$, where $M_k(T, T_1)$ is the space of forms of weight k. Then $M(T, T_1)$ is a Noetherian ring.

Recall the Bernoulli polynomials $B_r(x) \in \mathbb{Q}[x]$ defined by

$$\frac{t e^{tx}}{(e^t - 1)} = \sum_{r=0}^{\infty} \frac{B_r(x) t^r}{r!}.$$

For even $k \geq 2$, the normalized Eisenstein series $E_k(\tau)$ is given by

$$E_k(\tau) = \frac{-B_k}{k!} + \frac{2}{(k-1)!} \sum_{n=1}^{\infty} \sigma_{k-1}(n) q^n. \tag{4.1}$$

Also introduce

$$
\begin{aligned}
Q_k(\mu, \lambda, q_\tau) &= Q_k(\mu, \lambda, \tau) \\
&= \frac{1}{(k-1)!} \sum_{n \geq 0} \frac{\lambda (n + j/T)^{k-1} q_\tau^{n+j/T}}{1 - \lambda q_\tau^{n+j/T}} \\
&\quad + \frac{(-1)^k}{(k-1)!} \sum_{n \geq 1} \frac{\lambda^{-1} (n - j/T)^{k-1} q_\tau^{n-j/T}}{1 - \lambda^{-1} q_\tau^{n-j/T}} - \frac{B_k(j/T)}{k!} \quad (4.2)
\end{aligned}
$$

for $(\mu, \lambda) = \left(e^{\frac{2\pi i j}{T}}, e^{\frac{2\pi i l}{T_1}} \right)$ and $(\mu, \lambda) \neq (1, 1)$, when $k \geq 1$ and $k \in \mathbb{Z}$. Here $(n + j/T)^{k-1} = 1$ if $n = 0$, $j = 0$ and $k = 1$. Similarly, $(n - j/T)^{k-1} = 1$ if $n = 1$, $j = M$ and $k = 1$. We also define

$$Q_0(\mu, \lambda, \tau) = -1. \tag{4.3}$$

It is proved in [DLM3] that E_{2k}, Q_r are contained in $M(T, T_1)$ for $k \geq 2$ and $r \geq 0$.

Set $V(T, T_1) = M(T, T_1) \otimes_{\mathbb{C}} V$. Given $v \in V$ with $gv = \mu^{-1} v$, $hv = \lambda^{-1} v$ we define a vector space $O(g, h)$ which is a $M(T, T_1)$-submodule of $V(T, T_1)$ spanned by the following elements:

$$v[0] w, \; w \in V, \; (\mu, \lambda) = (1, 1) \tag{4.4}$$

$$v[-2] w + \sum_{k=2}^{\infty} (2k-1) E_{2k}(\tau) \otimes v[2k - 2] w, \; (\mu, \lambda) = (1, 1) \tag{4.5}$$

$$v, (\epsilon_v, \mu, \lambda) \neq (1, 1, 1) \tag{4.6}$$

$$\sum_{k=0}^{\infty} Q_k(\mu, \lambda, \tau) \otimes v[k-1]w, (\mu, \lambda) \neq (1, 1). \tag{4.7}$$

Definition 4.1. *Let* $\mathfrak{h}$ *denote the upper half plane. The space of* (g, h) *1-point functions* $C(g, h)$ *is defined to be the vector space consisting of functions*

$$S : V(T, T_1) \times \mathfrak{h} \to \mathbb{C}$$

such that

    (i) $S(v, \tau)$ *is holomorphic in* τ *for* $v \in V(T, T_1)$,
    (ii) $S(v, \tau)$ *is linear in* v *and for* $f \in M(T, T_1)$, $v \in V$,

$$S(f \otimes v, \tau) = f(\tau)S(v, \tau),$$

    (iii) $S(v, \tau) = 0$ *if* $v \in O(g, h)$,
    (iv) *If* $v \in V$ *with* $\sigma v = gv = hv = v$, *then*

$$S(L[-2]v, \tau) = \partial S(v, \tau) + \sum_{l=2}^{\infty} E_{2l}(\tau)S(L[2l-2]v, \tau). \tag{4.8}$$

Here ∂S *is the operator which is linear in* v *and satisfies*

$$\partial S(v, \tau) = \partial_k S(v, \tau) = \frac{1}{2\pi i}\frac{d}{d\tau}S(v, \tau) + kE_2(\tau)S(v, \tau) \tag{4.9}$$

for $v \in V_{[k]}$.

We have the following modular invariance result (see Theorem 5.4 of [DLM3]):

Theorem 4.2. *For* $S \in C(g, h)$ *and* $\gamma = \begin{pmatrix} a & b \\ c & d \end{pmatrix} \in \Gamma$, *we define*

$$S|\gamma(v, \tau) = S|_k\gamma(v, \tau) = (c\tau + d)^{-k}S(v, \gamma\tau) \tag{4.10}$$

for $v \in V_{[k]}$, *and extend linearly. Then* $S|\gamma \in C((g, h)\gamma)$.

Let g, h, σ, V be as before, and M be a simple g-twisted module. We now show how the graded $h\sigma$-trace function on g-twisted V-module M produces a (g, h) 1-point function.

From (2.11), we know that if M is a simple g-twisted module then there exists a complex number λ such that

$$M = \bigoplus_{n=0}^{\infty} M_{\lambda + \frac{n}{T}} \tag{4.11}$$

Now we define a $(h\sigma)g(h\sigma)^{-1}$-twisted V-module $(h\sigma \circ M, Y_{h\sigma \circ M})$ such that $h\sigma \circ M = M$ as vector spaces and

$$Y_{h\sigma \circ M}(v, z) = Y_M((h\sigma)^{-1}v, z).$$

Since g, h, σ commute each other, $h\sigma \circ M$ is, in fact, a simple g-twisted V-module again. The M is called $h\sigma$-*stable* if $h\sigma \circ M$ and M are isomorphic g-twisted V-modules. In this case, there is a linear map $\phi(h\sigma) : M \to M$ such that

$$\phi(h\sigma)Y_M(v, z)\phi(h\sigma)^{-1} = Y_M((h\sigma)v, z) \qquad (4.12)$$

for all $v \in V$. Note that $\phi(h\sigma)$ is unique up to a nonzero constant.

We now assume that M is $h\sigma$-stable. For homogeneous $v \in V$, we define the trace function T as follows:

$$T(v) = T_M(v, (g, h), q) = z^{\text{wt}v}\text{tr}_M Y_M(v, z)\phi(h\sigma)q^{L(0)-\frac{c}{24}} \qquad (4.13)$$

Here c is the central charge of V. Note that for $m \in \frac{1}{T}\mathbb{Z}$, $v(m)$ maps M_μ to $M_{\mu+\text{wt}v-m-1}$. Hence

$$T(v) = q^{\lambda-\frac{c}{24}}\sum_{n=0}^{\infty}\text{tr}_{M_{\lambda+\frac{n}{T}}}o(v)\phi(h\sigma)q^{\frac{n}{T}} = \text{tr}_M o(v)\phi(h\sigma)q^{L(0)-\frac{c}{24}}. \qquad (4.14)$$

In order to state the next theorem we need to recall the C_2-cofinite condition from [Z]. V is called C_2-cofinite if $V/C_2(V)$ is finite dimensional where $C_2(V) = \langle u_{-2}v | u, v \in V \rangle$.

Theorem 4.3. *Suppose that V is C_2-cofinite, $g, h \in \text{Aut}(V)$ commute and have finite orders. Let M be a simple g-twisted V-module such that M is $h\sigma$-stable. Then the trace function $T_M(v, (g, h), q)$ converges to a holomorphic function in the upper half plane $\mathfrak{h}$ where $q = e^{2\pi i \tau}$ and $\tau \in \mathfrak{h}$. Moreover, $T_M \in C(g, h)$.*

The proof of this theorem is similar to that of Theorem 4.3 of [DZ2] although the idea goes back to [Z] and [DLM3].

We also have:

Theorem 4.4. *Let $M^1, M^2, \ldots, M^s$ be the inequivalent simple $h\sigma$-stable g-twisted V-modules, then the corresponding trace functions $T_1, T_2, \ldots, T_s$ (4.13) are independent vectors of $C(g, h)$. Moreover, if V is g-rational, $T_1, T_2, \ldots, T_s$ form a basis of $C(g, h)$.*

The following theorem is an immediate consequence of Theorems 4.3 and 4.4.

Theorem 4.5. *Suppose that V is a C_2-cofinite vertex operator superalgebra and G a finite group of automorphisms of V. Assume that V is x-rational for each $x \in G$. Let $v \in V$ satisfy $\mathrm{wt}[v] = k$. Then the space of (holomorphic) functions in $\mathfrak{h}$ spanned by the trace functions $T_M(v, (g, h), \tau)$ for all choices of g, h in G and $h\sigma$-stable g-twisted V-modules M is a (finite-dimensional) $SL(2, \mathbb{Z})$-module such that*

$$T_M|\gamma(v, (g, h), \tau) = (c\tau + d)^{-k} T_M(v, (g, h), \gamma\tau),$$

where $\gamma \in SL(2, \mathbb{Z})$ acts on $\mathfrak{h}$ as usual.

More precisely, if $\gamma = \begin{pmatrix} a & b \\ c & d \end{pmatrix} \in SL(2, \mathbb{Z})$ then we have an equality

$$T_M\left(v, (g, h), \frac{a\tau + b}{c\tau + d}\right) = (c\tau + d)^k \sum_W \gamma_{M,W} T_W(v, (g^a h^c, g^b h^d), \tau),$$

where W ranges over the $g^a h^c$-twisted sectors which are $g^b h^d \sigma$-stable. The constants $\gamma_{M,W}$ depend on M, W and γ only.

Theorem 4.6. *Let V be a rational and C_2-cofinite $\mathbb{Z}$-graded VOSA. Let M^1, M^2,..., M^s be the collection of inequivalent simple σ-stable V-modules. Then the space spanned by*

$$T_i(v, \tau) = T_i(v, (1, 1), \tau) = \mathrm{tr}_{M^i} o(v) \phi(\sigma) q^{L(0) - \frac{c}{24}} \qquad (4.15)$$

admits a representation of the modular group. More precisely, for any $\gamma = \begin{pmatrix} a & b \\ c & d \end{pmatrix} \in \Gamma$ there exists a $s \times s$ invertible complex matrix (γ_{ij}) such that

$$T_i\left(v, \frac{a\tau + b}{c\tau + d}\right) = (c\tau + d)^n \sum_{j=1}^{s} \gamma_{ij} T_j(v, \tau)$$

for all $v \in V_{[n]}$. Moreover, the matrix (γ_{ij}) is independent of v.

Recall Definition 2.7. The Following corollary is a special case of Theorem 4.6.

Corollary 4.7. *Assume that V is a holomorphic and C_2-cofinite $\mathbb{Z}$-graded vertex operator superalgebra. Take $v \in V_{[k]}$. Then*

$$T(v, \tau) = tr_V o(v) \sigma q^{L(0) - \frac{c}{24}}$$

is a modular form on $SL(2, \mathbb{Z})$ of weight k.

Remark 4.8. *It is interesting to notice that the modular invariance result in Theorem 4.6 is different from that for a vertex operator algebra in [Z] and for a $\frac{1}{2}\mathbb{Z}$-graded vertex operator superalgebra in [DZ2]. In the case of vertex operator algebra, the space of the graded traces of simple modules is modular invariant [Z]. But for a $\frac{1}{2}\mathbb{Z}$-graded vertex operator superalgebra, the space of graded σ traces on the simple σ-twisted modules is modular invariant. In the present situation, the space of graded σ traces on the simple V-modules is modular invariant.*

One can also obtain results such as the number of inequivalent, $h\sigma$-stable simple g-twisted V-modules and rationality of central charges and conformal weights for rational vertex operator superalgebras as in [DLM3] and [DZ2].

5. An example

In this section we consider $\mathbb{Z}$-graded VOSA $V_{\mathbb{Z}\alpha}$ and its σ-twisted module $V_{\mathbb{Z}\alpha + \frac{1}{2}\alpha}$ to demonstrate the modular invariance directly.

We are working in the setting of Chapter 8 of [FLM3]. Let $L = \mathbb{Z}\alpha$ be a nondegenerate lattice of rank 1 with $\mathbb{Z}$-valued symmetric $\mathbb{Z}$-bilinear form $\langle \ , \ \rangle$ s.t. $\langle \alpha, \alpha \rangle = 1$. Set $M(1) = \mathbb{C}[\alpha(-n)|n > 0]$ and let $\mathbb{C}[L]$ be the group algebra of the abelian group L. Set $\mathbf{1} = 1 \otimes e^0 \in V_L$ and $\omega = \frac{1}{2}\alpha(-1)\alpha(-1)$.

Recall that a vertex operator (super)algebra is called *holomorphic* if it is rational and the only irreducible module is itself. We have the following theorem (see [B], [FLM3], [D1], [DLM1], [DM1]).

Theorem 5.1.

(i) $(V_L, Y, \mathbf{1}, \omega)$ is a holomorphic $\frac{1}{2}\mathbb{Z}$-graded vertex operator super-algebra with central charge $c = \mathrm{rank}(L) = 1$.

(ii) $(V_L)_{\bar{0}} = M(1) \otimes \mathbb{C}[2L]$ and $(V_L)_{\bar{1}} = M(1) \otimes \mathbb{C}[2L + \alpha]$.

(iii) $V_{L + \frac{1}{2}\alpha}$ is the unique irreducible σ-twisted module for V_L.

One can verify the next theorem easily.

Theorem 5.2.

(i) *If we let $\omega' = \frac{1}{2}\alpha(-1)^2 + \frac{1}{2}\alpha(-2)$, then $(V_L, Y, \mathbf{1}, \omega')$ is a holomorphic $\mathbb{Z}$-graded vertex operator superalgebra with central charge $c' = -2$. (ω' is a shift in the sense of [DM2].)*

(ii) $V_{L+\frac{1}{2}\alpha}$ *is the unique irreducible σ-twisted module for the $\mathbb{Z}$-graded vertex operator superalgebra V_L.*

We consider the group G to be the cyclic group generated by σ. It is straightforward to compute the following trace functions for the $\mathbb{Z}$-graded vertex operator superalgebra V_L :

$$T(\mathbf{1},(1,1),\tau) = tr_{V_{\mathbb{Z}\alpha}}\sigma q^{L(0)'-\frac{-2}{24}}$$

$$= q^{\frac{1}{12}}\sum_{n=0}^{\infty}P(n)q^n\sum_{s=-\infty}^{\infty}(-1)^s q^{\frac{s(s-1)}{2}}$$

$$= \eta(\tau)^{-1}\theta_1(q),$$

$$T(\mathbf{1},(1,\sigma),\tau) = tr_{V_{\mathbb{Z}\alpha}}q^{L'(0)-\frac{-2}{24}}$$

$$= q^{\frac{1}{12}}\sum_{n=0}^{\infty}P(n)q^n\sum_{s=-\infty}^{\infty}q^{\frac{s(s-1)}{2}}$$

$$= \eta(\tau)^{-1}\theta_2(q),$$

$$T(\mathbf{1},(\sigma,\sigma),\tau) = tr_{V_{\mathbb{Z}\alpha+\frac{1}{2}\alpha}}q^{L'(0)-\frac{-2}{24}}$$

$$= q^{\frac{1}{12}}\sum_{n=0}^{\infty}P(n)q^n\sum_{s=-\infty}^{\infty}q^{\frac{\left(s+\frac{1}{2}\right)\left(s-\frac{1}{2}\right)}{2}}$$

$$= \eta(\tau)^{-1}\theta_3(q),$$

$$T(\mathbf{1},(\sigma,1),\tau) = tr_{V_{\mathbb{Z}\alpha+\frac{1}{2}\alpha}}\sigma q^{L'(0)-\frac{-2}{24}}$$

$$= q^{\frac{1}{12}}\prod_{n=1}^{\infty}(1+q^n)\sum_{s=-\infty}^{\infty}(-1)^s q^{\frac{\left(s+\frac{1}{2}\right)\left(s-\frac{1}{2}\right)}{2}}$$

$$= \eta(\tau)^{-1}\theta_4(q),$$

where

$$\eta(\tau) = q^{\frac{1}{24}}\prod_{n\geq 1}(1-q^n)$$

$$\theta_1(q) = \sum_{n=-\infty}^{\infty}(-1)^n q^{\frac{1}{2}\left(n-\frac{1}{2}\right)^2} = 0$$

$$\theta_2(q) = \sum_{n=-\infty}^{\infty}q^{\frac{1}{2}\left(n-\frac{1}{2}\right)^2}$$

$$\theta_3(q) = \sum_{n=-\infty}^{\infty} q^{\frac{1}{2}n^2}$$

$$\theta_4(q) = \sum_{n=-\infty}^{\infty} (-1)^n q^{\frac{1}{2}n^2}.$$

Recall the transformation law for η functions

$$\eta(\tau + 1) = e^{\frac{\pi i}{12}} \eta(\tau), \quad \eta\left(-\frac{1}{\tau}\right) = (-i\tau)^{\frac{1}{2}} \eta(\tau)$$

$$\eta\left(\frac{\tau+1}{2}\right) = \frac{\eta(\tau)^3}{\eta(\frac{\tau}{2})\eta(2\tau)}$$

and relations

$$\theta_2(q) = 2\frac{\eta(2\tau)^2}{\eta(\tau)}$$

$$\theta_3(q) = \frac{\eta(\tau)^5}{\eta(2\tau)^2\eta(\frac{\tau}{2})^2}$$

$$\theta_4(q) = \frac{\eta(\frac{\tau}{2})^2}{\eta(\tau)}.$$

The modular transformation property for $T(\mathbf{1}, (g, h), \tau)$ for $g, h \in G$ can easily be verified and the result, of course, is the same as what Theorem 4.5 claimed. One can also compute the trace functions for the $\frac{1}{2}\mathbb{Z}$-graded vertex operator superalgebra V_L and notice that the sets of trace functions in the two cases are exactly the same. Recall from [DZ1]-[DZ2] the $\frac{1}{2}\mathbb{Z}$-graded vertex operator superalgebra $V\left(H, \mathbb{Z} + \frac{1}{2}\right)$ and the trace functions with group $G = \langle \sigma \rangle$ for any positive integer l. Since V_L and $V\left(H, \mathbb{Z} + \frac{1}{2}\right)$ with $l = 2$ are isomorphic $\frac{1}{2}\mathbb{Z}$-graded vertex operator superalgebras (the boson-fermion correspondence), one can use the modular invariance result for $V\left(H, \mathbb{Z} + \frac{1}{2}\right)$ obtained in [DZ2] to check the modular transformation property of the trace functions for the $\mathbb{Z}$-graded vertex operator superalgebra V_L.

Remark 5.3. *One can use another shift $\omega' = \frac{1}{2}\alpha(-1)^2 - \frac{1}{2}\alpha(-2)$ to get an isomorphic $\mathbb{Z}$-graded vertex operator superalgebra V_L and Theorem 5.2 is still valid. Moreover, the trace functions and the modular transformation property remain the same.*

References

[B]   R. E. Borcherds, Vertex algebras, Kac-Moody algebras, and the Monster, *Proc. Natl. Acad. Sci. USA* **83** (1986), 3068–3071.

[DVVV] R. Dijkgraaf, C. Vafa, E. Verlinde and H. Verlinde, The operator algebra of orbifold models, *Comm. Math. Phys.* **123** (1989), 485–526.

[D1]   C. Dong, Vertex algebras associated with even lattices, *J. Algebra* **160** (1993), 245–265.

[D2]   C. Dong, Twisted modules for vertex algebras associated with even lattices, *J. Algebra* **165** (1994), 90–112.

[DL]   C. Dong and J. Lepowsky, The algebraic structure of relative twisted vertex operators, *J. Pure Appl. Algebra* **110** (1996), 259–295.

[DLM1] C. Dong, H. Li and G. Mason, Regularity of rational vertex operator algebras, *Advances. in Math.* **132** (1997), 148–166.

[DLM2] C. Dong, H. Li and G. Mason, Twisted representations of vertex operator algebras, *Math. Ann.* **310** (1998), 571–600.

[DLM3] C. Dong, H. Li and G. Mason, Modular invariance of trace functions in orbifold theory and generalized moonshine, *Comm. Math. Phys.* **214** (2000), 1–56.

[DM1]   C. Dong and G. Mason, An orbifold theory of genus zero associated with the sporadic simple group M_{24}, *Comm. Math. Phys.* **164** (1994), 87–104.

[DM2]   C. Dong and G. Mason, Shifted vertex operator algebra, *Proc. Cambridge Philos. Soc.* **141** (2006), 67–80.

[DZ1]   C. Dong and Z. Zhao, Twisted representations of vertex operator superalgebras, *Comm. Contemp. Math.* **8** (2006), 101–121,

[DZ2]   C. Dong and Z. Zhao, Modularity in orbifold theory for vertex operator superalgebras, *Comm. Math. Phys.* **260** (2005), 227–256.

[FFR]   A. J. Feingold, I. B. Frenkel and J. F. X. Ries, *Spinor Construction of Vertex Operator Algebras, Triality and $E_8^{(1)}$*, Contemp. Math. **121**, Amer. Math. Soc., Providence, 1991.

[FLM1] I. B. Frenkel, J. Lepowsky and A. Meurman, A natural representation of the Fischer-Griess Monster with the modular function J as character, *Proc. Natl. Acad. Sci. USA* **81** (1984), 3256–3260.

[FLM2] I. B. Frenkel, J. Lepowsky and A. Meurman, Vertex operator calculus, in: *Mathematical Aspects of String Theory, Proc. 1986 Conference, San Diego.* ed. by S.-T. Yau, World Scientific, Singapore, 1987, 150–188.

[FLM3] I. B. Frenkel, J. Lepowsky and A. Meurman, *Vertex Operator Algebras and the Monster*, Pure and Applied Math., Vol. **134**, Academic Press, 1988.

[GSW] M. Green, J. Schwartz and E. Witten,. Superstring Theory, Vol. **1**, Cambridge University Press, 1987.

[G] R. Griess, The Friendly Giant, *Invent. Math.* **69** (1982), 1–102.

[H] G. Höhn, Self-dual vertex operator superalgebras and the Baby Monster, Bonn Mathematical Publications, 286. Universität Bonn, Mathematisches Institut, Bonn, 1996.

[KW] V. Kac and W. Wang, Vertex operator superalgebras and their representations, *Contemp. Math.* **175** (1994), 161–191.

[L1] J. Lepowsky, Calculus of twisted vertex operators, *Proc. Nat. Acad. Sci. USA* **82** (1985), no. 24, 8295–8299.

[L2] J. Lepowsky, Perspectives on vertex operators and the Monster, in: *Proc. 1987 Symposium on the Mathematical Heritage of Hermann Weyl, Duke Univ., Proc. Symp. Pure Math., American Math. Soc.*, **48** (1988).

[MS] G. Moore and N. Seiberg, Classical and quantum conformal field theory, *Comm. Math. Phys.* **123** (1989), 177–254.

[N] S. Norton, Generalized moonshine, *Proc. Symp. Pure. Math., American Math. Soc.* **47** (1987), 208–209.

[P] J. Polchinski, String Theory, Vol. **I, II**, Cambridge University Press, 1998.

[X] X. Xu, Introduction to Vertex Operator Superalgebras and Their Modules, *Mathematics and its Applications* Vol. 456, Kluwer Academic Publishers, Dordrecht, 1998.

[Z] Y. Zhu, Modular invariance of characters of vertex operator algebras, *J. Amer, Math. Soc.* **9** (1996), 237–302.

Twisted Modules for Vertex Operator Algebras

Benjamin Doyon*

Rudolf Peierls Centre for Theoretical Physics,
Oxford University, UK
Currrent address:
Department of Mathematical Sciences
Durham Unversity, UK

Abstract

This contribution is mainly based on joint papers with Lepowsky and Milas, and some parts of these papers are reproduced here. These papers further extended works by Lepowsky and by Milas. Following our joint papers, I explain the general principles of twisted modules for vertex operator algebras in their powerful formulation using formal series, and derive general relations satisfied by twisted and untwisted vertex operators. Using these, I prove new "equivalence" and "construction" theorems, identifying a set of sufficient conditions in order to have a twisted module for a vertex operator algebra, and a simple way of constructing the twisted vertex operator map. This essentially combines our general relations for twisted modules with ideas of Li (1996), who had obtained similar construction theorems using different relations. Then, I show how to apply these theorems in order to construct twisted modules for the Heisenberg vertex operator algebra. I obtain in a new way the explicit twisted vertex operator map, and in particular give a new derivation and expression for the formal operator Δ_x constructed some time ago by Frenkel, Lepowsky and Meurman. Finally, I reproduce parts of our joint papers. I use the untwisted relations in the Heisenberg vertex operator algebra in order to understand properties of a certain central extension of a Lie algebra of differential operators on the circle: the connection between the structure of the central term in Lie brackets and the Riemann Zeta function at negative integers. I then use the twisted relations in order to construct in a simple way a family of representations for this algebra based on twisted modules for the Heisenberg vertex operator algebra. As a simple consequence of the twisted

* The author gratefully acknowledges support from an EPSRC (UK) Postdoctoral Fellowship, grant GR/S91086/01.

144

relations, the construction involves the Bernoulli polynomials at rational values in a fundamental way.

1. Introduction

This contribution is mainly based on, and partly reproduces, the recent papers by the present author, Lepowsky and Milas [DLMi1], [DLMi2]. These works were a continuation of a series of papers of Lepowsky and Milas [L3], [L4], [M1]–[M3], stimulated by work of Bloch [Bl].

In those papers, we used the general theory of vertex operator algebras to study central extensions of classical Lie algebras and superalgebras of differential operators on the circle in connection with values of ζ–functions at the negative integers, and with the Bernoulli polynomials at rational values. Parts of the present contribution recall the main results of [DLMi1, DLMi2]: Using general principles of the theory of vertex operator algebras and their twisted modules, we obtain a bosonic, twisted construction of a certain central extension of a Lie algebra of differential operators on the circle, for an arbitrary twisting automorphism. The construction involves the Bernoulli polynomials in a fundamental way. This is explained through results in the general theory of vertex operator algebras, including an identity discovered in [DLMi1, DLMi2] which was called "modified weak associativity", and which is a consequence of the twisted Jacobi identity.

More precisely, we combine and extend methods from [L3], [L4], [M1]–[M3], [FLM1], [FLM2] and [DL2]. In those earlier papers, vertex operator techniques were used to analyze untwisted actions of the Lie algebra $\hat{\mathcal{D}}^+$, studied in [Bl], on a module for a Heisenberg Lie algebra of a certain standard type, based on a finite-dimensional vector space equipped with a nondegenerate symmetric bilinear form. Now consider an arbitrary isometry ν of period say p, that is, with $\nu^p = 1$. Then, it was shown in [DLMi1, DLMi2] that the corresponding ν–twisted modules carry an action of the Lie algebra $\hat{\mathcal{D}}^+$ in terms of twisted vertex operators, parametrized by certain quadratic vectors in the untwisted module. This extends a result from [FLM1], [FLM2], [DL2] where actions of the Virasoro algebra were constructed using twisted vertex operators.

Still following [DLMi1, DLMi2], we explicitly compute certain "correction" terms for the generators of the "Cartan subalgebra" of $\hat{\mathcal{D}}^+$ that naturally appear in any twisted construction. These correction terms are expressed in terms of special values of certain Bernoulli polynomials. They can in principle be generated, in the theory of vertex operator algebras, by the formal operator e^{Δ_x} [FLM1], [FLM2], [DL2] involved in the construction of a twisted action for a certain type of vertex operator algebra, the Heisenberg vertex operator algebra. We generate those correction terms in an easier way, using the modified weak associativity relation.

Then, the present contribution extends the works [DLMi1, DLMi2] described above by providing a detailed analysis of the modified weak associativity relation. We state and prove a new theorem (Theorem 5.1) about the equivalence of modified weak associativity and weak commutativity with the twisted Jacobi identity, and a new "construction" theorem (Theorem 5.5), where we identify a set of sufficient conditions in order to have a twisted module for a vertex operator algebra, and a simple way of constructing the twisted vertex operator map. The latter theorem essentially combines modified weak associativity with ideas of Li [Li1, Li2], where similar construction theorems were proven using different general relations of vertex operator algebras and twisted modules – there may be a "direct" path from Li's construction theorems to ours, but we haven't investigated this. The use of modified weak associativity seems to have certain advantages in the twisted case. As an illustration, we give a new proof that the ν-twisted Heisenberg Lie algebra modules mentioned above are also twisted modules for the Heisenberg vertex operator algebra. Using our theorems, we explicitly construct the twisted vertex operator map (Theorem 6.2). This gives a new and relatively simple derivation and expression for this map, and in particular for the formal operator Δ_x mentioned above. A consequence of this is that one minor technical assumption that had to be made in [DLMi1, DLMi2], about the action of the automorphism ν, can be taken away.

We should mention that in [KR] Kac and Radul established a relationship between the Lie algebra of differential operators on the circle and the Lie algebra $\widehat{\mathfrak{gl}}(\infty)$; for further work in this direction, see [AFMO], [KWY]. Our methods and motivation for studying Lie algebras of differential operators, based on vertex operator algebras, are new and very different, so we do not pursue their direction.

Although we will present many of the main results of [DLMi2] with some of the proofs, we refer the reader to this paper for a more extensive discussion of those results.

Acknowledgments. The author is grateful to J. Lepowsky and A. Milas for discussions and comments on the manuscript.

2. Vertex operator algebras, untwisted modules and twisted modules

In this section, we recall the definition of vertex operator algebras, (untwisted) modules and twisted modules. For the basic theory of vertex operator algebras and modules, we will use the viewpoint of [LL].

In the theory of vertex operator algebras, formal calculus plays a fundamental role. Here we recall some basic elements of formal calculus (cf. [LL]). Formal calculus is the calculus of formal doubly–infinite series of formal variables, denoted below by x, y, and by x_1, x_2, ..., y_1, y_2, ... The central object of formal calculus is the formal delta–function

$$\delta(x) = \sum_{n \in \mathbb{Z}} x^n$$

which has the property

$$\delta\left(\frac{x_1}{x_2}\right) f(x_1) = \delta\left(\frac{x_1}{x_2}\right) f(x_2)$$

for any formal series $f(x_1)$. The formal delta–function enjoys many other properties, two of which are:

$$x_2^{-1}\delta\left(\frac{x_1 - x_0}{x_2}\right) = x_1^{-1}\delta\left(\frac{x_2 + x_0}{x_1}\right) \tag{2.1}$$

and

$$x_0^{-1}\delta\left(\frac{x_1 - x_2}{x_0}\right) + x_0^{-1}\delta\left(\frac{x_2 - x_1}{-x_0}\right) = x_2^{-1}\delta\left(\frac{x_1 - x_0}{x_2}\right). \tag{2.2}$$

In these equations, binomial expressions of the type $(x_1 - x_2)^n$, $n \in \mathbb{Z}$ appear. Their meaning as formal series in x_1 and x_2, as well as the meaning of powers of more complicated formal series, is summarized in the "binomial expansion convention" – the notational device according to which binomial expressions are understood to be expanded in nonnegative integral powers of the second variable. When more elements of formal calculus are needed below, we shall recall them.

2.1. Vertex operator algebras and untwisted modules

We recall from [FLM2] the definition of the notion of vertex operator algebra, a variant of Borcherds' notion [Bo] of vertex algebra:

Definition 2.1. *A **vertex operator algebra** $(V, Y, \mathbf{1}, \omega)$, or V for short, is a $\mathbb{Z}$–graded vector space*

$$V = \coprod_{n \in \mathbb{Z}} V_{(n)}; \text{ for } v \in V_{(n)}, \text{ wt } v = n,$$

such that

$$V_{(n)} = 0 \text{ for } n \text{ sufficiently negative,}$$
$$\dim V_{(n)} < \infty \text{ for } n \in \mathbb{Z},$$

equipped with a linear map $Y(\cdot, x)$:

$$Y(\cdot, x) \;:\; V \to (\text{End } V)[[x, x^{-1}]]$$

$$v \mapsto Y(v, x) = \sum_{n \in \mathbb{Z}} v_n x^{-n-1}, \quad v_n \in \text{End } V, \tag{2.3}$$

where $Y(v, x)$ is called the vertex operator *associated with v, and two particular vectors, $\mathbf{1}, \omega \in V$, called respectively the* vacuum vector *and the* conformal vector, *with the following properties:*

truncation condition: *For every $v, w \in V$*

$$v_n w = 0 \tag{2.4}$$

for $n \in \mathbb{Z}$ sufficiently large;

vacuum property:

$$Y(\mathbf{1}, x) = 1_V \quad (1_V \text{ is the identity on } V); \tag{2.5}$$

creation property:

$$Y(v, x)\mathbf{1} \in V[[x]] \quad \text{and} \quad \lim_{x \to 0} Y(v, x)\mathbf{1} = v ; \tag{2.6}$$

Virasoro algebra conditions: Let

$$L(n) = \omega_{n+1} \text{ for } n \in \mathbb{Z}, \quad \text{i.e.,} \quad Y(\omega, x) = \sum_{n \in \mathbb{Z}} L(n) x^{-n-2} . \tag{2.7}$$

Then

$$[L(m), L(n)] = (m - n)L(m + n) + c_V \frac{m^3 - m}{12} \delta_{n+m,0} 1_V$$

for $m, n \in \mathbb{Z}$, where $c_V \in \mathbb{C}$ is the central charge (also called "rank" of V),

$$L(0)v = (\text{wt } v)v$$

for every homogeneous element v, and we have the $L(-1)$–derivative property:

$$Y(L(-1)u, x) = \frac{d}{dx} Y(u, x) ; \tag{2.8}$$

Jacobi identity:

$$x_0^{-1} \delta\left(\frac{x_1 - x_2}{x_0}\right) Y(u, x_1)Y(v, x_2) - x_0^{-1} \delta\left(\frac{x_2 - x_1}{-x_0}\right) Y(v, x_2)Y(u, x_1)$$

$$= x_2^{-1} \delta\left(\frac{x_1 - x_0}{x_2}\right) Y(Y(u, x_0)v, x_2) . \tag{2.9}$$

An important property of vertex operators is skew–symmetry, which is an easy consequence of the Jacobi identity (cf. [FHL]):

$$Y(u, x)v = e^{xL(-1)}Y(v, -x)u. \tag{2.10}$$

Another easy consequence of the Jacobi identity is the $L(-1)$–bracket formula:

$$[L(-1), Y(u, x)] = Y(L(-1)u, x). \tag{2.11}$$

Fix now a vertex operator algebra $(V, Y, \mathbf{1}, \omega)$, with central charge c_V.

Definition 2.2. *A ($\mathbb{Q}$-graded)* **module** *W for the vertex operator algebra V (or V–**module**) is a $\mathbb{Q}$-graded vector space,*

$$W = \coprod_{n \in \mathbb{Q}} W_{(n)}; \quad \text{for } v \in W_{(n)}, \text{ wt } v = n,$$

such that

$$W_{(n)} = 0 \quad \text{for } n \text{ sufficiently negative,}$$
$$\dim W_{(n)} < \infty \quad \text{for } n \in \mathbb{Q},$$

equipped with a linear map

$$Y_W(\cdot, x) : V \to (\text{End } W)[[x, x^{-1}]]$$
$$v \mapsto Y_W(v, x) = \sum_{n \in \mathbb{Z}} v_n^W x^{-n-1}, \quad v_n^W \in \text{End } W, \tag{2.12}$$

where $Y_W(v, x)$ is still called the vertex operator *associated with v, such that the following conditions hold:*
truncation condition: *For every $v \in V$ and $w \in W$*

$$v_n^W w = 0 \tag{2.13}$$

for $n \in \mathbb{Z}$ sufficiently large;
vacuum property:

$$Y_W(\mathbf{1}, x) = 1_W; \tag{2.14}$$

Virasoro algebra conditions: *Let*

$$L_W(n) = \omega_{n+1}^W \text{ for } n \in \mathbb{Z}, \quad \text{i.e.,} \quad Y_W(\omega, x) = \sum_{n \in \mathbb{Z}} L_W(n) x^{-n-2}.$$

We have

$$[L_W(m), L_W(n)] = (m - n)L_W(m + n) + c_V \frac{m^3 - m}{12} \delta_{m+n,0} 1_W,$$
$$L_W(0)v = (\text{wt } v)v$$

for every homogeneous element $v \in W$, and

$$Y_W(L(-1)u, x) = \frac{d}{dx} Y_W(u, x) ; \qquad (2.15)$$

Jacobi identity:

$$x_0^{-1} \delta \left(\frac{x_1 - x_2}{x_0} \right) Y_W(u, x_1) Y_W(v, x_2) - x_0^{-1} \delta \left(\frac{x_2 - x_1}{-x_0} \right) Y_W(v, x_2) Y_W(u, x_1)$$

$$= x_2^{-1} \delta \left(\frac{x_1 - x_0}{x_2} \right) Y_W(Y(u, x_0)v, x_2). \qquad (2.16)$$

From the Jacobi identity (2.16), one can derive the weak commutativity and weak associativity relations, respectively:

$$(x_1 - x_2)^{k(u,v)} Y_W(u, x_1) Y_W(v, x_2) = (x_1 - x_2)^{k(u,v)} Y_W(v, x_2) Y_W(u, x_1)$$

$$(2.17)$$

$$(x_0 + x_2)^{l(u,w)} Y_W(u, x_0 + x_2) Y_W(v, x_2)w = (x_0 + x_2)^{l(u,w)} Y_W(Y(u, x_0)v, x_2)w,$$

$$(2.18)$$

where $u, v \in V$ and $w \in W$, valid for large enough nonnegative integers $k(u, v)$ and $l(u, w)$, their minimum value depending respectively on u, v and on u, w. For definiteness, we will pick the integers $k(u, v)$ and $l(u, w)$ to be the smallest nonnegative integers for which the relations above are valid.

2.2. Twisted modules for vertex operator algebras

The notion of twisted module for a vertex operator algebra was formalized in [FFR] and [D] (see also the geometric formulation in [FrS]; see also [DLM]), summarizing the basic properties of the actions of twisted vertex operators discovered in [FLM1], [FLM2] and [L2]; the main nontrivial axiom in this notion is the twisted Jacobi identity of [FLM2] (and [L2]); cf. [FLM1].

A critical ingredient in formal calculus needed in the theory of twisted modules is the appearance of fractional powers of formal variables, like $x^{1/p}$, $p \in \mathbb{Z}_+$ (the positive integers). For the purpose of formal calculus, the object $x^{1/p}$ is to be treated as a new formal variable whose p–th power is x. The binomial expansion convention is applied as stated at the beginning of Section 2 to binomials of the type $(x_1 + x_2)^{1/p}$. From a geometrical point of view, these rules correspond to choosing a branch in the "orbifold structure" described (locally) by the twisted vertex operator algebra module.

We now fix a positive integer p and a primitive p–th root of unity

$$\omega_p \in \mathbb{C}. \tag{2.19}$$

We record here two important properties of the formal delta–function involving fractional powers of formal variables:

$$\delta(x) = \frac{1}{p} \sum_{r=0}^{p-1} \delta(\omega_p^r x^{1/p}) \tag{2.20}$$

and

$$x_2^{-1} \delta \left(\omega_p^r \left(\frac{x_1 - x_0}{x_2} \right)^{1/p} \right) = x_1^{-1} \delta \left(\omega_p^{-r} \left(\frac{x_2 + x_0}{x_1} \right)^{1/p} \right). \tag{2.21}$$

The latter formula can be found (in a slightly different form) in [Li2]. For the sake of completeness, we present here a proof.

Proof: The coefficient of x_0^0 in equation (2.21) is immediate. Consider some formal series $f(x) = \sum_{n \in \mathbb{C}} f_n x^n$, $f_n \in \mathbb{C}$. From the formula

$$(-1)^k (\partial/\partial x_1)^k \left(x_1^s x_2^{-s-1} \right) = (\partial/\partial x_2)^k \left(x_1^{s-k} x_2^{-s-1+k} \right)$$

for any $s \in \mathbb{C}$ and k a nonnegative integer, we find that

$$(-1)^k \left(\frac{\partial}{\partial x_1} \right)^k \left(x_2^{-1} f(x_1/x_2) \right) = \left(\frac{\partial}{\partial x_2} \right)^k \left(x_1^{-1} (x_1/x_2)^{1-k} f(x_1/x_2) \right). \tag{2.22}$$

With $f(x) = \delta \left(\omega_p^r x^{1/p} \right)$, we use the formal delta-function property to get

$$(x_1/x_2)^{1-k} \delta \left(\omega_p^r (x_1/x_2)^{1/p} \right) = \delta \left(\omega_p^r (x_1/x_2)^{1/p} \right)$$

and thus

$$(-1)^k \left(\frac{\partial}{\partial x_1} \right)^k \left(x_2^{-1} \delta \left(\omega_p^r (x_1/x_2)^{1/p} \right) \right) = \left(\frac{\partial}{\partial x_2} \right)^k \left(x_1^{-1} \delta \left(\omega_p^{-r} (x_2/x_1)^{1/p} \right) \right). \tag{2.23}$$

Summing over nonnegative integers k with the coefficients $x_0^k / k!$ on both sides, we obtain (2.21). ∎

Recall the vertex operator algebra $(V, Y, \mathbf{1}, \omega)$ with central charge c_V of the previous subsection. Fix an automorphism ν of period p of the vertex operator algebra V, that is, a linear automorphism of the vector space V preserving ω and $\mathbf{1}$ such that

$$\nu Y(v, x) \nu^{-1} = Y(\nu v, x) \text{ for } v \in V, \tag{2.24}$$

and

$$\nu^p = 1_V. \tag{2.25}$$

Definition 2.3. *A ($\mathbb{Q}$-graded) ν-**twisted** V-**module** M is a $\mathbb{Q}$-graded vector space,*

$$M = \coprod_{n \in \mathbb{Q}} M_{(n)}; \ \text{for} \ v \in M_{(n)}, \ \text{wt} \ v = n,$$

such that

$$M_{(n)} = 0 \ \text{for} \ n \ \text{sufficiently negative},$$
$$\dim M_{(n)} < \infty \ \text{for} \ n \in \mathbb{Q},$$

equipped with a linear map

$$Y_M(\cdot, x) \ : \ V \to (\text{End} \ M)\big[\big[x^{1/p}, x^{-1/p}\big]\big]$$
$$v \mapsto Y_M(v, x) = \sum_{n \in \frac{1}{p}\mathbb{Z}} v_n^\nu x^{-n-1}, \ \ v_n^\nu \in \text{End} \ M, \quad (2.26)$$

where $Y_M(v, x)$ is called the twisted vertex operator *associated with v, such that the following conditions hold:*
truncation condition: *For every $v \in V$ and $w \in M$*

$$v_n^\nu w = 0 \tag{2.27}$$

for $n \in \frac{1}{p}\mathbb{Z}$ sufficiently large;
vacuum property:

$$Y_M(\mathbf{1}, x) = 1_M; \tag{2.28}$$

Virasoro algebra conditions: *Let*

$$L_M(n) = \omega_{n+1}^\nu \ \text{for} \ n \in \mathbb{Z}, \quad \text{i.e.,} \quad Y_M(\omega, x) = \sum_{n \in \mathbb{Z}} L_M(n) x^{-n-2}.$$

We have

$$[L_M(m), L_M(n)] = (m-n) L_M(m+n) + c_V \frac{m^3 - m}{12} \delta_{m+n,0} 1_M,$$
$$L_M(0)v = (\text{wt} \ v)v \tag{2.29}$$

for every homogeneous element v, and

$$Y_M(L(-1)u, x) = \frac{d}{dx} Y_M(u, x) \ ; \tag{2.30}$$

Jacobi identity:

$$x_0^{-1}\delta\left(\frac{x_1-x_2}{x_0}\right)Y_M(u,x_1)Y_M(v,x_2) - x_0^{-1}\delta\left(\frac{x_2-x_1}{-x_0}\right)Y_M(v,x_2)Y_M(u,x_1)$$

$$= \frac{1}{p}x_2^{-1}\sum_{r=0}^{p-1}\delta\left(\omega_p^r\left(\frac{x_1-x_0}{x_2}\right)^{1/p}\right)Y_M(Y(\nu^r u,x_0)v,x_2). \tag{2.31}$$

Note that when restricted to the fixed–point subalgebra $\{u \in V \mid \nu u = u\}$, a twisted module becomes a *true* module: the twisted Jacobi identity (2.31) reduces to the untwisted one (2.16), by (2.20). This will enable us to construct natural representations of a certain infinite-dimensional algebra $\hat{\mathcal{D}}^+$ (see Section 7) on suitable twisted modules.

3. Heisenberg vertex operator algebra and its twisted modules

It is appropriate at this point to make these definitions more substantial by giving a simple but important example of a vertex operator algebra, and of some of its twisted modules.

3.1. Heisenberg vertex operator algebra

Following [FLM2], let $\mathfrak{h}$ be a finite-dimensional abelian Lie algebra (over $\mathbb{C}$) of dimension d on which there is a nondegenerate symmetric bilinear form $\langle\cdot,\cdot\rangle$. Let ν be an isometry of $\mathfrak{h}$ of period $p > 0$:

$$\langle\nu\alpha,\nu\beta\rangle = \langle\alpha,\beta\rangle, \quad \nu^p\alpha = \alpha$$

for all $\alpha, \beta \in \mathfrak{h}$. Consider the affine Lie algebra $\hat{\mathfrak{h}}$,

$$\hat{\mathfrak{h}} = \coprod_{n\in\mathbb{Z}}\mathfrak{h}\otimes t^n \oplus \mathbb{C}C,$$

with the commutation relations

$$[\alpha\otimes t^m, \beta\otimes t^n] = \langle\alpha,\beta\rangle m\delta_{m+n,0}C \quad (\alpha,\beta\in\mathfrak{h}, \ m,n\in\mathbb{Z})$$

$$[C,\hat{\mathfrak{h}}] = 0.$$

Set

$$\hat{\mathfrak{h}}^+ = \coprod_{n>0}\mathfrak{h}\otimes t^n, \qquad \hat{\mathfrak{h}}^- = \coprod_{n<0}\mathfrak{h}\otimes t^n.$$

The subalgebra

$$\hat{\mathfrak{h}}^+ \oplus \hat{\mathfrak{h}}^- \oplus \mathbb{C}C$$

is a Heisenberg Lie algebra. Form the induced (level–one) $\hat{\mathfrak{h}}$–module

$$S = \mathcal{U}(\hat{\mathfrak{h}}) \otimes_{\mathcal{U}(\hat{\mathfrak{h}}^+ \oplus \mathfrak{h} \oplus \mathbb{C}C)} \mathbb{C} \simeq S(\hat{\mathfrak{h}}^-) \quad \text{(linearly)},$$

where $\hat{\mathfrak{h}}^+ \oplus \mathfrak{h}$ acts trivially on $\mathbb{C}$ and C acts as 1; $\mathcal{U}(\cdot)$ denotes universal enveloping algebra and $S(\cdot)$ denotes the symmetric algebra. Then S is irreducible under the Heisenberg algebra $\hat{\mathfrak{h}}^+ \oplus \hat{\mathfrak{h}}^- \oplus \mathbb{C}C$. We will use the notation $\alpha(n)$ ($\alpha \in \mathfrak{h}$, $n \in \mathbb{Z}$) for the action of $\alpha \otimes t^n \in \hat{\mathfrak{h}}$ on S.

The induced $\hat{\mathfrak{h}}$-module S carries a natural structure of vertex operator algebra. This structure is constructed as follows (cf. [FLM2]). First, one identifies the vacuum vector as the element 1 in S: $\mathbf{1} = 1$. Consider the following formal series acting on S:

$$\alpha(x) = \sum_{n \in \mathbb{Z}} \alpha(n) x^{-n-1} \quad (\alpha \in \mathfrak{h}).$$

Then, the vertex operator map $Y(\cdot, x)$ is given by

$$Y(\alpha_1(-n_1) \cdots \alpha_j(-n_j)\mathbf{1}, x)$$
$$= \; :\frac{1}{(n_1 - 1)!} \left(\frac{d}{dx}\right)^{n_1-1} \alpha_1(x) \cdots \frac{1}{(n_j - 1)!} \left(\frac{d}{dx}\right)^{n_j-1} \alpha_j(x) : \quad (3.1)$$

for $\alpha_k \in \mathfrak{h}$, $n_k \in \mathbb{Z}_+$, $k = 1, 2, \ldots, j$, for all $j \in \mathbb{N}$, where $: \cdot : $ is the usual normal ordering, which brings $\alpha(n)$ with $n > 0$ to the right. Choosing an orthonormal basis $\{\bar{\alpha}_q | q = 1, \ldots, d\}$ of $\mathfrak{h}$, the conformal vector is $\omega = \frac{1}{2} \sum_{q=1}^{d} \bar{\alpha}_q(-1)\bar{\alpha}_q(-1)\mathbf{1}$. This implies in particular that the weight of $\alpha(-n)\mathbf{1}$ is n:

$$L(0)\alpha(-n)\mathbf{1} = n\alpha(-n)\mathbf{1} \quad (\alpha \in \mathfrak{h}, \; n \in \mathbb{Z}_+)$$

where we used

$$L(0) = \frac{1}{2} \sum_{n \in \frac{1}{p}\mathbb{Z}} \sum_{q=1}^{d} : \bar{\alpha}_q(n)\bar{\alpha}_q(-n) : .$$

The isometry ν on $\mathfrak{h}$ lifts naturally to an automorphism of the vertex operator algebra S, which we continue to call ν, of period p.

Then (cf. [FLM2]), the various properties of a vertex operator algebra are indeed satisfied by the quadruplet $(V, Y, \mathbf{1}, \omega)$ just defined.

3.2. Twisted modules

We now proceed as in [L1], [FLM1], [FLM2] and [DL2] to construct a space $S[\nu]$ that carries a natural structure of ν–twisted module for the vertex operator algebra S. In these papers, the twisted module structure was constructed assuming the minor hypothesis that ν preserves a rational lattice in $\mathfrak{h}$. We show in Section 6 that the space $S[\nu]$ is a twisted module, without the need for this minor assumption.

Consider a primitive p–th root of unity ω_p. For $r \in \mathbb{Z}$ set

$$\mathfrak{h}_{(r)} = \{\alpha \in \mathfrak{h} \mid \nu\alpha = \omega_p^r\alpha\} \subset \mathfrak{h}.$$

For $\alpha \in \mathfrak{h}$, denote by $\alpha_{(r)}$, $r \in \mathbb{Z}$, its projection on $\mathfrak{h}_{(r)}$. Define the ν-twisted affine Lie algebra $\hat{\mathfrak{h}}[\nu]$ associated with the abelian Lie algebra $\mathfrak{h}$ by

$$\hat{\mathfrak{h}}[\nu] = \coprod_{n\in\frac{1}{p}\mathbb{Z}} \mathfrak{h}_{(pn)} \otimes t^n \oplus \mathbb{C}C \tag{3.2}$$

with

$$[\alpha \otimes t^m, \beta \otimes t^n] = \langle \alpha, \beta\rangle m\delta_{m+n,0} C \left(\alpha \in \mathfrak{h}_{(pn)},\ \beta \in \mathfrak{h}_{(pm)},\ m, n \in \frac{1}{p}\mathbb{Z}\right)$$

$$[C, \hat{\mathfrak{h}}[\nu]] = 0. \tag{3.3}$$

Set

$$\hat{\mathfrak{h}}[\nu]^+ = \coprod_{n>0} \mathfrak{h}_{(pn)} \otimes t^n, \qquad \hat{\mathfrak{h}}[\nu]^- = \coprod_{n<0} \mathfrak{h}_{(pn)} \otimes t^n. \tag{3.4}$$

The subalgebra

$$\hat{\mathfrak{h}}[\nu]^+ \oplus \hat{\mathfrak{h}}[\nu]^- \oplus \mathbb{C}C \tag{3.5}$$

is a Heisenberg Lie algebra. Form the induced (level-one) $\hat{\mathfrak{h}}[\nu]$-module

$$S[\nu] = \mathcal{U}(\hat{\mathfrak{h}}[\nu]) \otimes_{\mathcal{U}(\hat{\mathfrak{h}}[\nu]^+\oplus\mathfrak{h}_{(0)}\oplus\mathbb{C}C)} \mathbb{C} \simeq S(\hat{\mathfrak{h}}[\nu]^-) \quad \text{(linearly)}, \tag{3.6}$$

where $\hat{\mathfrak{h}}[\nu]^+ \oplus \mathfrak{h}_{(0)}$ acts trivially on $\mathbb{C}$ and C acts as 1; $\mathcal{U}(\cdot)$ denotes universal enveloping algebra. Then $S[\nu]$ is irreducible under the Heisenberg algebra $\hat{\mathfrak{h}}[\nu]^+ \oplus \hat{\mathfrak{h}}[\nu]^- \oplus \mathbb{C}C$. We will use the notation $\alpha^\nu(n)$ $\left(\alpha \in \mathfrak{h}_{(pn)},\ n \in \frac{1}{p}\mathbb{Z}\right)$ for the action of $\alpha \otimes t^n \in \hat{\mathfrak{h}}[\nu]$ on $S[\nu]$.

Remark 3.1. *The special case where* $p = 1$ $(\nu = 1_\mathfrak{h})$ *corresponds to the* $\hat{\mathfrak{h}}$-*module S.*

The $\hat{\mathfrak{h}}[\nu]$-module $S[\nu]$ is naturally a ν–twisted module for the vertex operator algebra S. One first constructs the following formal series acting on $S[\nu]$:

$$\alpha^\nu(x) = \sum_{n\in\frac{1}{p}\mathbb{Z}} \alpha^\nu(n)x^{-n-1}, \tag{3.7}$$

as well as the formal series $W(v, x)$ for all $v \in S$:

$$W(\alpha_1(-n_1) \cdots \alpha_j(-n_j)\mathbf{1}, x)$$

$$= \: \frac{1}{(n_1 - 1)!} \left(\frac{d}{dx}\right)^{n_1 - 1} \alpha_1^v(x) \cdots \frac{1}{(n_j - 1)!} \left(\frac{d}{dx}\right)^{n_j - 1} \alpha_j^v(x) \: \quad (3.8)$$

where $\alpha_k \in \mathfrak{h}$, $n_k \in \mathbb{Z}_+$, $k = 1, 2, \ldots, j$, for all $j \in \mathbb{N}$. The twisted vertex operator map $Y_{S[v]}(\cdot, x)$ acting on $S[v]$ is then given by

$$Y_{S[v]}(v, x) = W(e^{\Delta_x} v, x) \quad (v \in S) \quad (3.9)$$

where Δ_x is a certain formal operator involving the formal variable x [FLM1], [FLM2], [DL2]. This operator is trivial on $\alpha(-n)\mathbf{1} \in S$ ($n \in \mathbb{Z}_+$), so that one has in particular

$$Y_{S[v]}(\alpha(-n)\mathbf{1}, x) = \frac{1}{(n - 1)!} \left(\frac{d}{dx}\right)^{n-1} \alpha^v(x). \quad (3.10)$$

One crucial (among others) role of the formal operator Δ_x is to make the fixed–point subalgebra $\{u \mid vu = u\}$ act according to a true module action. For instance, the conformal vector ω is in the fixed–point subalgebra, so that the vertex operator $Y_{S[v]}(\omega, x)$ generates a representation of the Virasoro algebra on the space $S[v]$. This representation of the Virasoro algebra was explicitly constructed in [DL2]. As one can see in the results of [DL2] and as will become clear below, the resulting representation $\mathrm{Res}_x \, x Y_{S[v]}(\omega, x)$ of the Virasoro generator $L(0)$ is not an (infinite) sum of normal-ordered products the type $\sum_{n \in \frac{1}{p}\mathbb{Z}} \: \alpha^v(n)\beta^v(-n) \:$; rather, there is an extra term proportional to the identity on $S[v]$, the so-called correction term, which appears because of the operator Δ_x. The correction term was calculated in [DL2] using the explicit action of e^{Δ_x} on ω. In the case of the period–2, $v = -1$ automorphism, this action is given by [FLM1], [FLM2]:

$$e^{\Delta_x}\omega = \omega + \frac{1}{16}(\dim \mathfrak{h})x^{-2},$$

and for general automorphism, the calculation was carried out in [DL2] (see also [FFR] and [FLM2]). These results are relevant, for instance, in the construction of the moonshine module [FLM2].

The calculation of the action of Δ_x on arbitrary elements of S is, however, a much more complicated task. Below we will derive some identities among twisted vertex operators. One of the important consequences of these identities, for us, will be to give a tool to explicitly construct the twisted vertex operators associated to elements of S from the knowledge of the twisted vertex operators associated to "simpler" elements, without requiring the explicit knowledge of Δ_x. In fact, these identities allow us to construct recursively twisted vertex operators associated to all elements of S and to compute Δ_x, only starting from the knowledge that Δ_x is trivial on $\alpha(-n)\mathbf{1} \in S$ ($n \in \mathbb{Z}_+$).

4. Commutativity and associativity properties

This section follows closely similar sections of [DLMi1] and [DLMi2], and reproduces the results and some of the proofs. We first recall the main commutativity and associativity properties of vertex operators in the context of modules ([FLM2], [FHL], [DL1], [Li1]; cf. [LL]), and then we derive other identities somewhat analogous to these. These other identities were stated and proven in [DLMi2], and the most important ones were stated in [DLMi1]. All these identities will be generalized to twisted modules, still following [DLMi1, DLMi2]. Note that taking the module to be the vertex operator algebra V itself, the relations below specialize to commutativity and associativity properties in vertex operator algebras. We will give the proofs of the simplest identities only, referring the reader to [DLMi2] for all the proofs. Throughout this and the next sections, we fix a vertex operator algebra V and a V-automorphism ν of period p, $\nu^p = 1_V$.

4.1. Formal commutativity and associativity for untwisted modules

We already stated the weak commutativity relation (2.17) and the weak associativity relation (2.18). They imply the main "formal" commutativity and associativity properties of vertex operators, which, along with the fact that these properties are equivalent to the Jacobi identity, can be formulated as follows (see [LL]):

Theorem 4.1. *Let W be a vector space (not assumed to be graded) equipped with a linear map $Y_W(\cdot, x)$ (2.12) such that the truncation condition (2.13) and the Jacobi identity (2.16) hold. Then for u, $v \in V$ and $w \in W$, there exist $k(u, v) \in \mathbb{N}$ and $l(u, w) \in \mathbb{N}$ and a (nonunique) element $F(u, v, w; x_0, x_1, x_2)$ of $W((x_0, x_1, x_2))$ such that*

$$x_0^{k(u,v)} F(u, v, w; x_0, x_1, x_2) \in W[[x_0]]((x_1, x_2)),$$
$$x_1^{l(u,w)} F(u, v, w; x_0, x_1, x_2) \in W[[x_1]]((x_0, x_2)) \tag{4.1}$$

and

$$Y_W(u, x_1)Y_W(v, x_2)w = F(u, v, w; x_1 - x_2, x_1, x_2),$$
$$Y_W(v, x_2)Y_W(u, x_1)w = F(u, v, w; -x_2 + x_1, x_1, x_2),$$
$$Y_W(Y(u, x_0)v, x_2)w = F(u, v, w; x_0, x_2 + x_0, x_2) \tag{4.2}$$

(where we are using the binomial expansion convention). Conversely, let W be a vector space equipped with a linear map $Y_W(\cdot, x)$ (2.12) such that the truncation condition (2.13) and the statement above hold, except that $k(u, v)$ ($\in \mathbb{N}$) and $l(u, w)$ ($\in \mathbb{N}$) may depend on all three of u, v and w. Then the Jacobi identity (2.16) holds.

It is important to note that since $k(u, v)$ can be (and typically is) greater than 0, the formal series $F(u, v, w; x_1 - x_2, x_1, x_2)$ and $F(u, v, w; -x_2 + x_1, x_1, x_2)$ are not in general equal. Along with (4.1), the first two equations of (4.2) represent formal commutativity, while the first and last equations of (4.2) represent formal associativity, as formulated in [LL] (see also [FLM2] and [FHL]). The twisted generalization of this theorem, written below, was proven in [DLMi2].

4.2. *Additional relations in untwisted modules*

From the equations in Theorem 4.1, we can derive a number of relations similar to weak commutativity and weak associativity but involving formal limit procedures (the meaning of such formal limit procedures is recalled below). Even though only one of these will be of use in the following sections, we state here for completeness of the discussion the two relations that are not "easy" consequences of weak commutativity and weak associativity. These relations were proven in [DLMi2]; we report the proofs here.

The first relation can be expressed as follows:

Theorem 4.2. *With W as in Theorem 4.1,*

$$\lim_{x_0 \to -x_2 + x_1} \left((x_0 + x_2)^{l(u,w)} Y_W(Y(u, x_0)v, x_2)w \right) = x_1^{l(u,w)} Y_W(v, x_2)Y(u, x_1)w$$

(4.3)

for u, v $\in V$.

The meaning of the formal limit

$$\lim_{x_0 \to -x_2 + x_1} \left((x_0 + x_2)^{l(u,w)} Y_W(Y(u, x_0)v, x_2)w \right) \qquad (4.4)$$

is that one replaces each power of the formal variable x_0 in the formal series $(x_0 + x_2)^{l(u,w)} Y_W(Y(u, x_0)v, x_2)w$ by the corresponding power of the formal series $-x_2 + x_1$ (defined using the binomial expansion convention). Notice again that the order of $-x_2$ and x_1 is important in $-x_2 + x_1$, according to the binomial expansion convention.

Proof of Theorem 4.2: Apply the limit $\lim_{x_0 \to -x_2 + x_1}$ to the expression

$$(x_0 + x_2)^{l(u,w)} Y_W(Y(u, x_0)v, x_2)w$$

written as in the right–hand side of the third equation of (4.2). This limit is well defined; indeed, the only possible problems are the negative powers of $x_2 + x_0$ in $F(u, v, w, x_0, x_2 + x_0, x_2)$, but they are cancelled out by the factor $(x_0 + x_2)^{l(u,w)}$. The resulting expression is read off the second relation of (4.2). ∎

Remark 4.2. *It is instructive to consider the following relation, deceptively similar to (4.3), but that is in fact an immediate consequence of weak associativity (2.18):*

$$\lim_{x_0 \to x_1 - x_2} \left((x_0 + x_2)^{l(u,w)} Y_W(Y(u, x_0)v, x_2)w \right) = x_1^{l(u,w)} Y_W(u, x_1) Y_W(v, x_2)w.$$

$$(4.5)$$

More precisely, it can be obtained by noticing that the replacement of x_0 by $x_1 - x_2$ independently in each factor in the expression as written on the left–hand side of (2.18) is well defined. We emphasize that, by contrast, the relation (4.3) cannot be obtained in such a manner. Indeed, although the formal limit procedure $\lim_{x_0 \to -x_2 + x_1}$ is of course well defined on the series on both sides of (2.18), one cannot replace x_0 by $-x_2 + x_1$ either in the factor $Y_W(u, x_0 + x_2) Y_W(v, x_2)w$ on the left–hand side or in the factor $Y_W(Y(u, x_0)v, x_2)w$ on the right–hand side of (2.18).

The second nontrivial relation, which we call *modified weak associativity*, will be important when generalized to twisted modules. It was first written in [DLMi1]. It is stated as:

Theorem 4.3. *With W as in Theorem 4.1,*

$$\lim_{x_1 \to x_2 + x_0} \left((x_1 - x_2)^{k(u,v)} Y_W(u, x_1) Y_W(v, x_2) \right) = x_0^{k(u,v)} Y_W(Y(u, x_0)v, x_2)$$

$$(4.6)$$

for $u, v \in V$.

Proof: Apply the limit $\lim_{x_1 \to x_2 + x_0}$ to the expression

$$(x_1 - x_2)^{k(u,v)} Y_W(u, x_1) Y_W(v, x_2)$$

written as in the right–hand side of the first equation of (4.2). This limit is well defined, since negative powers of $x_1 - x_2$ in $F(u, v, w; x_1 - x_2, x_1, x_2)$ are cancelled out by the factor $(x_1 - x_2)^{k(u,v)}$. The resulting expression is read off the third relation of (4.2). ∎

Remark 4.3. *Equation (4.6) can be written in the following form:*

$$\lim_{x_1 \to x_2 + x_0} \left(\left(\frac{x_1 - x_2}{x_0} \right)^{k(u,v)} Y_W(u, x_1) Y_W(v, x_2) \right) = Y_W(Y(u, x_0)v, x_2).$$

$$(4.7)$$

The factor $\left(\frac{x_1 - x_2}{x_0} \right)^{k(u,v)}$ *appearing in front of the product of two vertex opera-tors on the left–hand side is crucial in giving a well–defined limit, but when the limit is applied to this factor without the product of vertex operators, the result is simply* 1. *We will call such a factor a "resolving factor". Its power is apparent, in particular, in the proof of the main commutator formula (6.1) of [DLMi2]: it allows one to evaluate nontrivial limits of sums of terms with cancelling "singularities" in a straightforward fashion, evaluating the limit of each term independently. Its power will also be clear, in the present paper, when constructing the twisted vertex operator map* $Y_{S[v]}(\cdot, x)$ *and when studying the algebra* $\hat{D}^+$ *defined in Section 7.*

4.3. Formal commutativity and associativity for twisted modules

We derive below various commutativity and associativity properties of twisted vertex operators. In order to express some of these properties, we need one more element of formal calculus: a certain projection operator (see [DLMi2]). Consider the operator $P_{[[x_0, x_0^{-1}]]}$ acting on the space $\mathbb{C}\{x_0\}$ of formal series with any complex powers of x_0, which projects to the formal series with integral powers of x_0:

$$P_{[[x_0, x_0^{-1}]]} : \mathbb{C}\{x_0\} \to \mathbb{C}[[x_0, x_0^{-1}]] . \qquad (4.8)$$

We will extend the meaning of this notation in the obvious way to projections acting on formal series with coefficients lying in vector spaces other than $\mathbb{C}$, vector spaces which might themselves be spaces of formal series in other formal variables. Notice that when this projection operator acts on a formal series in x_0 with powers that are in $\frac{1}{p}\mathbb{Z}$, for instance on $f(x_0) \in \mathbb{C}[[x_0^{1/p}, x_0^{-1/p}]]$, it can be described by an explicit formula:

$$P_{[[x_0, x_0^{-1}]]} f(x_0) = \frac{1}{p} \sum_{r=0}^{p-1} \left(\lim_{x^{1/p} \to \omega_p^r x_0^{1/p}} f(x) \right) .$$

(See Remark 4.4 below for the meaning of formal limit procedures involv-ing fractional powers of formal variables.) We will also extend this projection notation to different kinds of formal series in obvious ways. For instance,

$$P_{x_0^{q/p}[[x_0, x_0^{-1}]]} : \mathbb{C}\{x_0\} \to \mathbb{C} x_0^{q/p} [[x_0, x_0^{-1}]].$$

Again, of course, we will extend the meaning of this notation to formal series with coefficients in vector spaces other than $\mathbb{C}$.

The twisted Jacobi identity (2.31) implies twisted versions of weak commutativity and weak associativity ($u, v \in V, w \in M$):

$$(x_2 - x_1)^k Y_M(v, x_2) Y_M(u, x_1) = (x_2 - x_1)^k Y_M(u, x_1) Y_M(v, x_2) \tag{4.9}$$

$$P_{[[x_0, x_0^{-1}]]} \left((x_0 + x_2)^l Y_M(u, x_0 + x_2) Y_M(v, x_2) w \right)$$

$$= (x_2 + x_0)^l \frac{1}{p} \sum_{r=0}^{p-1} \omega_p^{-lrp} Y_M(Y(v^r u, x_0) v, x_2) w. \tag{4.10}$$

These relations are valid for all large enough $k \in \mathbb{N}$ and $l \in \frac{1}{p}\mathbb{N}$, their minimum value depending respectively on u, v and on u, w. For definiteness, we will denote these minimum values by $k(u, v)$ and $l(u, w)$, respectively (they depend also on the module M; in particular, they differ from the integer numbers $k(u, v)$ and $l(u, w)$ used in the previous subsection in connection with the module W). As in the untwisted case, these relations imply the main "formal" commutativity and associativity properties of twisted vertex operators [Li2], which, along with with the fact that these properties are equivalent to the Jacobi identity, can be formulated as follows (it was first formulated in this form in [DLMi1]):

Theorem 4.4. *Let M be a vector space (not assumed to be graded) equipped with a linear map $Y_M(\cdot, x)$ (2.26) such that the truncation condition (2.27) and the Jacobi identity (2.31) hold. Then for $u, v \in V$ and $w \in M$, there exist $k(u, v) \in \mathbb{N}$ and $l(u, w) \in \frac{1}{p}\mathbb{N}$ and a (nonunique) element $F(u, v, w; x_0, x_1, x_2)$ of $M((x_0, x_1^{1/p}, x_2^{1/p}))$ such that*

$$x_0^{k(u,v)} F(u, v, w; x_0, x_1, x_2) \in M[[x_0]]((x_1^{1/p}, x_2^{1/p})),$$

$$x_1^{l(u,w)} F(u, v, w; x_0, x_1, x_2) \in M[[x_1^{1/p}]]((x_0, x_2^{1/p})) \tag{4.11}$$

and

$$Y_M(u, x_1) Y_M(v, x_2) w = F(u, v, w; x_1 - x_2, x_1, x_2),$$

$$Y_M(v, x_2) Y_M(u, x_1) w = F(u, v, w; -x_2 + x_1, x_1, x_2),$$

$$Y_M(Y(v^{-s} u, x_0) v, x_2) w = \lim_{x_1^{1/p} \to \omega_p^s (x_2 + x_0)^{1/p}} F(u, v, w; x_0, x_1, x_2) \tag{4.12}$$

for $s \in \mathbb{Z}$ (where we are using the binomial expansion convention). Conversely, let M be a vector space equipped with a linear map $Y_M(\cdot, x)$ (2.26) *such that the truncation condition (2.13) and the statement above hold, except that $k(u, v)$ ($\in \mathbb{N}$) and $l(u, w)$ ($\in \frac{1}{p}\mathbb{N}$) may depend on all three of u, v and w. Then the Jacobi identity (2.31) holds.*

This theorem, as well as (4.9) and (4.10), were proven in [DLMi2].

Remark 4.4. *Formal limit procedures involving fractional powers of formal variables like $x_1^{1/p}$ have the same meaning as in (4.4), but with $x_1^{1/p}$ being treated as a formal variable by itself. For instance, the formal limit procedure*

$$\lim_{x_1^{1/p} \to \omega_p^s (x_2 + x_0)^{1/p}} F(u, v, w; x_0, x_1, x_2)$$

above means that one replaces each integral power of the formal variable $x_1^{1/p}$ in the formal series $F(u, v, w; x_0, x_1, x_2)$ by the corresponding power of the formal series $\omega_p^s (x_2 + x_0)^{1/p}$ (defined using the binomial expansion convention).

Remark 4.5. *Note that this theorem, and in particular its proof in [DLMi2], illustrates the phenomenon, which arises again and again throughout the theory of vertex operator algebras, that formal calculus inherently involves just as much "analysis" as "algebra": in many relations there are integers that can be left unspecified, except for their minimum values, and the proof involves taking these integers "large enough". Recall that essentially the same issues arose for example in the use of formal calculus for the proof of the Jacobi identity for (twisted) vertex operators in [FLM2] (see Chapters 8 and 9). This is certainly not surprising, since we are using the Jacobi identity (for all twisting automorphisms) in order to prove, in a different approach, properties of (twisted) vertex operators.*

Along with (4.11), the first two equations of (4.12) represent what we call *formal commutativity* for twisted vertex operators, while the first and last equations of (4.12) represent *formal associativity* for twisted vertex operators. When specialized to the untwisted case $p = 1$ ($\nu = 1_V$), these two relations lead respectively to the usual formal commutativity and formal associativity for vertex operators, as described in (4.2).

4.4. Additional relations in twisted modules

As in the case of ordinary vertex operators, one can write other relations involving formal limit procedures. These relations were proven in [DLMi2]; we report the proofs here. Among them, two cannot be directly obtained from weak commutativity and weak associativity. One of these, the relation generalizing (4.3), is stated as follows:

Theorem 4.5. *With M as in Theorem 4.4,*

$$\lim_{x_0 \to -x_2+x_1} \left((x_2 + x_0)^l \frac{1}{p} \sum_{r=0}^{p-1} \omega_p^{-lrp} Y_M(Y(v^r u, x_0)v, x_2)w \right)$$

$$= P_{[[x_1, x_1^{-1}]]} \left(x_1^l Y_M(v, x_2) Y_M(u, x_1)w \right), \tag{4.13}$$

for all $l \in \frac{1}{p}\mathbb{Z}$, $l \geq l(u, w)$.

Proof: This is proved along the lines of the proof of Theorem 4.2, with some additions due to the fractional powers. One uses the third equation of (4.12) in order to rewrite the left–hand side of (4.13) as

$$\lim_{x_0 \to -x_2+x_1} \left((x_2 + x_0)^l \frac{1}{p} \sum_{r=0}^{p-1} \omega_p^{-lrp} \lim_{x_3^{1/p} \to \omega_p^{-r}(x_2+x_0)^{1/p}} F(u, v, w; x_0, x_3, x_2) \right).$$

The sum over r keeps only the terms in which $x_2 + x_0$ is raised to a power which has a fractional part equal to the negative of the fractional part of l. Multiplying by $(x_2 + x_0)^l$, for any $l \in \frac{1}{p}\mathbb{Z}$, $l \geq l(u, w)$, brings the remaining series to a series with finitely many negative powers of x_2 (as well as x_0), to which it is possible to apply the limit $\lim_{x_0 \to -x_2+x_1}$. This limit of course brings only integer powers of x_1, and the right–hand side of (4.13) can be obtained from the second equation of (4.12). ∎

Remark 4.6. *A relation similar to the last one, but that is a direct consequence of weak associativity (4.10), is*

$$\lim_{x_0 \to x_1-x_2} \left((x_2 + x_0)^l \frac{1}{p} \sum_{r=0}^{p-1} \omega_p^{-lrp} Y_M(Y(v^r u, x_0)v, x_2)w \right)$$

$$= P_{[[x_1, x_1^{-1}]]} \left(x_1^l Y_M(u, x_1) Y_M(v, x_2)w \right) \tag{4.14}$$

for all $l \in \frac{1}{p}\mathbb{Z}$, $l \geq l(u, w)$. This generalizes (4.5) (see the comments in Remark 4.2). It can be obtained by applying the formal limit involved in the left–hand side to both sides of (4.10).

The most important relation for our purposes, which was first stated in [DLMi1], generalizing (4.6) and which we call *modified weak associativity* for twisted vertex operators, is given by the following theorem:

Theorem 4.6. *With M as in Theorem 4.4,*

$$\lim_{x_1^{1/p} \to \omega_p^s(x_2+x_0)^{1/p}} \left((x_1 - x_2)^{k(u,v)} Y_M(u, x_1) Y_M(v, x_2) \right)$$

$$= x_0^{k(u,v)} Y_M(Y(v^{-s} u, x_0)v, x_2) \tag{4.15}$$

for u, v ∈ V and s ∈ Z.

Proof: The proof is a straightforward generalization of the proof of Theorem 4.3. ∎

Remark 4.7. *The specialization of Theorems 4.4 and 4.6 to the untwisted case $p = 1$ and $M = W$ gives, respectively, Theorems 4.1 and 4.3.*

Finally, we derive a simple relation, proved in [Li2], that specifies the structure of the formal series $Y_M(u, x)$.

Theorem 4.7. *With M as in Theorem 4.4,*

$$\lim_{x_1^{1/p} \to \omega_p^s x^{1/p}} Y_M(v^s u, x_1) = Y_M(u, x) \tag{4.16}$$

for $u \in V$ and $s \in Z$.

Proof: In the Jacobi identity (2.31), replace u by $v^s u$ and $x_1^{1/p}$ by $\omega_p^s x^{1/p}$. The right–hand side becomes

$$\frac{1}{p} x_2^{-1} \sum_{r=0}^{p-1} \delta \left(\omega_p^{r+s} \left(\frac{x - x_0}{x_2} \right)^{1/p} \right) Y_M(Y(v^{r+s} u, x_0)v, x_2),$$

which is independent of s, as is apparent if we make the shift in the summation variable $r \mapsto r - s$. Hence the left–hand side is also independent of s. Choosing $v = 1$ and using the vacuum property (2.28), this gives

$$\left(x_0^{-1} \delta \left(\frac{x - x_2}{x_0} \right) - x_0^{-1} \delta \left(\frac{x_2 - x}{-x_0} \right) \right) \lim_{x_1^{1/p} \to \omega_p^s x^{1/p}} Y_M(v^s u, x_1) \tag{4.17}$$

$$= \left(x_0^{-1} \delta \left(\frac{x - x_2}{x_0} \right) - x_0^{-1} \delta \left(\frac{x_2 - x}{-x_0} \right) \right) Y_M(u, x)$$

which, upon using (2.2) and taking Res_{x_2}, gives (4.16). ∎

From Theorem 4.7, we directly have the following corollary:

Corollary 4.8. *With M as in Theorem 4.4,*

$$Y_M(u, x) = \sum_{n \in Z + q/p} u_n x^{-n-1} \quad \text{for } u \in V, \ vu = \omega_p^q u, \ q \in Z.$$

5. Equivalence and construction theorems
from modified weak associativity

This section presents new results related to modified weak associativity. We will show, loosely speaking, that modified weak associativity (4.15) and weak commutativity (4.9) are equivalent to the Jacobi identity (2.31) ("equivalence theorem"). Then we will show that if the modified weak associativity for twisted vertex operators is valid for all pairs u, w (to be put in (4.15) instead of the ordered pair u, v) with $u \in U$ and $w \in V$, where U is a generating subset of a vertex operator algebra V, and if weak commutativity for twisted vertex operators is valid for all pairs u, v with u, $v \in U$, then both modified weak associativity and weak commutativity hold for the whole vertex operator algebra V ("construction theorem"). Similar construction theorems were proved by Li in the untwisted and twisted cases [Li1, Li2] (cf. [LL]), using the powerful idea of "local systems of (twisted) vertex operators." Here we start from similar ideas but we make use of modified weak associativity discovered in [DLMi1, DLMi2], in order to illustrate one of its applications. We find it instructive to give a direct proof of our construction theorem, although there may be a shorter route from Li's construction theorems. In the next section, our two theorems will allow us to construct in a relatively simple way the twisted vertex operator map for $S[\nu]$ – in particular, we will show how the use of modified weak associativity gives a new explicit form for the operator Δ_x – and to prove the twisted module structure for $S[\nu]$. We recall that throughout this section, V is a vertex operator algebra and ν is an automorphism of V.

5.1. Equivalence theorem

It is well known in the theory of vertex operator algebras that, under natural conditions, the weak commutativity relation (2.17) and the weak associativity relation (2.18) for untwisted modules are equivalent to the Jacobi identity (2.16). It is a simple matter to show that this statement is also true when weak associativity is replaced by modified weak associativity. We state this more generally for twisted modules (and for the twisted Jacobi identity (2.31)) in the following theorem.

Theorem 5.1. *Let M be a vector space (not assumed to be graded) equipped with a linear map $Y_M(\cdot, x)$ (2.26) such that the truncation condition (2.27) hold.*

If modified weak associativity (4.15) holds, for some $k'(u, v) \in \mathbb{N}$, and weak commutativity (4.9) holds, for a possibly different $k(u, v) \in \mathbb{N}$, then we may change $k'(u, v)$ (in particular, we may lower it) to $k'(u, v) = k(u, v)$, and the twisted Jacobi identity (2.31) holds. Also, the vector space M is a twisted module if, additionally, M is $\mathbb{Q}$ − graded and quasi-finite as in the first lines of Definition 2.3, with the $L(0)$-weight property (2.29), and the vacuum property (2.28) holds.

On the other hand, if the twisted Jacobi identity (2.31) holds, then modified weak associativity (4.15) and weak commutativity (4.9) hold.

Remark 5.2. *Note that this theorem can be specialized to $p = 1$, applying then to modules W. It can also be used to show that some vector space V is a vertex operator algebra; more precisely, in the definition 2.1, the Jacobi identity can be replaced by modified weak associativity and weak commutativity.*

Proof of Theorem 5.1. The last sentence of the theorem was proven already by proving modified weak associativity and weak commutativity above.

Weak commutativity (4.9), when both sides are applied on an element w of M, and the truncation condition immediately imply the first equation of (4.11), and the first two equations of (4.12). Then, the limit on the left-hand side of (4.15) with $k(u, v)$ replaced by $k''(u, v)$ certainly exists for all $k''(u, v) \geq k(u, v)$, and the result is the same for any $k''(u, v)$ that makes the limit exist, up to the obvious power of x_0 (because if all limits exist, then the product of the limits is the limit of the product). Since we know that this limit gives the right-hand side for some $k'(u, v)$, we may change it (and lower it if necessary) to $k'(u, v) = k(u, v)$. Then, writing the product of twisted vertex operators on the left-hand side of (4.15) as on the right-hand side of the first equation of (4.12), we see that (4.15) implies the third equation of (4.12). Hence, by Theorem 4.4, the twisted Jacobi identity holds. With the additional conditions stated in the theorem, all other parts of the definition (2.3) are satisfied (in particular, the fact that V is a vertex operator algebra implies that the Virasoro commutation relations are also satisfied, and the $L(-1)$-derivative property for vertex operator algebra (2.8) implies, using (4.15) with $v = 1$, the corresponding property for twisted modules (2.30)) and M is a twisted module for V. ∎

5.2. Construction theorem

Consider a generating subset $U \subset V$ of the vertex operator algebra V, defined as follows.

Definition 5.3. *A generating subset $U \subset V$ of a vertex operator algebra is a subset such that all elements of V can be written as linear combinations of elements of the form $u_m v_n \cdots 1$ for $u, v, \ldots \in U$ and $m, n, \ldots \in \mathbb{Z}$.*

Suppose we are able to define a twisted vertex operator map $Y_M(\cdot, x)$ as in (2.26) such that the truncation condition (2.27) holds for all $v \in V$, weak commutativity (4.9) holds for all $u, v \in U$, and modified weak associativity (4.15) holds for all $u \in U$ and $v \in V$. Theorem 5.5 below along with Theorem 5.1 tell us that this is enough to have the twisted Jacobi identity, and, with additional mild conditions (see Theorem 5.1) to have a twisted module.

Remark 5.4. *Theorem 5.5 of course requires us to already have a twisted vertex operator map $Y_M(\cdot, x)$ on the full vertex operator algebra V. On the other hand, once we have the map on U only (with the properties above), it is easy to extend it to a map on the set of symbols of the type $u_m v_n \cdots w$ for $u, v, \ldots, w \in U$ and $m, n, \ldots \in \mathbb{Z}$ by recursive use of modified weak associativity (4.15) with $u \in U$ and v an element in this set of symbols. However, the vertex operator algebra V is the span of this set of symbols with linear relations amongst them; these relations depend on the particular vertex operator algebra at hand. In order to have a twisted module, it is essential to verify that the map on this set of symbols is well-defined on V; that is, that it is in agreement with these linear relations. This is extremely nontrivial, and there does not seem to be yet a general theorem as to when that happens. Whatever these relations are, they have to imply those coming from the Jacobi identity (since V is a vertex operator algebra). It is possible to show that all linear relations coming from the Jacobi identity are satisfied by this construction. This is beyond the scope of the present paper, but we hope to clarify some of these issues in a future work.*

Note again that the theorem below and the issues discussed in the remark above are in close relation with results of Li [Li2]; cf. [LL].

Theorem 5.5. *Let M be a vector space (not assumed to be graded) equipped with a linear map $Y_M(\cdot, x)$ (2.26) such that the truncation condition (2.27) holds. Fix a generating subset $U \subset V$, as defined in Definition 5.3.*

If weak commutativity (4.9) is satisfied for all $u, v \in U$ (and fix $k(u, v) \in \mathbb{N}$ to be the lowest integer that can be taken both in (4.9) and in weak commutativity for the vertex operator algebra V), and if modified weak associativity (4.15) is satisfied for all $u \in U$ and for all $v \in V$ (for some $k'(u, v) \in \mathbb{N}$ that may be different from $k(u, v)$), then both weak commutativity and modified weak associativity are satisfied for all $u, v \in V$. Further, we may change $k'(u, v)$ (in particular, we may lower it) to $k'(u, v) = k(u, v)$ for $u, v \in V$, and these integers may be taken to satisfy the formula

$$k(u_{n_0+m}v, w) = k(w, u) + k(w, v) + m \tag{5.1}$$

for all $u, v, w \in V$ where n_0 is the highest integer such that $u_{n_0}v \neq 0$. The same integers may also be taken in weak commutativity for the vertex operator algebra V.

Proof: We start by showing weak commutativity (4.9), and the equation (5.1). Under the assumptions of the theorem, we have, for $u, v, w \in U$,

$$(x - x_2 - x_0)^{k(w,u)}(x - x_2)^{k(w,v)} Y_M(w, x)$$

$$\cdot \lim_{x_1^{1/p} \to \omega_p^s (x_2 + x_0)^{1/p}} \left((x_1 - x_2)^{k(u,v)} Y_M(u, x_1) Y_M(v, x_2) \right)$$

$$= \lim_{x_1^{1/p} \to \omega_p^s (x_2 + x_0)^{1/p}} \left((x - x_1)^{k(w,u)}(x - x_2)^{k(w,v)} Y_M(w, x) \right.$$

$$\left. \cdot (x_1 - x_2)^{k(u,v)} Y_M(u, x_1) Y_M(v, x_2) \right)$$

$$= \lim_{x_1^{1/p} \to \omega_p^s (x_2 + x_0)^{1/p}} \left((x_1 - x_2)^{k(u,v)} Y_M(u, x_1) Y_M(v, x_2) \right.$$

$$\left. \cdot (x - x_1)^{k(w,u)}(x - x_2)^{k(w,v)} Y_M(w, x) \right)$$

$$= \lim_{x_1^{1/p} \to \omega_p^s (x_2 + x_0)^{1/p}} \left((x_1 - x_2)^{k(u,v)} Y_M(u, x_1) Y_M(v, x_2) \right)$$

$$\cdot (x - x_2 - x_0)^{k(w,u)}(x - x_2)^{k(w,v)} Y_M(w, x) \qquad (5.2)$$

hence

$$(x - x_2 - x_0)^{k(w,u)}(x - x_2)^{k(w,v)} Y_M(w, x) Y_M(Y(v^{-s}u, x_0)v, x_2) \qquad (5.3)$$

$$= (x - x_2 - x_0)^{k(w,u)}(x - x_2)^{k(w,v)} Y_M(Y(v^{-s}u, x_0)v, x_2) Y_M(w, x) \ .$$

Both sides have finitely many negative powers of x_0. Take the lowest power:

$$(x - x_2)^{k(w,u)+k(w,v)} Y_M(w, x) Y_M((v^{-s}u)_{n_0}v, x_2)$$

$$= (x - x_2)^{k(w,u)+k(w,v)} Y_M((v^{-s}u)_{n_0}v, x_2) Y_M(w, x) \ . \qquad (5.4)$$

This proves weak commutativity for all pairs $u_{n_0}v, w$ with $u, v, w \in U$, and we may take $k(u_{n_0}v, w) = k(w, u) + k(w, v)$ (although we have not proven that this value is the minimum one for which weak commutativity is valid), where n_0 is the highest integer such that $u_{n_0}v \neq 0$.

The next power of x_0 in the equation (5.3) contains, on each side, two terms of the same type as those above: one with $(v^{-s}u)_{n_0+1}v$, the other with $(v^{-s}u)_{n_0}v$. The terms with $(v^{-s}u)_{n_0}v$ are multiplied by $(x - x_2)^{k(w,u)+k(w,v)-1}$, and those with $(v^{-s}u)_{n_0+1}v$, by $(x - x_2)^{k(w,u)+k(w,v)}$. Multiplying the resulting equation through by $x - x_2$, we can use the result (5.4) and we obtain weak commutativity for all pairs $u_{n_0+1}v, w$ with $u, v, w \in U$, where n_0 is again highest integer such that $u_{n_0}v \neq 0$. We can take $k(u_{n_0+1}v, w) = k(w, u) + k(w, v) + 1$. Repeating the process, we obtain weak commutativity for all pairs u_nv, w with $u, v, w \in U$ and for all $n \in \mathbb{Z}$, with $k(u_{n_0+m}v, w) = k(w, u) + k(w, v) + m$. Replacing v by u_nv for $u, v \in U$ in the argument above, and repeating, we obtain weak commutativity for all pairs v, w with $v \in V$ and $w \in U$. Finally, we can repeat the full argument above with $u, v \in U$ and $w \in V$. The induction from that gives us weak commutativity for all pairs v, w with $v \in V$ and $w \in V$, and in particular gives us (5.1).

Note that the same arguments can be used for the vertex operator $Y(\cdot, x)$, so that equation (5.1) may be taken to hold also for the $k(u, v)$ involved in weak commutativity for the vertex operator algebra V. This implies that the integers $k(u, v)$ involved in weak commutativity for $Y_M(\cdot, x)$ and those involved in weak commutativity for the vertex operator algebra V may both be taken to be given by (5.1) for all elements of the vertex operator algebra. We will do that for the rest of the proof.

Note also that weak commutativity (4.9) implies, through the arguments of the proof of Theorem 5.1, that in modified weak associativity (4.15) for $u \in U$ and $v \in V$, which holds by assumption with some integer $k'(u, v)$, we can change $k'(u, v)$ to $k'(u, v) = k(u, v)$. We will also do that for the rest of the proof.

Next we show modified weak associativity. First note the following lemma.

Lemma 5.6. *For $u, v, w \in V$ and $\phi \in M$, there exists a (nonunique) element $F(u, v, w, \phi; x_0, x_4, x_5, x_1, x_2, x_3)$ of $M((x_0, x_4, x_5, x_1^{1/p}, x_2^{1/p}, x_3^{1/p}))$ such that*

$$x_0^{k(u,v)} F(u, v, w, \phi; x_0, x_4, x_5, x_1, x_2, x_3) \in M[[x_0]]\left((x_4, x_5, x_1^{1/p}, x_2^{1/p}, x_3^{1/p})\right),$$

$$x_4^{k(u,w)} F(u, v, w, \phi; x_0, x_4, x_5, x_1, x_2, x_3) \in M[[x_4]]\left((x_0, x_5, x_1^{1/p}, x_2^{1/p}, x_3^{1/p})\right),$$

$$x_5^{k(v,w)} F(u, v, w, \phi; x_0, x_4, x_5, x_1, x_2, x_3) \in M[[x_5]]\left((x_0, x_4, x_1^{1/p}, x_2^{1/p}, x_3^{1/p})\right)$$

$$\tag{5.5}$$

and

$$Y_M(u, x_1)Y_M(v, x_2)Y_M(w, x_3)\phi = F(u, v, w, \phi; x_1 - x_2, x_1 - x_3,$$

$$x_2 - x_3, x_1, x_2, x_3) \tag{5.6}$$

where $k(u, v), k(u, w), k(v, w) \in \mathbb{N}$ can be taken as those that appear in the weak commutativity relation (4.9).

This is an immediate consequence of weak commutativity (4.9). Indeed, consider

$$(x_1 - x_2)^{k(u,v)}(x_1 - x_3)^{k(u,w)}(x_2 - x_3)^{k(v,w)} Y_M(u, x_1)Y_M(v, x_2)Y_M(w, x_3)\phi .$$

Thanks to the factor $(x_1 - x_2)^{k(u,v)}(x_1 - x_3)^{k(u,w)}(x_2 - x_3)^{k(v,w)}$, the twisted vertex operators may be written in any order. Looking at different orders and using the truncation property, we see the expression has finitely many negative powers of x_3, finitely many negative powers of x_2 and finitely many negative powers of x_1. This shows the lemma.

Now, take $u, v, \in U$ and $w \in V$, and again $\phi \in M$. From modified weak associativity for vertex operator algebra (4.6) (recall that the module W in that equation may be replaced by the vertex operator algebra itself V), which we will need only for the pair u, v, and from modified weak associativity for twisted modules, which is valid by assumption when, in (4.15), the (ordered) pair u, v is replaced by the pair v, w as well as when it is replaced by the pair $u, v_n w$, we have

$$Y_M(Y(Y(u, x_0)v, x_5)w, x_3)\phi$$

$$= \lim_{x_4 \to x_5 + x_0} \left(\frac{x_4 - x_5}{x_0} \right)^{k(u,v)} Y_M(Y(u, x_4)Y(v, x_5)w, x_3)$$

$$= \lim_{x_4 \to x_5 + x_0} \left(\frac{x_4 - x_5}{x_0} \right)^{k(u,v)} \sum_{n \in \mathbb{Z}} x_5^{-n-1} Y_M(Y(u, x_4)v_n w, x_3)\phi$$

$$= \lim_{x_4 \to x_5 + x_0} \left(\frac{x_4 - x_5}{x_0} \right)^{k(u,v)} \sum_{n \in \mathbb{Z}} x_5^{-n-1} \lim_{x_1^{\frac{1}{p}} \to (x_2 + x_4)^{\frac{1}{p}}} \left(\frac{x_1 - x_3}{x_4} \right)^{k(u, v_n w)}$$

$$\cdot Y_M(u, x_1) Y_M(v_n w, x_3)\phi$$

$$= \lim_{x_4 \to x_5 + x_0} \left(\frac{x_4 - x_5}{x_0} \right)^{k(u,v)} \sum_{n \in \mathbb{Z}} \lim_{x_1^{\frac{1}{p}} \to (x_3 + x_4)^{\frac{1}{p}}} \left(\frac{x_1 - x_3}{x_4} \right)^{k(u, v_n w)}$$

$$\cdot P_{x_5^{-n-1}} Y_M(u, x_1) Y_M(Y(v, x_5)w, x_3)\phi$$

where $P_{x_5^{-n-1}}$ projects onto the formal series with only the terms having the factor x_5^{-n-1}. Observe that the lowest power of x_5 is $-n_0 - 1$ where n_0 is the highest integer such that $v_{n_0} w \neq 0$. Continuing, we find

$$Y_M(Y(Y(u, x_0)v, x_5)w, x_3)\phi$$

$$= \lim_{x_4 \to x_5 + x_0} \left(\frac{x_4 - x_5}{x_0} \right)^{k(u,v)} \sum_{n \in \mathbb{Z}} \lim_{x_1^{\frac{1}{p}} \to (x_3 + x_4)^{\frac{1}{p}}} \left(\frac{x_1 - x_3}{x_4} \right)^{k(u, v_n w)}$$

$$\cdot P_{x_5^{-n-1}} \lim_{x_2^{\frac{1}{p}} \to (x_3 + x_5)^{\frac{1}{p}}} \left(\frac{x_2 - x_3}{x_5} \right)^{k(v,w)} Y_M(u, x_1) Y_M(v, x_2) Y_M(w, x_3)\phi$$

$$= \lim_{x_4 \to x_5 + x_0} \left(\frac{x_4 - x_5}{x_0} \right)^{k(u,v)} \sum_{n \in \mathbb{Z}} \lim_{x_1^{\frac{1}{p}} \to (x_3 + x_4)^{\frac{1}{p}}} \left(\frac{x_1 - x_3}{x_4} \right)^{k(u, v_n w)}$$

$$\cdot P_{x_5^{-n-1}} \lim_{x_2^{\frac{1}{p}} \to (x_3+x_5)^{\frac{1}{p}}} \left(\frac{x_2 - x_3}{x_5}\right)^{k(v,w)} F(u, v, w, \phi; x_1 - x_2, x_1 - x_3,$$

$$x_2 - x_3, x_1, x_2, x_3)$$

$$= \lim_{x_4 \to x_5+x_0} \left(\frac{x_4 - x_5}{x_0}\right)^{k(u,v)} \sum_{n \in \mathbb{Z}} \lim_{x_1^{\frac{1}{p}} \to (x_3+x_4)^{\frac{1}{p}}} \left(\frac{x_1 - x_3}{x_4}\right)^{k(u,v_n w)}$$

$$\cdot P_{x_5^{-n-1}} F(u, v, w, \phi; x_1 - x_3 - x_5, x_1 - x_3, x_5, x_1, x_3 + x_5, x_3).$$

Using the particular value of $k(u, v_n w)$ given by (5.1), we can continue evaluating the limits:

$$Y_M(Y(Y(u, x_0)v, x_5)w, x_3)\phi$$

$$= \lim_{x_4 \to x_5+x_0} \left(\frac{x_4 - x_5}{x_0}\right)^{k(u,v)} F(u, v, w, \phi; x_4 - x_5, x_4, x_5, x_3 + x_4, x_3 + x_5, x_3)$$

$$= F(u, v, w, \phi; x_0, x_5 + x_0, x_5, x_3 + x_5 + x_0, x_3 + x_5, x_3). \tag{5.7}$$

Now, consider the following formal series, and use similar arguments as those above in order to evaluate it:

$$\sum_{n \in \mathbb{Z}} x_0^{-n-1} \lim_{x_2^{1/p} \to (x_3+x_5)^{1/p}} \left(\frac{x_2 - x_3}{x_5}\right)^{k(u_n v,w)} Y_M(u_n v, x_2) Y_M(w, x_3)\phi$$

$$= \sum_{n \in \mathbb{Z}} \lim_{x_2^{1/p} \to (x_3+x_5)^{1/p}} \left(\frac{x_2-x_3}{x_5}\right)^{k(u_n v,w)} P_{x_0^{-n-1}} Y_M(Y(u, x_0)v, x_2) Y_M(w, x_3)\phi$$

$$= \sum_{n \in \mathbb{Z}} \lim_{x_2^{1/p} \to (x_3+x_5)^{1/p}} \left(\frac{x_2 - x_3}{x_5}\right)^{k(u_n v,w)}$$

$$\cdot P_{x_0^{-n-1}} \lim_{x_1^{1/p} \to (x_2+x_0)^{1/p}} \left(\frac{x_1 - x_2}{x_0}\right)^{k(u,v)} Y_M(u, x_1) Y_M(v, x_2) Y_M(w, x_3)\phi$$

$$= \sum_{n \in \mathbb{Z}} \lim_{x_2^{1/p} \to (x_3+x_5)^{1/p}} \left(\frac{x_2 - x_3}{x_5}\right)^{k(u_n v,w)}$$

$$\cdot P_{x_0^{-n-1}} \lim_{x_1^{1/p} \to (x_2+x_0)^{1/p}} \left(\frac{x_1 - x_2}{x_0}\right)^{k(u,v)} F(u, v, w, \phi; x_1 - x_2, x_1 - x_3,$$

$$x_2 - x_3, x_1, x_2, x_3)$$

$$= \sum_{n \in \mathbb{Z}} \lim_{x_2^{1/p} \to (x_3+x_5)^{1/p}} \left(\frac{x_2 - x_3}{x_5}\right)^{k(u_n v,w)}$$

$$\cdot P_{x_0^{-n-1}} F(u, v, w, \phi; x_0, x_2 + x_0 - x_3, x_2 - x_3, x_2 + x_0, x_2, x_3)$$

$$= F(u, v, w, \phi; x_0, x_5 + x_0, x_5, x_3 + x_5 + x_0, x_3 + x_5, x_3). \tag{5.8}$$

Comparing with (5.7), and taking any fixed power of x_0, we have

$$Y_M(Y(u_n v, x_5)w, x_3) = \lim_{x_2^{1/p} \to (x_3 + x_5)^{1/p}} \left(\frac{x_2 - x_3}{x_5}\right)^{k(u_n v, w)} Y_M(u_n v, x_2) Y_M(w, x_3)\phi.$$

(5.9)

This is modified weak associativity (4.15) for $s = 0$ and for the pairs $u_n v, w$, for all $n \in \mathbb{Z}$. It is a simple matter to compare, instead, the expression

$$Y_M(Y(v^{-s} Y(u, x_0)v, x_5)w, x_3)\phi$$

with the expression

$$\sum_{n \in \mathbb{Z}} x_0^{-n-1} \lim_{x_2^{1/p} \to \omega_p^s (x_3 + x_5)^{1/p}} \left(\frac{x_2 - x_3}{x_5}\right)^{k(u_n v, w)} Y_M(u_n v, x_2) Y_M(w, x_3)\phi$$

using similar steps, and using the fact that v is an automorphism of V. We obtain modified weak associativity (4.15) for arbitrary $s \in \mathbb{Z}$ and for the pairs $u_n v, w$, for all $n \in \mathbb{Z}$.

Repeating the argument with v replaced by $\tilde{u}_n v$ for all $\tilde{u} \in U$ and for all $n \in \mathbb{Z}$, and so on, we find modified weak associativity for all pairs v, w with $v \in V$ and $w \in V$. This proves the last part of the theorem. ∎

Remark 5.8. *As usual, it is possible to specialize the theorems above to the case $p = 1$ in order to obtain theorems applying to untwisted modules.*

Remark 5.9. *In the theorem above (in close relation with theorems of [Li2]), concerned with twisted modules for vertex operators algebras, we assumed the existence of a vertex operator algebra V. With slight adjustments, the theorem can be made into a construction theorem for vertex operator algebras themselves, in relation with constructions of [Li1] (a more complete construction theory, the representation theory of vertex operator algebras of [Li1], is explained at length in [LL], cf. Theorems 5.7.6, 5.7.11 for instance).*

6. Proof of the twisted module structure for $S[v]$ and construction of the twisted vertex operator map

The twisted module structure of $S[v]$, for the vertex operator S, was established in [L1], [FLM1], [FLM2] and [DL2] (assuming that v preserves a rational lattice in $\mathfrak{h}$). In this section, we present a new proof of the twisted module structure of $S[v]$ (which does not require this minor assumption), and we construct explicitly the twisted vertex operators using modified weak associativity (and in particular, we calculate the explicit form of Δ_x in a simple way).

Starting from (3.10), we will construct the twisted vertex operator $Y_{S[v]}(u, x)$ for all $u \in S$. As stated above, it is usual in studying vertex operator algebras to construct vertex operators associated to elements of the vertex operator algebra from vertex operator associated to "simpler" elements, by using associativity. However, for twisted vertex operators, the natural weak associativity (4.10), that is immediately obtained from the Jacobi identity, is somewhat complicated by the projection operator and hides the simple structure of the construction. On the other hand, the modified weak associativity (4.15) is simpler, especially when written in the form (specialized for convenience to $s = 0$)

$$Y_M(Y(u, x_0)v, x_2) = \lim_{x_1^{1/p} \to (x_2+x_0)^{1/p}} \left(\left(\frac{x_1 - x_2}{x_0} \right)^{k(u,v)} Y_M(u, x_1)Y_M(v, x_2) \right).$$

(6.1)

In order to better understand this formula, note that, as remarked in Remark 4.3, the pre-factor $\left(\frac{x_1-x_2}{x_0} \right)^{k(u,v)}$ gives exactly 1 when the limit procedure $\lim_{x_1^{1/p} \to (x_2+x_0)^{1/p}}$ is applied on it alone. The formula above cannot be simplified by replacing this pre-factor by 1, however, because the limit is not well defined on the other factor $Y_M(u, x_1)Y_M(v, x_2)$ alone. The construction principle will be to replace this product of vertex operators by a normal-ordered product plus extra terms. On the normal-ordered product, the limit is well defined, so that the pre-factor multiplying the normal-ordered product can be set to 1.

It is immediate to see that the set $U = \{1, \alpha(-1)1 | \alpha \in \mathfrak{h}\}$ is a generating subset for the vertex operator algebra S. Hence, in the proof of the twisted module structure of $S[v]$ using Theorems 5.5 and 5.1, we need first to prove weak commutativity (4.9) (and it turns out that it holds with $k(u, v) = 2$ – the integer $k(u, v) = 2$ is the one involved in the corresponding weak commutativity for the vertex operator algebra S) for the operators defined by (3.10) with $n = 1$. Then, we need to construct explicitly the map $Y_{S[v]}(\cdot, x)$ for all elements of S recursively by using modified weak associativity. Finally, we need to check modified weak associativity (4.15) for $u \in U$ and $v \in V$. All other requirements of Theorem 5.1 are immediate to see from the construction of Section 3.

Theorem 6.1. *The operators (3.10) with* $n = 1$ *satisfy weak commutativity (4.9) with* $k(u, v) = 2$.

Proof: We have

$$\alpha^v(x_1)\beta^v(x_2) = {}^{\bullet}_{\bullet}\alpha^v(x_1)\beta^v(x_2){}^{\bullet}_{\bullet} + h(\alpha, \beta, x_1, x_2)$$

(6.2)

with

$$h(\alpha, \beta, x_1, x_2) = \sum_{m \in \frac{1}{p}\mathbb{Z}, m > 0} x_1^{-m-1} x_2^{m-1} m \langle \alpha_{(mp)}, \beta \rangle$$

$$= \sum_{r=1}^{p} \left(\frac{x_2}{x_1}\right)^{\frac{r}{p}} \frac{1 - \frac{r}{p} + \frac{r}{p}\frac{x_1}{x_2}}{(x_1 - x_2)^2} \langle \alpha_{(r)}, \beta \rangle . \tag{6.3}$$

It is a simple matter to see that

$$(x_1 - x_2)^2 h(\alpha, \beta, x_1, x_2) = \sum_{r=1}^{p} \left(\frac{x_2}{x_1}\right)^{\frac{r}{p}} \left(1 - \frac{r}{p} + \frac{r}{p}\frac{x_1}{x_2}\right) \langle \alpha_{(r)}, \beta \rangle . \tag{6.4}$$

is equal to $(x_1 - x_2)^2 h(\beta, \alpha, x_2, x_1)$, using $\langle \beta_{(r)}, \alpha \rangle = \langle \alpha_{(-r)}, \beta \rangle$. Hence

$$(x_1 - x_2)^2 [\alpha^v(x_1), \beta^v(x_2)] = 0$$

which proves the theorem.   ∎

Theorem 6.2. *The space $S[v]$ has the structure of a twisted module for the vertex operator algebra S. The general form of the twisted vertex operator $Y_{S[v]}(u, x)$, for any element u of the vertex operator algebra S, is:*

$$Y_{S[v]}(\alpha_j(-n_j) \cdots \alpha_1(-n_1)\mathbf{1}, x)$$

$$= \sum_{J \subset \{1, \dots, j\}} f_{\{1, \dots, j\} \setminus J}(x) \; : \prod_{l \in J} \frac{1}{(n_l - 1)!} \left(\frac{d}{dx}\right)^{n_l - 1} \alpha_l^v(x) : \tag{6.5}$$

with $n_1, \dots, n_j \in \mathbb{Z}_+$, for some factors $f_I(x)$, where we just write the dependence on the index set I, but that really depend on the elements α_i and the integer numbers n_i for all $i \in I$. The set J on which we sum takes the values $\varnothing$ (the empty set), $\{1, \dots, j\}$ (if it is different from $\varnothing$), and all other proper subsets of $\{1, \dots, j\}$ (if any). The factors $f_I(x)$ are given by

$$f_I(x) = \begin{cases} 0 & |I| \text{ odd} \\ \displaystyle\sum_{s \in \text{Pairings}(I)} \prod_{l=1}^{|I|/2} g_{s_l}(x) & |I| \text{ even} \end{cases} \tag{6.6}$$

where $|I|$ is the cardinal of I, where $\text{Pairings}(I)$ is the set of all distinct sets $s = \{s_1, \dots, s_{|I|/2}\}$ of distinct (without any element in common) pairs $s_l = (i_l, i_l')$ (where the order of elements is not important) of elements $i_l \neq i_l'$ of I such that $\{i_1, \dots, i_{|I|/2}, i_1', \dots, i_{|I|/2}'\} = I$, and where

$$g_{(i,i')}(x) = g(\alpha_i, n_i, \alpha_{i'}, n_{i'}, x) \tag{6.7}$$

with

$$g(\alpha, m, \beta, n, x) = \mathrm{Res}_{x_0} \mathrm{Res}_{x_2} x_0^{-m} x_2^{-n} \sum_{r=1}^{p} \left(\frac{x+x_2}{x+x_0} \right)^{\frac{r}{p}} \frac{1 - \frac{r}{p} + \frac{r}{p} \frac{x+x_0}{x+x_2}}{(x_0 - x_2)^2} \langle \alpha_{(r)}, \beta \rangle.$$

(6.8)

Proof: Clearly, for $j = 0$ and $j = 1$ the form (6.5) is consistent with modified weak associativity and well defined on S, and we must have $f_\varnothing(x) = 1$ and $f_I(x) = 0$ if $|I| = 1$. Assume (6.5) to be valid for j replaced with $j - 1$. With k a nonnegative integer large enough and $n_1, \ldots, n_j \in \mathbb{Z}_+$, we have

$$Y_{S[\nu]}(\alpha_j(-n_j) \cdots \alpha_1(-n_1)\mathbf{1}, x)$$

$$= \mathrm{Res}_{x_0} x_0^{-n_j} Y_{S[\nu]}(Y(\alpha_j(-1)\mathbf{1}, x_0)\alpha_{j-1}(-n_{j-1}) \cdots \alpha_1(-n_1)\mathbf{1}, x)$$

$$= \mathrm{Res}_{x_0} x_0^{-n_j} \lim_{x_1^{1/p} \to (x+x_0)^{1/p}} \left(\left(\frac{x_1 - x}{x_0} \right)^k \right.$$

$$\left. \cdot Y_{S[\nu]}(\alpha_j(-1)\mathbf{1}, x_1) Y_{S[\nu]}(\alpha_{j-1}(-n_{j-1}) \cdots \alpha_1(-n_1)\mathbf{1}, x) \right)$$

$$= \mathrm{Res}_{x_0} x_0^{-n_j} \lim_{x_1^{1/p} \to (x+x_0)^{1/p}} \left(\left(\frac{x_1 - x}{x_0} \right)^k \right.$$

$$\left. \cdot \alpha_j^\nu(x_1) \sum_{J \subset \{1,\ldots,j-1\}} f_{\{1,\ldots,j-1\} \setminus J}(x) : \prod_{l \in J} \frac{1}{(n_l - 1)!} \left(\frac{d}{dx} \right)^{n_l - 1} \alpha_l^\nu(x) : \right).$$

(6.9)

Now, using the commutation relations (3.3), it is a simple matter to obtain

$$\alpha_j^\nu(x_1) : \prod_{l \in J} \left(\frac{d}{dx} \right)^{n_l - 1} \alpha_l^\nu(x) :$$

$$= : \alpha_j^\nu(x_1) \prod_{l \in J} \left(\frac{d}{dx} \right)^{n_l - 1} \alpha_l^\nu(x) :$$

$$+ \sum_{i \in J} \left(\frac{\partial}{\partial x} \right)^{n_i - 1} h(\alpha_j, \alpha_i, x_1, x) : \prod_{l \in J \setminus \{i\}} \left(\frac{d}{dx} \right)^{n_l - 1} \alpha_l^\nu(x) :$$

with $h(\alpha, \beta, x_1, x_2)$ defined in (6.3). Note that, comparing with (6.8), we have

$$g(\alpha, m, \beta, n, x) = \mathrm{Res}_{x_0} x_0^{-m} \lim_{x_1^{1/p} \to (x+x_0)^{1/p}} \left(\left(\frac{x_1 - x}{x_0} \right)^k \mathrm{Res}_{x_2} x_2^{-n} h(\alpha, \beta, x_1, x+x_2) \right).$$

(6.10)

Using the relation

$$\frac{1}{(n-1)!} \left(\frac{d}{dx} \right)^{n-1} f(x) = \mathrm{Res}_{x_2} x_2^{-n} f(x + x_2)$$

(6.11)

for formal series $f(x)$ with finitely many negative powers of x and for $n \in \mathbb{Z}_+$, we can now evaluate the limit and the residue on the right-hand side of the last equality of (6.9):

$$Y_{S[v]}(\alpha_j(-n_j) \cdots \alpha_1(-n_1)\mathbf{1}, x)$$

$$= \sum_{J \subset \{1,\dots,j-1\}} f_{\{1,\dots,j-1\}\setminus J}(x) \cdot$$

$$\left(: \frac{1}{(n_j - 1)!} \left(\frac{d}{dx}\right)^{n_j - 1} \alpha_j^v(x) \prod_{l \in J} \frac{1}{(n_l - 1)!} \left(\frac{d}{dx}\right)^{n_l - 1} \alpha_l^v(x):\right.$$

$$\left. + \sum_{i \in J} g(\alpha_j, n_j, \alpha_i, n_i, x) : \prod_{l \in J\setminus\{i\}} \frac{1}{(n_l - 1)!} \left(\frac{d}{dx}\right)^{n_l - 1} \alpha_l^v(x):\right).$$

(6.12)

This is still of the form (6.5). Moreover, it is simple to understand, from comparing (6.12) with (6.9), that the solution (6.6) is correct.

This construction certainly gives us a map $Y_{S[v]}(\cdot, x)$ on the vector space $S \simeq S(\hat{\mathfrak{h}}^-)$. We need to verify that this map satisfies modified weak associativity (4.15). This requires three steps.

First, we need to check that operators on the right-hand side of (6.5) are independent of the order of the pairs $(\alpha_1, n_1), \dots, (\alpha_j, n_j)$ for any positive integers $n_1, \dots, n_j$ and any elements $\alpha_1, \dots, \alpha_j$ of $\mathfrak{h}$. The sum over pairings has this symmetry, and we need to check that

$$g(\alpha, m, \beta, n, x) = g(\beta, n, \alpha, m, x). \tag{6.13}$$

This is not immediately obvious, because, by the binomial expansion convention, $(x_0 - x_2)^{-2} \neq (x_2 - x_0)^{-2}$). This symmetry can indeed be checked:

$$g(\alpha, m, \beta, n, x) - g(\beta, n, \alpha, m, x)$$

$$= \mathrm{Res}_{x_0}\mathrm{Res}_{x_2} x_0^{-m} x_2^{-n} \sum_{r=1}^{p}$$

$$\cdot \left(\frac{x + x_2}{x + x_0}\right)^{\frac{r}{p}} \left(1 - \frac{r}{p} + \frac{r}{p}\frac{x + x_0}{x + x_2}\right)$$

$$\times \left(\frac{1}{(x_0 - x_2)^2} - \frac{1}{(x_2 - x_0)^2}\right) \langle \alpha_{(r)}, \beta \rangle \tag{6.14}$$

$$= \mathrm{Res}_{x_0}\mathrm{Res}_{x_2} x_0^{-m} x_2^{-n} \sum_{r=1}^{p}$$

$$\cdot \left(\frac{x + x_2}{x + x_0}\right)^{\frac{r}{p}} \left(1 - \frac{r}{p} + \frac{r}{p}\frac{x + x_0}{x + x_2}\right) x_0^{-1} \frac{\partial}{\partial x_2} \delta\left(\frac{x_2}{x_0}\right) \langle \alpha_{(r)}, \beta \rangle$$

$$= 0 \tag{6.15}$$

where in the last step, we moved the derivative $\frac{\partial}{\partial x_2}$ towards the left using Leibniz's rule, and we used the formal delta-function property and the fact that $m, n \in \mathbb{Z}_+$.

Second, we need to check that modified weak associativity (4.15) with $Y_{S[v]}(Y(\alpha_j(-1)\mathbf{1}, x_0)\alpha_{j-1}(-n_{j-1}) \cdots \alpha_1(-n_1)\mathbf{1}, x)$ for its left-hand side is in agreement, at negative powers of x_0, with

$$Y_{S[v]}(\alpha_j(n_j)\alpha_{j-1}(-n_{j-1}) \cdots \alpha_1(-n_1)\mathbf{1}, x) =$$
$$\sum_{i=1}^{j-1} n_j \delta_{n_j, n_i} \langle \alpha_j, \alpha_i \rangle Y_{S[v]}(\alpha_{j-1}(-n_{j-1}) \cdots \widehat{\alpha_i(-n_i)} \cdots \alpha_1(-n_1)\mathbf{1}, x)$$
$$(6.16)$$

for $n_j \in \mathbb{N}$ (the nonnegative integers) where $\widehat{\alpha_i(-n_i)}$ means that the operator $\alpha_i(-n_i)$ is omitted. Repeating the derivation (6.9), (6.12), we see that this is equivalent to requiring (with the definition (6.8))

$$g(\alpha, -m, \beta, n) = m\delta_{m,n}\langle \alpha, \beta \rangle \tag{6.17}$$

for $m \in \mathbb{N}, n \in \mathbb{Z}_+$. This is a consequence of the fact that

$$\left(\frac{x + x_2}{x + x_0} \right)^s \frac{1 - s + s\frac{x+x_0}{x+x_2}}{(x_0 - x_2)^2} = \frac{1}{(x_0 - x_2)^2} + \mathbb{C}[[x_0, x_2, x^{-1}]] \tag{6.18}$$

for any $s \in \mathbb{C}$. The quantity

$$\left(\frac{x + x_2}{x + x_0} \right)^s \frac{1 - s + s\frac{x+x_0}{x+x_2}}{(x_0 - x_2)^2} - \frac{1}{(x_0 - x_2)^2}$$

obviously has only nonnegative powers of x_2 and nonpositive powers of x. On the other hand, it is equal to

$$\left(\frac{x + x_2}{x + x_0} \right)^s \frac{1 - s + s\frac{x+x_0}{x+x_2}}{(x_2 - x_0)^2} - \frac{1}{(x_2 - x_0)^2} +$$
$$\left(\left(\frac{x + x_2}{x + x_0} \right)^s \left(1 - s + s\frac{x + x_0}{x + x_2} \right) - 1 \right) x_0^{-1} \frac{\partial}{\partial x_2} \delta \left(\frac{x_2}{x_0} \right).$$

The first two terms obviously have only nonnegative powers of x_0. The third term can be evaluated using Leibniz's rule and gives zero. This completes the proof of (6.18).

Third, the structure of the construction of Section 3 shows that (4.15) is valid as well for $s \neq 0$ (equation (4.16) is satisfied).

The other requirements of Theorems 5.1 can be checked from the construction of Section 3, and with Theorems 5.5 and 6.1, this completes the proof. ∎

Finally, let us mention that formula (6.5) with (6.6) immediately leads to the following formula for the operator Δ_x introduced in (3.9):

$$\Delta_x = \sum_{q_1,q_2=1}^{d} \sum_{m,n\in\mathbb{Z}_+} \frac{\bar{\alpha}_{q_1}(m)\bar{\alpha}_{q_2}(n)}{mn} g(\bar{\alpha}_{q_1}, m, \bar{\alpha}_{q_2}, n, x) \qquad (6.19)$$

where we recall that $\bar{\alpha}_q$, $q = 1, \ldots, d$ form an orthonormal basis of $\mathfrak{h}$. This was first constructed, in a different form, in [FLM1] and [FLM2].

7. The Lie algebra $\hat{\mathcal{D}}^+$

We will now apply modified weak associativity and the results of the previous section, that $S[\nu]$, constructed in Section 3, is a twisted module for the vertex operator algebra S, in order to study a certain infinite-dimensional Lie algebra $\hat{\mathcal{D}}^+$ and its representations. This follows closely the results of [DLMi1] and [DLMi2], and does not give new results with respect to these works.

Let $\mathcal{D}$ be the Lie algebra of formal differential operators on $\mathbb{C}^\times$ spanned by $t^n D^r$, where $D = t\frac{d}{dt}$ and $n \in \mathbb{Z}$, $r \in \mathbb{N}$ (the nonnegative integers). This Lie algebra has an essentially unique one-dimensional central extension $\hat{\mathcal{D}} = \mathbb{C}c \oplus \mathcal{D}$ (denoted in the physics literature by $\mathcal{W}_{1+\infty}$).

The representation theory of the highest weight modules of $\hat{\mathcal{D}}$ was initiated in [KR], where, among other things, the complete classification problem of the so-called *quasi-finite* representations[1] was settled. The detailed study of the representation theory of certain subalgebras of $\hat{\mathcal{D}}$ having properties related to those of certain infinite–rank "classical" Lie algebras was initiated in [KWY] along the lines of [KR]. In [Bl] and [M2], related Lie algebras (and superalgebras) are considered from different viewpoints. As in [DLMi1, DLMi2], we will follow these lines and concentrate on the Lie subalgebra $\hat{\mathcal{D}}^+$ described in [Bl] and recalled below.

View the elements $t^n D^r$ ($n \in \mathbb{Z}$, $r \in \mathbb{N}$) as generators of the central extension $\hat{\mathcal{D}}$. They can be taken to satisfy the following commutation relations (cf. [KR]):

$$[t^m f(D), t^n g(D)] =$$
$$t^{m+n}(f(D+n)g(D) - g(D+m)f(D)) + \Psi(t^m f(D), t^n g(D))c,$$

where f and g are polynomials and Ψ is the 2–cocycle (cf. [KR]) determined by

---

1 These are representations with finite-dimensional homogeneous subspaces.

$$\Psi(t^m f(D), t^n g(D)) = -\Psi(t^n g(D), t^m f(D)) = \delta_{m+n,0} \sum_{i=1}^{m} f(-i) g(m-i), \ m > 0.$$

We consider the Lie subalgebra $\mathcal{D}^+$ of $\mathcal{D}$ generated by the formal differential operators

$$L_n^{(r)} = (-1)^{r+1} D^r (t^n D) D^r, \tag{7.1}$$

where $n \in \mathbb{Z}$, $r \in \mathbb{N}$ [Bl]. The subalgebra $\mathcal{D}^+$ has an essentially unique central extension (cf. [N]) and this extension may be obtained by restriction of the 2–cocycle Ψ to $\mathcal{D}^+$. Let $\hat{\mathcal{D}}^+ = \mathbb{C}c \oplus \mathcal{D}^+$ be the nontrivial central extension defined via the slightly normalized 2–cocycle $-\frac{1}{2}\Psi$, and view the elements $L_n^{(r)}$ as elements of $\hat{\mathcal{D}}^+$. This normalization gives, in particular, the usual Virasoro algebra bracket relations

$$[L_m^{(0)}, L_n^{(0)}] = (m - n) L_{m+n}^{(0)} + \frac{m^3 - m}{12} \delta_{m+n,0} \, c. \tag{7.2}$$

In [Bl] Bloch discovered that the Lie algebra $\hat{\mathcal{D}}^+$ can be defined in terms of generators that lead to a simplification of the central term in the Lie bracket relations. Oddly enough, if we let

$$\bar{L}_n^{(r)} = L_n^{(r)} + \frac{(-1)^r}{2} \zeta(-1 - 2r) \delta_{n,0} c, \tag{7.3}$$

then the central term in the commutator

$$[\bar{L}_m^{(r)}, \bar{L}_n^{(s)}] = \sum_{i=\min(r,s)}^{r+s} a_i^{(r,s)}(m, n) \bar{L}_{m+n}^{(i)} + \frac{(r+s+1)!^2}{2(2(r+s)+3)!} m^{2(r+s)+3} \delta_{m+n,0} \, c \tag{7.4}$$

is a pure monomial (here $a_i^{(r,s)}(m, n)$ are structure constants), in contrast to the central term in (7.2) and in other bracket relations that can be found from (7.1). As was announced in [L3], [L4] and shown in [DLMi2], in order to conceptualize this simplification (especially the appearance of zeta-values) one can construct certain infinite-dimensional projective representations of $\mathcal{D}^+$ using vertex operators.

Let us explain Bloch's construction [Bl]. Consider the Lie algebra $\hat{\mathfrak{h}}$ introduced in Section 3, and its induced (level-one) module S. Then the correspondence

$$L_n^{(r)} \mapsto \frac{1}{2} \sum_{q=1}^{d} \sum_{j \in \mathbb{Z}} j^r (n - j)^r {:} \bar{\alpha}_q(j) \bar{\alpha}_q(n - j) {:} \quad (n \in \mathbb{Z}), \ c \mapsto d, \tag{7.5}$$

where we recall that $\{\bar\alpha_q\}$ is an orthonormal basis of $\mathfrak{h}$ and $\colon\cdot\colon$ is the usual normal ordering, gives a representation of $\hat{\mathcal{D}}^+$. Let us denote the operator on the right–hand side of (7.5) by $L^{(r)}(n)$. In particular, the operators $L^{(0)}(m)$ ($m \in \mathbb{Z}$) give a well-known representation of the Virasoro algebra with central charge $c \mapsto d$,

$$[L^{(0)}(m), L^{(0)}(n)] = (m - n)L^{(0)}(m + n) + d\,\frac{m^3 - m}{12}\,\delta_{m+n,0},$$

and the construction (7.5) for those operators is the standard realization of the Virasoro algebra on a module for a Heisenberg Lie algebra (cf. [FLM2]).

Without going into any detail, let us mention that Bloch [Bl] also studied certain natural graded traces using this representation of $\bar{L}_0^{(r)}$, and found that, as in the well-known case of the Virasoro algebra $r = 0$, they possess nice modular properties.

The appearance of zeta–values in (7.3) can be conceptualized by the following heuristic argument [Bl]: Suppose that we remove the normal ordering in (7.5) and use the relation $[\bar\alpha_q(m), \bar\alpha_q(-m)] = m$ to rewrite $\bar\alpha_q(m)\bar\alpha_q(-m)$, with $m \geq 0$, as $\bar\alpha_q(-m)\bar\alpha_q(m)+m$. It is easy to see that the resulting expression contains an infinite formal divergent series of the form

$$1^{2r+1} + 2^{2r+1} + 3^{2r+1} + \cdots.$$

A heuristic argument of Euler's suggests replacing this formal expression by $\zeta(-1 - 2r)$, where ζ is the (analytically continued) Riemann ζ–function. The resulting (zeta–regularized) operator is well defined and gives the action of $\bar{L}_n^{(r)}$; such operators satisfy the bracket relations (7.4).

7.1. Realization in S: zeta function at negative integers

In order to understand the appearance of the zeta function at negative integers using the vertex operator algebra S, following [L3, L4, DLMi1, DLMi2], we need to introduce slightly different vertex operators. Consider a vertex operator algebra V. The *homogeneous vertex operators* are defined by

$$X(u, x) = Y(x^{L^{(0)}}u, x) \quad (u \in V).\tag{7.6}$$

The most important property of these operators, for us, is the homogenous version of modified weak associativity:

$$\lim_{x_1 \to e^y x_2} \left(\left(\frac{x_1}{x_2} - 1\right)^{k(u,v)} X(u, x_1)X(v, x_2)\right) = \left(e^y - 1\right)^{k(u,v)} X(Y[u, y]v, x_2)$$

$$\tag{7.7}$$

for $u, v \in V$, and $k(u, v)$ as in Theorem 4.3. Interestingly, in this relation, yet a new type of vertex operator appears:

$$Y[u, y] = Y(e^{yL(0)}u, e^y - 1) \quad (u \in V) . \tag{7.8}$$

This vertex operator map generates a vertex operator algebra that is *isomorphic* to V, and geometrically corresponding to a change to *cylindrical coordinates*. Those properties were proven in [Z1, Z2].

Consider the vertex operator algebra S. Recall that the Virasoro generators $L(n)$ acting on S are given by the operators on the right-hand side of (7.5) with $r = 0$. It is a simple matter to verify that $X(\alpha(-1)\mathbf{1}, x) = \alpha \langle x \rangle$, with

$$\alpha \langle x \rangle = \sum_{n \in \mathbb{Z}} \alpha(n) x^{-n} . \tag{7.9}$$

Consider now the following formal series, acting on S:

$$\bar{L}^{y_1, y_2} \langle x \rangle = X \left(\frac{1}{2} \sum_{q=1}^{d} Y[\bar{\alpha}_q(-1)\mathbf{1}, y_1 - y_2]\bar{\alpha}_q(-1)\mathbf{1}, e^{y_2} x \right) . \tag{7.10}$$

By (7.7), we have

$$\bar{L}^{y_1, y_2} \langle x_2 \rangle = \frac{1}{2} \lim_{x_1 \to x_2} \sum_{q=1}^{d} \left(\left(\frac{\frac{x_1}{x_2} e^{y_1 - y_2} - 1}{e^{y_1 - y_2} - 1} \right)^k \bar{\alpha}_q \langle e^{y_1} x_1 \rangle \bar{\alpha}_q \langle e^{y_2} x_2 \rangle \right) \tag{7.11}$$

for any fixed $k \in \mathbb{N}$, $k \geq 2$. Using

$$\sum_{q=1}^{d} \bar{\alpha}_q \langle e^{y_1} x_1 \rangle \bar{\alpha}_q \langle e^{y_2} x_2 \rangle = \sum_{q=1}^{d} {:}\bar{\alpha}_q \langle e^{y_1} x_1 \rangle \bar{\alpha}_q \langle e^{y_2} x_2 \rangle{:} - \frac{\partial}{\partial y_1} \left(\frac{1}{1 - \frac{x_2}{x_1} e^{-y_1 + y_2}} \right),$$

we immediately find that

$$\bar{L}^{y_1, y_2} \langle x \rangle = \frac{1}{2} \sum_{q=1}^{d} {:}\bar{\alpha}_q \langle e^{y_1} x \rangle \bar{\alpha}_q \langle e^{y_2} x \rangle{:} - \frac{1}{2} \frac{\partial}{\partial y_1} \left(\frac{1}{1 - e^{-y_1 + y_2}} \right) . \tag{7.12}$$

Defining the operators $\bar{L}^{r_1, r_2}(n)$, $r_1, r_2 \in \mathbb{N}$, $n \in \mathbb{Z}$ via

$$\bar{L}^{y_1, y_2} \langle x \rangle = \frac{1}{2} \frac{d}{(y_1 - y_2)^2} + \sum_{n \in \mathbb{Z}, r_1, r_2 \in \mathbb{N}} \bar{L}^{r_1, r_2}(n) x^{-n} \frac{y_1^{r_1} y_2^{r_2}}{r_1! r_2!} , \tag{7.13}$$

it is simple to see, using (7.5), that the correspondence

$$\bar{L}_n^{(r)} \mapsto \bar{L}^{r, r}(n), \quad c \mapsto d \tag{7.14}$$

for $n \in \mathbb{Z}$, $r \in \mathbb{N}$ gives a representation of the generators (7.3) of the algebra $\hat{\mathcal{D}}^+$. Recall that these generators were introduced by Bloch in order to simplify the central term in the commutation relations.

As was shown in [DLMi2], from the expression (7.11) of the formal series $L^{y_1,y_2}\langle x \rangle$, involving formal limits, it is a simple matter to compute the following commutators, first written in [L3]:

$$
\left[\bar{L}^{y_1,y_2}\langle x_1 \rangle, \bar{L}^{y_3,y_4}\langle x_2 \rangle \right]
$$
$$
= -\frac{1}{2}\frac{\partial}{\partial y_1}\left(\bar{L}^{-y_1+y_2+y_3,y_4}\langle x_2 \rangle \delta\left(\frac{e^{y_1}x_1}{e^{y_3}x_2} \right) + \bar{L}^{-y_1+y_2+y_4,y_3}\langle x_2 \rangle \delta\left(\frac{e^{y_1}x_1}{e^{y_4}x_2} \right) \right)
$$
$$
- \frac{1}{2}\frac{\partial}{\partial y_2}\left(\bar{L}^{y_1-y_2+y_3,y_4}\langle x_2 \rangle \delta\left(\frac{e^{y_2}x_1}{e^{y_3}x_2} \right) + \bar{L}^{y_1-y_2+y_4,y_3}\langle x_2 \rangle \delta\left(\frac{e^{y_2}x_1}{e^{y_4}x_2} \right) \right).
$$
$$(7.15)$$

As was announced in [L3, L4] and explained in [DLMi2], a simple analysis of this commutator shows that the central term in the commutators of the generators (7.14) is a pure monomial, as in (7.4). This gives a simple explanation of Bloch's phenomenon using the vertex operator algebras S. Moreover, the definition (7.10) says that the operators $L^{y_1,y_2}(x)$ represent on V the image of some fundamental algebra elements of V under the transformation to the cylinder. These fundamental elements, being closely related to the Virasoro element ω, can be expected, when transformed to the cylinder, to lead to graded traces with simple modular properties, in agreement with the observations of Bloch [Bl].

7.2. *Representations on* $S[\nu]$: *Bernoulli polynomials at rational values*

Following [DLMi1, DLMi2], we will now construct a representation of $\hat{\mathcal{D}}^+$ on $S[\nu]$. The property that a twisted module is a true module on the fixed-point subalgebra will be essential below in this construction. This property is guaranteed by the operator Δ_x that we calculated above (6.19). In order to have the correction terms for the representation of the algebra $\hat{\mathcal{D}}^+$ on the twisted space $S[\nu]$, one could apply e^{Δ_x} on the vectors generating the representation of the whole algebra $\hat{\mathcal{D}}^+$. This can be a complicated problem, mainly because generators of $\hat{\mathcal{D}}^+$ have arbitrarily large weights. In line with [DLMi1, DLMi2], below we will calculate the correction terms directly using the modified weak associativity relation for twisted operators, as well as the simple result (3.10). Hence in this argument, the explicit action of Δ_x on vectors generating the representation of the algebra $\hat{\mathcal{D}}^+$ is not of importance; all we need to know is that *there exists* such an operator Δ_x giving to the space $S[\nu]$ the properties of a twisted module for the vertex operator algebra S.

In parallel to the previous sub-section, we need to introduce *homogeneous twisted vertex operators*. Being given a vertex operator algebra V and a v-twisted V-module M, they are defined by

$$X_M(u, x) = Y_M(x^{L(0)}u, x) \quad (u \in V).$$ (7.16)

Again, the most important property of these operators, for us, is the homogenous version of modified weak associativity for twisted vertex operators:

$$\lim_{x_1^{1/p} \to \omega_p^s(e^y x_2)^{1/p}} \left(\left(\frac{x_1}{x_2} - 1 \right)^{k(u,v)} X_M(u, x_1) X_M(v, x_2) \right)$$

$$= (e^y - 1)^{k(u,v)} X_M(Y[v^{-s}u, y]v, x_2).$$ (7.17)

for $u, v \in V$, $s \in \mathbb{Z}$, and $k(u, v)$ as in Theorem 4.6. Recall the definition of $Y[u, y]$ in (7.8).

Consider the vertex operator algebra S and its twisted module $S[v]$. It is a simple matter to verify that $X_{S[v]}(\alpha(-1)\mathbf{1}, x) = \alpha^v\langle x \rangle$, with

$$\alpha^v\langle x \rangle = \sum_{n \in \frac{1}{p}\mathbb{Z}} \alpha^v(n)x^{-n}.$$ (7.18)

Consider now the following formal series, acting on $S[v]$:

$$\bar{L}^{v;y_1,y_2}\langle x \rangle = X_{S[v]} \left(\frac{1}{2} \sum_{q=1}^{d} Y[\bar{\alpha}_q(-1)\mathbf{1}, y_1 - y_2]\bar{\alpha}_q(-1)\mathbf{1}, e^{y_2}x \right).$$ (7.19)

Since the operator $\frac{1}{2} \sum_{q=1}^{d} Y[\bar{\alpha}_q(-1)\mathbf{1}, y_1 - y_2]\bar{\alpha}_q(-1)\mathbf{1}$ is in the fixed point subalgebra of S, it is immediate that these operators satisfy the same commutation relations as (7.15):

$$\left[\bar{L}^{v;y_1,y_2}\langle x_1 \rangle, \bar{L}^{v;y_3,y_4}\langle x_2 \rangle \right]$$

$$= -\frac{1}{2} \frac{\partial}{\partial y_1} \left(\bar{L}^{v;-y_1+y_2+y_3,y_4}\langle x_2 \rangle \delta \left(\frac{e^{y_1}x_1}{e^{y_3}x_2} \right) + \bar{L}^{v;-y_1+y_2+y_4,y_3}\langle x_2 \rangle \delta \left(\frac{e^{y_1}x_1}{e^{y_4}x_2} \right) \right)$$

$$- \frac{1}{2} \frac{\partial}{\partial y_2} \left(\bar{L}^{v;y_1-y_2+y_3,y_4}\langle x_2 \rangle \delta \left(\frac{e^{y_2}x_1}{e^{y_3}x_2} \right) + \bar{L}^{v;y_1-y_2+y_4,y_3}\langle x_2 \rangle \delta \left(\frac{e^{y_2}x_1}{e^{y_4}x_2} \right) \right).$$ (7.20)

Hence, defining the operators $\bar{L}^{v;r_1,r_2}(n)$, $r_1, r_2 \in \mathbb{N}$, $n \in \mathbb{Z}$ via

$$\bar{L}^{v;y_1,y_2}\langle x \rangle = \frac{1}{2} \frac{d}{(y_1 - y_2)^2} + \sum_{n \in \mathbb{Z}, \, r_1, r_2 \in \mathbb{N}} \bar{L}^{v;r_1,r_2}(n)x^{-n} \frac{y_1^{r_1} y_2^{r_2}}{r_1! r_2!}$$ (7.21)

(in particular, only integer powers of x appear in this expansion), we conclude that they satisfy the same commutation relations as the operators $\bar{L}^{r_1, r_2}(n)$ introduced in (7.13). Then, as in (7.14), we can expect that the correspondence

$$\bar{L}_n^{(r)} \mapsto \bar{L}^{v;r,r}(n), \quad c \mapsto d \tag{7.22}$$

for $n \in \mathbb{Z}$, $r \in \mathbb{N}$ gives a representation of the generators (7.3) of the algebra $\hat{\mathcal{D}}^+$ on $S[v]$. Bringing this expectation to a proof needs a little more analysis (in particular, one needs to show that the $L^{v;r_1,r_2}(n)$ are related to the $L^{v;r,r}(n)$ in the same way as the $L^{r_1,r_2}(n)$ are related to the $L^{r,r}(n)$), which is done in detail in [DLMi2].

Now, by (7.17) we have

$$\bar{L}^{v;y_1,y_2}\langle x_2 \rangle = \frac{1}{2} \lim_{x_1 \to x_2} \sum_{q=1}^{d} \left(\left(\frac{\frac{x_1}{x_2} e^{y_1 - y_2} - 1}{e^{y_1 - y_2} - 1} \right)^k \bar{\alpha}_q^v \langle e^{y_1} x_1 \rangle \bar{\alpha}_q^v \langle e^{y_2} x_2 \rangle \right) \tag{7.23}$$

for any fixed $k \in \mathbb{N}$, $k \geq 2$, which gives

$$\bar{L}^{v;y_1,y_2}\langle x \rangle = \frac{1}{2} \sum_{q=1}^{d} {:}\bar{\alpha}_q^v \langle e^{y_1} x \rangle \bar{\alpha}_q^v \langle e^{y_2} x \rangle {:} - \frac{1}{2} \frac{\partial}{\partial y_1} \left(\sum_{k=0}^{p-1} \frac{e^{\frac{k(-y_1 + y_2)}{p}} \dim \mathfrak{h}_{(k)}}{1 - e^{-y_1 + y_2}} \right) \tag{7.24}$$

using

$$\sum_{q=1}^{d} \bar{\alpha}_q^v \langle e^{y_1} x_1 \rangle \bar{\alpha}_q^v \langle e^{y_2} x_2 \rangle = \sum_{q=1}^{d} {:}\bar{\alpha}_q^v \langle e^{y_1} x_1 \rangle \bar{\alpha}_q^v \langle e^{y_2} x_2 \rangle {:}$$

$$- \frac{\partial}{\partial y_1} \left(\sum_{k=0}^{p-1} \frac{e^{\frac{k(-y_1 + y_2)}{p}} \dim \mathfrak{h}_{(k)}}{1 - \frac{x_2}{x_1} e^{-y_1 + y_2}} \right).$$

Evaluating the operators $\bar{L}^{v;r,r}(n)$ from (7.21), we conclude that the operators

$$\bar{L}^{v;r,r}(n) = \frac{1}{2} \sum_{q=1}^{d} \sum_{j \in \frac{1}{p}\mathbb{Z}} j^r (n-j)^r {:}\bar{\alpha}_q^v(j) \bar{\alpha}_q^v(n-j){:}$$

$$- \delta_{n,0} \frac{(-1)^r}{4(r+1)} \sum_{k=0}^{p-1} \dim \mathfrak{h}_{(k)} B_{2(r+1)}(k/p) \tag{7.25}$$

form a representation, on $S[v]$, of the generators (7.3) for the Lie algebra $\hat{\mathcal{D}}^+$. Notice the appearance of the Bernoulli polynomials. From our construction, this is seen to be directly related to general properties of homogeneous twisted vertex operators.

The next result is a simple consequence of the discussion above. It was shown in [DLMi2]. It describes the action of the "Cartan subalgebra" of $\hat{\mathcal{D}}^+$ on a highest weight vector of a canonical quasi-finite $\hat{\mathcal{D}}^+$–module; here we are using the terminology of [KR]. This corollary gives the "correction" terms referred to in the introduction.

Corollary 7.1. *Given a highest weight $\hat{\mathcal{D}}^+$–module W, let δ be the linear functional on the "Cartan subalgebra" of $\hat{\mathcal{D}}^+$ (spanned by $L_0^{(k)}$ for $k \in \mathbb{N}$) defined by*

$$L_0^{(k)} \cdot w = (-1)^k \delta \left(L_0^{(k)} \right) w,$$

where w is a generating highest weight vector of W, and let $\Delta(x)$ be the generating function

$$\Delta(x) = \sum_{k \geq 1} \frac{\delta(L_0^{(k)}) x^{2k}}{(2k)!}$$

(cf. [KR]). Then for every automorphism ν of period p as above,

$$\mathcal{U}(\hat{\mathcal{D}}^+) \cdot 1 \subset S[\nu]$$

is a quasi–finite highest weight $\hat{\mathcal{D}}^+$–module satisfying

$$\Delta(x) = \frac{1}{2} \frac{d}{dx} \sum_{k=0}^{p-1} \frac{(e^{\frac{kx}{p}} - 1) \dim \mathfrak{h}_{(k)}}{1 - e^x}. \tag{7.26}$$

References

[AFMO] H. Awata, M. Fukuma, Y. Matsuo and S. Odake, Representation theory of the $W_{1+\infty}$ algebra, Quantum field theory, integrable models and beyond (Kyoto, 1994), *Progr. Theoret. Phys. Suppl.* **118** (1995), 343–373.

[BPZ] A. A. Belavin, A. M. Polyakov and A. B. Zamolodchikov, Infinite conformal symmetries in two-dimensional quantum field theory, *Nucl. Phys.* **B241** (1984), 333–380.

[Bl] S. Bloch, Zeta values and differential operators on the circle, *J. Algebra* **182** (1996), 476–500.

[Bo] R. E. Borcherds, Vertex algebras, Kac-Moody algebras, and the Monster, *Proc. Natl. Acad. Sci. USA* **83** (1986), 3068–3071.

[D] C. Dong, Twisted modules for vertex algebras associated with even lattices, *J. Algebra* **165** (1994), 91–112.

[DL1] C. Dong and J. Lepowsky, *Generalized Vertex Algebras and Relative Vertex Operators*, Progress in Mathematics, Vol. **112**, Birkhäuser, Boston, 1993.

[DL2] C. Dong and J. Lepowsky, The algebraic structure of relative twisted vertex operators, *Journal of Pure and Applied Algebra*, **110** (1996), 259–295.

[DLM] C. Dong, H. Li and G. Mason, Modular invariance of trace functions in orbifold theory and generalized Moonshine, *Comm. Math. Phys.* **214** (2000), 1–56.

[DLMi1] B. Doyon, J. Lepowsky and A. Milas, Twisted modules for vertex operator algebras and Bernoulli polynomials, *I.M.R.N.* **44** (2003), 2391–2408.

[DLMi2] B. Doyon, J. Lepowsky and A. Milas, Twisted vertex operators and Bernoulli polynomials, to be published in *Comm. Cont. Math.* **8** (2006), 247–307.

[FFR] A. J. Feingold, I. B. Frenkel and J. F. X. Ries, *Spinor construction of vertex operator algebras, triality and $E_8^{(1)}$*, Contemporary Math., Vol. **121**, Amer. Math. Soc., Providence, 1991.

[FrS] E. Frenkel and M. Szczesny, Twisted modules over vertex algebras on algebraic curves, *Adv. Math.* **187** (2004), 195–227.

[FHL] I. B. Frenkel, Y.-Z. Huang and J. Lepowsky, On axiomatic approaches to vertex operator algebras and modules, preprint, 1989; *Memoirs Amer. Math. Soc.* **104**, 1993.

[FLM1] I. B. Frenkel, J. Lepowsky and A. Meurman, Vertex operator calculus, in: *Mathematical Aspects of String Theory*, Proc. 1986 Conf., San Diego, ed. by S.-T. Yau, World Scientific, Singapore, 1987, 150–188.

[FLM2] I. B. Frenkel, J. Lepowsky and A. Meurman, *Vertex Operator Algebras and the Monster*, Pure and Appl. Math., Vol. **134**, Academic Press, Boston, 1988.

[H1] Y.-Z. Huang, Applications of the geometric interpretation of vertex operator algebras, *Proc. 20th International Conference on Differential Geometric Methods in Theoretical Physics, New York, 1991*, ed. S. Catto and A. Rocha, World Scientific, Singapore, 1992, 333–343.

[H2] Y.-Z. Huang, *Two-dimensional Conformal Geometry and Vertex Operator Algebras*, Progress in Math., Vol. **148**, Birkhäuser, Boston, 1997.

[KR] V. Kac and A. Radul, Quasifinite highest weight modules over the Lie algebra of differential operators on the circle, *Comm. Math. Phys.* **157** (1993), 429–457.

[KWY] V. G. Kac, W. Wang and C. H. Yan, Quasifinite representations of classical Lie subalgebras of $\mathcal{W}_{1+\infty}$, *Adv. Math.* **139** (1998), 56–140.

[L1] J. Lepowsky, Calculus of twisted vertex operators, *Proc. Natl. Acad. Sci. USA* **82** (1985), 8295–8299.

[L2] J. Lepowsky, Perspectives on vertex operators and the Monster, in: *Proceedings of the Symposium on the Mathematical Heritage of Hermann Weyl*, Duke University, May, 1987, Proc. Symp. Pure Math., Vol. **48**, Amer. Math. Soc., Providence, 1988, 181–197.

[L3] J. Lepowsky, Vertex operator algebras and the zeta function, in: *Recent Developments in Quantum Affine Algebras and Related Topics*, ed. by N. Jing and K. C. Misra, Contemporary Math., Vol. 248, Amer. Math. Soc., Providence, 1999, 327–340.

[L4] J. Lepowsky, Application of a 'Jacobi Identity' for vertex operator algebras to zeta values and differential operators, *Lett. Math. Phys.* **53** (2000), 87–103.

[LL] J. Lepowsky and H. Li, *Introduction to Vertex Operator Algebras and Their Representations*, Progress in Mathematics, Vol. **227**, Birkhäuser, Boston, 2003.

[Li1] H. Li, Local systems of vertex operators, vertex superalgebras and modules, *Journal of Pure and Applied Algebra* **109** (1996), 143–195.

[Li2] H. Li, Local systems of twisted vertex operators, vertex operator superalgebras and twisted modules, in: *Moonshine, the Monster, and Related Topics (South Hadley, MA, 1994)*, Contemporary Math., Vol. **193**, Amer. Math. Soc., Providence, 1996, 203–236.

[M1] A. Milas, Correlation functions, vertex operator algebras and ζ–functions, Ph.D. thesis, Rutgers University, 2001.

[M2] A. Milas, Formal differential operators, vertex operator algebras and zeta–values, I, *Journal of Pure and Applied Algebra* **183** (2003), 129–190.

[M3] A. Milas, Formal differential operators, vertex operator algebras and zeta–values, II, *Journal of Pure and Applied Algebra* **183** (2003), 191–244.

[N] S.-H. Ng, The Lie bialgebra structures on the Witt and Virasoro algebras, Ph.D. thesis, Rutgers University, 1997.

[Z1] Y. Zhu, Vertex operators, elliptic functions and modular forms, Ph.D. thesis, Yale University, 1990.

[Z2] Y. Zhu, Modular invariance of characters of vertex operator algebras, *J. Amer. Math. Soc.* **9** (1996), 237–307.

Vertex Operators and Sporadic Groups

John F. Duncan*†

Abstract

In the 1980's, the work of Frenkel, Lepowsky and Meurman, along with that of Borcherds, culminated in the notion of vertex operator algebra, and an example whose full symmetry group is the largest sporadic simple group: the Monster. Thus it was shown that the vertex operators of mathematical physics play a role in finite group theory. In this article we describe an extension of this phenomenon by introducing the notion of enhanced vertex operator algebra, and constructing examples that realize other sporadic simple groups, including one that is not involved in the Monster.

1. Motivation

We begin not with the problem that motivates the article, but with motivation for the tools that will furnish the solution to this problem. The tools we have in mind are called *vertex operator algebras (VOAs)*; here follows one way to motivate the notion.

In mathematics there are various kinds of finite dimensional algebras that have proven to be significant or interesting in some respect. For example,

(1) semisimple Lie algebras (with invariant bilinear form)
(2) simple Jordan algebras (of type A, B, or C)
(3) the Chevalley algebra (see [Che54])
(4) the Griess algebra (see [Gri82])

* Harvard University, Department of Mathematics, One Oxford Street, Cambridge, MA 02138, U.S.A.
† Email: duncan@math.harvard.edu; homepage: http://math.harvard.edu/~jfd/

The items of this list are very different from each other in terms of their properties and structure theory. Perhaps the only thing they have in common (as algebras) is finite dimensionality.

Nonetheless, it turns out that there is such a process called *affinization* which associates a certain infinite dimensional algebra structure (let's say *affine algebra*), to each finite dimensional example in this list. Not only this, but the corresponding affine algebra has a distinguished representation (infinite dimensional) for which the action of the affine algebra extends in a natural way to a new kind of algebra structure called vertex operator algebra structure[1].

Thus we obtain objects of a common category for each distinct example here, and the notion of vertex operator algebra (VOA) furnishes a framework within which these distinct finite dimensional algebra structures may be unified.

2. VOAs

Let us now present a definition of the notion of VOA. For our purposes it is more natural to consider the larger category of super vertex operator algebras (SVOAs).

An SVOA is a quadruple $(U, Y, \mathbf{1}, \omega)$ where

- $U = U_{\bar{0}} \oplus U_{\bar{1}}$ is a super vector space over a field $\mathbb{F}$ say. (We will take $\mathbb{F}$ to be $\mathbb{R}$ or $\mathbb{C}$.)
- Y is a map $U \otimes U \rightarrow U((z))$, so that the image of the vector $u \otimes v$ under Y is a Laurent series with coefficients in U. This series is denoted $Y(u, z)v = \sum_n u_{(n)} v z^{-n-1}$, and the operator $Y(u, z)$ is called the *vertex operator* associated to u.
- $\mathbf{1} \in U_{\bar{0}}$ is called the *vacuum vector*, and is a kind of identity for U in the sense that we should have $Y(\mathbf{1}, z)u = u$ and $Y(u, z)\mathbf{1}|_{z=0}u = u$ for all $u \in U$.
- $\omega \in U_{\bar{0}}$ is called the *conformal element*, and is such that for $Y(\omega, z) = \sum L(n)z^{-n-2}$ the operators $L(n)$ should satisfy the relations

$$[L(m), L(n)] = (m - n)L(m + n) + \frac{m^3 - m}{12} c\, \delta_{m+n,0}\, \mathrm{Id} \qquad (1)$$

1 The VOA structure corresponding to a semisimple Lie algebra with invariant bilinear form is obtained in [FZ92] (see also [FLM88]). VOAs corresponding to the simple Jordan algebras of types A, B, and C are given in [Lam99]. An affinization of the Chevalley algebra is constructed in [FFR91], and the VOA corresponding to the Griess algebra is constructed in [FLM88].

for some scalar value $c \in \mathbb{F}$. In other words, the Fourier modes of $Y(\omega, z)$ should generate a representation of the Virasoro algebra, with some central charge c.

The main axiom for the vertex operators is the *Jacobi identity*, which states that

- for $\mathbb{Z}/2$–homogeneous $u, v \in U$ and arbitrary $a \in U$ we should have

$$z_0^{-1} \delta \left(\frac{z_1 - z_2}{z_0} \right) Y(u, z_1) Y(v, z_2)$$

$$- (-1)^{|u||v|} z_0^{-1} \delta \left(\frac{z_2 - z_1}{-z_0} \right) Y(v, z_2) Y(u, z_1)$$

$$= z_2^{-1} \delta \left(\frac{z_1 - z_0}{z_2} \right) Y(Y(u, z_0)v, z_2) \qquad (2)$$

where $|u|$ is 1 or 0 as u is odd or even, and similarly for $|v|$, and the expression $z_0^{-1}\delta((z_1 - z_2)/z_0)$, for example, denotes the formal power series

$$z_0^{-1} \delta \left(\frac{z_1 - z_2}{z_0} \right) = \sum_{\substack{n,k \in \mathbb{Z}, \\ k \geq 0}} (-1)^k \binom{n}{k} z_0^{-n-1} z_1^{n-k} z_2^k. \qquad (3)$$

From (2) and from the properties of $\mathbf{1}$ we see the extent to which the triple $(U, Y, \mathbf{1})$ behaves like a (super)commutative associative (super)algebra with identity, since the formal series (3) may be regarded as a "delta function supported at $z_1 - z_2 = z_0$ (and expanded in $|z_1| > |z_2|$)". The Jacobi identity (2) thus encodes, among other things, some sense in which the compositions $Y(u, z_1)Y(v, z_2)$, $Y(v, z_2)Y(u, z_1)$, and $Y(Y(u, z_1 - z_2)v, z_2)$ all coincide.

The structure in $(U, Y, \mathbf{1}, \omega)$ which has no analogue in the ordinary superalgebra case is that furnished by the conformal element ω. This structure furnished by ω (we will call it *conformal structure*) manifests in two important axioms.

- The action of $L(0)$ on U should be diagonalizable with eigenvalues in $\frac{1}{2}\mathbb{Z}$ and bounded from below. We write $U = \bigoplus_n U_n$ for the corresponding grading on U, and we call U_n the subspace of *degree n*.
- The operator $L(-1)$ should satisfy $Y(L(-1)u, z) = D_z Y(u, z)$ for all $u \in U$, where D_z denotes differentiation in z.

The conformal structure is essential for the construction of characters associated to an SVOA. Zhu has shown [Zhu90] that these characters (under certain finiteness conditions on the SVOA) span a representation of the Modular Group $PSL_2(\mathbb{Z})$.

We often write U in place of $(U, Y, \mathbf{1}, \omega)$. The scalar c of (1) is called the *rank* of U. Modules and module morphisms can be defined in a natural way. We say that an SVOA is *self-dual* in the case that it is irreducible as a module over itself, and has no other inequivalent irreducible modules.

3. Sporadic groups

The Classification of Finite Simple Groups (see [GLS94], [Sol01]) states that in addition to the

- cyclic groups of prime order
- alternating groups A_n for $n \geq 5$
- finite groups of Lie type (like $PSL_n(q)$, $PSU_n(q)$, $G_2(q)$, &c.)

there are exactly 26 other groups that are finite and simple, and can be included in none of the infinite families listed. These 26 groups are called the *sporadic groups*.

The largest of the sporadic groups is called the *Monster*[2], and was first constructed by Robert L. Griess, Jr., who obtained this result by explicitly constructing a certain commutative non-associative algebra (with invariant bilinear form) of dimension 196883, with the Monster group $\mathbb{M}$ as its full group of automorphisms. This algebra is named the *Griess algebra*.

The Monster group furnishes a setting in which the majority (but not all) of the other sporadic groups may be analyzed, since it involves 19 of the other sporadic simple groups. Here we say that a group G is *involved in the Monster* if G is the homomorphic image of some subgroup of the Monster; that is, if there is some H in $\mathbb{M}$ with normal subgroup N such that H/N is isomorphic to G. The sporadic groups that are involved in the Monster are called the *Monstrous* sporadic groups.

The remaining 6 sporadic groups not involved in the Monster are called the *non-Monstrous* sporadic groups, or more colorfully, the *Pariahs*.

4. Vertex operators and the Monster

It is a surprising and fascinating fact that the Griess algebra can appear in our list of finite dimensional algebras admitting affinization (see §1).

The fact that it does is due to the work of Frenkel, Lepowsky and Meurmann, who extended the notion of affinization (known to exist for certain Lie algebras) to the Griess algebra using vertex operators, obtaining

---

2 It is also called the *Friendly Giant*.

an infinite dimensional version: the affine Griess algebra. They constructed a distinguished infinite dimensional module over this affine Griess algebra called the Moonshine Module. They built upon the work of Borcherds [Bor86] so as to arrive at the notion of vertex operator algebra, and showed that the Moonshine Module admits such a structure. Finally, they showed that the full automorphism group of this structure is the Monster simple group.

In contrast to the situation with Lie algebras or Jordan algebras, the Griess algebra is an object which is hard to axiomatize. It is perhaps not clear that there is any reasonable category of algebras (in the orthodox sense) which includes the Griess algebra as an example (see [Con85]). The notion of VOA thus provides a remedy to this situation: a setting within which the Griess algebra can be axiomatized.

A related question is: "How might the Monster group be characterized?" Having found that such an extraordinary group is the symmetry group of some structure, we would like to be able recognize this structure as distinguished in its own right, so that our group might be defined to be just the group of automorphisms of this distinguished object. In particular, our structure should belong to some family of similar structures; a family equipped with invariants, sufficiently rich that they can distinguish our particularly interesting examples from all others, and sufficiently simple that we can communicate them easily.

This question of characterization can also be addressed (conjecturally, at least) within the theory of VOAs. Let us call to attention three invariants for VOAs:

- rank
- self-duality
- degree (vanishing conditions)

One may check that the Moonshine Module satisfies the following three properties.

- rank 24
- self-dual
- degree 1 subspace vanishes

It is a conjecture due to Frenkel, Lepowsky and Meurman that the Moonshine Module is uniquely determined by these properties. Modulo a proof of the conjecture, VOA theory thus provides a compelling definition of the Monster group: the automorphism group of the Moonshine Module, a beautifully characterized object in the category of VOAs.

5. Vertex operators and Monstrous groups

We have seen that vertex operators may be fruitfully applied to one of the sporadic groups, and we may wonder if there is anything they can do for the remaining 25. Let us formulate The Problem:

* Given a sporadic group G, find a VOA whose automorphism group is G, and characterize it.

The fact that 20 of the sporadic groups are involved in the Monster suggests that The Problem may have solutions, at least for G a Monstrous group. After all, if G is such a group, then a cover $\hat{G} = N.G$ say, of G, is a subgroup of the Monster, and in particular, acts on the Moonshine Module. This action is probably very reducible, but by choosing an appropriate irreducible subalgebra for example, we may well obtain an object – a VOA even – which serves as a reasonable analogue of the Moonshine Module for our new group $\hat{G}$. We may even find that the normal subgroup N acts trivially, so that our analogue of the Moonshine Module actually realizes G itself, and not a cover of G. This outline is extremely speculative, and certainly does not constitute an acceptable solution to The Problem, but it does at least give us somewhere to start, at least in the case that G is a Monstrous group.

It is important to mention that something like this has been carried out rigorously and successfully in at least one case: that in which G is the Baby Monster *BM*, and the group $\hat{G}$ is a double cover of G, the centralizer of a so-called $2A$ involution in the Monster. The precise method is due to Gerald Höhn [Höh96], and the result is a self-dual SVOA of rank $23\frac{1}{2}$ whose full automorphism group is a direct product $2 \times BM$ of the Baby Monster with a group of order 2.

6. Vertex operators and the Conway group

At this point we would like to describe a solution to The Problem for a specific Monstrous group G, which is nonetheless not along the lines just described in §5. The group we have in mind is the largest sporadic group of Conway, Co_1. The solution we have in mind is the object of the following Theorem.

Theorem 1 ([Dun07]). *Among nice rational $N = 1$ SVOAs, there is a unique one satisfying*

- *rank 12*
- *self-dual*
- *degree $1/2$ subspace vanishes*

Let us name this structure A_{Co}. We can see that it admits a convenient characterization. That A_{Co} is a solution to one of our problems is shown by the next Theorem.

Theorem 2 ([Dun07]). *The full automorphism group of A_{Co} is the sporadic group Co_1.*

We should go no further before addressing the new terminology that has arisen. The terms *nice* and *rational* refer to certain technical conditions on SVOAs, and it will be convenient to put aside their precise meaning, and refer the interested reader to the article [Dun07]. (One would expect such technical conditions to arise in a precise formulation of the uniqueness conjecture for the Moonshine Module.) Of more importance for our present purpose is the term $N = 1$ *SVOA*.

Definition. An $N = 1$ SVOA is a quadruple $(U, Y, \mathbf{1}, \{\omega, \tau\})$, such that $(U, Y, \mathbf{1}, \omega)$ is an SVOA, and τ is a distinguished vector of degree $3/2$ satisfying $\tau_{(0)}\tau = 2\omega$, and such that the Fourier coefficients of $Y(\tau, z)$ generate a representation of the Neveu–Schwarz Lie superalgebra on U.[3]

The Neveu–Schwarz superalgebra is a natural super-analogue of the Virasoro algebra. It is also known as the $N = 1$ Virasoro superalgebra. Thus an $N = 1$ SVOA is just like an ordinary SVOA except there is some extra structure: the role of the Virasoro algebra is now played by the $N = 1$ Virasoro superalgebra.

Let us consider for a moment longer, the difference between SVOA structure and $N = 1$ SVOA structure. Looking back at Theorem 1, experts will notice that the conclusion remains true if we drop the the the "$N = 1$" from "$N = 1$ SVOA" in the hypothesis. That is to say, there is indeed a unique self-dual SVOA with rank 12 that has vanishing degree $1/2$ subspace. In fact, it is a reasonably familiar object as SVOAs go: it is the lattice SVOA associated to the integral lattice D_{12}^+ (the unique self-dual integral lattice of rank 12 with no vectors of unit norm).

One may be surprised to see a sporadic group, or even a finite group here, since the SVOA underlying A_{Co} has infinite automorphism group. In fact, there is an action by Spin_{24} (faithful up to some subgroup of order 2). The crux of the matter is that

(1) for a suitably chosen vector in this Spin_{24}-module A_{Co} the fixing group is Co_1,

---

3 The definition of $N = 1$ SVOA here is almost identical to that of $N = 1$ *Neveu–Schwarz vertex operator superalgebra without odd formal variables ($N = 1$ NS-VOSA)* which was introduced earlier by Barron [Bar00].

(2) the precise choice is made for us by the $N = 1$ Virasoro superalgebra.

Let us also emphasize that the uniqueness result for A_{Co} furnishes a compelling definition of the Conway group: as the full automorphism group of A_{Co}, a well characterized object in the category of $N = 1$ SVOAs.

7. Enhanced SVOAs

With the example of A_{Co} in mind, and also with the suspicion that it may be interesting to consider other extensions of the Virasoro algebra, we formulate the notion of *enhanced SVOA*.

Roughly speaking, an enhanced SVOA is a quadruple $(U, Y, \mathbf{1}, \Omega)$ where Ω is a finite subset of U (the set of *conformal generators*) containing a vector ω for which $(U, Y, \mathbf{1}, \omega)$ is an SVOA. (We refer to [Dun06a, §2] for the precise definition.) The subSVOA of U generated by the elements of Ω is called the *conformal subSVOA*.

The rank of an enhanced SVOA is just the rank of the underlying SVOA. We say that an enhanced SVOA is self-dual just when it is self-dual as an SVOA. The automorphism group of an enhanced SVOA is the subgroup of the automorphism group of the underlying SVOA that fixes every conformal generator.

We see then that an ordinary SVOA is an enhanced SVOA with $\Omega = \{\omega\}$, and an $N = 1$ SVOA is an enhanced SVOA with $\Omega = \{\omega, \tau\}$, and conformal subSVOA a copy of the SVOA associated to the vacuum representation of the $N = 1$ Virasoro superalgebra. In order to show that there are other interesting examples of enhanced SVOA structure, we present the following result.

Theorem 3 ([Dun06b]). *There exists a self-dual enhanced SVOA A_{Suz} of rank* 12

$$A_{Suz} = (A_{Suz}, Y, \mathbf{1}, \{\omega, J, \nu, \mu\}) \tag{4}$$

with $\mathrm{Aut}(A_{Suz}) \cong 3 . Suz$.

It turns out that the conformal algebra in this example contains the direct product of a pair of $N = 1$ Virasoro superalgebras, at central charges 11 and 1, respectively. The SVOA underlying A_{Suz} coincides with that underlying A_{Co}, and taking the diagonal $N = 1$ Virasoro superalgebra generated by $\tau = \nu + \mu$ (with central charge 12), we recover the enhanced SVOA structure with automorphism group Co_1.

8. Beyond the Monster

We have seen already that vertex operators have a role to play in the analysis of several Monstrous sporadic groups. In the cases of Conway's group, Suzuki's group, the Monster, and the Baby Monster, precise theorems have been formulated. A very significant question is whether or not there is any application to sporadic groups beyond the Monster; that is, to pariahs. We observed at the very beginning that the notion of VOA can unify such disparate notions as 'semi-simple Lie algebra' and 'commutative non-associative algebra'.

The principal idea of this paper is that vertex operators do play a role in the representation theory of non-Monstrous sporadic groups.

At this point let us introduce the sporadic group of Rudvalis, *Ru*. This sporadic simple group is not involved in the Monster; it is one of the pariahs. It has order

$$145926144000 = 2^{14}.3^3.5^3.7.13.29 \approx \frac{3}{2} \times 10^{11}. \tag{5}$$

The largest maximal subgroup is of the form $^2F_4(2)$ and has index 4060 in *Ru*. (The Tits group has index two in this group.) The next largest maximal subgroup is a non-split extension of the form $2^6.G_2(2)$, and has index 188500. The smallest non-trivial irreducible representations of *Ru* have degree 378.

9. Vertex operators and Rudvalis's group

Consider The Problem for $G = Ru$. The main theorem we wish to present is the following.

Theorem 4 ([Dun06a],[Dun06b]). *There exists a self-dual enhanced SVOA A_{Ru} of rank* 28

$$A_{Ru} = (A_{Ru}, Y, \mathbf{1}, \{\omega, J, \nu, \varrho\}) \tag{6}$$

with $\mathrm{Aut}(A_{Ru}) \cong 7 \times Ru.$

(Compare this with the statement of Theorem 3.) We will now provide a description of the enhanced SVOA A_{Ru}.

At the level of SVOAs, we have an isomorphism

$$A_{Ru} \cong V_{D_{28}^+} \tag{7}$$

where D_{28}^+ denotes a self-dual integral lattice of rank 28 with no vectors of unit length, and with D_{28} as its even part. We have observed already that there are analogous statements for the SVOAs underlying the enhanced SVOAs A_{Co} and A_{Suz}.

$$A_{Co} \cong A_{Suz} \cong V_{D_{12}^+}, \quad \text{as SVOAs.} \tag{8}$$

In the case of the enhanced SVOA A_{Suz}, the conformal vectors j and ω have degree 1 and 2 respectively, and the two conformal vectors beyond these; viz. ν and μ, are both found in the degree $3/2$ subspace. In the case of A_{Ru}, the degree $3/2$ subspace is trivial. In fact the degree $1/2$, $3/2$, and $5/2$ subspaces are all trivial for A_{Ru}. The extra conformal vectors ν and ϱ for A_{Ru} are found in the degree $7/2$ subspace. This space is very large; the dimension is the number of vectors of square-length 7 in the lattice D_{28}^+.

$$\dim(A_{Ru})_{7/2} = 2^{28}/2 \approx 10^8 \tag{9}$$

The most effort in the construction of A_{Ru} goes into determining a precise description of the vectors ν and ϱ. It is a remarkable fact that the finite group eventually obtained has almost no other point-wise invariants[4] in its action on this 2^{27} dimensional space.

10. The 28 dimensional representation

It is important for our construction of A_{Ru} that the Rudvalis group admits a perfect double cover $2.Ru$ which has irreducible representations of degree 28 (writable over $\mathbb{Z}[\mathbf{i}]$). That the group $2.Ru$ preserves a lattice of rank 28 over $\mathbb{Z}[\mathbf{i}]$ was observed independently by Meurman[5], and by Conway and Wales [CW73] (see also [Con77] and [Wil84]), and this lattice is in fact self-dual when regarded as a lattice (of rank 56) over $\mathbb{Z}$. We choose to view this 28 dimensional representation in terms of the maximal of G_2-type: $2^6.G_2(2)$, which becomes $2^7.G_2(2)$ in the double cover $2.Ru$.

The action of the group $2^7.G_2(2)$ in the 28 dimensional representation can be understood in the following way in terms of the E_8 lattice. Let Λ denote a copy of the E_8 lattice, the unique self-dual even lattice of rank 8. (A *lattice* is a free $\mathbb{Z}$-module equipped with a bilinear form, and an *even lattice* is a lattice for which the square-norm of every vector is an even integer.) We may take

$$\Lambda = \left\{ \sum_{i \in \Pi} n_i h_i \mid \sum_{i \in \Pi} n_i \in 2\mathbb{Z}; \text{ all } n_i \in \mathbb{Z}, \text{ or all } n_i \in \mathbb{Z} + \tfrac{1}{2} \right\} \tag{10}$$

---

4 In fact, there is just one other invariant in addition to ν and ϱ. It turns out to be $j_{(0)}\nu$.
5 private communication

where the bilinear form is defined so that $\langle h_i, h_j \rangle = \delta_{ij}$. Then Λ supports a structure of non-associative algebra over $\mathbb{Z}$ which makes it a copy of the integral Cayley algebra, or what is the same, a maximal integral order in the Octonions. To see such a structure explicitly, we assume that the index set $\Pi = \{\infty, 0, 1, 2, 3, 4, 5, 6\}$ is a copy of the projective line over $\mathbb{F}_7$, and then offer the following defining relations (taken from [CCN+85])

$$1 = \tfrac{1}{2} \sum_{i \in \Pi} h_i \tag{11}$$

$$2h_i^2 = h_i - 1 \tag{12}$$

$$2h_\infty h_0 = 1 - h_3 - h_5 - h_6 \tag{13}$$

$$2h_0 h_\infty = 1 - h_2 - h_1 - h_4 \tag{14}$$

and also the images of these relations under the natural action of $L_2(7)$ on the indices. The sublattice of doubles 2Λ is an ideal in this algebra, and we may consider the quotient $\bar{\Lambda} = \Lambda/2\Lambda$ which becomes a copy of the Cayley algebra over the finite field $\mathbb{F}_2$; what we call the *binary Cayley algebra*. We should note that the automorphism group of this algebra is the finite group $G_2(2)$ (which contains the simple group $U_3(3)$ to index 2).

There are just $2^8 = 256$ elements in the binary Cayley algebra. We can count them.

	Type	Count	
zeroes		1	
identities		1	
involutions		63	(15)
square roots of 0		63	
idempotents		72	
cube roots of **1**		56	

There is a natural pairing on the elements of $\bar{\Lambda}$ obtained by sending x to the pair $\{x, \mathbf{1} + x\}$. This association pairs the zero with the identity, the involutions with the idempotents, and partitions the idempotents and cube roots into 36 and 28 pairs, respectively. The group $G_2(2)$ acts transitively on each of these different sets of pairs.

The 28 cube root pairs are in a sense the basis upon which the enhanced SVOA will be constructed. Let us denote them by Δ. A typical cube root of unity in $\bar{\Lambda}$ is $(h_i - h_j)$ for $i \neq j \in \Pi$. A typical involution in $\bar{\Lambda}$ is given by $(h_i + h_j)$ for $i \neq j \in \Pi$, and it is important that for any given involution there are exactly 24 cube roots (12 pairs of cube roots) that are not orthogonal to the chosen involution. The corresponding 12-subsets of Δ are called *dozens*.

We now introduce a complex vector space $\mathfrak{r}$ of dimension 28, with Hermitian form $(\cdot\,,\cdot)$ and an orthonormal basis $\{a_i\}_{i \in \Delta}$, indexed by our cube root pairs. The structures arising from the binary Cayley algebra described above allow us to define an action by the group $2.2^6.G_2(2)$ on this space. For example, the normal subgroup 2.2^6 is generated by the transformations that change sign on the coordinates of a given dozen. With a somewhat finer analysis of the geometry of $\bar{\Lambda}$ we can define the action of the rest of the group (cf. [Dun06a]); we call this group the *monomial group* and we denote it M. (Warning: the action of $G_2(2)$ cannot be realized as coordinate permutations. Instead we must write generators as coordinate permutations followed by multiplications by ± 1 or $\pm \mathbf{i}$ on particular coordinates. The group M is a non-split extension of $G_2(2)$.) The action of M on $\mathfrak{r}$ preserves the Hermitian form.

11. The conformal elements

Assume now that we have a Hermitian space $\mathfrak{r}$ and a unitary action on this space of the monomial group M of the shape $2.2^6.G_2(2)$. We assume further that M is regarded as a matrix group with respect to the basis $\{a_i\}$, and consists of monomial matrices (having one non-zero entry in each row and column). We set $\mathfrak{u} = \mathfrak{r} \oplus \mathfrak{r}^*$, where $\mathfrak{r}^*$ denotes the dual space to $\mathfrak{r}$, and is equipped with the induced Hermitian form. The space $\mathfrak{u}$ then comes equipped with a Hermitian form (obtained by taking direct sum of those associated to the summands) and also a bilinear form $\langle\cdot\,,\cdot\rangle$, induced by the canonical pairing $\mathfrak{r} \times \mathfrak{r}^* \to \mathbb{C}$. We define $\mathrm{Cliff}(\mathfrak{u})$ to be the Clifford algebra of $\mathfrak{u}$ defined with respect to this bilinear form.

$$\mathrm{Cliff}(\mathfrak{u}) = T(\mathfrak{u})/\langle u \otimes u + \langle u, u \rangle \mid u \in \mathfrak{u}\rangle \tag{16}$$

We define CM_X to be the module over $\mathrm{Cliff}(\mathfrak{u})$ spanned by a vector 1_X satisfying $u 1_X = 0$ whenever $u \in \mathfrak{r}^*$. We claim that the isomorphism

$$\mathrm{CM}_X \cong \bigoplus \wedge^n(\mathfrak{r}) 1_X \tag{17}$$

holds when these spaces are viewed as modules over $\mathrm{Cliff}(\mathfrak{r})$ (the subalgebra of $\mathrm{Cliff}(\mathfrak{u})$ generated by $\mathfrak{r} \hookrightarrow \mathrm{Cliff}(\mathfrak{u})$). Next we claim that the degree $7/2$ subspace of A_{Ru} may be naturally identified with the even part of CM_X.

$$(A_{Ru})_{7/2} \longleftrightarrow \mathrm{CM}_X^0 \cong \bigoplus \wedge^{2n}(\mathfrak{r}) \tag{18}$$

The Clifford algebra $\mathrm{Cliff}(\mathfrak{u})$ naturally contains a copy of the group $\mathrm{Spin}(\mathfrak{u})$, and the space CM_X^0 is an irreducible module for this group $\mathrm{Spin}(\mathfrak{u})$.

Recall that our goal is to define the elements v and ϱ. We now define v by setting

$$v = 1_X + a_\Delta 1_X \tag{19}$$

where $a_\Delta = a_\infty a_1 \cdots a_{27} \in \mathrm{Cliff}(\mathfrak{u})$ say. Let us write $\overline{M}$ for the copy of $G_2(2)$ obtained by replacing each non-zero entry with a 1 in each matrix in M. The vector ϱ will be expressed in terms of orbits of $\overline{M}$ on monomials $a_I 1_X$ for $I \subset \Delta$ with $|I| = 14$.

It turns out that there are 80 such orbits, but only 68 give rise to invariants for the monomial group M in $\wedge^{14}(\mathfrak{r})1_X \subset \mathrm{CM}_X^0$. The vector ϱ is a linear sum of these invariants, t_i say, where the coefficients r_i can be taken to lie in $\mathbb{Z}[\mathbf{i}]$.

$$\varrho = \sum_{i=1}^{68} r_i t_i \tag{20}$$

The orbits are paired under complementation, and the coefficients of invariants corresponding to complementary orbits are conjugate (up to sign), so that ultimately, we require to specify 34 values. We refer to [Dun06a, §5.2] for the details.

Finally, we take $A_{Ru} = (A_{Ru}, Y, \mathbf{1}, \{\omega, J, \nu, \varrho\})$ as in the statement of Theorem 4, and this completes the construction.

It follows from the isomorphism (7), with the lattice SVOA for D_{28}^+, that A_{Ru} is self-dual, and has rank 28. To prove that the automorphism group is of the stated form we prove first that it is finite, by showing that it is a reductive algebraic group with trivial Lie algebra (cf. [Dun06a, §5.4]). It follows that it has dimension 0, and hence, is finite. We can explicitly construct generators for $2.Ru$ acting on A_{Ru} and fixing the vectors $\{\omega, J, \nu, \varrho\}$, and we may employ an argument from [NRS01], to show that the only other symmetries possible are scalar multiples of the identity. It is then easy to check that such multiples must be 14th-roots of unity (since they must preserve $\varrho \in \wedge^{14}(\mathfrak{r})1_X$). Finally we obtain a central product $14 \circ 2.Ru$, but the central $\mathbb{Z}/2$ here acts trivially on A_{Ru}, and so the full automorphism group is just $7 \times Ru$.

12. The character

We conclude with consideration of the character of the enhanced SVOA A_{Ru}.

The action of the Rudvalis group Ru on A_{Ru} preserves a certain vector J of degree 1. The residue of the corresponding vertex operator $Y(J, z)$ is denoted $J(0)$, commutes with the Virasoro operator $L(0)$, and has diagonalizable action on A_{Ru}, thus giving rise to a grading by *charge*. It is natural then to consider the two variable series

$$\mathrm{tr}|_{A_{Ru}} p^{J(0)} q^{L(0)-c/24} \tag{21}$$

Table 1 The Character of A_{Ru}

	0	2	4	6	8
0	1				
1/2					
1	784	378			
3/2					
2	144452	92512	20475		
5/2					
3	11327232	8128792	2843568	376740	
7/2	40116600	30421755	13123110	3108105	376740
4	490068257	373673216	161446572	35904960	3108105
9/2	2096760960	1649657520	794670240	226546320	35904960
5	13668945136	10818453324	5284484352	1513872360	226546320
11/2	56547022140	45624923820	23757475560	7766243940	1513872360

which we call the *(2 variable) character* of A_{Ru}. Recall the Jacobi theta function given by

$$\vartheta_3(z|\tau) = \sum_{m \in \mathbb{Z}} e^{2izm + \pi i \tau m^2} \tag{22}$$

and also the Dedekind eta function

$$\eta(\tau) = q^{1/24} \prod_{m \geq 1} (1 - q^m) \tag{23}$$

written here according to the convention $q = e^{2\pi i \tau}$. Let us also convene to write $p = e^{2\pi i z}$. Then we have

Proposition 5 ([Dun06b]). *The character of A_{Ru} is given by*

$$\mathrm{tr}|_{A_{Ru}} p^{J(0)} q^{L(0) - c/24} = \frac{1}{2} \left(\frac{\vartheta_3(\pi z|\tau)^{28}}{\eta(\tau)^{28}} + \frac{\vartheta_3(\pi z + \pi/2|\tau)^{28}}{\eta(\tau)^{28}} \right)$$
$$+ \frac{1}{2} p^{14} q^{7/2} \left(\frac{\vartheta_3(\pi z + \pi\tau/2|\tau)^{28}}{\eta(\tau)^{28}} + \frac{\vartheta_3(\pi z + \pi\tau/2 + \pi/2|\tau)^{28}}{\eta(\tau)^{28}} \right) \tag{24}$$

The terms of lowest charge and degree in the character of A_{Ru} are recorded in Table 1. The column headed m is the coefficient of p^m (as a series in q), and the row headed n is the coefficient of $q^{n-c/24}$ (as a series in p). The coefficients of p^{-m} and p^m coincide, and all subspaces of odd charge vanish.

Many irreducible representations of *Ru* are visible in the entries of Table 1. For example, we have the following equalities, where the left hand sides are the dimensions of homogeneous subspaces of A_{Ru}, and the right hand sides indicate decompositions into irreducibles for the Rudvalis group.

$$378 = 378$$
$$784 = 1 + 783$$
$$20475 = 20475$$
$$92512 = (2)378 + 406 + 91350$$
$$144452 = (3)1 + (3)783 + 65975 + 76125$$
$$376740 = 27405 + 65975 + 75400 + 102400 \qquad (25)$$

References

[Bar00] Katrina Barron. $N = 1$ Neveu-Schwarz vertex operator superalgebras over Grassmann algebras and with odd formal variables. In *Representations and quantizations (Shanghai, 1998)*, pages 9–35. China High. Educ. Press, Beijing, 2000.

[Bor86] Richard Borcherds. Vertex algebras, Kac-Moody algebras, and the Monster. *Proceedings of the National Academy of Sciences, U.S.A.*, 83(10):3068–3071, 1986.

[CCN+85] J. H. Conway, R. T. Curtis, S. P. Norton, R. A. Parker, and R. A. Wilson. *Atlas of finite groups*. Oxford University Press, Eynsham, 1985. Maximal subgroups and ordinary characters for simple groups, With computational assistance from J. G. Thackray.

[Che54] Claude C. Chevalley. *The algebraic theory of spinors*. Columbia University Press, New York, 1954.

[Con77] J. H. Conway. A quaternionic construction for the Rudvalis group. In *Topics in group theory and computation (Proc. Summer School, University Coll., Galway, 1973)*, pages 69–81. Academic Press, London, 1977.

[Con85] J. H. Conway. A simple construction for the Fischer-Griess monster group. *Invent. Math.*, 79(3):513–540, 1985.

[CW73] J. H. Conway and D. B. Wales. Construction of the Rudvalis group of order 145, 926, 144, 000. *J. Algebra*, 27:538–548, 1973.

[Dun06a] John F. Duncan. Moonshine for Rudvalis's sporadic group I. arXiv:math.RT/0609449, September 2006.

[Dun06b] John F. Duncan. *Vertex operators, and three sporadic groups*. PhD thesis, Yale University, 2006.

[Dun07] John F. Duncan. Super-moonshine for Conway's largest sporadic simple group. *Duke Math. J.*, 139(2):255–315, 2007.

[FFR91] Alex J. Feingold, Igor B. Frenkel, and John F. X. Ries. *Spinor construction of vertex operator algebras, triality, and $E_8^{(1)}$*, volume 121 of *Contemporary Mathematics*. American Mathematical Society, Providence, RI, 1991.

[FLM88] Igor Frenkel, James Lepowsky, and Arne Meurman. *Vertex operator algebras and the Monster*, volume 134 of *Pure and Applied Mathematics*. Academic Press Inc., Boston, MA, 1988.

[FZ92] Igor B. Frenkel and Yongchang Zhu. Vertex operator algebras associated to representations of affine and Virasoro algebras. *Duke Math. J.*, 66(1):123–168, 1992.

[GLS94] Daniel Gorenstein, Richard Lyons, and Ronald Solomon. *The classification of the finite simple groups*, volume 40 of *Mathematical Surveys and Monographs*. American Mathematical Society, Providence, RI, 1994.

[Gri82] Robert L. Griess, Jr. The friendly giant. *Invent. Math.*, 69(1): 1–102, 1982.

[Höh96] Gerald Höhn. *Selbstduale Vertexoperatorsuperalgebren und das Babymonster*, volume 286 of *Bonner Mathematische Schriften [Bonn Mathematical Publications]*. Universität Bonn Mathematisches Institut, Bonn, 1996. Dissertation, Rheinische Friedrich-Wilhelms-Universität Bonn, Bonn, 1995.

[Lam99] Ching Hung Lam. On VOA associated with special Jordan algebras. *Comm. Algebra*, 27(4):1665–1681, 1999.

[NRS01] Gabriele Nebe, E. M. Rains, and N. J. A. Sloane. The invariants of the Clifford groups. *Designs, Codes and Cryptography*, 24: 99–122, 2001.

[Sol01] Ronald Solomon. A brief history of the classification of the finite simple groups. *Bull. Amer. Math. Soc. (N.S.)*, 38(3):315–352 (electronic), 2001.

[Wil84] Robert A. Wilson. The geometry and maximal subgroups of the simple groups of A. Rudvalis and J. Tits. *Proc. London Math. Soc. (3)*, 48(3):533–563, 1984.

[Zhu90] Yongchang Zhu. *Vertex Operator Algebras, Elliptic Functions, and Modular Forms*. PhD thesis, Yale University, 1990.

The Algebraic Meaning of Being a Hauptmodul

Terry Gannon

Department of Mathematical Sciences,
University of Alberta
Edmonton, Canada, T6G 2G1
tgannon@math.ualberta.ca

Abstract

The Conway–Norton conjectures unexpectedly related the Monster with certain special modular functions (Hauptmoduls). Their proof by Borcherds *et al* was remarkable for demonstrating the rich mathematics implicit there. Unfortunately Moonshine remained about as mysterious after the proof as before. In particular, a computer check — as opposed to a general conceptual argument — was used to verify the Monster functions equal the appropriate modular functions. This, the so-called 'conceptual gap', was eventually filled; we review the solution here. We conclude by speculating on the shape of a new proof of the Moonshine conjectures.

1. The conceptual gap

The main Conway–Norton conjecture [5] says:

Theorem 1. *There is an infinite-dimensional graded representation* $V = \oplus_{n=-1}^{\infty} V_n$ *of the Monster* $\mathbb{M}$, *such that the McKay–Thompson series*

$$T_g(\tau) := \mathrm{Tr}_V g \, q^{L_0 - 1} = \sum_{n \geq -1} c_n(g) \, q^n \tag{1.1}$$

equals the Hauptmodul J_g *for some subgroup* Γ_g *of* $SL_2(\mathbb{R})$.

Moreover, each coefficient $c_n(g)$ lies in $\mathbb{Z}$, and Γ_g contains the congruence subgroup $\Gamma_0(N)$ as a normal subgroup, where $N = h \, o(g)$ for some h dividing $\gcd(24, o(g))$ ($o(g)$ is the order of $g \in \mathbb{M}$). In his ICM (1998) talk [2], Borcherds outlined the proof of Theorem 1:

(i) Construction of the Frenkel–Lepowsky–Meurman Moonshine module $V^\natural$, which is to equal the space V;

204

(ii) Derivation of recursions for the McKay–Thompson coefficients $c_n(g)$ for $V^\natural$, such as

$$c_{4n+2}(g) = c_{2k+2}(g^2) + \sum_{j=1}^{k} c_j(g^2) c_{2k+1-j}(g^2) \qquad \forall k \geq 1 ; \quad (1.2)$$

(iii) From these recursions, prove $T_g = J_g$.

The original treatment of part (i) is [12]; see also the excellent review [24] and references therein. Borcherds derived (ii) by first constructing a Lie algebra out of $V^\natural$, and then computing its twisted denominator identities [1]. It was already known that Hauptmoduls satisfied these recursions, and that any function obeying all those recursions was uniquely determined by its first few coefficients. Thus establishing (iii) merely requires comparing finitely many coefficients of each T_g with J_g — in fact, comparing 5 coefficients for each of the 171 functions suffices [1]. In this way Borcherds accomplished (iii) and with it completed the proof of Theorem 1. The proof successfully established the mathematical richness of the subject, and for his work Borcherds deservedly received a Fields medal (math's highest honour) in 1998.

Our quick sketch hides the technical sophistication of the proof of (i) and (ii). Subsequent treatments of the vertex operator structure of the Moonshine module $V^\natural$ of (i) have been given in [11, 18, 26], and the derivations of the recursions (ii) have been simplified in [19, 20, 21]. But the biggest disappointment with the proof is hidden in the nearly trivial argument of (iii).

At the risk of sending shivers down Bourbaki's collective spine, the point of mathematics is surely not acquiring proofs (just as the point of theoretical physics is not careful calculations, and that of painting is not the creation of realistic scenes on canvas). The point of mathematics, like that of any intellectual discipline, is to find qualitative truths, to abstract out patterns from the inundation of seemingly isolated facts. An example is the notion of group. Another, dear to many of us, is the A-D-E meta-pattern: many different classifications (e.g. finite subgroups of SU_2, subfactors of small index, the simplest conformal field theories) fall unexpectedly into the same pattern. The conceptual explanation for the ubiquity of this meta-pattern — that is, the combinatorial fact common to its various manifestations — presumably involves the graphs with largest eigenvalues $|\lambda| \leq 2$.

Likewise, the real challenge of Monstrous Moonshine wasn't to prove Theorem 1; but rather to understand what the Monster has to do with modularity and genus-0. The first proof was due to Atkin, Fong and Smith [29], who by studying the first 100 coefficients of the T_g verified (without constructing it) that there existed a (possibly virtual) representation V of $\mathbb{M}$ obeying the Theorem. Their proof is forgotten because it didn't explain anything.

By contrast, the proof of Theorem 1 by Borcherds *et al* is clearly superior: it explicitly constructs $V = V^{\natural}$, and emphasises the remarkable mathematical richness saturating the problem. On the other hand, it also fails to explain modularity and the Hauptmodul property. The problem is step (iii): precisely at the point where we want to identify the algebraically defined T_g's with the topologically defined J_g's, a conceptually empty computer check of a few hundred coefficients is done. This is called the conceptual gap of Monstrous Moonshine, and it has an analogue in Borcherds' proof of Modular Moonshine [3] and in Höhn's proof of 'generalised Moonshine' for the Baby Monster [16]. Clearly preferable would be to replace the numerical check of [1] with a more general theorem.

Next section we review the standard definition of Hauptmodul. In Section 3 we describe the solution [9] to the conceptual gap: it replaces that topological definition of Hauptmodul with an algebraic one. We conclude the paper with some speculations. Even with the developments [11, 18, 19, 20, 21, 26] and especially [9] to the original proof of Theorem 1, the resulting argument still does a poor job explaining Monstrous Moonshine. Moonshine remains mysterious to this day. There is a lot left to do — for example establishing Norton's generalised Moonshine [28], or finding the Moonshine manifold [15]. But the greatest task for Moonshiners is to find a second independent proof of Theorem 1. It would (hopefully) clarify some things that the original proof leaves murky. In particular, we still don't know what really is so important about the Monster, that it has such a rich genus-0 moonshine. To what extent does Monstrous Moonshine determine the Monster? We turn to this open problem in Section 4.

2. The topological meaning of the Hauptmodul property

Just as a *periodic* function is a function on a compact *real* curve (i.e. on a circle), a *modular* function is a function on a compact *complex* curve. More precisely, let Σ be a compact surface. We can regard this as a complex curve, and thus put on it a complex analytic structure. Up to (biholomorphic or conformal) equivalence, there is a unique genus 0 surface (which we can take to be the Riemann sphere $\mathbb{C} \cup \{\infty\}$), but there is a continuum (moduli space) of inequivalent complex analytic structures which can be placed on a torus (genus 1), a double-torus, etc. For example, this moduli space for the torus can be naturally identified with $\mathbb{H}/SL_2(\mathbb{Z})$, where $\mathbb{H}$ is the upper half-plane $\{\tau \in \mathbb{C} \mid \operatorname{Im}(\tau) > 0\}$: for any $\tau \in \mathbb{H}$, we get the torus $\mathbb{C}/(\mathbb{Z} + \mathbb{Z}\tau)$; any torus is equivalent to one of that form; if $\tau' = (a\tau+b)/(c\tau+d)$ for $\begin{pmatrix} a & b \\ c & d \end{pmatrix} \in SL_2(\mathbb{Z})$, then the tori corresponding to τ and τ' are equivalent.

We learn from geometry and physics that we should study a space through the functions (fields) that live on it, which respect the relevant properties of the space. Therefore we should consider meromorphic functions $f : \Sigma \to \mathbb{C}$ (a meromorphic function is holomorphic everywhere, except for isolated finite poles; we would have preferred f to be holomorphic, but then f would be constant). These f are called *modular functions*, and are thus as important as complex curves. Clearly, they should be central to mathematics. Somewhat more surprising is that in particular they are fundamental to number theory, but that is another story.

We know all about meromorphic functions for $\mathbb{C}$: these include rational functions (i.e. quotients of polynomials), together with transcendental functions such as $\exp(z)$ and $\cos(z)$. These transcendental functions all have an essential singularity at ∞. In fact, the modular functions for the Riemann sphere are the rational functions in z. By contrast, the modular functions for the other compact surfaces will be rational functions in two generators, where those two generators satisfy a polynomial relation. For example, the modular functions for a torus are generated by the Weierstrass $\wp$-function and its derivative, and $\wp$ and $\wp'$ satisfy the cubic equation defining the torus. In this way, the sphere is distinguished from all other compact surfaces.

There are three possible geometries in two dimensions: Euclidean, spherical and hyperbolic. The most important of these, in any sense, is the hyperbolic one. The upper half-plane is a model for it: its 'lines' consist of vertical half-lines and semi-circles, its infinitesimal metric is $ds = |d\tau|/\mathrm{Im}(z)$, etc. As with Euclidean geometry, 'lines' are the paths of shortest distance. Just as the Euclidean plane $\mathbb{R}^2$ has a circular horizon (one infinite point for every angle θ), so does the hyperbolic plane $\mathbb{H}$, and it can be identified with $\mathbb{R} \cup \{i\infty\}$. The group of isometries (geometry-preserving transformations $\mathbb{H} \to \mathbb{H}$) is $SL_2(\mathbb{R})$, which acts on $\mathbb{H}$ as fractional linear transformations $\tau \mapsto \frac{a\tau+b}{c\tau+d}$, and sends the horizon to itself.

It turns out that any compact surface Σ can be realised (in infinitely many different ways) as the compactification of the space $\mathbb{H}/\Gamma$ of orbits, for some discrete subgroup Γ of $SL_2(\mathbb{R})$. The compactification amounts to including finitely many Γ-orbits of horizon points. For the most important example, $\mathbb{H}/SL_2(\mathbb{Z})$ can be identified with the sphere with one puncture, corresponding to the single compactification orbit $\mathbb{Q} \cup \{i\infty\}$. Those points in $\mathbb{Q} \cup \{i\infty\}$ are called cusps. The Γ of greatest interest in number theory, and to us, are those commensurable to $SL_2(\mathbb{Z})$: i.e. $\Gamma \cap SL_2(\mathbb{Z})$ is also an infinite discrete group with finite index in both Γ and $SL_2(\mathbb{Z})$. Their compactification points will again be the cusps $\mathbb{Q} \cup \{i\infty\}$. Examples of such groups are $SL_2(\mathbb{Z})$ itself, as well as its subgroups

$$\Gamma(N) = \{A \in SL_2(\mathbb{Z}) \mid A \equiv \pm I \ (\mathrm{mod}\ N)\}, \tag{2.1}$$

$$\Gamma_0(N) = \left\{ \begin{pmatrix} a & b \\ c & d \end{pmatrix} \in SL_2(\mathbb{Z}) \mid N \text{ divides } c \right\}, \tag{2.2}$$

$$\Gamma_1(N) = \left\langle \Gamma(N), \begin{pmatrix} 1 & 1 \\ 0 & 1 \end{pmatrix} \right\rangle. \tag{2.3}$$

For example, $\mathbb{H}/\Gamma_0(2)$ and $\mathbb{H}/\Gamma(2)$ are spheres with 2 and 3 punctures, respectively, while e.g. $\Gamma_0(24)$ is a torus with 7 punctures.

The modular functions for the compact surface $\Sigma = \overline{\mathbb{H}/\Gamma}$ are easy to describe: they are the meromorphic functions f on $\mathbb{H}$, which are also meromorphic at the cusps $\mathbb{Q} \cup \{i\infty\}$, and which obey the symmetry $f(\frac{a\tau+b}{c\tau+d}) = f(\tau)$ for all $\begin{pmatrix} a & b \\ c & d \end{pmatrix} \in \Gamma$. The precise definition of 'meromorphic at the cusps' isn't important here. For example, for any Γ obeying

$$\begin{pmatrix} 1 & t \\ 0 & 1 \end{pmatrix} \in \Gamma \text{ iff } t \in \mathbb{Z}, \tag{2.4}$$

a meromorphic function $f(\tau)$ with symmetry Γ will have a Fourier expansion $\sum_{n=-\infty}^{\infty} a_n q^n$ for $q = e^{2\pi i \tau}$ (q is a local coordinate for $\tau = i\infty$); then f is meromorphic at the cusp $i\infty$ iff all but finitely many a_n, for $n < 0$, are nonzero.

We say a group Γ is genus-0 when $\Sigma = \overline{\mathbb{H}/\Gamma}$ is a sphere. For these Γ, there will be a uniformising function $f_\Gamma(\tau)$ identifying Σ with the Riemann sphere $\mathbb{C} \cup \{\infty\}$. That is, f_Γ will be the mother-of-all modular functions; i.e., it is a modular function for Γ with the property that any other modular function $f(\tau)$ for Γ can be written uniquely as a rational function $poly(f_\Gamma(\tau))/poly(f_\Gamma(\tau))$. This function f_Γ is not quite unique ($SL_2(\mathbb{R})$ permutes these generating functions). In genus > 0, two (non-canonical) generating functions will be needed.

The groups Γ we are interested in are genus-0, contain some $\Gamma_0(N)$, and obey (2.4). We call such Γ *genus-0 groups of Moonshine-type*. Cummins [8] has classified all of these — there are precisely 6486 of them. For these groups (and more generally a group containing a $\Gamma_1(N)$ rather than a $\Gamma_0(N)$) there is a canonical choice of generator f_Γ: we can always choose it uniquely so that it has a q-expansion of the form $q^{-1} + \sum_{n=1}^{\infty} a_n q^n$. This choice of generator is called the *Hauptmodul*, and will be denoted $J_\Gamma(\tau)$. Some examples are

$$J_{\Gamma(1)}(\tau) = q^{-1} + 196884\,q + 21493760\,q^2 + 864299970\,q^3 + \cdots \tag{2.5}$$

$$J_{\Gamma_0(2)}(\tau) = q^{-1} + 276q - 2048q^2 + 11202q^3 - 49152q^4 + 184024q^5 + \cdots \tag{2.6}$$

$$J_{\Gamma_0(13)}(\tau) = q^{-1} - q + 2q^2 + q^3 + 2q^4 - 2q^5 - 2q^7 - 2q^8 + q^9 + \cdots \tag{2.7}$$

$$J_{\Gamma_0(25)}(\tau) = q^{-1} - q + q^4 + q^6 - q^{11} - q^{14} + q^{21} + q^{24} - q^{26} + \cdots . \tag{2.8}$$

Of course $J_{\Gamma(1)}$ is the famous J-function. Exactly 616 of these Hauptmoduls have integer coefficients (171 of which are the McKay–Thompson series); the remainder have cyclotomic integer coefficients.

3. The algebraic meaning of the Hauptmodul property

The conceptual gap will be bridged only when we can directly relate the definition of a Hauptmodul (which is inherently topological), with the 'replicability' recursions coming from the twisted denominator identities.

The easiest way to produce functions invariant with respect to some symmetry, is to average over the group. For example, given any function $f(x)$, the average $f(x) + f(-x)$ is invariant under $x \leftrightarrow -x$. When the group is infinite, a little more subtlety is required but the same idea can work.

For example, take $\Gamma = \mathrm{SL}_2(\mathbb{Z})$ and let p be any prime. Then

$$\Gamma \begin{pmatrix} p & 0 \\ 0 & 1 \end{pmatrix} \Gamma = \{A \in M_{2 \times 2}(\mathbb{Z}) \mid \det(A) = p\}$$

$$= \Gamma \begin{pmatrix} p & 0 \\ 0 & 1 \end{pmatrix} \cup \bigcup_{k=0}^{p-1} \Gamma \begin{pmatrix} 1 & k \\ 0 & p \end{pmatrix}. \tag{3.1}$$

This means that, for any modular function $f(\tau)$ of $\mathrm{SL}_2(\mathbb{Z})$, $f(p\tau)$ will no longer be $\mathrm{SL}_2(\mathbb{Z})$-invariant, but

$$s_f^{(p)}(\tau) := f(p\tau) + \sum_{k=0}^{p-1} f\left(\frac{\tau + k}{p}\right) \tag{3.2}$$

is. Considering now f to be the Hauptmodul J, we thus obtain that $s_J^{(p)}(\tau) = P(J(\tau))/Q(J(\tau))$, for polynomials P, Q. By considering poles and the surjectivity of J, we see that Q must be constant, and hence that $s_J^{(p)}$ must be a polynomial in J. The same will hold for $s_{J^k}^{(p)}$.

This implies that there is a monic polynomial $F_p(x, y)$ of degree $p + 1$ in x, y, such that

$$F_p(J(\tau), J(p\tau)) = F_p\left(J(\tau), J\left(\frac{\tau + k}{p}\right)\right) = 0 \qquad \forall k = 0, \ldots, p - 1,$$

$$\tag{3.3}$$

or equivalently

$$F_p(J(\tau), Y) = (J(p\tau) - Y) \prod_{k=0}^{p-1} \left(J\left(\frac{\tau+k}{p}\right) - Y \right). \tag{3.4}$$

For example,

$$F_2(x, y) = (x^2 - y)(y^2 - x) - 393768\,(x^2 + y^2) - 42987520\,xy$$
$$- 40491318744\,(x + y) + 12098170833256. \tag{3.5}$$

There is nothing special about p being prime; for a composite number m, the sum in e.g. (3.2) becomes a sum over $\begin{pmatrix} m/d & k \\ 0 & d \end{pmatrix}$, for all divisors d of m and all $0 \le k < d$. Write $\mathcal{A}_m$ for the set of all these pairs (d, k). Note that its cardinality $\|\mathcal{A}_m\|$ is $\psi(m) = m \prod_{p|m}(1 + 1/p)$.

Definition 1. Let $h(\tau) = q^{-1} + \sum_{n=1}^{\infty} a_n q^n$. We say that $h(\tau)$ satisfies a modular equation of order $m > 1$, if there is a monic polynomial $F_m(x, y) \in \mathbb{C}[x, y]$ such that F_m is of degree $\psi(m)$ in both x and y, and

$$F_m(h(\tau), Y) = \prod_{(d,k)\in\mathcal{A}_m} \left(h\left(\frac{m\tau}{d^2} + \frac{k}{d}\right) - Y \right). \tag{3.6}$$

It is unnecessary to assume that the series h converges; it is enough to require that (3.6) holds formally at the level of q-series. An easy consequence of this definition is that $F_m(x, y) = F_m(y, x)$.

We've learnt above that the Hauptmodul J satisfies a modular equation of all orders $m > 1$. In fact similar reasoning verifies, more generally, that:

Proposition 1. (a) If $J_\Gamma(\tau)$ is the Hauptmodul of some genus-0 group Γ of Moonshine-type, with rational coefficients, then J_Γ satisfies a modular equation for all m coprime to N.

(b) [10] Likewise, any McKay–Thompson series $T_g(\tau)$ (1.1) satisfies a modular equation for any m coprime to the order of the element $g \in \mathbb{M}$.

Recall there are 616 such J_Γ, and 171 such T_g. The N in part (a) is the level of any congruence group $\Gamma_0(N)$ contained in Γ. Part (b) involves showing that the recursions such as (1.2) imply the modular equation property q-coefficient-wise.

Note that the Hauptmodul property of J_Γ plays a crucial role in the proof that they satisfy modular equations. Could the converse of Proposition 1(a) hold? Such a converse combined with Proposition 1(b) would fill the conceptual gap.

Unfortunately, that is too naive. In particular, $h(\tau) = q^{-1}$ also satisfies a modular equation for any $m > 1$: take $F_m(x, y) = (x^m - y)(y^m - x)$. Using

Tchebychev polynomials, it is easy to show that $h(\tau) = q^{-1} + q$ (which is essentially cosine) likewise satisfies a modular equation for any m.

However, Kozlov, in a thesis directed by Meurman, proved the following remarkable fact:

Theorem 2. [22] *If $h(\tau) = q^{-1} + \sum_{n=1}^{\infty} a_n q^n$ satisfies a modular equation for all $m > 1$, then either $h = J$, $h(\tau) = q^{-1}$, or $h(\tau) = q^{-1} \pm q$.*

His proof breaks down when we no longer have all those modular equations, but it gives us confidence to hope that modular equations could provide an algebraic interpretation of what it means to be a Hauptmodul. Indeed that is the case!

Theorem 3. [9] *Suppose a formal series $h(\tau) = q^{-1} + \sum_{n=1}^{\infty} a_n q^n$ satisfies a modular equation for all $m \equiv 1$ (mod K). Then $h(\tau)$ is holomorphic throughout $\mathbb{H}$. Write*

$$\Gamma_h := \left\{ \begin{pmatrix} a & b \\ c & d \end{pmatrix} \in \mathrm{SL}_2(\mathbb{R}) \mid h\left(\frac{a\tau + b}{c\tau + d}\right) = h(\tau) \, \forall \tau \in \mathbb{H} \right\}. \tag{3.7}$$

(a) *If $\Gamma_h \neq \left\{ \pm \begin{pmatrix} 1 & n \\ 0 & 1 \end{pmatrix} \mid n \in \mathbb{Z} \right\}$, then h is a Hauptmodul for Γ_h, and Γ_h obeys (2.4) and contains $\Gamma_0(N)$ for some $N | K^{\infty}$.*

(b) *If $\Gamma_h = \left\{ \pm \begin{pmatrix} 1 & n \\ 0 & 1 \end{pmatrix} \mid n \in \mathbb{Z} \right\}$, and the coefficients a_n of h are algebraic integers, then $h(z) = q^{-1} + \xi q$ where $\xi = 0$ or $\xi^{\gcd(K, 24)} = 1$.*

By '$N | K^{\infty}$' we mean any prime dividing N also divides K. Of course part (a) implies that Γ_h is genus-0 and of moonshine-type. When $h = T_g$, $K = o(g)$ works (see Prop.1(b)), and all coefficients are integers, and so Theorem 3 establishes the Hauptmodul property and fills the conceptual gap. The proof of Theorem 3 is difficult: if $h(\tau_1) = h(\tau_2)$, then it is fairly easy to prove that locally there is an invertible holomorphic map α sending an open disc about τ_1 onto one about τ_2; the hard part of the proof is to show that α extends to a globally invertible map $\mathbb{H} \to \mathbb{H}$ (and hence lies in Γ_h).

The converse of Theorem 3 is also true:

Proposition 2. [9] *If $h(\tau) = q^{-1} + \sum_{n=1}^{\infty} a_n q^n$ is a Hauptmodul for a group Γ_h of moonshine-type, and the coefficients a_n all lie in the cyclotomic field $\mathbb{Q}[\xi_N]$, then there exists a generalised modular equation for any order m coprime to N. Moreover, the field generated over $\mathbb{Q}$ by all coefficients a_n will be a Galois extension of $\mathbb{Q}$, with Galois group of exponent 2.*

The exponent 2 condition means that that field is generated over $\mathbb{Q}$ by a number of square-roots of rationals. We write ξ_N for the root of unity $\exp[2\pi i/N]$.

The condition that all a_n lie in the cyclotomic field should be automatically satisfied. By a 'generalised modular equation' of order $m > 1$, we mean that there is a polynomial $F_m(x, y) \in \mathbb{Q}[\xi_N][x, y]$ such that F_m is monic of degree $\psi(m)$ in both x and y, and

$$F_m((\sigma_m.h)(\tau), Y) = \prod_{(d,k)\in\mathcal{A}_m} \left(h\left(\frac{m\tau}{d^2} + \frac{k}{d}\right) - Y \right). \qquad (3.8)$$

We also have the symmetry condition $F_m(x, y) = (\sigma_m.F_m)(y, x)$. Here, $\sigma_m \in \mathrm{Gal}(\mathbb{Q}[\xi_N]/\mathbb{Q}) \cong \mathbb{Z}_N^*$ is the Galois automorphism sending ξ_N to ξ_N^m; it acts on h and F_m coefficient-wise. The beautiful relation of modular functions to cyclotomic fields and their Galois groups is classical and is reviewed in e.g. Chapter 6 of [23].

Proposition 2 explains why Proposition 1 predicts more modular equations than Theorem 3 assumes: if all coefficients a_n in Theorem 3 are rational, then indeed we'd get that h would satisfy an ordinary modular equation for all m coprime to n.

The lesson of Moonshine is that we probably shouldn't completely ignore the exceptional functions in Theorem 3(b). It is tempting to call those 25 functions *modular fictions* (following John McKay). So a question could be:

Question 1. *What is the question in e.g. vertex algebras, for which the modular fictions are the answer?*

Theorem 3 requires many more modular equations than is probably necessary. In particular, the computer experiments in [4] show that if h has integer coefficients and satisfies modular equations of order 2 and 3, then f is either a Hauptmodul, or a modular fiction. Cummins has made the following conjecture:

Conjecture 1. [7] *Let p_1, p_2 be distinct primes, and $a_i \in \mathbb{C}$. Suppose $h(\tau) = q^{-1} + \sum_{n=1}^{\infty} a_n q^n$ satisfies modular equations of order p_1, p_2. Then*

(a) *If $\Gamma_h \neq \left\{ \pm \begin{pmatrix} 1 & n \\ 0 & 1 \end{pmatrix} \mid n \in \mathbb{Z} \right\}$, then h is a Hauptmodul for Γ_h, and Γ_h obeys (2.4) and contains $\Gamma_1(N)$ for some N coprime to p_1, p_2.*

(b) *If $\Gamma_h = \left\{ \pm \begin{pmatrix} 1 & n \\ 0 & 1 \end{pmatrix} \mid n \in \mathbb{Z} \right\}$, then $h(z) = q^{-1} + \xi q$ where $\xi = 0$ or $\xi^{\gcd(p_1-1, p_2-1)} = 1$.*

We are far from proving this. However, if h obeys a modular equation of order m for all m with the property that all prime divisors p of m obey $p \equiv 1$

(mod K) for some fixed K, then h is either a Hauptmodul for Γ_h containing some $\Gamma_1(N)$, or h is 'trivial' (see [7] for details and a proof). The converse again is true (again provided the a_i are cyclotomic).

It would be interesting to apply similar arguments to fill the related conceptual gaps of Modular Moonshine [3] and Baby Moonshine [16]. Modular equations have many uses in number theory, besides these in Moonshine — see e.g. [6] for important applications to class field theory. Modular equations are also closely related to the notion of replicable functions (see e.g. [27]).

4. The meaning of moonshine

As mentioned earlier, the greatest open challenge for Monstrous Moonshine is to find a second independent proof. In this section we offer some thoughts on what this proof may involve; see [14] for details.

A powerful guide to Monstrous Moonshine has been rational conformal field theory (RCFT). Modularity arises in RCFT through the conjunction of two standard pictures:

(1) *canonical quantisation* presents us with a state space V, carrying a representation of the symmetries of the theory, a Hamiltonian operator H, etc. In RCFT, we take graded traces such as $\mathrm{Tr}_V q^H$, defining the coefficients of our q-expansions.

(2) The *Feynman picture* interprets the amplitudes using path integrals. In RCFT this permits us to interpret these graded traces as functions (sections) over moduli spaces, and hence they carry actions by the relevant mapping class groups such as $\mathrm{SL}_2(\mathbb{Z})$. This gives us modularity.

In Monstrous Moonshine, canonical quantisation is successfully abstracted into the language of vertex operator algebras (VOAs). The present proof of the Conway–Norton conjectures however ignores the Feynman side, and with it the lesson from RCFT that modularity is ultimately topological. Perhaps this is where to search for a second more conceptual proof. After all, the proof of the modularity of VOA characters [30] — perhaps the deepest result concerning VOAs — follows exactly this RCFT intuition. Let's briefly revisit the RCFT treatment of characters.

In an RCFT, with 'chiral algebra' (i.e. VOA) V, the character of a 'sector' (i.e. V-module) M is really a 'one-point function' on the torus. Fix a torus $\mathbb{C}/(\mathbb{Z} + \mathbb{Z}\tau)$, a local parameter $z \in \mathbb{C}$ at the marked point (which we can take to be 0), and a 'field insertion' v (which can belong to any V-module, but for now we'll take $v \in V$). The local parameter z is needed for sewing surfaces together at the marked points (a fundamental process in RCFT). For convenience assume that $Hv = kv$ (this eigenvalue $k \in \mathbb{Q}$ is called the 'conformal weight' of v). The character is given by

$$\chi_M(\tau, v, z) := \mathrm{Tr}_M Y(v, e^{2\pi i z}) q^{H-c/24} = e^{-2k\pi i z} \mathrm{Tr}_M o(v) q^{H-c/24}, \quad (4.1)$$

where c is the 'central charge' ($c = 24$ in Monstrous Moonshine), and $o(v)$ is an endomorphism commuting with H (also called L_0). We get a moduli space $\widehat{\mathcal{M}}_{1,1}$ of 'extended tori', i.e. tori with a choice of local parameter z at 0, and what naturally acts on these χ_M is the mapping class group $\widehat{\Gamma}_{1,1}$ of $\widehat{\mathcal{M}}_{1,1}$.

This 'extended' moduli space $\widehat{\mathcal{M}}_{1,1}$ is larger than the usual moduli space $\mathcal{M}_{1,1} = \mathbb{H}/SL_2(\mathbb{Z})$ of a torus with one marked point, and the mapping class group $\widehat{\Gamma}_{1,1}$ is larger than the familiar mapping class group $\Gamma_{1,1} = SL_2(\mathbb{Z})$. In fact, $\widehat{\Gamma}_{1,1}$ can be naturally identified with the braid group

$$\mathcal{B}_3 = \langle \sigma_1, \sigma_2 \mid \sigma_1\sigma_2\sigma_1 = \sigma_2\sigma_1\sigma_2 \rangle, \quad (4.2)$$

and acts on the characters by

$$\sigma_1.\chi_M(\tau, v, z) = e^{-2\pi i k/12} \chi_M(\tau + 1, v, z), \quad (4.3)$$

$$\sigma_2.\chi_M(\tau, v, z) = e^{-2\pi i k/12} \chi_M\left(\frac{\tau}{1-\tau}, \frac{v}{(1-\tau)^k}, z\right). \quad (4.4)$$

Thus in RCFT it is really $\mathcal{B}_3$ and not $SL_2(\mathbb{Z})$ which acts. This is usually ignored because we specialise χ_M, and more fundamentally because typically we consider only insertions $v \in V$, and what results is a true action of the modular group $SL_2(\mathbb{Z})$. But taking v from other V-modules is equally fundamental in the theory, and for those insertions we only get a projective action of $SL_2(\mathbb{Z})$ (though again a true action of $\mathcal{B}_3$).

This is just a hint of a much more elementary phenomenon. Recall that a modular form f for $\Gamma := SL_2(\mathbb{Z})$ is a holomorphic function $f : \mathbb{H} \to \mathbb{C}$, which is also holomorphic at the cusps, and which obeys

$$f\left(\frac{a\tau + b}{c\tau + d}\right) = \mu\begin{pmatrix} a & b \\ c & d \end{pmatrix}(c\tau + d)^k f(\tau) \quad \forall \begin{pmatrix} a & b \\ c & d \end{pmatrix} \in \Gamma, \quad (4.5)$$

for some $k \in \mathbb{Q}$ (called the *weight*) and some function μ (called the *multiplier*) with modulus $|\mu| = 1$. For example, the Eisenstein series

$$E_k(\tau) = \sum_{(m,n)\in\mathbb{Z}^2} {}' (m\tau + n)^{-k} \quad (4.6)$$

for even $k > 2$ is a modular form of weight k with trivial multiplier μ, but the Dedekind eta

$$\eta(\tau) = q^{1/24} \prod_{n=1}^{\infty}(1 - q^n) \quad (4.7)$$

is a modular form of weight $k = 1/2$ with a nontrivial multiplier, given by

$$\mu \begin{pmatrix} a & b \\ c & d \end{pmatrix} = \exp\left(\pi i \left(\frac{a+d}{12c} - \frac{1}{2} - \sum_{i=1}^{c-1} \frac{i}{c} \left(\frac{di}{c} - \left\lfloor \frac{di}{c} \right\rfloor - \frac{1}{2} \right) \right) \right) \quad (4.8)$$

when $c > 0$.

$\mathbb{H}$ can be regarded as a homogeneous space $SL_2(\mathbb{R})/SO_2(\mathbb{R})$. Nowadays we are taught to lift a modular form f from $\mathbb{H}$ to $SL_2(\mathbb{R})$:

$$\phi_f \begin{pmatrix} a & b \\ c & d \end{pmatrix} := f\left(\frac{ai+b}{ci+d} \right) (ci+d)^{-k} \mu \begin{pmatrix} a & b \\ c & d \end{pmatrix}^* . \quad (4.9)$$

We've sacrificed the implicit $SO_2(\mathbb{R})$-invariance and explicit Γ-covariance of f, for explicit $SO_2(\mathbb{R})$-covariance and explicit Γ-invariance of ϕ_f. This is significant, because compact Lie groups like the circle $SO_2(\mathbb{R})$ are much easier to handle than infinite discrete groups like $SL_2(\mathbb{Z})$. The result is a much more conceptual and powerful picture.

Thus a modular form should be regarded as a function on the orbit space $X := \Gamma \backslash \mathbb{H}$. Remarkably, this 3-space X can be naturally identified with the complement of the trefoil! We are thus led to ask:

Question 2. *Do modular forms for* $SL_2(\mathbb{Z})$ *see the trefoil?*

An easy calculation shows that the fundamental group $\pi_1(X)$ is in fact the braid group $\mathcal{B}_3$! It is a central extension of $SL_2(\mathbb{Z})$ by $\mathbb{Z}$. In particular, the quotient of $\mathcal{B}_3$ by its centre $\langle (\sigma_1 \sigma_2 \sigma_1)^2 \rangle$ is $PSL_2(\mathbb{Z})$; the isomorphism $\mathcal{B}_3/\langle (\sigma_1 \sigma_2 \sigma_1)^4 \rangle \cong SL_2(\mathbb{Z})$ is defined by the (reduced and specialised) Burau representation

$$\sigma_1 \mapsto \begin{pmatrix} 1 & 1 \\ 0 & 1 \end{pmatrix}, \qquad \sigma_2 \mapsto \begin{pmatrix} 1 & 0 \\ -1 & 1 \end{pmatrix} . \quad (4.10)$$

Through this map, which is implicit in (4.3) and (4.4), $\mathcal{B}_3$ acts on modular forms, and the multiplier μ can be lifted to $\mathcal{B}_3$. For example, the multiplier of the Dedekind eta becomes

$$\mu(\beta) = \xi_{24}^{\deg \beta} \qquad \forall \beta \in \mathcal{B}_3 , \quad (4.11)$$

where 'deg β' denotes the crossing number or degree of a braid. This is vastly simpler than (4.8)!

In hindsight it isn't so surprising that the multiplier is simpler as a function of braids than of 2×2 matrices. The multiplier μ will be a true representation of $SL_2(\mathbb{Z})$ iff the weight k is integral; otherwise it is only a projective representation. And the standard way to handle projective representations is to centrally extend. Of course number theorists know this, but have preferred using the minimal necessary extension; as half-integer weights are the most common, they typically only look at a $\mathbb{Z}_2$-extension of $SL_2(\mathbb{Z})$ called the metaplectic group $Mp_2(\mathbb{Z})$. But unlike $\mathcal{B}_3$, $Mp_2(\mathbb{Z})$ isn't much different from the modular group and the multipliers don't simplify much when lifted to $Mp_2(\mathbb{Z})$. At least in the context of modular forms, *the braid group can be regarded as the universal central extension of the modular group, and the universal symmetry of its modular forms.*

Topologically, $SL_2(\mathbb{R})$ is the interior of the solid torus, so its universal covering group $\widetilde{SL_2(\mathbb{R})}$ will be the interior of the solid helix, and a central extension by $\pi_1 \cong \mathbb{Z}$ of $SL_2(\mathbb{R})$. $\widetilde{SL_2(\mathbb{R})}$ can be realised [25] as the set of all pairs $\left(\begin{pmatrix} a & b \\ c & d \end{pmatrix}, n \right)$ where $\begin{pmatrix} a & b \\ c & d \end{pmatrix} \in SL_2(\mathbb{R})$ and $n \equiv 0, 1, 2, 3 \pmod 4$ depending on whether $c = 0$ and $a > 0$, $c < 0$, $c = 0$ and $a < 0$, or $c > 0$, respectively. The group operation is $(A, m)(B, n) = (AB, m + n + \tau)$, where $\tau \in \{0, \pm 1\}$ is called the Maslov index. Just as $SL_2(\mathbb{Z})$ is the set of all integral points in $SL_2(\mathbb{R})$, the braid group $\mathcal{B}_3$ is the set of all integral points in $\widetilde{SL_2(\mathbb{R})}$.

Incidentally, similar comments apply when $SL_2(\mathbb{Z})$ is replaced with other discrete groups — e.g. for $\Gamma(2)$ the relevant central extension is the pure braid group $\mathcal{P}_3$. It would be interesting to topologically identify the central extension for all the genus-0 groups Γ_g of Monstrous Moonshine.

So far we have only addressed the issue of modularity. A more subtle question in Moonshine is the relation of the Monster to the genus-0 property. Our best attempt at answering this is that the Monster is probably the largest exceptional 6-transposition group [28]. This relates to Norton's generalised Moonshine through the notion of quilts (see e.g. [17]). The relation of '6' to genus-0 is that $\mathbb{H}/\Gamma(n)$ is genus-0 iff $n < 6$, while $\mathbb{H}/\Gamma(6)$ is 'barely' genus 1. The notion of quilts, and indeed the notion of generalised Moonshine and orbifolds in RCFT, is related to braids through the right action of $\mathcal{B}_3$ on any $G \times G$ (for any group G) given by

$$(g, h).\sigma_1 = (g, gh), \qquad (g, h).\sigma_2 = (gh^{-1}, h). \tag{4.12}$$

Limited space has forced us to be very sketchy here. For more on all these topics, see [14]. We suggest that the braid group $\mathcal{B}_3$ and related central extensions may play a central role in a new, more conceptual proof of the Monstrous Moonshine conjectures.

Acknowledgements. This paper was written at the University of Hamburg, whom I warmly thank for their hospitality. My research was supported in part by the von Humboldt Foundation and by NSERC.

References

[1] R. E. Borcherds, 'Monstrous moonshine and monstrous Lie superalgebras', *Invent. Math.* **109** (1992) 405–444.

[2] R. E. Borcherds, 'What is moonshine?', *Proceedings of the International Congress of Mathematicians (Berlin, 1998)* (Documenta Mathematica, Bielefeld 1998) 607–615.

[3] R. E. Borcherds, 'Modular moonshine III', *Duke Math. J.* **93** (1998) 129–154.

[4] H. Cohn and J. McKay, 'Spontaneous generation of modular invariants', *Math. of Comput.* **65** (1996) 1295–1309.

[5] J. H. Conway and S. P. Norton, 'Monstrous moonshine', *Bull. London Math. Soc.* **11** (1979) 308–339.

[6] D. Cox, *Primes of the Form $x^2 + ny^2$* (Wiley, New York 1989).

[7] C. J. Cummins, 'Modular equations and discrete, genus-zero subgroups of SL(2, $\mathbb{R}$) containing $\Gamma(N)$', *Canad. Math. Bull.* **45** (2002) 36–45.

[8] C. J. Cummins, 'Congruence subgroups of groups commensurable with $PSL(2, \mathbb{Z})$ of genus 0 and 1', *Experiment. Math.* **13** (2004) 361–382.

[9] C. J. Cummins and T. Gannon, 'Modular equations and the genus zero property', *Invent. Math.* **129** (1997) 413–443.

[10] C. J. Cummins and S. P. Norton, 'Rational Hauptmodul are replicable', *Canad. J. Math.* **47** (1995) 1201–1218.

[11] L. Dolan, P. Goddard, and P. Montague, 'Conformal field theory of twisted vertex operators', *Nucl. Phys.* **B338** (1990) 529–601.

[12] I. Frenkel, J. Lepowsky, and A. Meurman, *Vertex Operator Algebras and the Monster* (Academic Press, San Diego 1988).

[13] T. Gannon, 'Monstrous Moonshine: The first twenty-five years', *Bull. London Math. Soc.* **38** (2006) 1–33.

[14] T. Gannon, *Moonshine beyond the Monster* (Cambridge University Press, Cambridge, 2006).

[15] F. Hirzebruch, T. Berger, and R. Jung, *Manifolds and Modular Forms*, 2nd edn (Aspects of Math, Vieweg, Braunschweig 1994).

[16] G. Höhn, 'Generalized moonshine for the Baby Monster', Preprint (2003).

[17] T. Hsu, *Quilts: Central Extensions, Braid Actions, and Finite Groups*, Lecture Notes in Math **1731** (Springer, Berlin 2000).

[18] Y.-Z. Huang, 'A nonmeromorphic extension of the moonshine module vertex operator algebra', in: *Moonshine, the Monster and Related Topics (Mount Holyoke, 1994)*, Contemp. Math. **193** (American Mathematical Society, Providence, 1996) 123–148.

[19] E. Jurisich, 'Generalized Kac-Moody Lie algebras, free Lie algebras and the structure of the monster Lie algebra', *J. Pure Applied Algebra* **126** (1998) 233-266.

[20] E. Jurisich, J. Lepowsky, and R.L. Wilson, 'Realizations of the Monster Lie algebra', *Selecta Math. (NS)* **1** (1995) 129–161.

[21] V. G. Kac and S.-J. Kang, 'Trace formula for graded Lie algebras and Monstrous Moonshine', *Representations of Groups (Banff, 1994)* (American Mathematical Society, Providence 1995) 141–154.

[22] D. N. Kozlov, 'On completely replicable functions and extremal poset theory', MSc thesis, Univ. of Lund, Sweden 1994.

[23] S. Lang, *Elliptic Functions*, 2nd edn (Springer, New York 1997).

[24] J. Lepowsky, 'Some developments in vertex operator algebra theory, old and new', Preprint (arxiv: math/0706.4072).

[25] G. Lion and M. Vergne, *The Weil Representation, Maslov Index and Theta Series* (Birkhäuser, Boston 1980).

[26] M. Miyamoto, 'A new construction of the Moonshine vertex operator algebras over the real number field', *Ann. Math.* **159** (2004) 535–596.

[27] S. P. Norton, 'More on Moonshine', *Computational Group Theory (Durham, 1982)* (Academic Press, New York 1984) 185–193.

[28] S. P. Norton, 'Generalized moonshine', *The Arcata Conference on Representations of Finite Groups (Arcata, 1986)*, Proc. Symp. Pure Math. **47** (American Mathematical Society, Providence 1987) 208–209.

[29] S. D. Smith, 'On the head characters of the Monster simple group', *Finite Groups – Coming of Age (Montréal, 1982)*, Contemp. Math. **45** (American Mathematical Society, Providence 1996).

[30] Y. Zhu, 'Modular invariance of characters of vertex operator algebras', *J. Amer. Math. Soc.* **9** (1996) 237–302.

Borcherds' Proof of the Conway-Norton Conjecture

Elizabeth Jurisich

Department of Mathematics
The College of Charleston
Charleston SC 29464
jurisiche@cofc.edu

Abstract

We give a summary of R. Borcherds' solution (with some modifications) to the following part of the Conway-Norton conjectures: Given the Monster $\mathbb{M}$ and Frenkel-Lepowsky-Meurman's moonshine module $V^\natural$, prove the equality between the graded characters of the elements of $\mathbb{M}$ acting on $V^\natural$ (i.e., the McKay-Thompson series for $V^\natural$) and the modular functions provided by Conway and Norton. The equality is established using the homology of a certain subalgebra of the monster Lie algebra, and the Euler-Poincaré identity.

1. Introduction

In this paper we present a summary of R. Borcherds' proof of part of the Conway-Norton "monstrous moonshine" conjectures: the proof that the McKay-Thompson series of the Monster simple group acting on the known structure $V^\natural$ [FLM88] do indeed correspond to the Hauptmoduls presented in Conway and Norton [CN79]. Interested readers should certainly consult primary sources, a few of which are [B86], [FLM84], [FLM88], and [B92]. The simplification of the original proof, presented in this paper, can be found in [J98] and [JLW95]. See also Borcherds' survey articles about moonshine [B94] and [B98]. A brief overview of the historical development of the subject can be found in [FLM88]. What the reader will find in this paper is an outline of the proof itself, with references to particular results needed to establish the equality between the McKay-Thompson series for $V^\natural$ and the Laurent expansions of the relevant modular functions.

Given a group G and a $\mathbb{Z}$-graded G-module $V = \coprod_{n \in \mathbb{Z}} V_{[n]}$, with $\dim V_{[n]} < \infty$ for all $n \in \mathbb{Z}$ truncated so $V_{[k]} = 0$ for $k < N$ for some fixed $N \in \mathbb{Z}$, the

McKay-Thompson series for $g \in G$ acting on V is defined to be the graded trace

$$T_g(q) = \sum_{n>N} \text{tr}(g|V_{[n]})q^n.$$

The moonshine conjectures of Conway and Norton in [**CN79**] include the conjecture that there should be an infinite-dimensional representation of the (not yet constructed) Fischer-Griess Monster simple group $\mathbb{M}$ such that the McKay-Thompson series T_g for $g \in \mathbb{M}$ acting on V have coefficients that are equal to the coefficients of the q-series expansions of certain modular functions. After the construction of $\mathbb{M}$ [**Gr82**], a "moonshine module" $V^{\natural}$ for the Monster simple group was constructed [**FLM84**], [**FLM88**] and many of its properties, including the determination of some of its McKay-Thompson series, were proven. The critical example involves the modular function $j(\tau)$, a generator for the field of functions invariant under the action of $SL_2(\mathbb{Z})$ on the upper half plane. Let $q = e^{2\pi i \tau}$. Then the normalized q-series expansion denoted by $J(q) = j(q) - 744$ is one of the Hauptmoduls that occur in the moonshine correspondence [**CN79**]. Results of [**FLM88**] include the vertex operator algebra structure of $V^{\natural}$, with the conformal grading $V^{\natural} = \coprod_{i \geq 0} V_i^{\natural}$, as an $\mathbb{M}$-module, and the graded dimension correspondence $\dim V_i^{\natural} \leftrightarrow c(i - 1)$ to the coefficients of the modular function $J(q) = \sum_{n \geq -1} c(n)q^n$. In other words, shifting the grading by defining $V_{[i-1]}^{\natural} = V_i^{\natural}$, we have $T_e(q) = J(q)$ as formal series, where $e \in \mathbb{M}$ is the identity element.

After the above results, the nontrivial problem of computing the rest of the McKay-Thompson series of Monster group elements acting on $V^{\natural}$ remained. Borcherds showed in [**B92**] that the McKay-Thompson series are the expected modular functions. The argument can be summarized as follows: Borcherds establishes a product formula

$$p(J(p) - J(q)) = \prod_{i=1,2,\dots,\,j=-1,1,\dots} (1 - p^i q^j)^{c(ij)}.$$

This formula is used in the proof, but first note that the formula leads to recursion formulas for the coefficients of the q-series expansion of $J(q)$, and hence to recursions for the dimensions of the homogeneous components of $V^{\natural}$. The approach of [**B92**] is to establish a product formula involving all of the McKay-Thompson series $T_g(q)$ of elements of $g \in \mathbb{M}$ acting on $V^{\natural}$, analogous to the above identity for $T_e(q) = J(q)$. The more general product formula in turn leads to a set of recursion formulas that determine the coefficients of the series, given a (large) set of initial data. For example, to determine $\sum \text{tr}(g|V_{i+1}^{\natural})q^i = \sum_{i \geq -1} c_g(i)q^i$ it is sufficient to compute four of the first five coefficients of $\sum \text{tr}(g^k | V_{i+1}^{\natural}) q^i$, $k \in \mathbb{Z}$, $g \in \mathbb{M}$. One also determines that the Hauptmoduls listed in [**CN79**] satisfy the same recursion relations and initial data as the McKay-Thompson series for $V^{\natural}$.

The crucial product identity for the McKay-Thompson series is obtained by Borcherds from the Euler-Poincaré identity for the homology groups of a particular Lie algebra, the "monster" Lie algebra. This Lie algebra is constructed using the tensor product of the vertex operator algebra $V^{\natural}$ and a vertex algebra associated with a two-dimensional Lorentzian lattice. The monster Lie algebra m constructed in [**B92**] is an infinite-dimensional $\mathbb{Z} \times \mathbb{Z}$-graded Lie algebra. The Lie algebra m is then shown to be a generalized Kac-Moody algebra, or a Borcherds algebra. The "No-ghost" theorem of string theory is used as a step in establishing an isomorphism between the $\mathbb{Z} \times \mathbb{Z}$-homogeneous components of m and the weight spaces of $V^{\natural}$. The product formula for $p(J(p) - J(q))$ is interpreted as the denominator formula for the Lie algebra m and used to determine the simple roots. The results pertaining to the homology groups of Lie algebras of [**GL76**] are then extended to include the class of Borcherds algebras and applied to a subalgebra $\mathfrak{n}^-$ of m to obtain the desired family of identities. Of course, the recursions and initial data must also be established for the Hauptmoduls; see [**Koi**] and [**Fer96**].

In this paper we will actually discuss a modification of Borcherds proof, in which it is not necessary to generalize the homology results of [**GL76**]. We shall compute the homology groups as in [**J98**], [**JLW95**] with respect to a smaller subalgebra $\mathfrak{u}^- \subset \mathfrak{n}^-$. We shall use $\mathfrak{u}^-$ because it is a free Lie algebra, and therefore computing the homology groups is straightforward. The Euler-Poincaré identity applied to the subalgebra $\mathfrak{u}^-$ and the trivial $\mathfrak{u}^-$-module $\mathbb{C}$ leads to recursions sufficient to establish the correspondence between the McKay-Thompson series for $V^{\natural}$ and the Hauptmoduls specified by Conway and Norton [**CN79**].

2. Vertex operator algebras

We begin by recalling the definition of vertex operator algebra and vertex algebra. The following definition is a variant of Borcherds' original definition in [**B86**]. For a detailed discussion the reader can consult [**FLM88**], [**FHL93**], [**DL93**].

Definition 1. A *vertex operator algebra*, $(V, Y, \mathbf{1}, \omega)$, consists of a vector space V, distinguished vectors called the *vacuum vector* $\mathbf{1}$ and the *conformal vector* ω, and a linear map $Y(\cdot, z) : V \to (\text{End } V)[[z, z^{-1}]]$ which is a generating function for operators v_n, i.e., for $v \in V$, $Y(v, z) = \sum_{n \in \mathbb{Z}} v_n z^{-n-1}$, satisfying the following conditions:

(V1) $V = \coprod_{n \in \mathbb{Z}} V_n$; for $v \in V_n$, $n = \text{wt}(v)$
(V2) $\dim V_n < \infty$ for $n \in \mathbb{Z}$
(V3) $V_n = 0$ for n sufficiently small
(V4) If $u, v \in V$ then $u_n v = 0$ for n sufficiently large
(V5) $Y(\mathbf{1}, z) = 1$
(V6) $Y(v, z)\mathbf{1} \in V[[z]]$ and $\lim_{z \to 0} Y(v, z)\mathbf{1} = v$, i.e., the *creation property* holds
(V7) The following *Jacobi identity* holds:

$$z_0^{-1} \delta\left(\frac{z_1 - z_2}{z_0}\right) Y(u, z_1) Y(v, z_2) - z_0^{-1} \delta\left(\frac{z_2 - z_1}{-z_0}\right) Y(v, z_2) Y(u, z_1)$$
$$= z_2^{-1} \delta\left(\frac{z_1 - z_0}{z_2}\right) Y(Y(u, z_0)v, z_2). \tag{2.1}$$

The following conditions relating to the vector ω also hold;

(V8) The operators ω_n generate a Virasoro algebra i.e., if we let $L(n) = \omega_{n+1}$ for $n \in \mathbb{Z}$ then

$$[L(m), L(n)] = (m - n)L(m + n) + (1/12)(m^3 - m)\delta_{m+n,0}(\text{rank } V) \tag{2.2}$$

(V9) If $v \in V_n$ then $L(0)v = (\text{wt } v)v = nv$
(V10) $\frac{d}{dz} Y(v, z) = Y(L(-1)v, z)$.

Definition 2. A *vertex algebra* (with conformal vector) $(V, Y, \mathbf{1}, \omega)$ is a vector space V with all of the above properties except for **V2** and **V3**.

For a vertex algebra V, and $v \in V$ with $\text{wt } v = n$, let $(-z^{-2})^{L(0)} v = (-z^{-2})^n v$. This action extends linearly to all of V.

Definition 3. A bilinear form on a vertex algebra V is *invariant* if for $v, u, w \in V$
$$(Y(v, z)u, w) = (u, Y(e^{zL(1)}(-z^{-2})^{L(0)}v, z^{-1})w).$$

We note that an invariant form satisfies $(u, v) = 0$ unless $\text{wt}(u) = \text{wt}(v)$ for u, v homogeneous elements of V.

The tensor product of vertex algebras is also a vertex algebra [**FHL93**], [**DL93**]. Given two vertex algebras $(V, Y, \mathbf{1}_V, \omega_V)$ and $(W, Y, \mathbf{1}_W, \omega_W)$ the vacuum of $V \otimes W$ is $\mathbf{1}_V \otimes \mathbf{1}_W$ and the conformal vector ω is given by $\omega_V \otimes \mathbf{1}_W + \mathbf{1}_V \otimes \omega_W$. If the vertex algebras V and W both have invariant forms in the sense of Definition 3 then it follows from the definition of the tensor product that the form on $V \otimes W$ given by the product of the forms on V and W is also invariant.

One large and important class of vertex algebras are those associated with even lattices. Although the moonshine module $V^\natural$ is not a vertex operator algebra associated with a lattice (it is a far more complicated object), it is constructed using the vertex operator algebra associated to the Leech lattice. The vertex algebra used in the proof of the moonshine correspondence is the tensor product of the moonshine module and a vertex algebra associated with a two- dimensional Lorentzian lattice.

Given an even lattice L the vertex algebra V_L [**B86**] associated to the lattice has underlying vector space

$$V_L = S(\hat{\mathfrak{h}}_{\mathbb{Z}}^-) \otimes \mathbb{C}\{L\}.$$

(We are using the notation and constructions in [**FLM88**].) Here we take $\mathfrak{h} = L \otimes_{\mathbb{Z}} \mathbb{C}$, and $\hat{\mathfrak{h}}_{\mathbb{Z}}^-$ is the negative part of the Heisenberg algebra (with c central) defined by

$$\hat{\mathfrak{h}}_{\mathbb{Z}} = \coprod_{n \in \mathbb{Z}} \mathfrak{h} \otimes t^n \oplus \mathbb{C}c \subset \mathfrak{h} \otimes \mathbb{C}[[t]] \oplus \mathbb{C}c,$$

so that

$$\hat{\mathfrak{h}}_{\mathbb{Z}}^- = \coprod_{n < 0} \mathfrak{h} \otimes t^n.$$

The symmetric algebra on $\hat{\mathfrak{h}}_{\mathbb{Z}}^-$ is denoted $S(\hat{\mathfrak{h}}_{\mathbb{Z}}^-)$. Let $\hat{L}$ be a central extension of L by a group of order 2, i.e.,

$$1 \to \langle \kappa | \kappa^2 = 1 \rangle \to \hat{L} \xrightarrow{-} L \to 1,$$

with commutator map given by $\kappa^{\langle \alpha, \beta \rangle}$, α, $\beta \in L$. Define $\mathbb{C}\{L\}$ to be the induced module $\mathrm{Ind}_{\langle \kappa \rangle}^{\hat{L}} \mathbb{C}$, where κ acts on $\mathbb{C}$ as multiplication by -1.

If $a \in \hat{L}$ denote by $\iota(a)$ the element $a \otimes 1 \in \mathbb{C}\{L\}$. We will use the notation $\alpha(n) = \alpha \otimes t^n \in S(\hat{\mathfrak{h}}_{\mathbb{Z}}^-)$. The vector space V_L is spanned by elements of the form:

$$\alpha_1(-n_1)\alpha_2(-n_2)\dots\alpha_k(-n_k)\iota(a) \tag{2.3}$$

where $a \in \hat{L}, \alpha_i \in \mathfrak{h}$ and $n_i \in \mathbb{N}$. The space V_L, equipped with $Y(v, z)$ as defined in [**FLM88**] satisfies properties **V1** and **V4** $-$ **V10**, so is a vertex algebra with conformal vector ω.

A vertex algebra V_L constructed from an even lattice L automatically has an invariant bilinear form [**B86**], which can be defined using the contragredient module V_L' [**FHL93**].

3. Construction of the monster Lie algebra from the moonshine module.

The moonshine module $V^\natural$, a graded $\mathbb{M}$-module and vertex operator algebra, is constructed in [**FLM88**]. The following results describing the structure and properties of $V^\natural$ appear in Corollary 12.5.4 and Theorem 12.3.1 of [**FLM88**], part of which we restate here for the convenience of the reader. The invariance of the form in the sense of Definition 3 follows from the construction and results of [**Li94**]; see [**J98**]. Recall that $J(q)$ denotes the Laurent, or q-series, expansion of the modular function $j(\tau)$, normalized so that the coefficient of q^0 is zero.

Theorem 1.

(i) *The graded dimension of the moonshine module $V^\natural$ is $J(q)$.*

(ii) *$V^\natural$ is a vertex operator algebra of rank 24.*

(iii) *$\mathbb{M}$ acts in a natural way as automorphisms as of the vertex operator algebra $V^\natural$, i.e.,*

$$gY(v,z)g^{-1} = Y(gv,z)$$

for $g \in \mathbb{M}, v \in V^\natural$

(iv) *There is an invariant positive definite hermitian form $(\cdot,\cdot)$ on $V^\natural$ which is also invariant under $\mathbb{M}$.*

In [**B92**] the monster Lie algebra is constructed using $V^\natural$ and the vertex algebra associated with a Lorentzian lattice as follows. Let $\Pi_{1,1} = \mathbb{Z} \oplus \mathbb{Z}$ be the rank two Lorentzian lattice with bilinear form $\langle \cdot, \cdot \rangle$ given by the matrix $\begin{pmatrix} 0 & -1 \\ -1 & 0 \end{pmatrix}$. The vertex algebra $V_{\Pi_{1,1}}$ has a conformal vector, and is given the structure of a trivial $\mathbb{M}$-module. Since $V_{\Pi_{1,1}}$ is a vertex algebra associated with an even lattice it has an invariant bilinear form, which we consider as $\mathbb{M}$-invariant under the trivial group action. Note that $V_{\Pi_{1,1}}$ is not a vertex operator algebra, because it does not satisfy conditions (**V2**) or (**V3**). For example, the weight of an element of the form (2.3) is $\sum_{i=1}^{k} n_i + \frac{1}{2}\langle a, a \rangle, n_i > 0 \in \mathbb{Z}$, $a \in \Pi_{1,1}$, which can be less than zero and arbitrarily large in absolute value.

Lemma 3.1. *The tensor product $V = V^\natural \otimes V_{\Pi_{1,1}}$ is a vertex algebra with conformal vector, and an invariant bilinear form, which is also $\mathbb{M}$-invariant.*

Given a vertex operator algebra V, or a vertex algebra V with conformal vector ω and therefore an action of the Virasoro algebra, let

$$P_i = \{v \in V | L(0)v = iv, L(n)v = 0 \text{ if } n > 0\}.$$

Thus P_i consists of the lowest weight vectors for the Virasoro algebra of conformal weight i. (P_1 is called the *physical space*). Let $u \in P_0$, then $\text{wt}L(-1)u = \text{wt}\omega + \text{wt}u - 1 = 1$ and $L_{-1}P_0 \subset P_1$.

Lemma 3.2. *The space $P_1/L(-1)P_0$ is a Lie algebra with bracket given by*

$$[u + L(-1)P_0, v + L(-1)P_0] = u_0v + L(-1)P_0.$$

For $u, v \in P_1$.

Proof. Let $u, v \in P_1$. By formula (8.8.7) of [**FLM88**] (see [**B86**])

$$Y(u, z)v = e^{zL(-1)}Y(v, -z)u. \tag{3.1}$$

Taking coefficients of z^{-1} on both sides of (3.1) yields

$$u_0v = -v_0u + \sum_{k=1}^{\infty}(-1)^{k-1}L(-1)^k v_k u \in -v_0u + L(-1)P_0.$$

Thus the bracket is anti-symmetric. Let $u, v, w \in P_1$ The Jacobi identity (**V7**) implies

$$u_0v_0w - v_0u_0w = (u_0v)_0w$$

$$(u_0(v_0w)) - (v_0(u_0w)) - ((u_0v)_0w) = 0.$$

Since we have shown anti-symmetry (modulo $L(-1)P_0$) the above is equivalent to the usual Lie algebra Jacobi identity for $[\cdot, \cdot]$ on $P_1/L(-1)P_0$. □

We can now give the definition of Borcherds' monster Lie algebra. The tensor product $V = V^\natural \otimes V_{\Pi_{1,1}}$ is a vertex algebra with conformal vector, and invariant bilinear form. This form induces a bilinear form on the Lie algebra $P_1/L_{-1}P_0$. Note that if $u, v, w \in P_1$, then by invariance, and the fact that $L(1)u = 0$

$$(Y(u, z)v, w) = (v, Y(e^{zL(1)}(-z^{-2})^{L(0)}u, z^{-1})w)$$

$$= -(v, Y(u, z^{-1})z^{-2}w) = -(v, \sum_{n \in \mathbb{Z}} u_n z^{n-1}). \tag{3.2}$$

Taking the coefficient of z^{-1} we have for $u, v, w \in P_1/L(-1)P_0$

$$(u_0v, w) = -(v, u_0w),$$

and so the form on $P_1/L(1)P_0$ is invariant in the usual Lie algebra sense.

In addition to the weight grading, the vertex algebra $V^\natural \otimes V_{\Pi_{1,1}}$ is graded by the lattice $\Pi_{1,1}$. For u, v elements of degree $r, s \in \Pi_{1,1}$, the invariant form satisfies $(u, v) = 0$ unless $r = s$.

Let $N(\cdot, \cdot)$ denote the nullspace of the bilinear form on P_1 so $N(\cdot, \cdot) = \{u \in P_1 \mid (u, v) = 0 \quad \forall v \in P_1\}$. Since, for $u = L(-1)v, v \in P_0, w \in P_1$, it is immediate that $(L(-1)v, w) = (v, L(1)w) = 0$, we see $L(-1)P_0 \subset N(\cdot, \cdot)$. This in conjunction with Lemma 3.2 ensures the following is a Lie algebra.

Definition 4. The *monster Lie algebra* $\mathfrak{m}$ is defined by

$$\mathfrak{m} = P_1/N(\cdot, \cdot).$$

The monster Lie algebra is graded by the Lorentzian lattice $\Pi_{1,1}$ by construction. Elements of $\mathfrak{m}$ can be written as $\sum u \otimes ve^r$, where $u \in V^\natural$ and $ve^r = v\iota(e^r) \in V_{\Pi_{1,1}}$. Here, a section of the map $\hat{\Pi}_{1,1} \overset{\sim}{\to} \Pi_{1,1}$ has been chosen so that $e^r \in \hat{\Pi}_{1,1}$ satisfies $\overline{e^r} = r \in \Pi_{1,1}$. There is a grading of $\mathfrak{m}$ by the lattice defined by $\deg(u \otimes ve^r) = r$. It follows from the construction that the Lie algebra $\mathfrak{m}$ has a Lie invariant bilinear form, whose radical is zero.

In order to establish the equality between the coefficients of the McKay-Thompson series for $V^\natural$ and the given Hauptmoduls, it is necessary to determine the dimensions of the components of $\mathfrak{m}$ of degrees $r \in \Pi_{1,1}$. Borcherds [B92] computes the dimensions by using Theorem 2 below, which uses the No-ghost theorem of string theory. For a proof of the No-ghost theorem see [GT72], [B92], or the appendix of [J98] for one written more algebraically.

Theorem 2. *Let V be a vertex operator algebra with the following properties:*

  i. *V has a symmetric invariant nondegenerate bilinear form.*
 ii. *The central element of the Virasoro algebra acts as multiplication by 24.*
iii. *The weight grading of V is an $\mathbb{N}$-grading of V, i.e., $V = \coprod_{n=0}^{\infty} V_n$, and $\dim V_0 = 1$.*
 iv. *V is acted on by a group G preserving the above structure; in particular the form on V is G-invariant.*

Let $P_1 = \{u \in V \otimes V_{\Pi_{1,1}} | L(0)u = u, L(i)u = 0, i > 0\}$. The group G acts on $V \otimes V_{\Pi_{1,1}}$ via the trivial action on $V_{\Pi_{1,1}}$. Let P_1^r denote the subspace of P_1 of degree $r \in \Pi_{1,1}$. Then the quotient of P_1^r by the nullspace of its bilinear form is isomorphic as a G-module with G-invariant bilinear form to $V_{1-\langle r,r \rangle/2}$ if $r \neq 0$ and to $V_1 \oplus \mathbb{C}^2$ if $r = 0$.

Applying Theorem 2 to $V = V^\natural \otimes V_{\Pi_{1,1}}$ we see that the monster Lie algebra has $(m, n) \in \mathbb{Z} \times \mathbb{Z}$ homogeneous subspaces isomorphic to the weight spaces $V^\natural_{mn+1}$ when $(m, n) \neq (0, 0)$, that is, $\mathfrak{m}_{(m,n)} = V^\natural_{[mn]}$. We have shown:

$$\mathfrak{n}^+ = \coprod_{m>0, n\geq -1} \mathfrak{m}_{(m,n)},$$

with

$$\mathfrak{m}_{(m,n)} \simeq V^\natural_{mn+1}$$

and similarly for $\mathfrak{n}^-$.

4. The structure of the monster Lie algebra

A crucial step in [**B92**] is to identify $\mathfrak{m} = P_1/N(\cdot, \cdot)$ with a Lie algebra given by a generalization of a Cartan matrix. This allows one to be able to compute the homology groups of the trivial module $\mathbb{C}$ with respect to an appropriate subalgebra of $\mathfrak{m}$, as can be done for symmetrizable Kac-Moody Lie algebras [**GL76**], [**Liu92**].

The Lie algebra $\mathfrak{g}(A)$ associated to a symmetrizable matrix A is introduced in [**K90**] and [**Mo67**], but the systematic study of the case where A satisfies conditions **B1-B3** below was carried out by Borcherds. Borcherds algebras have many properties in common with symmetrizable Kac-Moody algebras such as an invariant bilinear form and a root lattice grading. One notable difference is that there may be simple imaginary roots in the root lattice. This is a desirable property in the monster case $\mathfrak{m}$ because we wish to associate a root grading to the hyperbolic $\mathbb{Z} \times \mathbb{Z}$-grading inherited from $V_{\Pi_{1,1}}$.

We review the construction of the Borcherds algebra $\mathfrak{g}(A)$ of [**B88**]. Let I be a (finite or) countable index set and let $A = (a_{ij})_{i,j \in I}$ be a matrix with entries in $\mathbb{C}$, satisfying the following conditions:

(B1) A is symmetric.

(B2) If $i \neq j$ $(i, j \in I)$, then $a_{ij} \leq 0$.

(B3) If $a_{ii} > 0$ $(i \in I)$, then $2a_{ij}/a_{ii} \in \mathbb{Z}$ for all $j \in I$.

Let $\mathfrak{g}'(A)$ be the Lie algebra with generators $h_i, e_i, f_i, i \in I$, and the following defining relations: For all $i, j, k \in I$,

$$[h_i, h_j] = 0, [e_i, f_j] - \delta_{ij}h_i = 0,$$

$$[h_i, e_k] - a_{ik}e_k = 0, [h_i, f_k] + a_{ik}f_k = 0$$

and Serre relations

$$(\operatorname{ad} e_i)^{-2a_{ij}/a_{ii}+1}e_j = 0, (\operatorname{ad} f_i)^{-2a_{ij}/a_{ii}+1}f_j = 0$$

for all $i \neq j$ with $a_{ii} > 0$, and finally

$$[e_i, e_j] = 0, [f_i, f_j] = 0$$

whenever $a_{ij} = 0$.

Let $\mathfrak{h} = \sum_{i \in I} \mathbb{C}h_i$, $\mathfrak{n}^{\pm}$ the subalgebra generated by the elements e_i (resp. the f_i) for $i \in I = \langle e_i \rangle$. As in the Kac-Moody case, the simple roots $\alpha_i \in (\mathfrak{h})^*$ are defined to satisfy $(\alpha_i, \alpha_j) = a_{ij}$. Also as in the Kac-Moody case, we may have linearly dependent simple roots α_i and we extend the Lie algebra as in [**GL76**] and [**Le79**] by an appropriate abelian Lie algebra $\mathfrak{d}$ of degree derivations, chosen so that the simple roots are linearly independent in $(\mathfrak{h} \ltimes \mathfrak{d})^*$.

Definition 5. The Lie algebra $\mathfrak{g}(A) = g'(A) \ltimes \mathfrak{d}$ is the *Borcherds* or *generalized Kac-Moody* (Lie) algebra associated to the matrix A. Any Lie algebra of the form $\mathfrak{g}(A)/\mathfrak{c}$ where $\mathfrak{c}$ is a central ideal is also called a Borcherds algebra.

Versions of the following theorem appear in [**B88**] and [**B95**], see also [**J98**]. This theorem allows us to recognize a Lie algebra associated to a matrix A satisfying **B1** − **B3**.

Theorem 3. *Let $\mathfrak{g}$ be a Lie algebra satisfying the following conditions:*

(i) $\mathfrak{g}$ *can be $\mathbb{Z}$-graded as* $\bigsqcup_{i \in \mathbb{Z}} \mathfrak{g}_i$, $\mathfrak{g}_i$ *is finite dimensional if $i \neq 0$, and $\mathfrak{g}$ is diagonalizable with respect to $\mathfrak{g}_0$.*

(ii) $\mathfrak{g}$ *has an involution η which maps $\mathfrak{g}_i$ onto $\mathfrak{g}_{-i}$ and acts as -1 on noncentral elements of $\mathfrak{g}_0$, in particular, $\mathfrak{g}_0$ is abelian.*

(iii) $\mathfrak{g}$ *has a Lie algebra-invariant bilinear form $(\cdot, \cdot)$, invariant under η, such that $\mathfrak{g}_i$ and $\mathfrak{g}_j$ are orthogonal if $i \neq -j$, and such that the form $(\cdot, \cdot)_0$, defined by $(x, y)_0 = -(x, \eta(y))$ for $x, y \in \mathfrak{g}$, is positive definite on $\mathfrak{g}_m$ if $m \neq 0$.*

(iv) $\mathfrak{g}_0 \subset [\mathfrak{g}, \mathfrak{g}]$.

Then there is a central extension $\hat{\mathfrak{g}}$ of a Borcherds algebra and a homomorphism, π, from $\hat{\mathfrak{g}}$ onto $\mathfrak{g}$, such that the kernel of π is in the center of $\hat{\mathfrak{g}}$.

The theorem is proven by inductively constructing a set of generators of $\mathfrak{g}_n$, $n \in \mathbb{Z}$, consisting of $\mathfrak{g}_0$ weight vectors, using the form $(x, y)_0$. Proofs can be found in [**B91**], see [**J98**] for the theorem stated exactly as above. An alternative characterization of Borcherds algebra can be found in [**B92**].

Theorem 4. *The Lie algebra $\mathfrak{m} = P_1/N(\cdot, \cdot)$ is a Borcherds algebra.*

Proof. The abelian subalgebra $\mathfrak{m}_{(0,0)}$ is spanned by elements of the form $1 \otimes \alpha(-1)\iota(1)$ where $\alpha \in \Pi_{1,1} \otimes_{\mathbb{Z}} \mathbb{C}$. Note that $\mathfrak{m}_{(0,0)}$ is two-dimensional. In order to apply Theorem 3, grade $\sum_{(m,n) \in \Pi_{1,1}} m_{(m,n)}$ by $i = 2m + n \in \mathbb{Z}$. With this grading, $\mathfrak{m}$ satisfies condition (i) of Theorem 3.

There is an involution η is on the vertex algebra $V_{\Pi_{1,1}}$, determined by $\eta(\alpha) = -\alpha$ for $\alpha \in \Pi_{1,1}$. Extend the involution to $V^{\natural} \otimes V_{\Pi_{1,1}}$ by taking $\eta(\sum u \otimes v) = \sum(u \otimes \eta v)$ for $u \in V^{\natural}$, $v \in V_{\Pi_{1,1}}$. The invariant form given by Lemma 3.1 is the required non-degenerate invariant bilinear form, satisfying condition (iii) in Theorem 3. Let $a = e^{(1,1)}$, $b = e^{(1,-1)}$. Condition (iv) follows from the fact that $\mathfrak{m}_{(0,0)}$ is two-dimensional and that the elements $[u \otimes \iota(a), v \otimes \iota(a^{-1})]$ and $[\iota(b), \iota(b^{-1})]$ for $u, v \in V_2^{\natural}$ are two linearly independent vectors in $\mathfrak{m}_{(0,0)}$. Thus the Lie algebra $\mathfrak{m}$ is the homomorphic image of some Borcherds algebra $\mathfrak{g}(A)$ associated to a matrix. $\square$

By computing the action of $a \in \mathfrak{g}_0 = \mathfrak{m}_{(0,0)}$ on $v \in \mathfrak{m}_r$, $r \in \Pi_{1,1}$ one obtains $[a, v] = \langle \alpha, r \rangle v$ and Borcherds identifies the elements of $\Pi_{1,1}$ with the root lattice of $\mathfrak{m}$. The following is Theorem 7.2 of [**B92**].

Theorem 5. *The simple roots of the monster Lie algebra* $\mathfrak{m}$ *are the vectors* $(1, n)$, $n = -1$ *or* $n > 0$, *each with multiplicity* $c(n)$.

This theorem is proven [**B92**] by identifying the product formula

$$p(J(p) - J(q)) = \prod_{i=1,2,\ldots, \, j=-1,1,\ldots} (1 - p^i q^j)^{c(ij)}$$

with the denominator identity for the Borcherds algebra $\mathfrak{m}$.

Since, by definition of $\mathfrak{m}$ the radical of the invariant form on $\mathfrak{m}$ is zero, the kernel of the homomorphism in Theorem 3 is in the center of $\mathfrak{g}(A)$. We construct a symmetric matrix B, determined by the root lattice $\Pi_{1,1}$ and the multiplicities given by Theorem 2. We have the Lie algebra $\mathfrak{m}$ is isomorphic to $\mathfrak{g}(B)/\mathfrak{c}$, where $\mathfrak{g}(B)$ is the Borcherds algebra associated to the following matrix B and $\mathfrak{c}$ is the full center of $\mathfrak{g}(B)$:

$$
B = \left(
\begin{array}{ccc|ccc|c}
2 & 0 & \cdots \ 0 & -1 & \cdots & -1 & \cdots \\
\hline
0 & -2 & \cdots \ -2 & -3 & \cdots & -3 & \\
\vdots & \vdots & \ddots \ \vdots & \vdots & \ddots & \vdots & \cdots \\
0 & -2 & \cdots \ -2 & -3 & \cdots & -3 & \\
\hline
-1 & -3 & \cdots \ -3 & -4 & \cdots & -4 & \\
\vdots & \vdots & \ddots \ \vdots & \vdots & \ddots & \vdots & \cdots \\
-1 & -3 & \cdots \ -3 & -4 & \cdots & -4 & \\
\hline
\vdots & & \vdots & & \vdots & &
\end{array}
\right).
$$

In this summary of Borcherds' proof of part of the moonshine conjectures we are able to bypass the part of the argument of [**B92**] that requires a more extensive development of the theory of Borcherds algebras including generalizing the results of [**GL76**]. Instead, we will use Theorem 6 below, proven in [**J98**]. Given a vector space U, let $L(U)$ denote the free Lie algebra generated by a basis of U. Let $J \subset I$ be the set $\{i \in I \,|\, a_{ii} > 0\}$. Note that the matrix $(a_{ij})_{i,j \in J}$ is a generalized Cartan matrix. Let $\mathfrak{g}_J$ be the Kac-Moody algebra associated to this matrix. Then $\mathfrak{g}_J = \mathfrak{n}_J^+ \oplus \mathfrak{h}_J \oplus \mathfrak{n}_J^-$, and $\mathfrak{g}_J$ is isomorphic to the subalgebra of $\mathfrak{g}(A)$ generated by $\{e_i, f_i\}$ with $i \in J$.

Theorem 6. *Let A be a matrix satisfying conditions* **B1-B3**. *Let J and $\mathfrak{g}_J$ be as above. Assume that if $i, j \in I \setminus J$ and $i \neq j$ then $a_{ij} < 0$. Then*

$$\mathfrak{g}(A) = \mathfrak{u}^+ \oplus (\mathfrak{g}_J + \mathfrak{h}) \oplus \mathfrak{u}^-,$$

where $\mathfrak{u}^- = L(\coprod_{j \in I \backslash J} \mathcal{U}(\mathfrak{n}_J^-) \cdot f_j)$ *and* $\mathfrak{u}^+ = L(\coprod_{j \in I \backslash J} \mathcal{U}(\mathfrak{n}_J^+) \cdot e_j)$. *The* $\mathcal{U}(\mathfrak{n}_J^-) \cdot f_j$ *for* $j \in I \backslash J$ *are integrable highest weight* $\mathfrak{g}_J$-*modules, and the* $\mathcal{U}(\mathfrak{n}_J^+) \cdot e_j$ *are integrable lowest weight* $\mathfrak{g}_J$-*modules.*

Note that the conditions on the a_{ij} given in the theorem are equivalent to the statement that the Lie algebra has no mutually orthogonal imaginary simple roots. This is the case for the monster Lie algebra $\mathfrak{m}$.

The structure of $\mathfrak{m}$ can now be summarized. There are natural isomorphisms

$$\mathfrak{m}_{(m,n)} \cong V_{mn+1}^\natural \text{ as an } M\text{-module for } (m, n) \neq (0, 0),$$

$$\mathfrak{m}_{(0,0)} \cong \mathbb{C} \oplus \mathbb{C}, \text{ a trivial } M\text{-module.}$$

It follows from the definition of $\mathfrak{m}$ that

$$\mathfrak{m}_{(-1,1)} \oplus \mathfrak{m}_{(0,0)} \oplus \mathfrak{m}_{(1,-1)} \cong \mathfrak{gl}_2.$$

Applying Theorem 6 to the above realization of $\mathfrak{m}$ by generators and relations gives

$$\mathfrak{m} = \mathfrak{u}^+ \oplus \mathfrak{gl}_2 \oplus \mathfrak{u}^-,$$

with $\mathfrak{u}^- = L(U)$ and $\mathfrak{u}^+ = L(U')$. Where $L(U), L(U')$ are free Lie algebras over vector spaces that are direct sums of $\mathfrak{gl}_2$-modules.

$$U = \coprod_{i>0} W_i \otimes V_{i+1}^\natural \text{ and } U' = \coprod_{i>0} W_i' \otimes V_{i+1}^\natural.$$

For $i > 0$, $V_{i+1}^\natural$ is (as usual) the weight $i + 1$ component of $V^\natural$, W_i denotes the (unique up to isomorphism) irreducible highest weight $\mathfrak{gl}_2$-module of dimension i on which z acts as $i + 1$ and W_i', $i > 0$, denotes the irreducible lowest weight module.

5. The homology computation and recursion formulas

We are now ready to establish the recursion relations for the coefficients of the McKay-Thompson series $\sum_{i>0} \text{Tr}(g|V_{i+1}^\natural)q^i = \sum_{i \in \mathbb{Z}} c_g(i)q^i$. What follows is a summary of what has appeared in [**JLW95**]. See [**Ka94**] and [**KK95**] for similar computations.

To compute the homology of the free Lie algebra $L(U)$ for a vector space U, with coefficients in the trivial module (as in [**CE56**]), consider the following exact sequence is a $\mathcal{U}(L(U)) = T(U)$-free resolution of the trivial module:

$$0 \to T(U) \otimes U \xrightarrow{\mu} T(U) \xrightarrow{\epsilon} \mathbb{C} \to 0$$

where μ is the multiplication map and ϵ is the augmentation map. One obtains:

$$H_0(L(U), \mathbb{C}) = \mathbb{C}$$

$$H_1(L(U), \mathbb{C}) = U \cong L(U)/[L(U), L(U)]$$

$$H_n(L(U), \mathbb{C}) = 0 \text{ for } n \geq 2.$$

Let p, q and t be commuting formal variables. The variables p^{-1} and q^{-1} will be used to index the $\mathbb{Z} \oplus \mathbb{Z}$-grading of our vector spaces. All of the $\mathbb{M}$-modules we encounter are finite-dimensionally $\mathbb{Z} \oplus \mathbb{Z}$-graded with grading suitably truncated and will be identified with formal series in $R(\mathbb{M})[[p, q]]$. Definitions and results from [**Kn73**] about the λ-ring $R(\mathbb{M})$ of finite-dimensional representations of $\mathbb{M}$ are applicable to formal series in $R(\mathbb{M})[[p, q]]$. We summarize the results of, for example, [**Kn73**] that we use below.

The representation ring $R(\mathbb{M})$ is a λ-ring [**Kn73**] with the λ operation given by exterior powers, so $\lambda^i V = \bigwedge^i V$ for $V \in R(\mathbb{M})$.

In the following discussion we let $W, V \in R(\mathbb{M})$. The operation $\bigwedge^i$ satisfies

$$\bigwedge^i (W \oplus V) = \sum_{n=0}^{i} \bigwedge^n (W) \otimes \bigwedge^{i-n}(V).$$

Define

$$\bigwedge_t(W) = \bigwedge^0(W) + \bigwedge^1(W)t + \bigwedge^2(W)t^2 + \cdots.$$

Then

$$\bigwedge_t(V \oplus W) = \bigwedge_t(V) \cdot \bigwedge_t(W). \tag{5.1}$$

The Adams operations $\Psi^k : R(\mathbb{M}) \to R(\mathbb{M})$ are defined for $W \in R(\mathbb{M})$ by:

$$\tfrac{d}{dt} \log \bigwedge_t(W) = \sum_{n \geq 0} (-1)^n \Psi^{n+1}(W)t^n. \tag{5.2}$$

For a class function $f : \mathbb{M} \to \mathbb{C}$, define

$$(\Psi^k f)(g) = f(g^k).$$

for all $g \in \mathbb{M}$.

Now let W be a finite-dimensionally $\mathbb{Z} \oplus \mathbb{Z}$- graded representation of $\mathbb{M}$ such that $W_{(\gamma_1, \gamma_2)} = 0$ for $\gamma_1, \gamma_2 > 0$. We shall write $W = \sum_{(\gamma_1, \gamma_2) \in \mathbb{N}^2} W_{(-\gamma_1, -\gamma_2)} p^{\gamma_1} q^{\gamma_2}$, identifying the graded space and formal series. We extend the definition of Ψ^k to formal series $W \in R(\mathbb{M})[[p, q]]$ by defining $\Psi^k(p) = p^k$, $\Psi^k(q) = q^k$ and in general,

$$\Psi^k \left(\sum_{(\gamma_1,\gamma_2)\in\mathbb{N}^2} W_{(-\gamma_1,-\gamma_2)} p^{\gamma_1} q^{\gamma_2} \right) = \sum_{(\gamma_1,\gamma_2)\in\mathbb{N}^2} \Psi^k \left(W_{(-\gamma_1,-\gamma_2)} \right) p^{k\gamma_1} q^{k\gamma_2}.$$

Recall the structure of $\mathfrak{u}^-$ and $U = H_1(\mathfrak{u}^-)$ as $\mathbb{Z} \oplus \mathbb{Z}$-graded M-modules. We index the grading by p^{-1} and q^{-1}; then write $\mathfrak{u}^-$ and $U = H_1(\mathfrak{u}^-)$ as elements of $R[M][[p, q]]$:

$$\mathfrak{u}^- = \sum_{(m,n)} V^{\natural}_{mn+1} p^m q^n \tag{5.3}$$

and

$$U = \sum_{(m,n)} V^{\natural}_{m+n} p^m q^n, \tag{5.4}$$

where here and below the sums are over all pairs (m, n) such that $m, n > 0$.
    Define

$$H_t(\mathfrak{u}^-) = \sum_{i=0}^{\infty} H_i(\mathfrak{u}^-) t^i$$

and let $H(\mathfrak{u}^-)$ denote the alternating sum $H_t(\mathfrak{u}^-)|_{t=-1}$. Recall the Euler-Poincaré identity:

$$\bigwedge_{-1}(\mathfrak{u}^-) = H(\mathfrak{u}^-). \tag{5.5}$$

Taking log of both sides of (5.5) results in the formal power series identity in $R(M)[[p, q]] \otimes \mathbb{Q}$:

$$\log \bigwedge_{-1}(\mathfrak{u}^-) = \log H(\mathfrak{u}^-), \tag{5.6}$$

where we have

$$\log H(\mathfrak{u}^-) = \log(1 - H_1(\mathfrak{u}^-)) = -\sum_{n=1}^{\infty} \frac{1}{n} H_1(\mathfrak{u}^-)^n.$$

Formally integrating (5.2), with $W = \mathfrak{u}^-$, gives

$$\log \bigwedge_t(\mathfrak{u}^-) = -\sum_{n\geq 0} \Psi^{n+1}(\mathfrak{u}^-) \frac{(-t)^{n+1}}{n+1}.$$

Then setting $t = -1$ gives:

$$-\log \bigwedge_{-1}(\mathfrak{u}^-) = \sum_{k=1}^{\infty} \frac{1}{k} \Psi^k(\mathfrak{u}^-).$$

Since $H_1(\mathfrak{u}^-) = U$, equation (5.6) gives

$$\sum_{k=1}^{\infty} \frac{1}{k} \Psi^k \left(\sum_{(m,n)} V^{\natural}_{mn+1} p^m q^n \right) = \sum_{k=1}^{\infty} \frac{1}{k} \left(\sum_{(m,n)} V^{\natural}_{m+n} p^m q^n \right)^k \quad (5.7)$$

We say $k|(i, j)$ if $k(m, n) = (i, j)$ for some $(m, n) \in \mathbb{Z} \oplus \mathbb{Z}$. For $(i, j) \in \mathbb{Z}_+ \oplus \mathbb{Z}_+$ we define

$$P(i, j) = \left\{ a = (a_{rs})_{r,s \in \mathbb{Z}_+} \mid a_{rs} \in \mathbb{N}, \sum_{(r,s) \in \mathbb{Z}_+ \oplus \mathbb{Z}_+} a_{rs}(r, s) = (i, j) \right\}.$$

We will use the notation $|a| = \sum a_{rs}$, $a! = \prod a_{rs}!$. Expanding both sides of equation (5.7)

$$\sum_{(i,j)} \sum_{k|(i,j)} \frac{1}{k} \Psi^k (V^{\natural}_{ij/k^2+1}) p^i q^j = \sum_{(i,j)} \sum_{a \in P(i,j)} \frac{(|a|-1)!}{a!} \prod_{r,s \in \mathbb{Z}_+} (V^{\natural}_{r+s})^{a_{rs}} p^i q^j.$$

$$(5.8)$$

Then taking the trace of an element $g \in \mathbb{M}$ on both sides of the identity (5.8)

$$\sum_{(i,j)} \sum_{k|(i,j)} \frac{1}{k} \Psi^k (c_g(ij/k^2)) p^i q^j$$

$$= \sum_{(i,j)} \sum_{a \in P(i,j)} \frac{(|a|-1)!}{a!} \prod_{r,s \in \mathbb{Z}_+} c_g(r+s-1)^{a_{rs}} p^i q^j.$$

Equating the coefficients of $p^i q^j$ and applying Möbius inversion yields the recursion formulas:

$$c_g(ij) = \sum_{\substack{k>0 \\ k(m,n)=(i,j)}} \frac{1}{k} \mu(k) \left(\sum_{a \in P(m,n)} \frac{(|a|-1)!}{a!} \prod_{r,s \in \mathbb{Z}_+} c_{g^k}(r+s-1)^{a_{rs}} \right).$$

$$(5.9)$$

The coefficients of any replicable function are determined by the first 23 coefficients [**CN95**], but in the case of the above McKay-Thompson series we can use a smaller set of coefficients.

An examination of the formula (5.9) shows that $c_g(n)$ is determined by expressions of lower level except when $n = 1, 2, 3, 5$. Thus the values of the $c_g(n)$ are determined by the $c_h(1)$, $c_h(2)$, $c_h(3)$, $c_h(5)$, $h \in \mathbb{M}$, and the above recursions.

As in [**B92**], we conclude that since both the McKay-Thompson series for $V^{\natural}$ and the modular functions of [**CN79**] satisfy (5.9), all that is necessary to prove that these functions are the same is to check the initial data listed above. For the modular functions, see [**Koi**] and [**Fer96**]. For the relevant initial data about the graded traces of the actions of the elements of the Monster on $V^\natural$, the main theorem of [**FLM88**] is needed, as used in [**B92**].

References

[**B86**] R. E. Borcherds, *Vertex algebras, Kac-Moody algebras, and the Monster*, Proc. Nat. Acad. Sci. U.S.A. **83** (1986), no. 10, 3068–3071.

[**B88**] ———, *Generalized Kac-Moody algebras*, J. Algebra **115** (1988), no. 2, 501–512.

[**B91**] ———, *Central extensions of generalized Kac-Moody algebras*, J. Algebra **140** (1991), no. 2, 330–335.

[**B92**] ———, *Monstrous moonshine and monstrous Lie superalgebras*, Invent. Math. **109** (1992), no. 2, 405–444.

[**B94**] ———, *Sporadic groups and string theory*, First European Congress of Mathematics, Vol. **I** (Paris, 1992), Progr. Math., vol. 119, Birkhäuser, Basel, 1994, pp. 411–421.

[**B95**] ———, *A characterization of generalized Kac-Moody algebras*, J. Algebra **174** (1995), no. 3, 1073–1079.

[**B98**] ———, *What is Moonshine?*, Proceedings of the International Congress of Mathematicians, Vol. I (Berlin, 1998), no. Extra Vol. **I**, 1998, pp. 607–615.

[**CE56**] H. Cartan and S. Eilenberg, *Homological algebra*, Princeton University Press, Princeton, N. J., 1956.

[**CN79**] J. H. Conway and S. P. Norton, *Monstrous moonshine*, Bull. London Math. Soc. **11** (1979), no. 3, 308–339.

[**CN95**] C. J. Cummins and S. P. Norton, *Rational Hauptmoduls are replicable*, Canad. J. Math. **47** (1995), no. 6, 1201–1218.

[**DL93**] C. Dong and J. Lepowsky, *Generalized vertex algebras and relative vertex operators*, Progress in Mathematics, vol. 112, Birkhäuser Boston Inc., Boston, MA, 1993.

[**Fer96**] C. R. Ferenbaugh, *Replication formulae for n\h-type Hauptmoduls*, J. Algebra **179** (1996), no. 3, 808–837.

[**FHL93**] I. B. Frenkel, Y.-Z. Huang, and J. Lepowsky, *On axiomatic approaches to vertex operator algebras and modules*, Mem. Amer. Math. Soc. **104** (1993), no. 494.

[**FLM84**] I. Frenkel, J. Lepowsky, and A. Meurman, *A natural representation of the Fischer-Griess Monster with the modular function J as* *character*, Proc. Nat. Acad. Sci. U.S.A. **81** (1984), no. 10, Phys. Sci., 3256–3260.

[FLM88] _____, *Vertex operator algebras and the Monster*, Pure and Applied Mathematics, vol. 134, Academic Press Inc., Boston, MA, 1988.

[GL76] H. Garland and J. Lepowsky, *Lie algebra homology and the Macdonald-Kac formulas*, Invent. Math. **34** (1976), no. 1, 37–76.

[Gr82] R. L. Griess, Jr., *The friendly giant*, Invent. Math. **69** (1982), no. 1, 1–102.

[GT72] P. Goddard and C. B. Thorn, *Compatibility of the dual pomeron with unitarity and the absence of ghosts in the dual resonance model*, Phys. Lett. B **40** (1972), no. 2, 235–238.

[JLW95] E. Jurisich, J. Lepowsky, and R. L. Wilson, *Realizations of the Monster Lie algebra*, Selecta Math. (N.S.) **1** (1995), no. 1, 129–161.

[J98] E. Jurisich, *Generalized Kac-Moody Lie algebras, free Lie algebras and the structure of the Monster Lie algebra*, J. Pure Appl. Algebra **126** (1998), no. 1-3, 233–266.

[K90] V. Kac, *Infinite-dimensional Lie algebras*, third ed., Cambridge University Press, Cambridge, 1990.

[Ka94] S.-J. Kang, *Generalized Kac-Moody algebras and the modular function j*, Math. Ann. **298** (1994), no. 2, 373–384.

[KK95] V. G. Kac and S.-J. Kang, *A trace formula for graded Lie algebras and Monstrous Moonshine*, Sūrikaisekikenkyūsho Kōkyūroku (1995), no. 904, 116–129, Moonshine and vertex operator algebra (Kyoto, 1994).

[Kn73] D. Knutson, *λ-rings and the representation theory of the symmetric group*, Springer-Verlag, Berlin, 1973, Lecture Notes in Mathematics, Vol. **308**.

[Koi] M. Koike, *On replication formulas and Hecke operators*, Nagoya University preprint.

[Le79] J. Lepowsky, *Generalized Verma modules, loop space cohomology and Macdonald-type identities*, Ann. Sci. École Norm. Sup. (4) **12** (1979), no. 2, 169–234.

[Li94] H.-S. Li, *Symmetric invariant bilinear forms on vertex operator algebras*, J. Pure Appl. Algebra **96** (1994), no. 3, 279–297.

[Liu92] L. Liu, *Kostant's formula for Kac-Moody Lie algebras*, J. Algebra **149** (1992), 155–178.

[Mo67] R. V. Moody, *Lie algebras associated with generalized Cartan matrices*, Bull. Amer. Math. Soc. **73** (1967), 217–221.

On the Connection of Certain Lie Algebras with Vertex Algebras

Haisheng Li

Department of Mathematical Sciences,
Rutgers University, Camden, NJ 08102
E-mail address: hli@camden.rutgers.edu

Abstract

In this paper we survey the connection of certain infinite-dimensional Lie algebras, including twisted and untwisted affine Lie algebras, toroidal Lie algebras and quantum torus Lie algebras, with vertex algebras.

Introduction

Vertex (operator) algebras are a new class of algebraic structures and they have deep connections with numerous fields. In mathematics, vertex algebras have been a vibrant research area. On the other hand, as the algebraic counterpart of chiral algebras, vertex operator algebras together with their representations provide a solid foundation for the study of conformal field theory in physics.

Though vertex algebras are highly non-classical, they have connections with classical algebras such as Lie algebras, associative algebras and groups. In particular, vertex algebras are often constructed and studied by using classical (infinite-dimensional) Lie algebras. For example, those vertex operator algebras associated to (untwisted) affine Kac-Moody Lie algebras (including infinite-dimensional Heisenberg Lie algebras) and the Virasoro Lie algebra (cf. [FZ], [DL], [Li1], [LL]) are among the important examples. These two families of vertex operator algebras underline the algebraic study of the physical Wess-Zumino-Novikov-Witten model and the minimal models in conformal field theory, respectively. On the other hand, twisted affine Lie algebras (see [K1]) can be also associated with vertex operator algebras in terms of twisted modules (see [FLM], [Li2]). In the theory of Lie algebras,

1991 *Mathematics Subject Classification.* Primary 17B69, 17B67; Secondary 17B68.
Key words and phrases. Affine Lie algebra, toroidal Lie algebra, vertex algebra.
The author was supported in part by an NSA Grant.

by generalizing the loop-realization of untwisted affine Lie algebras, one has toroidal Lie algebras, which are perfect central extensions of multi-loop Lie algebras. Toroidal Lie algebras can be also associated with vertex algebras and their modules (see [BBS]). There is another family of Lie algebras called quantum torus Lie algebras, which are central extensions of (associative) quantum tori viewed as Lie algebras. Recently, in [Li3] we introduced a new notion of quasi module for vertex algebras and we proved that certain quantum torus Lie algebras can be also associated to vertex algebras in terms of quasi modules. Affine Lie algebras, toroidal Lie algebras and quantum torus Lie algebras belong to a larger family of Lie algebras called extended affine Lie algebras (see [S1,2], [AABGP]). Extended affine Lie algebras generalize affine Kac-Moody Lie algebras in a certain way and they also very much resemble affine Kac-Moody Lie algebras. A general theory and many examples of extended affine Lie algebras were given in [AABGP]. We believe that every extended affine Lie algebra can be associated with vertex algebras in terms of quasi modules in a similar way to the way that quantum torus Lie algebras were associated with vertex algebras in [Li3]. This is a survey and it does not contain any essentially new result.

We would like to thank the organizers for their hard work in organizing this excellent workshop and we thank S. Berman for many discussions on toroidal Lie algebras.

1. Affine Lie algebras and vertex algebras

In this section we present the connection of (untwisted) affine Lie algebras and toroidal Lie algebras with vertex algebras.

For convenience we first give a definition of a vertex algebra. Recall that a general (or abstract) non-associative algebra is a vector space A equipped with a bilinear operation, which is equivalent to a linear map from A to End A (through left multiplication).

Definition 1.1. A *vertex algebra* is a vector space V equipped with infinitely many bilinear operations parametrized by integers, which are equivalent to infinitely many linear maps:

$$V \to \text{End } V; \ v \mapsto v_n. \tag{1.1}$$

For each v, the infinitely many associated left multiplications are written in terms of the generating function as:

$$Y(v, x) = \sum_{n \in \mathbb{Z}} v_n x^{-n-1} \in (\text{End } V)[[x, x^{-1}]], \tag{1.2}$$

which is called the *vertex operator associated to v*. The following are the axioms:

(V1) For any $u, v \in V$, $u_n v = 0$ for n sufficiently large.

(V2) There exists a special vector $\mathbf{1}$, called the *vacuum vector*, such that for $v \in V$, $n \in \mathbb{Z}$,

$$\mathbf{1}_n v = \delta_{n,-1} v, \tag{1.3}$$

$$v_n \mathbf{1} = 0 \quad \text{if } n \geq 0 \quad \text{and} \quad v_{-1} \mathbf{1} = v. \tag{1.4}$$

(V3) For any $u, v \in V$, there exists a nonnegative integer k such that

$$(x_1 - x_2)^k Y(u, x_1) Y(v, x_2) = (x_1 - x_2)^k Y(v, x_2) Y(u, x_1). \tag{1.5}$$

(V4) For any $u, v, w \in V$, there exists a nonnegative integer l such that

$$(x_0 + x_2)^l Y(u, x_0 + x_2) Y(v, x_2) w = (x_0 + x_2)^l Y(Y(u, x_0)v, x_2) w, \tag{1.6}$$

where

$$Y(u, x_0 + x_2) = \sum_{n \in \mathbb{Z}} u_n (x_0 + x_2)^{-n-1} = \sum_{n \in \mathbb{Z}} \sum_{i \geq 0} \binom{-n-1}{i} u_n x_0^{-n-1-i} x_2^i$$

is a formal series in x_0 and x_2.

The axioms (V3) and (V4) together are equivalent to the Jacobi identity

$$x_0^{-1} \delta \left(\frac{x_1 - x_2}{x_0} \right) Y(u, x_1) Y(v, x_2) w - x_0^{-1} \delta \left(\frac{x_2 - x_1}{-x_0} \right) Y(v, x_2) Y(u, x_1) w$$

$$= x_2^{-1} \delta \left(\frac{x_1 - x_0}{x_2} \right) Y(Y(u, x_0)v, x_2) w, \tag{1.7}$$

where $\delta(x) = \sum_{n \in \mathbb{Z}} x^n$ and

$$x_0^{-1} \delta \left(\frac{x_1 - x_2}{x_0} \right) = \sum_{n \in \mathbb{Z}} x_0^{-1-n} (x_1 - x_2)^n = \sum_{n \in \mathbb{Z}} \sum_{i \geq 0} \binom{n}{i} (-1)^i x_0^{-1-n} x_1^{n-i} x_2^i.$$

Note that for a non-associative algebra A, an A-module is a vector space U equipped with a linear map from A to End U. For a vertex algebra V, a V-*module* is a vector space W equipped with infinitely many linear maps parametrized by integers, which together amount to a linear map

$$Y_W(\cdot, x) : V \to (\text{End } W)[[x, x^{-1}]],$$

$$v \mapsto Y_W(v, x) = \sum_{n \in \mathbb{Z}} v_n x^{-n-1},$$

such that for $v \in V$, $w \in W$, $v_n w = 0$ for n sufficiently large, $Y_W(\mathbf{1}, x) = 1_W$ (the identity operator on W), and such that (V3) and (V4) with the obvious modification hold.

Note that linear operators on a vector space give rise to classical associative algebras and Lie algebras. Similarly, there is a general construction of vertex algebras by using vertex operators.

Let W be any vector space. Set

$$\mathcal{E}(W) = \mathrm{Hom}(W, W((x))). \tag{1.8}$$

In particular, $\mathcal{E}(W)$ contains the identity operator on W, denoted by 1_W. For $\phi(x), \psi(x) \in \mathcal{E}(W)$, $n \in \mathbb{Z}$, define

$$\psi(x)_n \phi(x) = \mathrm{Res}_{x_1} \left((x_1 - x)^n \phi(x_1)\psi(x) - (-x + x_1)^n \psi(x)\phi(x_1) \right). \tag{1.9}$$

A subspace U of $\mathcal{E}(W)$ is said to be *closed* if

$$\phi(x)_n \psi(x) \in U \quad \text{for all } \phi(x), \psi(x) \in \mathcal{E}(W), \ n \in \mathbb{Z}.$$

A subset S of $\mathcal{E}(W)$ is said to be *local* if for any $\phi(x), \psi(x) \in S$, there exists a nonnegative integer k such that

$$(x_1 - x_2)^k \phi(x_1)\psi(x_2) = (x_1 - x_2)^k \psi(x_1)\phi(x_2). \tag{1.10}$$

We have the following result due to [Li1] (cf. [LL]):

Theorem 1.2. *Let W be any vector space. For any local subset S of $\mathcal{E}(W)$, there exists a (unique) smallest closed local subspace $\langle S \rangle$ of $\mathcal{E}(W)$, containing S and 1_W, and $\langle S \rangle$ is a vertex algebra with W as a faithful module with $Y_W(\phi(x), x_0) = \phi(x_0)$.*

Let $\mathfrak{g}$ be a (possibly infinite-dimensional) Lie algebra equipped with a (possibly degenerate) symmetric invariant bilinear form $\langle \cdot, \cdot \rangle$. Associated to $(\mathfrak{g}, \langle \cdot, \cdot \rangle)$ we have the (untwisted) affine Lie algebra

$$\hat{\mathfrak{g}} = \mathfrak{g} \otimes \mathbb{C}[t, t^{-1}] \oplus \mathbb{C}c, \tag{1.11}$$

where c is central and

$$[a \otimes t^m, b \otimes t^n] = [a, b] \otimes t^{m+n} + m\langle a, b \rangle \delta_{m+n,0} c \tag{1.12}$$

for $a, b \in \mathfrak{g}$, $m, n \in \mathbb{Z}$.

For $a \in \mathfrak{g}$, $n \in \mathbb{Z}$, we use $a(n)$ for $a \otimes t^n$ and for its corresponding operator on any $\hat{\mathfrak{g}}$-module. For $a \in \mathfrak{g}$, form the generating function

$$a(x) = \sum_{n \in \mathbb{Z}} a(n)x^{-n-1} \in \hat{\mathfrak{g}}[[x, x^{-1}]]. \tag{1.13}$$

For $a, b \in \mathfrak{g}$, we have

$$[a(x_1), b(x_2)] = [a, b](x_2)x_2^{-1}\delta(x_1/x_2) + \langle a, b \rangle c \frac{\partial}{\partial x_2} x_2^{-1}\delta(x_1/x_2). \tag{1.14}$$

As $(x_1 - x_2)^m \left(\frac{\partial}{\partial x_2}\right)^n x_2^{-1} \delta(x_1/x_2) = 0$ for $m > n \geq 0$, we have

$$(x_1 - x_2)^2 [a(x_1), b(x_2)] = 0. \tag{1.15}$$

A $\hat{\mathfrak{g}}$-module W is said to be *restricted* if for any $a \in \mathfrak{g}$, $w \in W$, $a(n)w = 0$ for n sufficiently large, that is, for any $a \in \mathfrak{g}$, $a(x) \in \mathcal{E}(W)$. Then for any restricted $\hat{\mathfrak{g}}$-module W, $U_W = \{a(x) \mid a \in \mathfrak{g}\}$ is a local subspace of $\mathcal{E}(W)$. A $\hat{\mathfrak{g}}$-module W is said to be of *level* $\ell \in \mathbb{C}$ if c acts on W as scalar ℓ. We have (see [Li1], [LL]):

Proposition 1.3. *Let W be any restricted $\hat{\mathfrak{g}}$-module of level $\ell \in \mathbb{C}$. Then U_W is a local subspace of $\mathcal{E}(W)$ and U_W generates a vertex algebra V_W with W as a module. Furthermore, V_W is naturally a $\hat{\mathfrak{g}}$-module of level ℓ with $a(n)$ acting as $a(x)_n$ for $a \in \mathfrak{g}$, $n \in \mathbb{Z}$ and c acting as scalar ℓ, and V_W as a $\hat{\mathfrak{g}}$-module is generated by 1_W satisfying the following relation*

$$a(n)1_W = 0 \quad \text{for } a \in \mathfrak{g}, \ n \in \mathbb{Z}.$$

Set $\hat{\mathfrak{g}}_{\geq 0} = \mathfrak{g} \otimes \mathbb{C}[t] \oplus \mathbb{C}c$, a Lie subalgebra of $\hat{\mathfrak{g}}$. Let ℓ be any complex number. Denote by $\mathbb{C}_\ell$ the 1-dimensional $\hat{\mathfrak{g}}_{\geq 0}$-module with $\mathfrak{g}[t]$ acting trivially and with c acting as scalar ℓ. Form the induced $\hat{\mathfrak{g}}$-module

$$V_{\hat{\mathfrak{g}}}(\ell, 0) = U(\hat{\mathfrak{g}}) \otimes_{U(\hat{\mathfrak{g}}_{\geq 0})} \mathbb{C}_\ell. \tag{1.16}$$

Set $\mathbf{1} = 1 \otimes 1 \in V_{\hat{\mathfrak{g}}}(\ell, 0)$. In view of the P-B-W theorem, we may and we should identify $\mathfrak{g}$ as a subspace of $V_{\hat{\mathfrak{g}}}(\ell, 0)$ through the map $a \mapsto a(-1)\mathbf{1}$.

The following result was obtained in [FZ], [Li1] (cf. [DL]):

Theorem 1.4. *For any complex number ℓ, there exists a unique vertex algebra structure on $V_{\hat{\mathfrak{g}}}(\ell, 0)$ with $\mathbf{1}$ as the vacuum vector such that*

$$Y(a, x) = a(x) \quad \text{for } a \in \mathfrak{g}.$$

The following result was obtained in [Li1] (cf. [FZ], [LL]):

Theorem 1.5. *Let W be any restricted $\hat{\mathfrak{g}}$-module of level $\ell \in \mathbb{C}$. There exists a unique module structure Y_W on W for $V_{\hat{\mathfrak{g}}}(\ell, 0)$ such that*

$$Y_W(a, x) = a(x) \quad \text{for } a \in \mathfrak{g}.$$

On the other hand, for any $V_{\hat{\mathfrak{g}}}(\ell, 0)$-module (W, Y_W), W is a restricted $\hat{\mathfrak{g}}$-module of level ℓ with $a(x)$ acting as $Y_W(a, x)$ for $a \in \mathfrak{g}$.

Remark 1.6. For the well known Virasoro Lie algebra $Vir = \oplus_{n \in \mathbb{Z}} L_n \oplus \mathbb{C}c$, one has similar results (cf. [FZ], [Li1], [LL]). We here just mention that the generating function

$$L(x) = \sum_{n \in \mathbb{Z}} L(n) x^{-n-2}$$

satisfies the locality relation $(x_1 - x_2)^4 [L(x_1), L(x_2)] = 0$.

Remark 1.7. There is a notion of conformal Lie algebra (see [K2]), or equivalently, a notion of vertex Lie algebra (see [P], [DLM]), which unifies (untwisted) affine Lie algebras and the Virasoro Lie algebra. For each vertex Lie algebra, one has an honest Lie algebra which can be defined in terms of mutually local generating functions. Furthermore, one obtains similar results (see [P], [DLM]).

Next we consider toroidal Lie algebras (cf. [AABGP]). Let $\mathfrak{g}$ be a Lie algebra equipped with a symmetric invariant bilinear form $\langle \cdot, \cdot \rangle$ as before. (Usually, $\mathfrak{g}$ is a finite-dimensional simple Lie algebra equipped with the Killing form suitably normalized.) Fix a positive integer r. We have the following multi-loop Lie algebra

$$L^{(r)}(\mathfrak{g}) = \mathfrak{g} \otimes \mathbb{C}[t_1^{\pm 1}, \ldots, t_r^{\pm 1}]. \tag{1.17}$$

For $\mathbf{n} = (n_1, \ldots, n_r) \in \mathbb{Z}^r$, set $\mathbf{t^n} = t_1^{n_1} \cdots t_r^{n_r}$. Endow $L^{(r)}(\mathfrak{g})$ with the symmetric bilinear form defined by

$$\langle a \otimes \mathbf{t^m}, b \otimes \mathbf{t^n} \rangle = \langle a, b \rangle \delta_{\mathbf{m+n}, 0} \tag{1.18}$$

for $a, b \in \mathfrak{g}$, $\mathbf{m}, \mathbf{n} \in \mathbb{Z}^r$, which is invariant. For $1 \le i \le r$, set

$$d_i = 1 \otimes t_i \frac{d}{dt_i} \in \mathrm{Der}(L^{(r)}(\mathfrak{g})). \tag{1.19}$$

The toroidal Lie algebra associated with $\mathfrak{g}$ and r is a perfect central extension of the multi-loop Lie algebra $L^{(r)}(\mathfrak{g})$:

$$T^{(r)}(\mathfrak{g}) = \mathfrak{g} \otimes \mathbb{C}[t_1^{\pm 1}, \ldots, t_r^{\pm 1}] \oplus \left(\oplus_{i=1}^r \mathbb{C}c_i \right) \oplus \left(\oplus_{i=1}^r \mathbb{C}d_i \right), \tag{1.20}$$

where $\oplus_{i=1}^r \mathbb{C}c_i$ is central and

$$[a \otimes \mathbf{t^m}, b \otimes \mathbf{t^n}] = [a, b] \otimes \mathbf{t^{m+n}} + \sum_{i=1}^r m_i \langle a, b \rangle \delta_{\mathbf{m+n}, 0} c_i \tag{1.21}$$

for $a, b \in \mathfrak{g}$, $\mathbf{m}, \mathbf{n} \in \mathbb{Z}^r$. Set

$$T^{(r)}(\mathfrak{g})' = \mathfrak{g} \otimes \mathbb{C}[t_1^{\pm 1}, \ldots, t_r^{\pm 1}] \oplus \left(\oplus_{i=1}^r \mathbb{C}c_i \right), \tag{1.22}$$

which is a Lie subalgebra of $T^{(r)}(\mathfrak{g})$ and which is called the *core* of the toroidal Lie algebra. (If $[\mathfrak{g}, \mathfrak{g}] = \mathfrak{g}$ and if $\langle \cdot, \cdot \rangle$ is nondegenerate, then $T^{(r)}(\mathfrak{g})'$ is the derived subalgebra of $T^{(r)}(\mathfrak{g})$.)

For $a \in \mathfrak{g}$, $\mathbf{p} = (p_1, \ldots, p_{r-1}) \in \mathbb{Z}^{r-1}$, form the generating function

$$a(\mathbf{p}, x) = \sum_{m \in \mathbb{Z}} (a \otimes \mathbf{t}^{(\mathbf{p}, m)}) x^{-m-1}.$$

In terms of the generating functions the relations (1.21) can be rewritten as

$$
\begin{aligned}
[a(\mathbf{p}, x_1), b(\mathbf{q}, x_2)] = {}& [a, b](\mathbf{p} + \mathbf{q}, x_2) x_1^{-1} \delta\left(\frac{x_2}{x_1}\right) \\
& + \langle a, b \rangle \delta_{\mathbf{p}+\mathbf{q},0} \frac{\partial}{\partial x_2} x_1^{-1} \delta\left(\frac{x_2}{x_1}\right) c_r \\
& + \langle a, b \rangle \delta_{\mathbf{p}+\mathbf{q},0} x_1^{-1} \delta\left(\frac{x_2}{x_1}\right) \sum_{i=1}^{r-1} p_i c_i x_2^{-1} \quad (1.23)
\end{aligned}
$$

for $a, b \in \mathfrak{g}$, $\mathbf{p}, \mathbf{q} \in \mathbb{Z}^{r-1}$. It follows that

$$(x_1 - x_2)^2 [a(\mathbf{p}, x_1), b(\mathbf{q}, x_2)] = 0. \quad (1.24)$$

We say that a $T^{(r)}(\mathfrak{g})$-module W is a *restricted module* if for any $w \in W$ and for any $a \in \mathfrak{g}$, $\mathbf{p} \in \mathbb{Z}^{r-1}$, $(a \otimes \mathbf{t}^{(\mathbf{p}, m)})w = 0$ for m sufficiently large. Then for a restricted $T^{(r)}(\mathfrak{g})$-module W, we have

$$a(\mathbf{p}, x) \in \mathcal{E}(W) \quad \text{for } a \in \mathfrak{g}, \ \mathbf{p} \in \mathbb{Z}^{r-1} \quad (1.25)$$

and the set $\{a(\mathbf{p}, x) \mid a \in \mathfrak{g}, \ \mathbf{p} \in \mathbb{Z}^{r-1}\}$ is a local subset of $\mathcal{E}(W)$.

We say that a $T^{(r)}(\mathfrak{g})$-module W is of *level* $\ell \in \mathbb{C}$ if c_r acts as scalar ℓ.

Remark 1.8. Let W be a restricted $T^{(r)}(\mathfrak{g})$-module of level ℓ, on which $c_1, \ldots, c_r$ act trivially. In view of (1.23), the space spanned by the set $\{1_W\} \cup \{a(\mathbf{p}, x) \mid a \in \mathfrak{g}, \ \mathbf{p} \in \mathbb{Z}^{r-1}\}$ is a "closed" subspace of $\mathcal{E}(W)$ in a certain sense.

Set

$$T^{(r)}(\mathfrak{g})^{\geq 0} = \mathfrak{g} \otimes \mathbb{C}[t_1^{\pm 1}, \ldots, t_{r-1}^{\pm 1}, t_r] + \left(\oplus_{i=1}^r \mathbb{C}c_i\right) + \left(\oplus_{i=1}^r \mathbb{C}d_i\right), \quad (1.26)$$

$$T^{(r)}(\mathfrak{g})^- = \mathfrak{g} \otimes t_r^{-1} \mathbb{C}[t_1^{\pm 1}, \ldots, t_{r-1}^{\pm 1}, t_r^{-1}]. \quad (1.27)$$

Let ℓ be a complex number. Denote by $\mathbb{C}_\ell$ the 1-dimensional $T^{(r)}(\mathfrak{g})^{\geq 0}$-module on which c_r acts as ℓ and $\mathfrak{g} \otimes \mathbb{C}[t_1^{\pm 1}, \ldots, t_{r-1}^{\pm 1}, t_r] + \left(\oplus_{i=1}^{r-1} \mathbb{C}c_i\right) + \left(\oplus_{i=1}^r \mathbb{C}d_i\right)$ acts trivially. Form the induced $T^{(r)}(\mathfrak{g})$-module

$$V_{T^{(r)}(\mathfrak{g})}(\ell, 0) = U(T^{(r)}(\mathfrak{g})) \otimes_{U(T^{(r)}(\mathfrak{g})^{\geq 0})} \mathbb{C}_\ell, \quad (1.28)$$

which is a restricted $T^{(r)}(\mathfrak{g})$-module of level ℓ. Set $\mathbf{1} = 1 \otimes 1 \in V_{T^{(r)}(\mathfrak{g})}(\ell, 0)$. For $a \in \mathfrak{g}$, $\mathbf{p} = (p_1, \ldots, p_{r-1}) \in \mathbb{Z}^{r-1}$, set

$$a[\mathbf{p}] = a \otimes t_1^{p_1} \cdots t_{r-1}^{p_{r-1}} \in L^{(r-1)}(\mathfrak{g}). \tag{1.29}$$

Now we identify the multi-loop algebra $L^{(r-1)}(\mathfrak{g})$ as a subspace of $V_{T^{(r)}(\mathfrak{g})}(\ell, 0)$ through the linear map $a[\mathbf{p}] \mapsto (a \otimes \mathbf{t}^{(\mathbf{p}, -1)})\mathbf{1}$.

Proposition 1.9. *For any complex number ℓ, there exists a (unique) vertex algebra structure on $V_{T^{(r)}(\mathfrak{g})}(\ell, 0)$ with $\mathbf{1}$ as the vacuum vector such that*

$$Y(a[\mathbf{p}], x) = a(\mathbf{p}, x) \quad for \; a \in \mathfrak{g}, \; \mathbf{p} \in \mathbb{Z}^{r-1}.$$

Furthermore, on any restricted $T^{(r)}(\mathfrak{g})$-module W of level ℓ, on which $c_1, \ldots, c_{r-1}$ act trivially, there exists a (unique) $V_{T^{(r)}(\mathfrak{g})}(\ell, 0)$-module structure Y_W with

$$Y_W(a[\mathbf{p}], x) = a(\mathbf{p}, x) \quad for \; a \in \mathfrak{g}, \; \mathbf{p} \in \mathbb{Z}^{r-1}.$$

Proof. Notice that $L^{(r-1)}(\mathfrak{g})$ is a Lie algebra equipped with a symmetric invariant bilinear form $\langle \cdot, \cdot \rangle$ as defined in (1.18). Then we have a (generalized) affine Lie algebra $\widehat{L^{(r-1)}(\mathfrak{g})}$. Clearly, $\widehat{L^{(r-1)}(\mathfrak{g})}$ is isomorphic to the quotient Lie algebra of $T^{(r)}(\mathfrak{g})'$ by the $(r-1)$-dimensional central ideal $\oplus_{i=1}^{r-1} \mathbb{C} c_i$. We see that $V_{T^{(r)}(\mathfrak{g})}(\ell, 0) \simeq V_{\widehat{L^{(r-1)}(\mathfrak{g})}}(\ell, 0)$ as an $\widehat{L^{(r-1)}(\mathfrak{g})}$-module. Then it follows immediately from Theorems 1.4 and 1.5. $\qquad\square$

Next we consider the general case with restricted $T^{(r)}(\mathfrak{g})$-modules with nontrivial actions of $c_1, \ldots, c_{r-1}$. First we have the toroidal Lie algebra $T^{(r-1)}(\mathfrak{g})$ and its subalgebra

$$T^{(r-1)}(\mathfrak{g})' = \mathfrak{g} \otimes \mathbb{C}[t_1^{\pm 1}, \ldots, t_{r-1}^{\pm 1}] \oplus \left(\oplus_{i=1}^{r-1} \mathbb{C} c_i \right).$$

Extend the symmetric invariant bilinear form $\langle \cdot, \cdot \rangle$ on (the multi-loop Lie algebra) $L^{(r-1)}(\mathfrak{g})$ to $T^{(r-1)}(\mathfrak{g})'$ such that c_i for $i = 1, \ldots, r-1$ lie in the kernel. With the Lie algebra $T^{(r-1)}(\mathfrak{g})'$ equipped with this symmetric invariant bilinear form (which is degenerate), we have a (generalized) affine Lie algebra $\widehat{T^{(r-1)}(\mathfrak{g})'}$. From Theorem 1.4, for any complex number ℓ, we have a vertex algebra $V_{\widehat{T^{(r-1)}(\mathfrak{g})'}}(\ell, 0)$ with $T^{(r-1)}(\mathfrak{g})'$ as a subspace.

Proposition 1.10. *Let ℓ be any complex number. For any restricted $T^{(r)}(\mathfrak{g})$ -module W of level ℓ, there exists a (unique) $V_{\widehat{T^{(r-1)}(\mathfrak{g})'}}(\ell, 0)$-module structure Y_W with*

$$Y_W(a[\mathbf{p}], x) = a(\mathbf{p}, x) \quad for \; a \in \mathfrak{g}, \; \mathbf{p} \in \mathbb{Z}^{r-1}$$

and $Y_W(c_i, x) = c_i x^{-1}$ for $i = 1, \ldots, r - 1$, where we abuse c_i for elements of $V_{\widehat{T^{(r-1)}(\mathfrak{g})'}}(\ell, 0)$ and for elements of $T^{(r)}(\mathfrak{g})$. On the other hand, let (W, Y_W) be a $V_{\widehat{T^{(r-1)}(\mathfrak{g})'}}(\ell, 0)$-module such that $Y_W(c_i, x) \in x^{-1}(\mathrm{End}\ W)$ for $i = 1, \ldots, r - 1$. Then W is a restricted $T^{(r)}(\mathfrak{g})$-module of level ℓ where

$$a(\mathbf{p}, x) = Y_W(a[\mathbf{p}], x) \quad \text{for } a \in \mathfrak{g}, \ \mathbf{p} \in \mathbb{Z}^{r-1}$$

and $c_i x^{-1} = Y_W(c_i, x)$ for $i = 1, \ldots, r - 1$.

Proof. Notice that $\sum_{i=1}^{r-1} \mathbb{C}(c_i \otimes \mathbb{C}[t, t^{-1}])$ is central in the generalized affine Lie algebra $\widehat{T^{(r-1)}(\mathfrak{g})'}$. In view of (1.23), the Lie algebra $T^{(r)}(\mathfrak{g})'$ is isomorphic to the quotient Lie algebra of $\widehat{T^{(r-1)}(\mathfrak{g})'}$ modulo the central ideal $\sum_{i=1}^{r-1} \sum_{n \neq 0} \mathbb{C}(c_i \otimes t^n)$. Then it follows immediately from Theorem 1.5. □

Remark 1.11. In [BBS], certain Lie algebras, which are closely related to toroidal Lie algebras, were studied in terms of vertex operator algebras and their modules.

2. Twisted affine Lie algebras and vertex algebras

Affine Kac-Moody Lie algebras consist of untwisted affine Lie algebras and twisted affine Lie algebras (see [K1]). As we have seen in Section 2, untwisted affine Lie algebras are associated with vertex algebras and their modules. On the other hand, twisted affine Lie algebras are associated with vertex algebras and their twisted modules.

First we review the definition of the notion of twisted module for a vertex algebra and a conceptual result on vertex algebras and twisted modules. Let V be a vertex algebra and let σ be an automorphism of V of period r (finite). Set $\omega_r = \exp(2\pi\sqrt{-1}/r)$, a primitive r-th root of unity. Then $V = \coprod_{j=0}^{r-1} V^j$, where $V^j = \{v \in V \mid \sigma(v) = \omega_r^j v\}$.

A σ-*twisted V-module* ([L], [FLM], [FFR], [D]) is a vector space W equipped with a linear map

$$Y_W : V \to \mathrm{Hom}(W, W((x^{\frac{1}{r}}))) \subset (\mathrm{End}\ W)[[x^{\frac{1}{r}}, x^{-\frac{1}{r}}]]$$

such that $Y_W(\mathbf{1}, x) = 1_W$ and for $u \in V^j$, $v \in V$ with $0 \le j \le r - 1$ the following twisted Jacobi identity holds

$$x_0^{-1}\delta\left(\frac{x_1 - x_2}{x_0}\right) Y_W(u, x_1)Y_W(v, x_2) - x_0^{-1}\delta\left(\frac{x_2 - x_1}{-x_0}\right) Y_W(v, x_2)Y_W(u, x_1)$$

$$= x_2^{-1}\delta\left(\frac{x_1 - x_0}{x_2}\right)\left(\frac{x_1 - x_0}{x_2}\right)^{-\frac{j}{r}} Y_W(Y(u, x_0)v, x_2). \quad (2.1)$$

As a simple consequence we have

$$x^{\frac{i}{r}} Y_W(u, x) \in \operatorname{Hom}(W, W((x))) \tag{2.2}$$

for $u \in V^j$ with $0 \le j \le r - 1$.

Now let W be a plain vector space and let r be any positive integer. Set

$$\mathcal{E}(W, r) = \operatorname{Hom}(W, W((x^{1/r}))) \subset (\operatorname{End} W)[[x^{\frac{1}{r}}, x^{-\frac{1}{r}}]]. \tag{2.3}$$

Notice that $\mathcal{E}(W) = \mathcal{E}(W, 1)$. We define local subspaces of $\mathcal{E}(W, r)$ in the same way as we defined local subspaces of $\mathcal{E}(W)$ in Section 2. Notice that $(\operatorname{End} W)[[x^{\frac{1}{r}}, x^{-\frac{1}{r}}]]$ is naturally $\mathbb{Z}/r\mathbb{Z}$-graded as

$$(\operatorname{End} W)[[x^{\frac{1}{r}}, x^{-\frac{1}{r}}]] = \coprod_{j=0}^{r-1} x^{-\frac{j}{r}} (\operatorname{End} W)[[x, x^{-1}]]. \tag{2.4}$$

The subspace $\mathcal{E}(W, r)$ is $\mathbb{Z}/r\mathbb{Z}$-graded and

$$\mathcal{E}(W, r) = \coprod_{j=0}^{r-1} \mathcal{E}(W, r)^j = \coprod_{j=0}^{r-1} x^{-\frac{j}{r}} \mathcal{E}(W). \tag{2.5}$$

We define an order r automorphism σ of $\mathcal{E}(W, r)$ by

$$\sigma(a(x)) = \omega_r^j a(x) \tag{2.6}$$

for $a(x) \in \mathcal{E}(W, r)^j = x^{-\frac{j}{r}} \mathcal{E}(W)$, $0 \le j \le r - 1$. Under this definition, we have (cf. (2.2))

$$x^{\frac{j}{r}} a(x) \in \mathcal{E}(W) = \operatorname{Hom}(W, W((x)))$$

for $a(x) \in \mathcal{E}(W, r)^j$, $0 \le j \le r - 1$.

Let $a(x), b(x)$ be a local pair in $\mathcal{E}(W, r)$ with homogeneous $a(x) \in \mathcal{E}(W, r)^j$ for some $0 \le j \le r - 1$. We define $a(x)_n b(x) \in \mathcal{E}(W, r)$ for $n \in \mathbb{Z}$ by

$$a(x)_n b(x) = \operatorname{Res}_{x_1} \sum_{i \ge 0} \binom{-\frac{j}{r}}{i} x^{-\frac{j}{r} - i} x_1^{\frac{j}{r}}$$
$$\times \left((x_1 - x)^{n+i} a(x_1) b(x) - (-x + x_1)^{n+i} b(x) a(x_1) \right). \tag{2.7}$$

Remark 2.1. As in [Li2], the elements $a(x)_n b(x) \in \mathcal{E}(W, r)$ for $n \in \mathbb{Z}$ can also be defined in terms of the generating function

$$Y_{\mathcal{E}}(a(x), x_0) b(x) = \sum_{n \in \mathbb{Z}} a(x)_n b(x) x_0^{-n-1}$$

by

$$Y_{\mathcal{E}}(a(x), x_0)b(x) = \text{Res}_{x_1} \left(\frac{x_1 - x_0}{x}\right)^{\frac{i}{r}} \cdot X \left(= \text{Res}_{x_1} \left(\frac{x + x_0}{x_1}\right)^{-\frac{i}{r}} \cdot X\right),$$

(2.8)

where

$$X = x_0^{-1}\delta\left(\frac{x_1 - x}{x_0}\right) a(x_1)b(x) - x_0^{-1}\delta\left(\frac{x - x_1}{-x_0}\right) b(x)a(x_1).$$

Remark 2.2. The elements $a(x)_n b(x) \in \mathcal{E}(W, r)$ for $n \in \mathbb{Z}$ can also be defined in terms of the generating function $Y_{\mathcal{E}}(a(x), x_0)b(x)$ by

$$x_0^k Y_{\mathcal{E}}(a(x), x_0)b(x) = \text{Res}_{x_1}(x + x_0)^{-\frac{i}{r}} \left((x_1 - x)^k x_1^{\frac{i}{r}} a(x_1)b(x)\right)\big|_{x_1 = x + x_0},$$

(2.9)

where k is any nonnegative integer such that

$$(x_1 - x_2)^k a(x_1)b(x_2) = (x_1 - x_2)^k b(x_2)a(x_1).$$

This definition uses the associative algebra aspect of vertex algebras (cf. [Li3]).

The following results were obtained in [Li2]:

Theorem 2.3. *Let W be any vector space and let r be any positive integer. Every maximal $\mathbb{Z}/r\mathbb{Z}$-graded local subspace U of $\mathcal{E}(W, r)$ is a vertex algebra with σ as an order r automorphism. Furthermore, W is a faithful σ-twisted U-module with $Y_W(a(x), x_0) = a(x)$.*

Corollary 2.4. *Let W be any vector space and let r be any positive integer. Every $\mathbb{Z}/r\mathbb{Z}$-graded local subspace S of $\mathcal{E}(W, r)$ generates a vertex algebra $\langle S \rangle$ with σ as an order r automorphism. Furthermore, W is a faithful σ-twisted $\langle S \rangle$-module with $Y_W(a(x), x_0) = a(x)$.*

Let $\mathfrak{g}$ be a Lie algebra equipped with a symmetric invariant bilinear form $\langle \cdot, \cdot \rangle$ as in Section 2. Recall from Section 2 that we have the (untwisted) affine Lie algebra $\hat{\mathfrak{g}}$ and for any complex number ℓ, we have a vertex algebra $V_{\hat{\mathfrak{g}}}(\ell, 0)$ which is also a $\hat{\mathfrak{g}}$-module of level ℓ. Let σ be an order r automorphism of $\mathfrak{g}$, preserving the bilinear form. Extend σ to an order r automorphism of the affine Lie algebra $\hat{\mathfrak{g}}$ by $\sigma(c) = c$ and

$$\sigma(a \otimes t^n) = \sigma(a) \otimes t^n \quad \text{for } a \in \mathfrak{g}, \ n \in \mathbb{Z}.$$

Furthermore, σ gives rise to an order r automorphism, denoted also by σ, of $V_{\hat{\mathfrak{g}}}(\ell, 0)$ as a $\hat{\mathfrak{g}}$-module. It follows that σ is an order r automorphism of the vertex algebra $V_{\hat{\mathfrak{g}}}(\ell, 0)$.

Consider the Lie algebra

$$\hat{\mathfrak{g}}[r] = \mathfrak{g} \otimes \mathbb{C}[t^{\frac{1}{r}}, t^{-\frac{1}{r}}] \oplus \mathbb{C}c, \qquad (2.10)$$

where c is central and for $a, b \in \mathfrak{g}$, $m, n \in \mathbb{Z}$,

$$[a \otimes t^{m/r}, b \otimes t^{n/r}] = [a, b] \otimes t^{(m+n)/r} + \frac{m}{r}\langle a, b \rangle \delta_{m+n,0}c.$$

Clearly, the linear map ψ from $\hat{\mathfrak{g}}$ to $\hat{\mathfrak{g}}[r]$, defined by $\psi(a \otimes t^m) = a \otimes t^{m/r}$ and $\psi(c) = \frac{1}{r}c$, is an isomorphism of Lie algebras.

Extend σ to an automorphism of $\hat{\mathfrak{g}}[r]$ by $\sigma(c) = 1$ and

$$\sigma(a \otimes t^{m/r}) = \omega_r^{-m}(\sigma(a) \otimes t^{m/r}) \quad \text{for } a \in \mathfrak{g}, \ m \in \mathbb{Z}. \qquad (2.11)$$

The twisted affine Lie algebra $\hat{\mathfrak{g}}[\sigma]$ (see [FLM]) is defined to be the σ-fixed point Lie subalgebra of $\hat{\mathfrak{g}}[r]$. That is,

$$\hat{\mathfrak{g}}[\sigma] = \coprod_{j=0}^{r-1} \mathfrak{g}_j \otimes t^{\frac{j}{r}}\mathbb{C}[t, t^{-1}] \oplus \mathbb{C}c, \qquad (2.12)$$

where $\mathfrak{g}_j = \{a \in \mathfrak{g} \mid \sigma(a) = \omega_r^j a\}$. For $a \in \mathfrak{g}_j$ form the generating function

$$a(\sigma, x) = \sum_{n \in \mathbb{Z}}(a \otimes t^{\frac{j}{r}+n})x^{-\frac{j}{r}-n-1} \in x^{-\frac{j}{r}}\hat{\mathfrak{g}}[\sigma][[x, x^{-1}]]. \qquad (2.13)$$

For $a \in \mathfrak{g}_j$, $b \in \mathfrak{g}_k$ we have

$$[a(\sigma, x_1), b(\sigma, x_2)]$$

$$= x_1^{-1}\delta\left(\frac{x_2}{x_1}\right)\left(\frac{x_2}{x_1}\right)^{\frac{j}{r}}[a, b](\sigma, x_2) + \langle a, b \rangle c\frac{\partial}{\partial x_2}x_1^{-1}\delta\left(\frac{x_2}{x_1}\right)\left(\frac{x_2}{x_1}\right)^{\frac{j}{r}}. \qquad (2.14)$$

Just as with the untwisted case the following local relation holds:

$$(x_1 - x_2)^2[a(\sigma, x_1), b(\sigma, x_2)] = 0.$$

A $\hat{\mathfrak{g}}[\sigma]$-module W is said to be *restricted* if for any $a \in \mathfrak{g}$, $w \in W$, $a(m/r)w = 0$ for m sufficiently large and a $\hat{\mathfrak{g}}[\sigma]$-module W is said to be of level $\ell \in \mathbb{C}$ if c acts on W as scalar ℓ. If W is a restricted $\hat{\mathfrak{g}}[\sigma]$-module, then $a(\sigma, x)$ for homogeneous $a \in \mathfrak{g}$ are homogeneous elements of $\mathcal{E}(W, r)$ and they form a local subset. We have (see [Li2]):

Theorem 2.5. *Let W be any restricted $\hat{\mathfrak{g}}[\sigma]$-module of level ℓ. There exists a unique σ-twisted module structure Y_W on W for the vertex algebra $V_{\hat{\mathfrak{g}}}(\ell, 0)$ such that*

$$Y_W(a, x) = a(\sigma, x) \quad for \ a \in \mathfrak{g}.$$

On the other hand, for any σ-twisted $V_{\hat{\mathfrak{g}}}(\ell, 0)$-module (W, Y_W), W is a restricted $\hat{\mathfrak{g}}[\sigma]$-module of level ℓ with $a(\sigma, x)$ acting as $Y_W(a, x)$ for $a \in \mathfrak{g}$.

Remark 2.6. Just as Theorem 1.5 can be used to associate restricted modules for toroidal Lie algebra $L^{(r)}(\mathfrak{g})$ with vertex algebras and modules in Section 2, Theorem 2.5 can be used to associate restricted modules for certain twisted toroidal Lie algebras with vertex algebras and twisted modules.

3. Quantum torus Lie algebras and vertex algebras

Quantum torus Lie algebras are a family of infinite-dimensional Lie algebras which are closely related to toroidal Lie algebras. In [Li3] quantum torus Lie algebras were associated with vertex algebras and quasi modules.

First we present the Lie algebras of 2-dimensional quantum torus, following [BGT]. Let q be a nonzero complex number. Consider the following twisted group algebra of $\mathbb{Z}^2$

$$\mathbb{C}_q[t_0^{\pm 1}, t_1^{\pm 1}] = \mathbb{C}[t_0^{\pm 1}, t_1^{\pm 1}] \tag{3.1}$$

as a vector space, where for $m, n, r, s \in \mathbb{Z}$,

$$(t_0^m t_1^n)(t_0^r t_1^s) = q^{nr} t_0^{m+r} t_1^{n+s}. \tag{3.2}$$

Let A be an associative algebra equipped with a (possibly degenerate) symmetric invariant bilinear form $\langle \cdot, \cdot \rangle$ in the sense that

$$\langle ab, c \rangle = \langle a, bc \rangle \quad \text{for } a, b, c \in A. \tag{3.3}$$

Consider the tensor product associative algebra

$$A_q[t_0^{\pm 1}, t_1^{\pm 1}] = A \otimes \mathbb{C}_q[t_0^{\pm 1}, t_1^{\pm 1}]. \tag{3.4}$$

Naturally, $A_q[t_0^{\pm 1}, t_1^{\pm 1}]$ is a Lie algebra with the commutator map as the Lie bracket, which is denoted by $[\cdot, \cdot]_{loop}$. For $a, b \in A$, $m, n, r, s \in \mathbb{Z}$, we have

$$[a \otimes t_0^m t_1^n, b \otimes t_0^r t_1^s]_{loop} = q^{nr}(ab \otimes t_0^{m+r} t_1^{n+s}) - q^{ms}(ba \otimes t_0^{m+r} t_1^{n+s}). \tag{3.5}$$

Furthermore, consider the following two-dimensional central extension of the Lie algebra $A_q[t_0^{\pm 1}, t_1^{\pm 1}]$:

$$\widehat{A_q}[t_0^{\pm 1}, t_1^{\pm 1}] = A_q[t_0^{\pm 1}, t_1^{\pm 1}] \oplus \mathbb{C}c_0 \oplus \mathbb{C}c_1 = A \otimes \mathbb{C}_q[t_0^{\pm 1}, t_1^{\pm 1}] \oplus \mathbb{C}c_0 \oplus \mathbb{C}c_1, \tag{3.6}$$

where c_0, c_1 are central and

$$[a \otimes t_0^m t_1^n, b \otimes t_0^r t_1^s] = q^{nr}(ab \otimes t_0^{m+r} t_1^{n+s}) - q^{ms}(ba \otimes t_0^{m+r} t_1^{n+s})$$
$$+ \langle a, b \rangle q^{nr} \delta_{m+r,0} \delta_{n+s,0}(mc_0 + nc_1). \quad (3.7)$$

For $a \in A$, $n \in \mathbb{Z}$, set

$$X(a, n, x) = \sum_{m \in \mathbb{Z}} (a \otimes t_0^m t_1^n) x^{-m-1} \in \widehat{A_q}[t_0^{\pm 1}, t_1^{\pm 1}][[x, x^{-1}]]. \quad (3.8)$$

Then

$$[X(a, n, x_1), X(b, s, x_2)] = q^{-n} X(ab, n + s, q^{-n} x_2) x_1^{-1} \delta \left(\frac{q^{-n} x_2}{x_1} \right)$$
$$- X(ba, n + s, x_2) x_1^{-1} \delta \left(\frac{q^s x_2}{x_1} \right) + \langle a, b \rangle \delta_{n+s,0} \frac{\partial}{\partial x_2} x_1^{-1} \delta \left(\frac{q^{-n} x_2}{x_1} \right) c_0$$
$$+ \langle a, b \rangle \delta_{n+s,0} n \left(c_1 x_2^{-1} \right) x_1^{-1} \delta \left(\frac{q^{-n} x_2}{x_1} \right). \quad (3.9)$$

Consequently, we have

$$(x_1 - q^{-n} x_2)^2 (x_1 - q^s x_2)[X(a, n, x_1), X(b, s, x_2)] = 0. \quad (3.10)$$

For convenience let us just use $\widehat{A_q}$ for the Lie algebra $\widehat{A_q}[t_0^{\pm 1}, t_1^{\pm 1}]$. An $\hat{A}_q$-module on which c_0 acts as a scalar $\ell \in \mathbb{C}$ is said to be *of level* ℓ. An $\hat{A}_q$-module W is said to be *restricted* if for any $a \in A$, $n \in \mathbb{Z}$ and for any $w \in W$, $X(a, n, x)w \in W((x))$. That is, $X(a, n, x)$ acting on W is an element of $\mathcal{E}(W)$. It was proved in [Li3] that restricted $\hat{A}_q$-modules of a fixed level can be associated with vertex algebras. This follows from a conceptual result. To state this result we shall need the following notion:

Definition 3.1. Let V be a vertex algebra. A *quasi V-module* is a vector space W equipped with a linear map

$$Y_W(\cdot, x) : V \to \text{Hom}(W, W((x))) \subset (\text{End } W)[[x, x^{-1}]]$$

such that $Y_W(\mathbf{1}, x) = 1_W$ and for any $u, v \in V$, there exists a nonzero polynomial $p(x_1, x_2)$ such that

$$x_0^{-1} \delta \left(\frac{x_1 - x_2}{x_0} \right) p(x_1, x_2) Y_W(u, x_1) Y_W(v, x_2)$$
$$- x_0^{-1} \delta \left(\frac{x_2 - x_1}{-x_0} \right) p(x_1, x_2) Y_W(v, x_2) Y_W(u, x_1)$$
$$= x_2^{-1} \delta \left(\frac{x_1 - x_0}{x_2} \right) p(x_1, x_2) Y_W(Y(u, x_0)v, x_2). \quad (3.11)$$

Let $\mathbb{C}(x_1, x_2)$ be the field of rational functions in variables x_1 and x_2. We define ι_{x_1, x_2} to be the algebra homomorphism from $\mathbb{C}(x_1, x_2)$ to $\mathbb{C}((x_1))((x_2))$ such that $\iota_{x_1, x_2}(x_1^m x_2^n) = x_1^m x_2^n$ for $m, n \in \mathbb{Z}$ and

$$\iota_{x_1, x_2}(x_1 + \alpha x_2)^n = \sum_{i \geq 0} \binom{n}{i} \alpha^i x_1^{n-i} x_2^i$$

for $\alpha \in \mathbb{C}$, $n \in \mathbb{Z}$.

Now, let W be any vector space. A subset S of $\mathcal{E}(W)$ is said to be *quasi local* if for any $\phi(x), \psi(x) \in S$,

$$p(x_1, x_2)\phi(x_1)\psi(x_2) = p(x_1, x_2)\psi(x_2)\phi(x_1) \qquad (3.12)$$

for some nonzero homogeneous polynomial $p(x_1, x_2)$.

If $a(x), b(x)$ are quasi local, we define $a(x)_n b(x)$ for $n \in \mathbb{Z}$ in terms of the generating function

$$Y_{\mathcal{E}}(a(x), x_0)b(x) = \sum_{n \in \mathbb{Z}} (a(x)_n b(x)) x_0^{-n-1}$$

by

$$Y_{\mathcal{E}}(a(x), x_0)b(x) = \iota_{x, x_0}(1/p(x_0 + x, x)) \, (p(x_1, x)a(x_1)b(x)) \,|_{x_1 = x + x_0}.$$
$$(3.13)$$

One can show that this definition is independent of the choice of the polynomial $p(x_1, x_2)$. A quasi local subspace U of $\mathcal{E}(W)$ is said to be *closed* if

$$a(x)_n b(x) \in U \quad \text{for } a(x), b(x) \in U, \ n \in \mathbb{Z}.$$

The following was proved in [Li3]:

Theorem 3.2. *Let W be any vector space. Every maximal quasi local subspace U of $\mathcal{E}(W)$ is a vertex algebra with 1_W as the vacuum vector and W is a faithful quasi U-module with $Y_W(\alpha(x), x_0) = \alpha(x_0)$. Any quasi local subset S of $\mathcal{E}(W)$ generates a vertex algebra with W as a faithful quasi module.*

It is clear that Theorem 3.2 is applicable to to the case with W a restricted $\hat{A}_q$-module of level ℓ. Moreover, we can also determine the corresponding vertex algebra. To present the corresponding vertex algebra we need another algebra. The following was obtained in [Li3]:

Lemma 3.3. *Define a non-associative algebra with the underlying vector space*

$$\mathbb{C}_*[t_0^{\pm 1}, t_1^{\pm 1}] = \mathbb{C}[t_0^{\pm 1}, t_1^{\pm 1}], \qquad (3.14)$$

equipped with the multiplication

$$(t_0^n t_1^m)(t_0^s t_1^r) = \delta_{n+m,r} t_0^{n+s} t_1^m \quad for \ m, n, r, s \in \mathbb{Z}. \tag{3.15}$$

Endow this non-associative algebra with a bilinear form defined by

$$\langle t_0^n t_1^m, t_0^s t_1^r \rangle = \delta_{n+s,0} \delta_{m+n,r} \quad for \ m, n, r, s \in \mathbb{Z}. \tag{3.16}$$

Then this algebra is associative and the bilinear form is symmetric and associative.

Let A be an associative algebra equipped with a symmetric invariant bilinear form as before. Now, we have an associative algebra $A \otimes \mathbb{C}_*[t_0^{\pm 1}, t_1^{\pm 1}]$ equipped with a symmetric invariant bilinear form $\langle \cdot, \cdot \rangle$. The bilinear form is also invariant when $A \otimes \mathbb{C}_*[t_0^{\pm 1}, t_1^{\pm 1}]$ is viewed as a Lie algebra. For convenience, we set

$$A_* = A \otimes \mathbb{C}_*[t_0^{\pm 1}, t_1^{\pm 1}]. \tag{3.17}$$

Form the (generalized) affine Lie algebra

$$\widehat{A_*} = (A \otimes \mathbb{C}_*[t_0^{\pm 1}, t_1^{\pm 1}]) \otimes \mathbb{C}[t, t^{-1}] \oplus \mathbb{C}\mathbf{k}. \tag{3.18}$$

Recall that for any complex number ℓ, we have a vertex algebra $V_{\widehat{A_*}}(\ell, 0)$ associated with the affine Lie algebra $\widehat{A_*}$. The following result was obtained in [Li3]:

Theorem 3.4. *Let A be an associative algebra equipped with a symmetric invariant bilinear form $\langle \cdot, \cdot \rangle$ and let q be a nonzero complex number which is not a root of unity. Let $\widehat{A_q}$ be the quantum torus Lie algebra defined in (3.6). Let A_* be the Lie algebra defined in (3.17) equipped with the symmetric invariant bilinear form. Let W be any restricted $\widehat{A_q}$-module of level $\ell \in \mathbb{C}$, on which the central element c_1 acts as zero. Then there exists a unique quasi $V_{\widehat{A_*}}(\ell, 0)$-module structure on W with $Y_W(a \otimes t_0^m t_1^n, x) = X(a, m, q^n x)$ for $a \in A, \ m, n \in \mathbb{Z}$.*

Next we give a new connection between twisted affine Lie algebras and vertex algebras in terms of quasi modules. Note that the twisted affine Lie algebra $\hat{\mathfrak{g}}[\sigma]$ can also be realized as a subalgebra of $\hat{\mathfrak{g}}$ (see [K1]). First extend σ to be an automorphism of $\hat{\mathfrak{g}}$ by

$$\sigma(c) = c, \quad \sigma(a \otimes t^n) = \sigma(a) \otimes \omega_r^{-n} t^n$$

for $a \in \mathfrak{g}, \ n \in \mathbb{Z}$. Denote by $\hat{\mathfrak{g}}^\sigma$ the σ-fixed point Lie subalgebra of $\hat{\mathfrak{g}}$. Then $\hat{\mathfrak{g}}^\sigma$ is isomorphic to $\hat{\mathfrak{g}}[\sigma]$ with c corresponding to c/r.

Let $a \in \mathfrak{g}$ with $\sigma a = \omega_r^j a$. Set

$$a_\sigma(x) = \sum_{n \in \mathbb{Z}} (a \otimes t^{j+nr}) x^{-j-nr-1}. \tag{3.19}$$

We have

$$(x_1^r - x_2^r)^2 a_\sigma(x_1) b_\sigma(x_2) = (x_1^r - x_2^r)^2 b_\sigma(x_2) a_\sigma(x_1). \tag{3.20}$$

The following was proved in [Li3]:

Theorem 3.5. *Let σ be an order r automorphism of $\mathfrak{g}$ preserving the bilinear form and let W be any restricted $\hat{\mathfrak{g}}^\sigma$-module of level ℓ/r. There exists a unique quasi module structure Y_W on W for $V_{\hat{\mathfrak{g}}}(\ell, 0)$ such that $Y_W(a, x) = a_\sigma(x)$ for $a \in \mathfrak{g}$.*

Remark 3.6. Recall that the Lie algebra $\hat{\mathfrak{g}}^\sigma$ is isomorphic to $\hat{\mathfrak{g}}[\sigma]$ with c corresponding to c/r. Consequently, the category of restricted $\hat{\mathfrak{g}}^\sigma$-modules of level ℓ/r is canonically isomorphic to the category of restricted $\hat{\mathfrak{g}}[\sigma]$-modules of level ℓ. Combining Theorems 2.5 and 3.5 we have that on every σ-twisted $V_{\hat{\mathfrak{g}}}(\ell, 0)$-module there exists a canonical quasi $V_{\hat{\mathfrak{g}}}(\ell, 0)$-module structure. In view of this, we expect that for a general vertex operator algebra V, there is a canonical connection between twisted V-modules and quasi modules. Closely related to this is a beautiful work [BDM], in which Barron, Dong and Mason established a correspondence between V-modules and twisted modules for $V^{\otimes n}$ with respect to permutation automorphisms.

4. Extended affine Lie algebras and vertex algebras

Extended affine Lie algebras are a class of Lie algebras which were defined (see [AABGP]) as Lie algebras equipped with a nondegenerate symmetric invariant bilinear form and equipped with a Cartan subalgebra such that the associated root system satisfies a set of conditions. This class of Lie algebras contains finite-dimensional simple Lie algebras, affine Kac-Moody Lie algebras, toroidal Lie algebras, and some quantum torus Lie algebras as well. On the one hand, this class is general enough to contain many interesting examples other than affine Kac-Moody Lie algebras, and on the other hand the structures of this class of Lie algebras are simple enough to be completely determined even though there are still lots to be done.

We have seen that affine Lie algebras and toroidal Lie algebras have been associated to vertex algebras in terms of modules while (special) quantum torus Lie algebras are associated to vertex algebras in terms of quasi modules. Extended affine Lie algebras resemble affine Lie algebras very much and those examples constructed in [AABGP] are Lie subalgebras of Lie algebras of (higher dimensional) quantum tori. In view of this we believe that the following is true:

Conjecture 4.1. *Every extended affine Lie algebra can be associated to vertex algebras in terms of quasi modules.*

To prove this conjecture, first one needs to obtain a realization of extended affine Lie algebras in terms of generating functions. Then one just follows the example of [Li3] for quantum torus Lie algebras.

References

[AABGP] B. N. Allison, S. Azam, S. Berman, Y. Gao and A. Pianzola, Extended affine Lie algebras and their root systems, *Memoirs Amer. Math. Soc.* **126**, 1997.

[BDM] K. Barron, C. Dong and G. Mason, Twisted sectors for tensor product vertex operator algebras associated to permutation groups, *Commun. Math. Phys.* **227** (2002), 349–384.

[BB] S. Berman and Y. Billig, Irreducible representations for toroidal Lie algebras, *J. Algebra* **221** (1999), 188–231.

[BBS] S. Berman, Y. Billig and J. Szmigielski, Vertex operator algebras and the representation theory of toroidal algebras, Contemporary Math. **297**, Amer. Math. Soc., Providence, 2002, 1–26.

[BGT] S. Berman, Y. Gao and S. Tan, A unified view of some vertex operator constructions, *Israel J. Math.* **134** (2003), 29–60.

[D] C. Dong, Twisted modules for vertex algebras associated with even lattices, *J. Algebra* **165** (1994), 91–112.

[DL] C. Dong and J. Lepowsky, *Generalized Vertex Algebras and Relative Vertex Operators*, Progress in Math., Vol. 112, Birkhäuser, Boston, 1993.

[DLM] C. Dong, H.-S. Li and G. Mason, Vertex Lie algebra, vertex Poisson algebras and vertex algebras, in: *Recent Developments in Infinite-Dimensional Lie Algebras and Conformal Field Theory (Charlottesville, VA, 2000)*, Contemporary Math. **297**, Amer. Math. Soc., Providence, 2002, 69–96.

[FFR] A. J. Feingold, I. B. Frenkel and J. F. X. Ries, *Spinor Construction of Vertex Operator Algebras, Triality, and $E_8^{(1)}$*, Contemporary Math. **121**, 1991.

[FLM]   I. Frenkel, J. Lepowsky and A. Meurman, *Vertex Operator Algebras and the Monster*, Pure and Appl. Math., **Vol. 134**, Academic Press, Boston, 1988.

[FZ]   I. Frenkel and Y.-C. Zhu, Vertex operator algebras associated to representations of affine and Virasoro algebras, *Duke Math. J.* **66** (1992), 123–168.

[G-K-K]   M. Golenishcheva-Kutuzova and V. Kac, Γ-conformal algebras, *J. Math. Phys.* **39** (1998), 2290–2305.

[G-K-L]   M. Golenishcheva-Kutuzova and D. Lebedev, Vertex operator representation of some quantum tori Lie algebras, *Commun. Math. Phys.* **148** (1992), 403–416.

[K1]   V. G. Kac, *Infinite-dimensional Lie Algebras*, 3rd ed., Cambridge Univ. Press, Cambridge, 1990.

[K2]   V. G. Kac, *Vertex Algebras for Beginners*, University Lecture Series, Vol. 10, Amer. Math. Soc., 1997.

[L]   J. Lepowsky, Calculus of twisted vertex operators, *Proc. Natl. Acad. Sci. USA* **82** (1985), 8295–8299.

[LL]   J. Lepowsky and H.-S. Li, *Introduction to Vertex Operator Algebras and Their Representations*, Progress in Math. Vol. 227, Birkhäuser, Boston, 2004.

[Li1]   H.-S. Li, Local systems of vertex operators, vertex superalgebras and modules, preprint 1993; *J. Pure Appl. Alg.* **109** (1996), 143–195.

[Li2]   H.-S. Li, Local systems of twisted vertex operators, vertex superalgebras and twisted modules, Contemporary Math. **193**, Amer. Math. Soc., Providence, 1996, 203–236.

[Li3]   H.-S. Li, A new construction of vertex algebras and their quasi modules, *Advances in Math.* **202** (2006), 232–286.

[MRY]   R. V. Moody, S. Eswara Rao and T. Yokonuma, Toroidal Lie algebras and vertex representations, *Geom. Dedicata* **35** (1990), 283–307.

[EM]   S. Eswara Rao and R. V. Moody, Vertex representations for n-toroidal Lie algebras and a generalization of the Virasoro algebra, *Commun. Math. Phys.* **159** (1994), 239–264.

[P]   M. Primc, Vertex algebras generated by Lie algebras, *J. Pure Applied Algebra* **135** (1999), 253–293.

[S1]   K. Saito, Extended affine root systems 1 (Coxeter transformations), Publ. RIMS., Kyoto University **21** (1985), 75–179.

[S2]   K. Saito, Extended affine root systems 2 (flat invariants), Publ. RIMS., Kyoto University **26** (1990), 15–78.

Vertex Operators and Arithmetic: How a Single Photon Illuminates Number Theory

Geoffrey Mason*

Department of Mathematics, UC Santa Cruz

1. Introduction

This paper is based on a talk I gave at the Spitalfields Day, held under the auspices of the London Mathematical Society during the Edinburgh Conference on Moonshine. As such, it is intended for nonspecialists.

Consider the following confluence of ideas:

- As closed strings move in space-time they sweep out a worldsheet which carries the structure of a Riemann surface.
- Riemann surfaces occur as quotients $\Gamma \setminus \mathcal{H}$ of the complex upper half-plane $\mathcal{H}$ by arithmetic groups $\Gamma \subseteq SL(2, \mathbf{R})$.
- The modular forms associated to $\Gamma \setminus \mathcal{H}$ carry arithmetic information manifested in the coefficients of the corresponding Fourier series.
- Vertex operator algebra theory may be construed as an algebraicization of aspects of the theory of elementary particles (bosonic strings) and their interactions.
- *Monstrous Moonshine* relates certain distinguished elliptic modular functions (so-called hauptmoduln) to the Monster simple group **M**.
- **M** is the automorphism group of a particular vertex operator algebra, the *Moonshine Module $V^{\natural}$*, which models bosons in the critical dimension $c = 24$ as a $\mathbf{Z}_2$-orbifold.

The reader who is not conversant with all of these ideas should have no fear. The statements above are intended only as background, and to suggest several points. The first is the *inevitability* of connections between elliptic modular forms, vertex operator algebras and finite groups; the second is the intimate connection with ideas from theoretical physics (string

* Research supported by the NSF and the Committee on Research, UC Santa Cruz

255

theory and conformal field theory). There is also the contrast between general statements versus interesting special cases. Indeed, the subject of vertex operator algebras arose from the very special and quite remarkable Moonshine conjectures of Conway and Norton [CN] which concerned the Monster, and the equally remarkable work of Frenkel-Lepowsky-Meurman [FLM] and Borcherds [B1], [B2], which showed how these conjectures could be understood (indeed, solved) by making use of a theory of vertex operator algebras.

In this paper I will explain how modular forms naturally arise in the theory of vertex operator algebras. This has nothing to do with groups, but rather is a manifestation of the relations that exist among the first four items listed above. It is interesting to compare the advent of modularity with the rise of field theory from classical and quantum mechanics and the parallel development of vertex operator algebras from Lie algebras. We take the simplest possible model, a single free boson. This already contains all (level 1) modular forms in a sense that we will explain. Following this, we briefly discuss the situation for more general modular forms, specifically Siegel modular forms. This is a relatively new topic, but the results that I will mention (obtained in collaboration with Michael Tuite) suggest that vertex operator algebras are capable of describing classes of automorphic forms far more general than the elliptic modular forms which have been the main focus of attention heretofore.

Thanks are due to the referee for helpful suggestions.

2. In the Beginning...

The passage $CM \Rightarrow QM$ from classical mechanics to quantum mechanics may briefly be described as follows. A state of a (classical) particle can be described by a pair of *canonically conjugate coordinates*, namely position q and momentum p. This means that they satisfy the relations

$$\{p, p\} = \{q, q\} = 0, \quad \{p, q\} = -\{q, p\} = 1 \tag{1}$$

where $\{,\}$ is the *Poisson bracket*. One may include independent states with canonical conjugates $p_i, q_i, i = 1, 2, \ldots$ by setting

$$\{p_i, p_j\} = \{q_i, q_j\} = 0, \tag{2}$$

$$\{p_i, q_j\} = -\{q_j, p_i\} = \delta_{i,j}. \tag{3}$$

There is no need to linger over this; we have only to be aware that QM involves 'substitutions'

$$p_i, q_i \Rightarrow \text{operators on } \textit{Fock space},$$

$$\text{Poisson brackets} \Rightarrow \text{Lie bracket of operators}.$$

This means that p_i and q_i are promoted to operators on some linear space (the Fock space) in such a way that relations (1)–(3) are satisfied when Poisson brackets are replaced by Lie brackets of the corresponding operators. One can see easily that Fock space for independent states may be taken to be the *tensor product* of the individual Fock spaces. In physics, Fock space is synonymous with *Hilbert space*, but for what we wish to say nothing more than a linear space is required.

The basic example. Recall that a (complex) *Heisenberg Lie algebra* is a Lie algebra H defined as a central extension

$$0 \to Z \to H \to A \to 0$$

with A abelian and $Z = [H, H] = Z(H)$ being 1-dimensional. Up to isomorphism, there is a unique Heisenberg Lie algebra of any (odd) finite dimension and a unique one of countably infinite dimension. It is apparent that replacing Poisson brackets by Lie brackets in (1)–(3) yields a Heisenberg Lie algebra. So our Fock space will afford a representation of a Heisenberg Lie algebra.

The Fock space we take is a weighted polynomial algebra

$$V = \mathbf{C}[x_1, x_2, \ldots] \cong \bigotimes_{i=1}^{\infty} \mathbf{C}[x_i] \tag{4}$$

with $\deg x_i = i, i = 1, 2, \ldots$. It is indeed a tensor product of Fock spaces for the individual states. The operators corresponding to p_i and q_i are

$$a_i = \frac{\partial}{\partial x_i}, b_i = \text{multiplication by } x_i$$

respectively, and the *Heisenberg canonical commutator relations* (CCR) are satisfied:

$$[a_i, a_j] = [b_i, b_j] = 0, \quad [a_i, b_j] = \delta_{i,j}\text{Id}_V.$$

This all goes back to Heisenberg and his conception of matrix mechanics. The following reformulation has a more modern feel. Namely, we can organize the CCR into a representation π of the Heisenberg algebra in the guise of an *affine Lie algebra* $\hat{L}(A)$. To explain this, introduce a 1-dimensional Lie algebra

$$A = \mathbf{C}a, \ \langle a, a \rangle = 1,$$

equipped with a symmetric bilinear form $\langle\ ,\ \rangle$ and normalized basis vector a. The loop algebra $L(A)$ is an abelian Lie algebra

$$L(A) = \bigoplus_{n \in \mathbf{Z}} A \otimes t^n \cong \mathbf{C}[t, t^{-1}].$$

The affine algebra $\hat{L}(A)$ is the corresponding Lie algebra defined via

$$0 \longrightarrow \mathbf{C}K \longrightarrow \hat{L}(A) \longrightarrow L(A) \longrightarrow 0$$

where K is a central element and

$$[u \otimes t^m, v \otimes t^n] = m \langle u, v \rangle \delta_{m+n,0} K, \ u, v \in A.$$

This is not quite a Heisenberg Lie algebra – the center of $\hat{L}(A)$ contains $a \otimes t^0$ in addition to K. But by setting

$$\pi\left(\frac{a \otimes t^n}{\sqrt{in}}\right) = \begin{cases} \frac{\partial}{\partial x_n} & n \geq 1 \\ x_n & n \leq -1, \end{cases}$$

$$\pi(a \otimes t^0) = 0,$$

$$\pi(K) = \mathrm{Id}_V,$$

we obtain a representation π of $\hat{L}(A)$ on Fock space which pushes down to a representation of the corresponding Heisenberg Lie algebra and which is nothing more than a reorganization of the CCR.

It is useful to notice that π is an *induced* representation, corresponding to a *Verma* module over $\hat{L}(A)$. To see this, let **C1** be a 1-dimensional module for the abelian Lie subalgebra $\hat{L}(A)^{\geq} = \mathbf{C}K \oplus_{n \geq 0} A \otimes t^n$ where K acts as multiplication by 1 and $A \otimes t^n$ acts as $0, n \geq 0$. With $\mathcal{U}$ denoting universal enveloping algebra, we have

$$V \cong Ind_{\mathcal{U}(\hat{L}(A)^{\geq})}^{\mathcal{U}(\hat{L}(A))} \mathbf{1}. \tag{5}$$

In this context we call **1** the *vacuum vector*. The virtue of this point-of-view will become apparent later.

3. First Arithmetic Intermezzo

Fock space (4) decomposes into homogeneous spaces

$$V = \bigoplus_{n \geq 0} V_n, \tag{6}$$

where V_n is spanned by homogeneous polynomials of degree n. The monomial $x_1^{e_1} x_2^{e_2} \ldots$ of degree n corresponds to a *partition* λ of n in the usual way, that is

$$\lambda \vdash \overbrace{1 \ldots 1}^{e_1} \overbrace{2 \ldots 2}^{e_2} \ldots.$$

It follows that

$$\dim V_n = p(n) = \text{number of (unrestricted) partitions of } n.$$

The *partition function* of V is

$$\begin{aligned}
\text{ch}_V(q) &= \sum_{n \geq 0} \dim V_n \, q^n \\
&= \sum_{n \geq 0} p(n) q^n \\
&= 1 + q + 2q^2 + \ldots
\end{aligned}$$

The arithmetic version of the tensor product in (4) is well-known and due to Euler:

$$\begin{aligned}
\sum_{n \geq 0} p(n) q^n &= \prod_{n=1}^{\infty} \text{ch}_{C[x_n]}(q) \\
&= \prod_{n=1}^{\infty} (1 + q^n + q^{2n} + \ldots) \\
&= \prod_{n=1}^{\infty} (1 - q^n)^{-1}.
\end{aligned}$$

One can usefully refine the partition function of V by setting

$$\begin{aligned}
Z_V(q) &= q^{-1/24} \text{ch}_V(q) \\
&= q^{-1/24} \prod_{n=1}^{\infty} (1 - q^n)^{-1},
\end{aligned} \tag{7}$$

which is also called the partition function. The inverse is the *Dedekind eta function*

$$\eta(q) = q^{1/24} \prod_{n=1}^{\infty} (1 - q^n).$$

When we interpret q as a complex variable $q = e^{2\pi i \tau}$ with $\tau \in \mathcal{H}$, the eta-function is a holomorphic modular form of weight $1/2$ on $\Gamma = SL(2, \mathbb{Z})$. Without going into details, this means that there are functional identities

$$\eta\left(\frac{a\tau + b}{c\tau + d}\right) = (\text{24th root of unity})(c\tau + d)^{1/2}\eta(\tau)$$

for all $\begin{pmatrix} a & b \\ c & d \end{pmatrix} \in \Gamma$.

The connection between modular forms and QM that we have sketched above is somewhat tenuous. It can be strengthened considerably once we make the move to field theory.

4. ... There was Light

Quantum field theory is a generalization of QM. Recall the representation π of the Heisenberg algebra $\hat{L}(A)$ on V. Set

$$a(n) = \pi(a \otimes t^n),$$

and form a sort of generating function

$$a(z) = \sum_{n \in \mathbb{Z}} a(n)z^{-n-1} \in \text{End}V[[z, z^{-1}]].$$

It is the most basic example of a *quantum field* or *vertex operator*. As we saw in Section 2, V has a basis of states indexed by partitions. Indeed, there is a natural basis adapted to the induced module structure (5), namely

$$v = v_\lambda = a(-n_1)a(-n_2)...a(-n_k).\mathbf{1} \tag{8}$$

where $n_1 \geq n_2 \geq ... \geq n_k \geq 1$ are the parts of the partition λ.

Now comes the basic idea. Define

$$Y(v_\lambda, z) = {}^{\circ}_{\circ} \prod_{i=1}^{k} \frac{1}{(n_i - 1)!} \left(\frac{\partial}{\partial z}\right)^{n_i - 1} a(z) {}^{\circ}_{\circ} \tag{9}$$

$$= \sum_{n \in \mathbb{Z}} v_\lambda(n)z^{-n-1}.$$

What does this mean? Ignoring the *normal ordering* symbol $\circ\!\!\circ \;\; \circ\!\!\circ$ for now, we are defining a sequence of operators $v_\lambda(n)$ on Fock space, the *modes* of v_λ, associated to each of the basis states v_λ, using (formal) partial derivatives of $a(z)$. Formally, $v_\lambda(n)$ is the operator which appears as the coefficient of z^{-n-1} in (9), and as such it is a sum of terms of the form $(\text{constant})a(n_1)...a(n_k)$. For example, if $\lambda \vdash 1^2$ then

$$Y(v_{1^2}, z) = \; {}^{\circ}_{\circ}\, a(z)^2 \, {}^{\circ}_{\circ}$$

$$= \; {}^{\circ}_{\circ} \sum_{m,n \in \mathbf{Z}} a(m)a(n-m-1)z^{-n-1} \, {}^{\circ}_{\circ},$$

$$v_{1^2}(n) = \sum_{m \in \mathbf{Z}} {}^{\circ}_{\circ}\, a(m)a(n-m-1) \, {}^{\circ}_{\circ}\,.$$

For $u \in V$ we would like to define $v_{1^2}(n)u$ via the formalism

$$v_{1^2}(n)u = \sum_{m \in \mathbf{Z}} {}^{\circ}_{\circ}\, a(m)a(n-m-1) \, {}^{\circ}_{\circ}\, u. \tag{10}$$

Without normal ordering, (10) is generally not well-defined for $n = 1$, meaning that it will not reduce to a *finite* sum of nonzero states in Fock space. It *is* well-defined if we set

$$ {}^{\circ}_{\circ}\, a(m)a(n) \, {}^{\circ}_{\circ} = \begin{cases} a(n)a(m), & m+n = 0 \le m, \\ a(m)a(n), & \text{otherwise.} \end{cases}$$

(Note that $a(m)a(n) = a(n)a(m)$ if $n+m \ne 0$.) With this device, (10) achieves legitimacy, because all but a finite number of the normally ordered operators appearing as summands will annihilate any fixed state u. When we have a normally ordered product of more than two modes $a(n)$s, the rule is the same: operate first with those modes satisfying $n \ge 0$.

We define $Y(v, z)$ for an arbitrary state $v \in V$ by extending (9) linearly. This produces a linear map, known as the *state-field correspondence* in QFT:

$$V \longrightarrow \text{End}(V)[[z, z^{-1}]]$$

$$v \mapsto Y(v, z).$$

$Y(v, z)$ is the quantum field, or vertex operator, associated to state v.

We distinguish three special states in V, two of which we have already seen: $\mathbf{1}$, a, and $\omega = 1/2a(-1)^2\mathbf{1} \in V_2$. From what we have said so far, it follows that

$$V_0 = \mathbf{C1}, \quad Y(\mathbf{1}, z) = \text{Id}$$

$$V_1 = \mathbf{C}a, \quad Y(a, z) = a(z).$$

The importance of the state ω, often called the *conformal vector* or *Virasoro vector*, is less clear at this point, however its properties are central to our enterprise. It is traditional to write the field corresponding to ω in the form

$$Y(\omega, z) = \sum_{n \in \mathbf{Z}} L(n) z^{-n-2}$$

(note the shift in grading). It is a remarkable fact that the modes $L(n)$ satisfy the relations defining the *Virasoro algebra Vir* of central charge $c = 1$. That is,

$$[L(m), L(n)] = (m - n) L(m + n) + \frac{m^3 - m}{12} \delta_{m+n,0} \mathrm{Id}_V.$$

Furthermore, the homogeneous subspaces V_n of V are eigenspaces for $L(0)$:

$$V_n = \{v \in V | L(0)v = nv\}. \tag{11}$$

The quadruple $(V, Y, \mathbf{1}, \omega)$ is the basic example of a *vertex operator algebra* (on the sphere), called the *Heisenberg VOA*, or *free bosonic theory* of rank 1. It models the annihilation and creation of (an infinite set of) bosonic states from the vacuum. The distinguished state a might correspond to a photon. The linear span of all modes is a huge Lie algebra

$$\mathcal{L} = \langle v(n) \mid v \in V, n \in Z \rangle$$

of operators on V. This follows from the following beautiful identity [B1]:

$$[u(m), v(n)] = \sum_{i \geq 0} \binom{m}{i} (u(i)v)(m + n - i). \tag{12}$$

As we explained before, for given u, v we have $u(i)v = 0$ for all large enough i, so that the right-hand-side of (12) is a well-defined finite sum.

It is well known that the Virasoro algebra (more precisely, the *Witt algebra* Wi, which is the central quotient of Vir), acts on $\hat{L}(A)$ as derivations. Indeed, the modes $a(n)$ and $L(n)$ span a Lie subalgebra of $\mathcal{L}$ which is an extension of Wi by $\hat{L}(A)$.

5. Second Arithmetic Intermezzo

So far we have seen that the passage from CM to QM and then QFT corresponds to a series of increasingly complicated Lie algebras:

$$CM \Rightarrow QM \Rightarrow QFT$$
$$L(A) \Rightarrow \hat{L}(A) \Rightarrow \mathcal{L}.$$

We have also seen that the eta-function arises rather naturally in QM. We will now explain how QFT, aka $\mathcal{L}$ or the Heisenberg VOA V, produces a large number of holomorphic modular forms on $SL(2, \mathbb{Z})$ - all of them, in fact.

Recall the grading (6) on V. If $v \in V_k$ and

$$Y(v, z) = \sum_{n \in \mathbf{Z}} v(n) z^{-n-1}$$

then one knows that

$$v(m) : V_n \longrightarrow V_{n+k-m-1}.$$

In particular, the *zero mode* $o(v) = v(k-1)$ satisfies

$$o(v) : V_n \longrightarrow V_n,$$

that is $o(v)$ has weight zero as an operator. So we can take the *graded trace*, or *(1-point) correlation function*

$$Z(v, q) = \mathrm{Tr}_V o(v) q^{L(0)-1/24}$$
$$= q^{-1/24} \sum_{n \geq 0} \mathrm{Tr}_{V_n} o(v) q^n,$$

where as in (7) we have included an extra term $q^{-1/24}$. Extending Z to V by linearity, we obtain a map

$$Z : V \longrightarrow q^{-c/24} \mathbb{C}[[q]],$$

which should be thought of as the *character* of V.

The discussion in Section 3 shows that $Z(\mathbf{1}, \tau) = \eta(\tau)^{-1}$ is the partition function, or graded dimension of V. We will now describe the full image of Z, for which we need some standard notation for modular forms (see [S], for example, for details).

$$E_{2k}(q) = \frac{-B_{2k}}{(2k)!} + \frac{2}{(2k-1)!} \sum_{n=1}^{\infty} \sigma_{2k-1}(n) q^n \quad \text{(Eisenstein series)},$$

where B_k is a Bernoulli number and $\sigma_{2k-1}(n)$ a power sum. For example,

$$E_2(q) = -1/12 + 2(q + 3q^2 + 4q^3 + ...)$$
$$720 E_4(q) = 1 + 240(q + 9q^2 + 28q^3 + ...)$$
$$-30240 E_6(q) = 1 - 504(q + 33q^2 + 244q^3 + ...)$$

For $k \geq 2$, $E_{2k}(q)$ is a *holomorphic modular form* of weight $2k$. That is, taking $q = e^{2\pi i \tau}$ as before, $E_{2k}(\tau)$ is holomorphic throughout $\mathcal{H}$ and

$$E_{2k}(\gamma \tau) = (c\tau + d)^{2k} E_{2k}(\tau), \quad \gamma = \begin{pmatrix} a & b \\ c & d \end{pmatrix} \in SL(2, \mathbf{Z}).$$

Note that E_2 is *not* a modular form (loc. cit.) $E_4(q)$ and $E_6(q)$ generate the algebra of all holomorphic modular forms on $SL(2, \mathbb{Z})$, and together with $E_2(q)$ they freely generate the algebra of *quasi-modular forms* [KZ]

$$Q = \mathbf{C}[E_2(q), E_4(q), E_6(q)].$$

We then have [DMN]:

$$Z(v, q) = \frac{f(q)}{\eta(q)} \text{ for some } f(q) \in Q,$$

every $f(q) \in Q$ occurs in this way.

In other words, there is a linear surjection

$$Z : V \longrightarrow \frac{1}{\eta(q)} Q. \tag{13}$$

It is an interesting feature that, up to a factor of $\eta(q)^{-1}$, the correlation functions are quasi-modular, but not necessarily modular.

There is an integral grading on quasimodular forms in which E_{2k} has weight $2k$:

$$Q = \bigoplus_{k \geq 0} Q_{2k}.$$

It would be natural to think that (13) is a graded map, i.e. if $v \in V_k$ then $Z(v, q) = f(q)/\eta(q)$ with $f(q) \in Q_k$. This is true if $v = \mathbf{1}$ (when $f(q) = 1$), but is generally incorrect. It is important to understand how to restore a grade-preserving map. We explain the mechanism that achieves this in the next Section.

6. QFT on the Cylinder

Consider the conformal map

$$z \mapsto e^z - 1$$

mapping the complex plane to an infinite cylinder. This induces a VOA *on the cylinder*, utilized by physicists for some time and introduced into mathematics by Zhu [Z]. The idea is to retain the space V but adjust the state-field correspondence as follows. For $v \in V_k$, set

$$Y[v, z] = e^{kz} Y(v, e^z - 1) = \sum_{n \in \mathbf{Z}} v[n] z^{-n-1},$$

$$\tilde{\omega} = \omega - \frac{1}{24} \mathbf{1}.$$

Then we obtain a new vertex operator algebra $(V, Y[\ ], \mathbf{1}, \tilde{\omega})$ which is *isomorphic*[1] to the original one. What we mainly need from this is that the new conformal vector $\tilde{\omega}$ corresponds to a field $\sum_n L[n] z^{-n-2}$ and that the mode $L[0]$ produces a new grading on V

$$V = \oplus_{n \geq 0} V_{[n]}, \quad V_{[n]} = \{v \in V \mid L[0]v = nv\}$$

which is isomorphic to, but distinct from, the $L(0)$-grading (11). It turns out that (13) is a graded map as long as V is endowed with the $L[0]$-grading. Thus, there is a surjection for each $k \geq 0$,

$$Z : V_{[k]} \longrightarrow \frac{1}{\eta(q)} Q_k.$$

There is an *explicit* version of this map which is useful. The analog of (8) in the cylindrical basis of V is

$$v[\lambda] = a[-1]^{e_1} a[-2]^{e_2} ... a[-r]^{e_r} \mathbf{1} \in V_{[k]},$$

for a partition $\lambda \vdash k = \sum i e_i$. Attached to λ is the *labelled set* Φ_λ of size k consisting of e_1 elements (nodes) with label 1, e_2 nodes with label 2, etc. Set

$$F(\Phi_\lambda) = \{\varphi \in \text{Symm}(\Phi_\lambda) \mid \varphi \text{ a fixed-point-free involution}\}.$$

(*Symm* denotes symmetric group.) Then [MT1]

$$Z(v[\lambda], q) = \frac{1}{\eta(q)} \sum_{\varphi \in F(\Phi)} \prod_{(r,s)} (-1)^{r+1} \frac{(r+s-1)!}{(r-1)!(s-1)!} E_{r+s}(q),$$

where (r, s) ranges over the orbits of φ acting on Φ_λ and where we are identifying a node with its label.

7. Compact Bosons

We have been looking at the Heisenberg, or free bosonic CFT on the sphere (genus 0) and on the torus (genus 1), and have found that the correlation functions are essentially quasimodular forms, but are not always true modular forms. In this Section we discuss a class of vertex operator algebras for which the correlation functions are modular. These are[2] the *rational conformal field theories*, or rational vertex operator algebras.

---

1 Since we have not actually defined what a VOA is, this is somewhat prosaic
2 In fact, the implied assertion that the correlation functions of a rational vertex operator algebra are modular remains open at this time.

The Heisenberg VOA is not rational, however there is a closely associated VOA that is. Suppose that

$$\langle \, , \rangle : L \times L \longrightarrow \mathbb{Z}$$

is a positive-definite *even* lattice of rank l. Then there is a VOA V_L with Fock space defined by

$$V_L \cong V^{\otimes l} \otimes \mathbb{C}[L]. \tag{14}$$

Physically, this corresponds to *compactifying* l free bosons so that their momenta lie on L. Rather than moving in $\mathbb{R}$, they are constrained to the torus $\mathbb{R}^l/L$. For more information on *lattice theories*, see [FLM]. A special case of this construction may be familiar: if we take $l = 1$ and $L = \sqrt{2}\mathbb{Z}$ then L is just the root lattice of type A_1 and V_L can be identified with the corresponding level 1 affine Kac-Moody Lie algebra [K].

In (14), $\mathbb{C}[L]$ is the group algebra of L with basis $e^\alpha, \alpha \in L$. Inflicting this element with degree $\langle \alpha, \alpha \rangle/2$ and giving (14) the natural tensor product grading, we find that the partition function of V_L satisfies

$$Z_{V_L}(q) = Z_V(q) ch_{\mathbb{C}[L]}(q) = \theta_L(\tau)/\eta(\tau)^l$$

where

$$ch_{\mathbb{C}[L]}(q) = \theta_L(\tau) = \sum_{\alpha \in L} q^{\langle \alpha, \alpha \rangle/2}$$

is the *theta function* of L.

In contrast to V, the partition function Z_{V_L} of V_L is a meromorphic modular function of weight *zero* on a congruence subgroup of $SL(2, \mathbb{Z})$ [S]. As in the case of the Heisenberg VOA discussed before, one can describe the space of correlation functions spanned by all $Z_{V_L}(v, q), v \in V_L$, and to give explicit formulas. In particular, when we transfer V_L from the sphere to the cylinder, one finds for $v \in (V_L)_{[k]}$ that $Z_{V_L}(v, q)$ is a modular function of weight k on a congruence subgroup of $SL(2, \mathbb{Z})$. For further details see [DMN] and [MT1].

8. Genus 2

We have so far focused on *elliptic* modular functions, i.e. functions invariant under a (congruence subgroup of) $SL(2, \mathbb{Z})$. However, one expects that there are similar connections with other arithmetic groups, and in particular with the symplectic groups $Sp(2g, \mathbb{Z})$. We will consider only the case $g = 2$. In this Section we recall some relevant arithmetic facts. See [F], for example, for further details.

If X is a compact Riemann surface of genus 2 then the first homology group is isomorphic to $\mathbb{Z}^4$ and X admits 2 linearly independent holomorphic differentials v_1, v_2. We integrate along a suitable basis for $H_1(X)$ (the a-cycles and b cycles), using the former to normalize and the latter to obtain the period matrix Ω for X:

$$\int_{a_i} v_j = \delta_{i,j}, \quad \int_{b_i} v_j = \Omega_{ij},$$

$$\Omega = \begin{pmatrix} \Omega_{11} & \Omega_{12} \\ \Omega_{21} & \Omega_{22} \end{pmatrix}.$$

Ω lies in the genus 2 *Siegel upper half-space*

$$\mathcal{H}^{(2)} = \{\Omega \in M_2(\mathbb{C}) \mid \Omega = \Omega^t, \operatorname{Im}\Omega \text{ positive-definite}\}.$$

There is a left action

$$Sp(4, \mathbb{Z}) \times \mathcal{H}^{(2)} \longrightarrow \mathcal{H}^{(2)}$$

$$\begin{pmatrix} A & B \\ C & D \end{pmatrix} \circ \Omega = (A\Omega + B)(C\Omega + D)^{-1}$$

$F(\Omega)$ is a *weight k holomorphic Siegel modular form on* $\Gamma \subseteq Sp(4, \mathbf{Z})$ if $F(\Omega)$ is holomorphic and

$$F(\gamma \circ \Omega) = \det(C\Omega + D)^k F(\Omega), \quad \gamma \in \Gamma.$$

Example: Genus 2 Theta function. For an even lattice L of rank l, its genus 2 theta function is defined as

$$\Theta_L^{(2)}(\Omega) = \sum_{\alpha,\beta \in L} q^{\langle\alpha,\alpha\rangle/2} r^{\langle\alpha,\beta\rangle} s^{\langle\beta,\beta\rangle/2},$$

$$(q = e^{2\pi i \Omega_{11}}, r = e^{2\pi i \Omega_{12}}, s = e^{2\pi i \Omega_{22}}).$$

This is a Siegel modular form of weight $l/2$ on a finite index subgroup of $Sp(4, \mathbb{Z})$.

9. Genus 2 Partition Function

For a VOA W such as $V^{\otimes l}$ or V_L, we can make the following definition:

$$Z_W^{(2)}(q_1, \epsilon, q_2) = \sum_{n \geq 0} \epsilon^n \sum_{u,v \in V_{[n]}} Z_W(u, q_1) \, G_{u,v}^{-1} \, Z_W(v, q_2).$$

Here, q_1, q_2 and ϵ are (for the moment) formal variables, and Z_W is the correlation function already discussed. In the inner sum, u and v range independently over a basis of $V_{[n]}$. If $\dim W_{[n]} = d$, $G_{u,v}$ is a certain invertible $d \times d$ matrix whose (u, v)-entry is a complex number $\langle u, v \rangle$ given by a (normalized) invariant bilinear form on W as in [Li]. By way of example, in the case of the free boson VOA V we can rewrite the definition in terms of symmetric groups S_k as follows:

$$Z_V^{(2)}(q_1, \epsilon, q_2) = \sum_{k \geq 0} \sum_{g \in S_k} Z(g, q_1)\, Z(g, q_2) \frac{\epsilon^k}{k!}, \tag{15}$$

where $Z(g, \tau) = Z(v[\lambda], \tau)$ is the correlation function determined by $\lambda \vdash k$ whenever g lies in the conjugacy class of S_k determined by λ.

What we really wish to do is construct a genus 2 partition function as a function of $\Omega \in \mathcal{H}^{(2)}$, but it does not seem to be possible to do this directly. The fact that we could do so at genus 1 is a happy coincidence, and a misleading one at that. One must proceed indirectly, first studying the function (15) and then relating it to the Siegel upper half-space. Michael Tuite and the author have established the following result:

Let $q_i = e^{2\pi i \tau_i}$, $\tau_i \in \mathcal{H}$. There is a domain $\mathcal{D} \subseteq \mathcal{H} \times \mathbb{C} \times \mathcal{H}$ and a holomorphic map $F : \mathcal{D} \to \mathcal{H}^{(2)}$ such that the following holds:

(a) $Z_W(\tau_1, \epsilon, \tau_2)$ is holomorphic on $\mathcal{D}$

(b) If L is an even lattice of rank l then there is an identity of formal power series and holomorphic functions

$$\frac{Z_{V_L}^{(2)}(\tau_1, \epsilon, \tau_2)}{Z_{V^{\otimes l}}^{(2)}(\tau_1, \epsilon, \tau_2)} = \Theta_L^{(2)}(\Omega), \tag{16}$$

where $(\tau_1, \epsilon, \tau_2) \in \mathcal{D}$ and $\Omega = F(\tau_1, \epsilon, \tau_2)$

Compare (16) with the genus 1 case:

$$\frac{Z_{V_L}(\tau)}{Z_{V^{\otimes l}}(\tau)} = \theta_L(\tau).$$

The idea of the proof, which is long and quite difficult, is to interpret the formula for $Z^{(2)}$ in terms of *sewing* a pair of genus 1 punctured Riemann surfaces described by the data encoded in the triple $(\tau_1, \epsilon, \tau_2)$. Sewing produces a compact genus 2 Riemann surface which has a period matrix Ω. This determines the map F. The proof of (16) makes use of the explicit formulas for the genus 1 correlation functions that we mentioned in Section 7. For further details, see [T] and the forthcoming [MT2], [MT3].

References

[B1] R. Borcherds, Vertex algebras, Kac-Moody algebras, and the Monster, *Proc. Natl. Acad. Sci. USA* **83** (1986), 3068-3071.

[B2] R. Borcherds, Monstrous moonshine and monstrous Lie superalgebras, Inv. Math. **109** (1992), 405–444.

[CN] J. Conway and S. Norton, Monstrous Moonshine, Bull. Lond. Math. Soc. **11**(1979), 308–339.

[DMN] C. Dong, G. Mason and K, Nagatomo, Quasi-modular forms and trace functions associated to free boson and lattice vertex operator algebras, I.M.R.N. **8**(2001), 409–427.

[F] E. Freitag, *Siegelsche Modulfunktionen*, Springer, New York, 1983.

[FLM] I. Frenkel, J. Lepowsky and A. Meurman, *Vertex Operator Algebras and the Monster*, Pure and Applied Math., Vol. **134**, Academic Press, 1988.

[K] V. Kac, *Infinite-dimensional Lie Algebras*, 3rd. ed., CUP, Cambridge, 1990.

[KZ] M. Kaneko and D. Zagier: A generalized Jacobi theta function and quasimodular forms, *The Moduli Space of Curves (Texel Island, 1994)*, Progr. in Math.129, Birkhauser, Boston, 1995.

[Li] H. Li, Symmetric invariant bilinear forms on vertex operator algebras, J. Pure and Appl. Alg. **96**(1994), 279–297.

[MT1] G. Mason and M. Tuite: Torus Chiral n-Point Functions for Free Boson and Lattice Vertex Operator Algebras, Commun.Math.Phys. **235**(2003), 47–68.

[MT2] G. Mason and M. Tuite: On Genus Two Riemann Surfaces Formed from Sewn Tori, Commun. Math. Phys. **270**(2007), 587–634.

[MT3] G. Mason and M. Tuite: The Genus Two Partition Function for Free Bosonic and Lattice Vertex Operator Algebras, in preparation.

[S] B. Schoeneberg: *Elliptic Modular Functions*, Springer, New York, 1974.

[T] M. Tuite: Genus two meromorphic conformal field theory, *CRM Proceedings and Lecture Notes* **30** (2001), 231–251.

[Z] Y. Zhu: Modular Invariance of characters of vertex operator algebras. J. Amer.Math.Soc.**9** (1996), 237–302.

Rational Vertex Operator Algebras
and their Orbifolds

Geoffrey Mason*

Department of Mathematics, UC Santa Cruz

1. Introduction

The *Conway-Norton conjectures* about the Monster-modular connection, later established by Borcherds [B] following the work of Frenkel-Lepowsky-Meurman [FLM], set the stage for an intensive study of the origins of the relations between finite groups and modular functions. Norton also introduced *generalized moonshine* which associates q-expansions to *pairs of commuting elements* in M. His conjecture that the nonconstant functions that one obtains in this way are also hauptmoduln remains open.

By now it is clear that the principal mathematical idea underlying the general study of such phenomena is that of a *vertex operator algebra*. For an extensive class of vertex operator algebras, so-called[1] *rational orbifold models*, one expects a theory completely parallel to Monstrous Moonshine whereby one associates modular functions to automorphisms of finite order. Furthermore, this theory *naturally accommodates* generalized moonshine. From this perspective, the Conway-Norton phenomena would be a particularly interesting sporadic example of a general theory, just as the Monster itself is a particularly interesting sporadic example of a finite simple group.

In the spirit of the Edinburgh Conference, this paper is mainly devoted to a review of some of the main results currently available concerning the structure of rational vertex operator algebras and their orbifolds. We sometimes use the Frenkel-Lepowsky-Meurman Moonshine Module $V^\natural$ [FLM] to illustrate the ideas. We also suggest open problems - many of them well-known - whose solution would contribute to a more complete theory. At the end of the paper we announce some new results (joint work with Chongying Dong) which solve the problem of generalized moonshine for a large class of orbifolds.

* Research supported by the NSF and the Committee on Research, UC Santa Cruz
1 A discussion of the term *orbifold* is given at the beginning of Section 4.

Our discussion is brief throughout, and by no means complete. We generally assume the reader to be familiar with the basic facts about the theory of vertex operator algebras ([FHL], [FLM], [LL]).

2. Rational Vertex Operator Algebras

We denote a vertex operator algebra (VOA) as a $\mathbb{Z}$-graded complex linear space

$$V = \bigoplus V_n.$$

There is a notion of *representation* for VOAs, and hence a module category

$$V\text{-Mod.}$$

For background we refer to [FHL], [LL]. The module category V-Mod is generally difficult to deal with. For example, although there is a general notion of *dual module* (loc. cit), the *adjoint* module V may not be self-dual. We will not discuss this further here (cf. [DM1]), but it suggests that one should not expect a nice theory unless the class of vertex operator algebras is suitably restricted.

One has to deal with several types of V-modules. The most basic are the simple modules. They have a grading of the general shape

$$M = \bigoplus_{n \geq 0} M_{\lambda+n}$$

for some (a priori complex) number λ, the *conformal weight* of M. In addition to the simples, there is the idea of an *admissible* or $\mathbb{N}$-*gradable* module. We will not give the definition here (cf. [DLM1], [Z]), but merely note that they arise naturally from Zhu's idea of constructing V-modules via the so-called Zhu algebra $A(V)$ [Z].

Definition[DLM2]: V is *rational* in case V-Mod is *semisimple* in the sense that every admissible module over V is completely reducible.

Definition[Zhu]: Let $C_2(V)$ be the subspace of V spanned by $a(-2)b$, $a, b \in V$. Call V C_2-*cofinite* if

$$\dim V/C_2(V) < \infty.$$

The relation between these two definitions is not entirely understood. First we have

Suppose that V is either *rational* or C_2-*cofinite*.

Then V has only finitely many simple modules.

This is proved in [Z] under the hypothesis of C_2-cofiniteness. Indeed, it is not hard to see that the finiteness of dim $V/C_2(V)$ implies that of dim $A(V)$, and this in turn implies the finiteness of the number of simple V-modules. If V is rational then it is shown in [DLM2] that $A(V)$ is a *semisimple* algebra, and hence is necessarily of finite dimension. Physicists have known for some time that there are VOAs which are not rational but which have only finitely many simple modules (logarithmic field theories). It seems likely that such theories are C_2-cofinite, though this seems not to have been checked. If so, then one can expect a large number of C_2-cofinite, irrational VOAs. Still, one can hope that the following is true:

Problem 1: Prove that

$$V \text{ rational} \Rightarrow V \ C_2\text{-cofinite.}$$

We refer the reader to [DLM1], [ABD], [GN] for further background.

For many purposes one needs to know that a given V is *both* rational and C_2-cofinite. It is therefore of some importance to resolve Problem 1, so that the C_2- assumption can be dispensed with in the presence of rationality. In many ways, the 'best' type of VOA, which we call *strongly rational* or SRVOAs, satisfies these and other conditions:

Definition: V is *strongly rational* if it satisfies

   (*a*) V is rational

   (*b*) V is C_2-cofinite

   (*c*) $V = \mathbf{C1} \oplus V_1 \oplus V_2 \dots$

   (*d*) V is *self-dual* i.e. $V \cong V'$ as V-modules.

Conditions (c) and (d) are *independent* of (a) and (b): one can construct examples of VOAs satisfying (a) and (b) for which any combination of (c) or its negation together with (d) or its negation hold [DM1]. The class of strongly rational vertex operator algebras enjoy many of the properties generally assumed in the physics literature.

From now on we assume that V is simple SRVOA of central charge c. We review some basic properties of such V:

   1. V admits a *unique nondegenerate, invariant bilinear form*

$$\langle \, , \, \rangle : V \otimes V \longrightarrow \mathbb{C}$$

   normalized so that $\langle \mathbf{1}, \mathbf{1} \rangle = -1$. In part, this amounts to a reformulation of property (d) of a SRVOA. Uniqueness follows from properties (c), (d) and Li's theory [Li1]. See also [FHL].

2. The pair $(V_1, \langle \, , \, \rangle)$ consisting of the weight 1 subspace of V together with (the restriction of) $\langle \, , \rangle$ is a Lie algebra equipped with a nondegenerate, symmetric, invariant bilinear form, where

$$[a, b] = a(0)b, \quad \langle a, b \rangle \mathbf{1} = a(1)b, \quad a, b \in V_1.$$

These facts are elementary.

3. V_1 is a *reductive* Lie algebra, that is

$$V_1 = T \oplus L_1 \oplus \ldots \oplus L_k,$$

where T is abelian, each L_i a simple Lie algebra, and the restriction of $\langle \, , \rangle$ to T and each L_i is non-degenerate. It is worth noting that ideas concerning modular-invariance underlie the proofs [DM2].

4. Each of the affine Lie algebras $\hat{L}_i$ determined by L_i is *integrable*. To explain this, for $u, v \in L_i$ we have

$$[u(m), v(n)] = [u, v](m + n) + m \langle u, v \rangle \delta_{m+n,0}$$
$$= [u, v](m + n) + m l_i(u, v) \delta_{m+n,0}.$$

The first equality comes from the VOA axioms, the second arises from the theory of Kac-Moody Lie algebras [K], the notation being that $(\, ,)$ is the non-degenerate form on L_i normalized so that $(\alpha, \alpha) = 2$ for a long root α in the root system determined by L_i, and l_i is the *level* of L_i. Integrability is the assertion that each level is a *positive integer*. This is proved in [DM3].

Now we come to the question of modular-invariance. Let the (finitely many) simple V-modules be $M^1, \ldots, M^r$, with conformal weights $\lambda_1, \ldots, \lambda_r$ and graded dimensions

$$Z_{M^i}(q) = \mathrm{Tr}_{M^i} q^{L(0)-c/24} = q^{-c/24+\lambda} \sum_{n \geq 0} \dim M^i_{\lambda_i+n} q^n.$$

Let $q = e^{2\pi i \tau}$ with τ in the complex upper half-plane $\mathcal{H}$, so that we may think of the graded dimension as a function $Z_{M^i}(\tau)$ on $\mathcal{H}$.

5. Each $Z_{M^i}(\tau)$ is holomorphic in $\mathcal{H}$. For $\gamma \in SL(2, \mathbb{Z})$ there are scalars $\rho_{ij}(\gamma)$ such that

$$Z_{M^i}\left(\frac{a\tau+b}{c\tau+d}\right) = \sum_{j=1}^{r} \rho_{ij}(\gamma) Z_{M^j}(\tau), \quad \gamma = \begin{pmatrix} a & b \\ c & d \end{pmatrix}.$$

The matrices $\rho(\gamma)$ afford an r-dimensional representation of $SL(2, \mathbb{Z})/\pm I$. The main ideas for proving this are in Zhu's fundamental paper [Z].

6. The central charge c and each of the conformal weights λ_i are *rational* numbers. This is proved in [DLM2], extending an argument of Anderson and Moore [AM].

As a consequence of the last result, there is a positive integer N such that for each i, $\lambda_i - c/24 = n_i/N$, $n_i \in \mathbb{Z}$, and

$$Z_{M^i}(q) \in q^{n_i/N}\mathbb{Z}[[q]].$$

Problem 2 (Modular-invariance): Prove that the kernel of the representation ρ contains the principal congruence subgroup $\Gamma(N)$ of level N. Equivalently, prove that each $Z_{M^i}(q)$ is a modular function (weight zero) on a congruence subgroup of $SL(2, \mathbb{Z})$.

3. Automorphism Groups

We continue to assume in this Section that V is a SRVOA, although all four of the conditions (a)-(d) will not always be required in what follows. The automorphism group of V is defined in the usual way:

$$\text{Aut}\,V = \{g \in GL(V) \mid gv(n)g^{-1} = (gv)(n),\ g\omega = \omega\}.$$

We set $\mathcal{G} = \text{Aut}\,V$. It acts on each V_n and preserves $\langle\ ,\ \rangle$, so that restriction provides a sequence of orthogonal representations

$$\mathcal{G} \longrightarrow O(V_n).$$

For $g \in \mathcal{G}$, the graded trace of g is defined in the obvious way:

$$Z_V(g, q) = \text{Tr}_V g\, q^{L(0)-c/24} = q^{-c/24} \sum_{n \geq 0} \text{Tr}_{V_n} g\, q^n.$$

Example: $\text{Aut}\,V^\natural$ is the Monster simple group M [FLM]. The orthogonal representations of M are rational-valued, so that $V_n^\natural$ is a *rational-valued M-module*. Hence, for $g \in M$,

$$Z(g, q) \in q^{-1}\mathbf{Z}[[q]].$$

The nature of $\mathcal{G}$ is still not well-understood. The best general result obtained so far [DG] is that

$$\mathcal{G} \text{ is an algebraic group.}$$

The process of exponentiating elements of a (semisimple) Lie algebra to obtain automorphism in the corresponding Lie group works perfectly well for V too. More precisely, there is a normal subgroup

$$\mathcal{L} = \langle \exp(u(0)) \mid u \in V_1 \rangle \trianglelefteq \mathcal{G},$$

which we call the *linear* automorphism group of V. The theory of algebraic groups [H] together with the structure of V_1 described in the previous Section provide a clear picture of the group-theoretic structure of $\mathcal{L}$ in case V is an RSVOA. What is still lacking is a solution to

Problem 4: Prove that $\mathcal{L} = \mathcal{G}^o$, i.e. $\mathcal{L}$ is the connected component of the identity of $\mathcal{G}$.

A consequence of this (assuming its truth!) is that $\mathcal{L}$ has finite index in $\mathcal{G}$, and from this one further deduces that $\mathcal{G}$ is a finite group if, and only if, $V_1 = 0$. If we consider the Moonshine Module $V^\natural$, for example, this would imply (because $V_1^\natural = 0$) that $\mathrm{Aut} V^\natural$ is finite - something that even now requires some effort to prove.

4. Rational orbifolds

Like many things, 'orbifold' means different things to different people, and the different meanings are not necessarily consistent. Orbifold theory within the context of VOAs has come to mean the general study of the pair (V, G) where V is a VOA and $G \subseteq \mathcal{G}$ a group of automorphisms. We shall allow ourselves to call the fixed-point space

$$V^G = \{v \in V \mid gv = v, \ \forall g \in G\}$$

an *orbifold*, or *orbifold model*. It is, of course, a subVOA of V. Orbifold theory in this sense includes the study of twisted sectors, fusion rules, and intertwining algebra defined by V and G, among other things, and we shall look at some of these ideas below. One of the first major constructions in orbifold theory was the work of Frenkel-Lepowsky-Meurman concerning a $\mathbb{Z}_2$-orbifold of the Leech lattice VOA. We refer the reader to [FLM] for an account of this, where it is also explained how this relates to a $\mathbb{Z}_2$-orbifold of the Leech torus in the topological sense.

Now let us return to the case when V is a SRVOA. One is then particularly interested in the *closed* subgroups of $\mathcal{G}$ and their orbifolds in the above sense. In particular, we want to know when V^G is a rational VOA for a closed subgroup $G \subseteq \mathcal{G}$.

By a theorem in [DM4], every irreducible unitary representation of G occurs as a constituent of the restriction to V_n, for at least one (and therefore infinitely many) n. Furthermore, V decomposes into a direct sum of simple V^G-modules indexed by the irreducible representations of G, with inequivalent irreducible G-modules giving rise to inequivalent V^G-modules. The upshot is that if G is not finite then there are infinitely many inequivalent simple V^G-modules contained in V, so that the orbifold V^G *cannot* be rational. This leaves the case of finite G, for which the following is a well-known open problem:

Problem 5: If G is a finite group, prove that

$$V \text{ rational } \Rightarrow V^G \text{ rational.}$$

One also needs to know about C_2-cofiniteness. Affirmative solutions to Problems 1 and 5 would take care of this, but one can also ask

Problem 6: If G is a finite group, prove that

$$V \ C_2\text{-cofinite } \Rightarrow V^G \ C_2\text{-cofinite.}$$

If solutions to these problems are available (and there are very few choices of V and G which have been checked thus far), then modular-invariance of rational orbifolds is subsumed within the general problem of modular-invariance for rational VOAs. However, when we have a group acting on V, the structure of the orbifold V^G and its module category has additional features which are not present in general. These concern the *twisted sectors*, which we take up next.

5. Twisted Sectors

Let V be a VOA, $G \subseteq \mathcal{G}$ a finite group of automorphisms of V, and $g \in G$ an automorphism of order N. A (simple) g-*twisted sector* is a graded space

$$V(g) = \bigoplus_{n \geq 0} V(g)_{\lambda(g) + \frac{n}{N}}$$

for some scalar $\lambda(g)$ (the *conformal weight*), where V acts by 'twisted' operators. Naturally, the graded dimension of $V(g)$ is defined to be

$$Z_{V(g)}(q) = \mathrm{Tr}_{V(g)} q^{L(0) - c/24}$$
$$= q^{-c/24 + \lambda(g)} \sum_{n \geq 0} \dim V_{\lambda(g) + \frac{n}{N}} q^{n/N}.$$

We forgo technical definitions here (cf. [L], [DLM3]), observing only that the idea of twisted sectors has no good analog in classical representation theory. In a more leisurely account we would have introduced general g-twisted sectors, which form a category $V(g)$-Mod. If $g = 1$ (the 'untwisted' case), a g-twisted module is nothing but a V-module, and from this perspective the theory of twisted modules is a generalization of the representation theory of V. Thus it becomes important to know when V is g-*rational*, which means that $V(g)$-Mod is completely reducible in a sense which parallels the definition in the untwisted case (cf. Section 2). For more background and results in this direction, see [DLM2], [DLM3]. The main problem in this direction is the following:

Problem 7: Prove that V rational $\Rightarrow V$ g-rational.

It is known [DLM3] that, as in the untwisted case, g-rationality implies the finiteness of the number of inequivalent simple g-twisted modules. One can therefore refine Problem 7 by asking for the *number* of simple g-twisted modules. For more on this, see [DLM2].

One of the main applications of twisted modules is to the representation theory of orbifolds. It is an immediate consequence of the definitions that a g-twisted module is an (untwisted) module over the orbifold VOA V^G. Thus one can look for simple V^G-modules by decomposing twisted modules. The basic problem here is

Problem 8: Prove that *every* simple V^G-module is contained in a g-twisted module for *some* $g \in G$.

This suggests, for example, that the graded dimension of a simple g-twisted module should also be a modular function. We abandon the general development of rational orbifolds at this point in order to take up a special case where many of the basic ideas and conjectures show up in simpler form.

6. Modular-invariance in holomorphic orbifolds

Definition: V is *holomorphic* if it is rational and if, in addition, the adjoint module V is the *unique simple V-module*.

From now on, V is assumed to be both holomorphic and an SRVOA, with $G \subseteq \mathcal{G}$ a finite group of automorphisms. We call V^G a *holomorphic orbifold*.

Example The Moonshine module $V^\natural$ satisfies these conditions.

Working with holomorphic VOAs rather than general rational VOAs is rather like working with modular forms of level 1 rather than level N. This is particularly apropos with regard to questions of modular-invariance. For example, if V is holomorphic then the representation ρ which figures in Problem 2 is 1-dimensional, in which case modular-invariance is clear. Hence we have [Z]

If V is a holomorphic VOA then the graded dimension $Z_V(q)$ is a

modular function of weight zero on $SL(2, \mathbb{Z})$ (possibly with a character).

For this to even make sense, we must set $q = e^{2\pi i \tau}$ with τ in the complex upper half-plane $\mathbb{H}$, and identify $Z_V(q)$ with the corresponding function $Z_V(\tau)$ on $\mathbb{H}$. Implicit in the last displayed assertion is the fact that $Z_V(\tau)$ is holomorphic throughout $\mathbb{H}$. Similar comments apply to the additional q-expansions that occur below. The most famous example is of course that of $V^\natural$, where

$$Z_{V^\natural}(q) = q^{-1} + 0 + 196884q + \ldots$$

is the modular function $J(q)$ with constant term zero [FLM].

The theory of twisted sectors is also better understood in the case of holomorphic VOAs. In particular [DLM2], there is a *unique* simple g-twisted sector $V(g)$ for each finite order automorphism g. There is an interesting relation between the graded dimension of $V(g)$ and the graded trace of g (the formal definition is given below):

$$Z_V(g, -1/\tau) = (\text{constant})Z_{V(g)}(\tau).$$

This is proved in [DLM3], and shows that the modularity of $Z_V(g, q)$ is *equivalent* to that of $Z_{V(g)}(q)$. In the case of the Moonshine Module $V^\natural$ for example, each $Z_{V^\natural}(g, q)$ is modular thanks to Borcherds solution of the Conway-Norton Moonshine Conjectures [B]. Therefore, each simple twisted sector $V^\natural(g), g \in M$, has a graded dimension which is a modular function. Note that explicit constructions of the twisted sectors are known in only a few cases. To be most effective, one would like to know

Problem 9: Show that the constant above is always 1.

This would establish a remarkable fact - that the graded dimension of the g-twisted sector just the S-transform of the graded trace of g on V. Problem 9 remains open even in the case of $V^\natural$ and general $g \in M$, though it is known in some cases.

Example: (Identify elements of M as in [CN]; $\eta(\tau)$ is the Dedekind eta-function.)
(a) $2A \in M; \lambda(2A) = 1/2$,

$$Z(2A, \tau) = q^{-1} + 4372q + \ldots$$
$$Z_{V^\natural(2A)}(q) = q^{-1/2} + 4372q^{1/2} + \ldots$$

(b) $2B \in M : \lambda(2B) = 1$,

$$Z(2B, \tau) = 24 + \frac{\eta(\tau)^{24}}{\eta(2\tau)^{24}}$$

$$Z_{V^\natural(2B)}(q) = 24 + 2^{12}\frac{\eta(\tau)^{24}}{\eta(\tau/2)^{24}}.$$

The uniqueness of $V(g)$ can be exploited. It is a general fact [DM5] that if $g, h \in G$ then h induces a functor $V(g)$-Mod $\rightarrow V(hgh^{-1})$-Mod. In particular, the unicity of $V(g)$ leads to a *projective* action of the centralizer $C_G(g)$ on $V(g)$. Thus if $gh = hg$ then there is a graded trace

$$Z_{V(g)}(h, q) = \text{Tr}_{V(g)} h q^{L(0)-c/24}$$
$$= q^{-c/24+\lambda(g)} \sum_{n \geq 0} \text{Tr}_{V_{\lambda(g)+\frac{n}{N}}} h \, q^{n/N}$$

which is well-defined up to an overall nonzero scalar. Setting $Z(g, h, q) = Z_{V(g)}(h, q)$, there is the following conjecture which generalizes the relation between $Z_{V(g)}(q)$ and $Z_V(g, q)$ which we discussed above:

Problem 10: For each $\gamma = \begin{pmatrix} a & b \\ c & d \end{pmatrix} \in SL(2, \mathbb{Z})$, prove that

$$Z(g, h, \frac{a\tau + b}{c\tau + d}) = (\text{constant}) Z(g^a h^c, g^b h^d, \tau),$$

where the constant is a *root of unity*.

In the case of $V^{\natural}$, this is precisely Norton's generalized moonshine conjecture mentioned in the Introduction. Note that unlike Problem 9 (corresponding to the case $\gamma = S$) the constant cannot always be taken to be 1. On the other hand, all of the remaining difficulty in the problem resides in the nature of the constant, since the equality (with no restrictions on the constant) is known to be true [DLM3]. It is a consequence of Problem 10 that each $Z(g, h, \tau)$ is a modular function of weight zero. Assuming an affirmative solution to Problem 8, this in turn is equivalent to the modularity of the graded dimensions of the simple modules over a holomorphic orbifold.

Finally, we mention unpublished work of the author and Chongying Dong. A holomorphic *linear* orbifold is a holomorphic orbifold V^G with a finite group $G \subseteq \mathcal{L}$. Roughly, we can show that much of the program outlined above can be *proved* for holomorphic linear orbifolds, and in particular Problems 9 and 10 have affirmative solutions. Indeed, there is a symmetric function $\zeta(\cdot|\cdot)$: $G \times G \longrightarrow \mathbf{C}^*$ such that

$$\zeta(g \mid \cdot) \text{ is a character of } C_G(g)$$
$$Z(g, h, S\tau) = \zeta(g|h) Z(h, g^{-1}, \tau)$$

This corresponds *precisely* to the formalism of [DVVV] in one of the first papers on orbifolds. For the Moonshine Module $V^{\natural}$ the linear automorphism group is trivial, so we find nothing new in this case. On the other hand, for the holomorphic VOA V_{E_8} corresponding to the E_8 root lattice, $\mathcal{L}$ is the full automorphism group. In this case, then, we have a complete proof of modular-invariance for commuting pairs of elements. Given a linear orbifold, G arises from exponentiating elements of V_1 and one can make use of the connection with Lie algebras in the proofs. Furthermore, there are fairly explicit descriptions of the twisted sectors for elements in $\mathcal{L}$ based on a construction of Li [Li2]. There are connections with some modular-invariance results of Miyamoto [M].

This completes our survey, though there is much more that one could (and perhaps should) say. In particular, we have omitted discussion of the ideas of Bantay [Ba] for proving modular-invariance results using permutation orbifolds and modular data. Huang has recently announced some important results [Hu] concerning the categorical nature of V-Mod which bear on these issues.

References

[ABD] T. Abe, G. Buhl and C. Dong, Rationality, regularity and C_2-cofiniteness, Trans. Amer. Math. Soc. **356**(2004), 3391–3402.

[AM] G. Anderson and G. Moore, Rationality in conformal field theory, Comm. Math. Phys. **117**(1988), 441–450.

[B] R. Borcherds, Monstrous moonshine and monstrous Lie superalgebras, Inv. Math. **109**(1992), 405–444.

[Ba] P. Bantay, Permutation orbifolds and their applications, in Vertex Operator Algebras in Mathematics and Physics, Fields Inst. Comm. 39, **Amer. Math. Soc.**, 2003.

[DG] C. Dong and R. Griess, Automorphism groups and derivation algebras of finitely generated vertex operator algebras, Mich. Math. J. **50**(2002), 227–239.

[DLM1] C. Dong, H. Li and G. Mason, Regularity of rational vertex operator algebras, Adv. Math. **132** (1997), 148–166.

[DLM2] C. Dong, H. Li and G. Mason, Modular invariance of trace functions in orbifold theory, Comm. Math. Phys. **214** (2000), 1–56.

[DLM3] C. Dong, H. Li and G. Mason, Twisted representations of vertex operator algebras, Math. Ann. **310** (1998), 389–397.

[DM1] C. Dong and G. Mason, Shifted vertex operator algebras, Math. Proc. Cambs. Phil. Soc. **141** (2006), 67–80.

[DM2] C. Dong and G. Mason, Rational Vertex Operator Algebras and the Effective Central Charge, Int. Math. Res. Not. **No.56** (2004), 2989–3008.

[DM3] C. Dong and G. Mason, Integrability of C_2-cofinite vertex operator algebras, Int. Math. Res. Not. **Vol. 2006**, Art. ID 80468, 1–15.

[DM4] C. Dong and G. Mason, On Quantum Galois Theory, Duke. Math. J. Vol. 86, **No.2** (1997), 305–321.

[DM5] C. Dong and G. Mason, Non-abelian orbifolds and the boson-fermion correspondence, Comm. Math. Phys. **163** (1994), 523–559. 1

[DVVV] R. Dijkgraaf, C. Vafa, E. Verlinde and H. Verlinde, The operator algebra of orbifold models, Comm. Math. Phys. **123** (1989), 485–526.

[FHL] I. Frenkel, Y-Z Huang and J. Lepowsky, On axiomatic approaches to vertex operator algebras and modules, Mem. Amer. Math. Soc. **104**, 1993.

[FLM] I. Frenkel, J. Lepowsky and A. Meurman, Vertex Operator Algebras and the Monster, **Academic Press**, San Diego, 1988.

[GN] M. Gaberdiel and A. Neitzke, Rationality, quasirationality and finite W-algebras, Comm. Math. Phys. **238** (2003), 305–331.

[Hu] Y-Z Huang, Vertex operator algebras, the Verlinde conjecture, and modular tensor categories, Proc. Natl. Acad. Sci. USA **102**, No. 15(2005), 5352–5356.

[H] J. Humphreys, Linear Algebraic Groups, Graduate Texts in Mathematics, **Springer**, New York, 1975.

[Li1] H. Li, Symmetric invariant bilinear forms on vertex operator algebras, J. Pure and Appl. Alg. **109** (1994), 279–297.

[Li2] H. Li, Local systems of twisted vertex operators, vertex operator superalgebras and twisted Modules, Contemp. Math. **193** (1996), 203–236.

[L] J. Lepowsky, Calculus of twisted vertex operators, Proc. Nat. Acad. Sci. **82** (1985), 8295–8299.

[LL] J. Lepowsky and H. Li, Introduction to Vertex Operator Algebras and Their Representations, Progress in Mathematics, **Birkhäuser**, Boston, 2004.

[M] M. Miyamoto, A modular invariance on the theta functions defined on vertex operator algebras, Duke. Math. J. **101** (2000), 221–236.

[Z] Y-Z Zhu, Modular invariance of characters of vertex operator algebras, J. Amer. Math. Soc. **9** (1996), 237–302.

Quasi-finite Algebras Graded by Hamiltonian and Vertex Operator Algebras

Atsushi Matsuo*, Kiyokazu Nagatomo† and Akihiro Tsuchiya‡

Abstract

A general notion of a quasi-finite algebra is introduced as an algebra graded by the set of all integers equipped with topologies on the homogeneous subspaces satisfying certain properties. An analogue of the regular bimodule is introduced and various module categories over quasi-finite algebras are described. When applied to the current algebras (universal enveloping algebras) of vertex operator algebras satisfying Zhu's C_2-finiteness condition, our general consideration derives important consequences on representation theory of such vertex operator algebras. In particular, the category of modules over such a vertex operator algebra is shown to be equivalent to the category of modules over a finite-dimensional associative algebra.

Introduction

In order to construct conformal field theories on Riemann surfaces associated with a vertex operator algebra V and to obtain their properties such as the finite-dimensionality of the space of conformal blocks, factorization of the blocks along the boundaries of the moduli space of Riemann surfaces and the fusion functors or the tensor product of V-modules, we need first to impose an appropriate finiteness condition on V and second to study the structure of the abelian category of V-modules to some extent.

* Graduate School of Mathematical Sciences, University of Tokyo, Komaba, Tokyo 153-8914, Japan
† Department of Pure and Applied Mathematics, Graduate School of Information Science and Technology, Osaka University, Toyonaka, Osaka 560-0043, Japan
‡ Graduate School of Mathematics, Nagoya University, Furo-cho, Nagoya 464-8602, Japan
    Partially supported by Grant-in-Aid for Scientific Research No. 14740005 (Atsushi Matsuo), 16340007 and 16634001 (Kiyokazu Nagatomo) and 14204003 (Akihiro Tsuchiya) from the Ministry of Education, Science, Sports and Culture, Japan.

One of the candidates of such a finiteness condition is the one introduced by Y.-C. Zhu ([Zhu]), usually called the C_2-finiteness (or C_2-cofiniteness), saying that a certain quotient space $V/C_2(V)$ is finite-dimensional. We will call this condition *Zhu's finiteness condition* in the rest of the paper. This condition was used in [Zhu] in an essential way to the proof of the modular invariance of characters of V-modules, as well as the condition that the category of V-modules is semisimple. The modular invariance is a part of the characteristic properties of rational conformal field theories.

Another important ingredient in Zhu's derivation of the modular invariance is the use of an associative algebra $A(V)$, called Zhu's algebra, and a functor from the category of V-modules to the category of $A(V)$-modules. The fundamental result of Zhu is as follows: the functor sends an irreducible V-module to an irreducible $A(V)$-module in such a way that the equivalence classes of irreducibles in the two categories are in one-to-one correspondence. In particular, the number of irreducible classes is finite if $A(V)$ is finite-dimensional. The last property actually follows from Zhu's finiteness condition mentioned above.

However, the above-mentioned functor need not give us an equivalence of categories. Specifically, if we include the cases when the category of V-modules is not semisimple, the algebra $A(V)$ is not enough to understand the properties of the category of V-modules in general. In this regard, it seems to the authors that detailed analysis of the structure of the abelian category of V-modules has not yet been done. (See [DLM3] for some related results.)

The purpose of the present paper is to fill this gap by making use of the universal enveloping algebra associated with the vertex operator algebra.

The universal enveloping algebra associated with V, which we will denote by $\mathbb{U} = \mathbb{U}(V)$, is a certain topological algebra first considered by Frenkel and Zhu in [FrZ]. The presence of a topology is inevitable as the universal defining relations of vertex operator algebras contain infinite sums. We will call $\mathbb{U}$ the *current algebra* for short in the rest of the paper. More precisely, the current algebra $\mathbb{U}$ is an associative algebra with unity graded by the set of integers equipped with a separated linear topology defined by a certain sequence $I_0(\mathbb{U}), I_1(\mathbb{U}), \ldots$ of open left ideals such that each homogeneous subspace $\mathbb{U}(d)$ is complete with respect to the relative topology. Note that the quotient space $Q_n(\mathbb{U}) = \mathbb{U}/I_n(\mathbb{U})$ is a discrete space, which inherits a grading from $\mathbb{U}$. We will denote its subspace of degree d by $Q_n(\mathbb{U})(d)$. The algebra $\mathbb{U}$ is important in that there is a one-to-one correspondence between V-modules of certain type and continuous discrete $\mathbb{U}$-modules. (See Section 6 for the precise statement.)

The authors noticed while trying to understand the category of V-modules by means of $\mathbb{U}$ that the finiteness property should better be imposed on $\mathbb{U}$ rather than on V as far as the properties of the category as an abelian category are concerned. Thus we arrived at the concept of *quasi-finiteness*, which is defined as follows: $\mathbb{U}$ is quasi-finite if and only if the spaces $Q_n(\mathbb{U})(d)$ are finite-dimensional for all integers d and nonnegative integers n. As we will show in Theorem 9.2.1, Zhu's finiteness condition actually implies the quasi-finiteness of $\mathbb{U}$.

Let us assume that $\mathbb{U}$ is quasi-finite. Then the spaces $Q_n(\mathbb{U})(d)$ have the generalized eigenspace decompositions with respect to the action of the Virasoro operator L_0, which we will call the Hamiltonian of $\mathbb{U}$. By using this, we can construct a series of finite-dimensional algebras $\mathbb{U}_n$ and functors from the category of continuous discrete left $\mathbb{U}$-modules to the category of left $\mathbb{U}_n$-modules which give rise to equivalences of categories when n is sufficiently large. Therefore, under the quasi-finiteness, the category of continuous discrete left $\mathbb{U}$-modules as an abelian category is completely described by the properties of the finite-dimensional algebra $\mathbb{U}_n$.

In application to conformal field theories on Riemann surfaces, we have to enlarge the algebra $\mathbb{U}$ and to take into account the concept of duality of $\mathbb{U}$-modules. Therefore we consider two different topologies on $\mathbb{U}$, one is the original left linear one and the other a right linear one, and take the filter-wise completion $\mathcal{U}(V)$ and $\mathcal{U}(V)^\vee$ with respect to the two topologies. Then we may consider the left $\mathcal{U}(V)$-modules, the right $\mathcal{U}(V)$-modules, the left $\mathcal{U}(V)^\vee$-modules, and the right $\mathcal{U}(V)^\vee$-modules. We can now formulate the concept of quasi-finiteness for each of the four and can establish the equivalences or the dualities among them. We will call this type of results the *finiteness theorems*.

As the argument above deriving the equivalences and the dualities of categories only uses quite general properties of $\mathbb{U}$ and L_0, we now postulate them: we will call a topological algebra $\mathbb{A}$ with a distinguished element h a *quasi-finite algebra graded by Hamiltonain* if it shares the same properties with $\mathbb{U}$ and L_0 as mentioned above. (See Section 2 for the precise definition.)

The present paper is divided into two parts: Part I consists in explaining general theory of quasi-finite algebras graded by Hamiltonian and categories of modules over them and Part II in proving that Zhu's finiteness condition on V implies the quasi-finiteness of $\mathbb{U}$.

The plan of Part I is as follows. We will begin by describing in Section 1 the concept of compatible degreewise topological algebras and the associated topological filtered algebras which are modeled on the topological features of $\mathbb{U}$. In Section 2, we define the concept of quasi-finite algebras graded by Hamiltonian and give some consequences. In particular, for a pair $(\mathbb{A}, h)$ of a quasi-finite algebra $\mathbb{A}$ and a Hamiltonian h, we introduce an analogue of the regular bimodule and define certain finite-dimensional algebras $\mathbb{A}_n$. We will show that the regular bimodule is a dense subspace of the algebra (Theorem 2.5.4). Section 3 is devoted to the equivalence of categories between the category of continuous discrete $\mathbb{A}$-modules, which we will call *exhaustive modules*, and the category of $\mathbb{A}_n$-modules (Theorem 3.3.4). In Section 4, we will introduce a notion of *coexhaustive modules* which is a *dual* notion of exhaustive modules and establish an equivalence between the category of exhaustive modules and the category of coexhaustive modules (Theorem 4.3.3). The duality of modules will be formulated in Section 5 as the duality between quasi-finite exhaustive left modules and quasi-finite coexhaustive right modules. We will then summarize various equivalences and the dualities for quasi-finite modules (Theorem 5.5.1).

The proof of quasi-finiteness of the current algebra $\mathbb{U}$ in Part II under Zhu's finiteness condition will be done by considering a certain filtration G on $\mathbb{U}$, which was introduced in [NaT], and a certain universal Poisson algebra $\mathbb{S} = \mathbb{S}(\mathfrak{p})$ associated with $\mathfrak{p} = V/C_2(V)$. We will construct a surjective homomorphism of Poisson algebra from $\mathbb{S}$ to the degreewise completion $\widetilde{\mathrm{gr}}^{\mathrm{G}}\mathbb{U}$ of $\mathrm{gr}^{\mathrm{G}}\mathbb{U}$. The quasi-finiteness of $\mathbb{U}$ then follows from that of $\mathbb{S}$. We will call $\mathbb{S}$ the *Poisson current algebra*.

In Section 6, we will describe the construction and some properties of the current algebra $\mathbb{U}$ associated with a vertex operator algebra V. In Section 7, we will consider the filtration G on $\mathbb{U}$ and show that the associated graded algebra has a structure of a Poisson algebra. In Section 8, we will construct the Poisson current algebra $\mathbb{S}$ associated with any Poisson algebra $\mathfrak{p}$ and we will show that $\mathbb{S}$ is quasi-finite if $\mathfrak{p}$ is finite-dimensional (Theorem 8.3.3). In the final section, we will consider the case when $\mathfrak{p} = V/C_2(V)$ and construct the surjective homomorphism $\psi : \mathbb{S} \to \widetilde{\mathrm{gr}}^{\mathrm{G}}\mathbb{U}$ of Poisson algebras (Theorem 9.1.2). We will then combine these results to show that Zhu's finiteness condition on V implies quasi-finiteness of the current algebra $\mathbb{U}$ (Theorem 9.2.1).

The finiteness theorems mentioned above will give us not only a nice conceptual understanding of the role of Zhu's finiteness condition in representation theory of vertex operator algebras but also a foundation in the strategy mentioned at the beginning of this introduction of constructing the spaces of conformal blocks on general Riemann surfaces and of showing their expected properties. This will be developed in our forthcoming paper, which will serve as the continuation of the previous paper [NaT] by two of the authors.

Acknowledgement. The results of this paper were presented in part at 'Moonshine - the First Quarter Century and Beyond, Edinburgh, July 2004'. The authors grateful to the organizers the workshop. They also thank the organizers of 'Tensor Categories in Mathematics and Physics, Vienna, June, 2004', 'International Conference on Infinite Dimensional Lie Theory, Beijing, July 2004' and 'Perspectives arising from vertex algebra theory, Osaka, November 2004'.

The authors thank Dr. T. Arakawa for discussion. A.M. thanks Profs. T. Miwa and S. Loktev and Dr. M. Rosellen for conversation. K.N. thanks Prof. N. Kawanaka for his continuous encouragements and dedicates the paper to him on the occasion of his sixtieth birthday.

Part I

Quasi-finite Algebras Graded by Hamiltonian

1. Linear topologies on graded algebras

In this section, we introduce compatible degreewise topological algebras and some related notions.

1.1. Preliminaries

Throughout the paper, we will work over an algebraically closed field $\mathbf{k}$ of characteristic zero. A vector space always means a vector space over $\mathbf{k}$ and the scalar multiplication of a vector space is denoted by juxtaposition.

An algebra means an associative algebra over $\mathbf{k}$ with unity. The multiplication of an algebra A is denoted by the dot $\cdot$ and the unity by 1_A or simply by 1. For subsets S and T of A, we denote by $S \cdot T$ the linear span of the elements of the form $s \cdot t$ with $s \in S$ and $t \in T$. The action of A on an A-module M is denoted again by the dot. We always assume that the unity 1_A acts by the identity operator.

We endow the field $\mathbf{k}$ with the discrete topology. Let $I_0, I_1, \ldots, I_n, \ldots$ be a decreasing sequence of linear subspaces of a vector space V. A *linear topology defined by* I_n means a topology on V such that for each $v \in V$ the set $\{v + I_n \mid n = 0, 1, 2, \ldots\}$ forms a fundamental system of open neighborhoods of v.

Throughout the paper, a topology on a vector space always means a linear topology given in this way. Such a space is usually called a linearly topologized vector space in the literatures.

Let V be a vector space with the linear topology defined by I_n. Any subspace U which contains I_n for some n is open and closed, and the quotient topology on V/U is the discrete topology. The closure of a subspace U is given by $\bigcap_n (U + I_n)$ and U is dense in V if and only if the composite $U \to V \to V/I_n$ of canonical maps is surjective for any n.

The completion of V with respect to the linear topology defined by I_n is the projective limit

$$\hat{V} = \varprojlim_n V/I_n \qquad (1.1.1)$$

endowed with the projective limit topology, namely the relative topology induced from $\prod_{n=0}^{\infty} V/I_n$ with the product topology of the discrete topologies on V/I_n. Let $\hat{I}_n$ be the closure of the image of I_n under the canonical map $V \to \hat{V}$. Then $\hat{I}_n$ agrees with the kernel of the canonical map $\hat{V} \to V/I_n$ and the topology on $\hat{V}$ is the linear topology defined by $\hat{I}_n$.

The space V is said to be complete if the canonical map $V \to \hat{V}$ is a homeomorphism. Thus a complete space is separated (i.e. Hausdorff) in our convention. If V is complete then the closure of a subspace U is given by

$$\hat{U} = \varprojlim_n (U + I_n)/I_n. \qquad (1.1.2)$$

and for a closed subspace F the quotient space V/F with the quotient topology is also complete.

We refer the reader to [Bou], [EGA] and [Mac] for linear topologies.

1.2. Compatible degreewise topological algebras

Let $\mathbb{A}$ be an algebra and suppose given a grading

$$\mathbb{A} = \bigoplus_d \mathbb{A}(d) \qquad (1.2.1)$$

indexed by integers such that $\mathbb{A}(d) \cdot \mathbb{A}(e) \subset \mathbb{A}(d + e)$. We simply call such an $\mathbb{A}$ a *graded algebra*. Let us set

$$F_p \mathbb{A} = \bigoplus_{d \leq p} \mathbb{A}(d), \quad F_p^{\vee} \mathbb{A} = \bigoplus_{d \geq -p} \mathbb{A}(d) \qquad (1.2.2)$$

where p is an integer. We call the filtration F the *associated filtration* and $F^{\vee}$ the *opposite filtration*. Note that the subspaces $F_0 \mathbb{A}$ and $F_0^{\vee} \mathbb{A}$ are subalgebras of $\mathbb{A}$.

Let $\mathbb{A} = \bigoplus_d \mathbb{A}(d)$ be a graded algebra and suppose given a linear topology on each $\mathbb{A}(d)$. In such a situation, we will say that $\mathbb{A}$ is endowed with a *degreewise topology*. We assume that the multiplication maps $\mathbb{A}(d) \times \mathbb{A}(e) \to \mathbb{A}(d + e)$ are continuous. We will say that $\mathbb{A}$ is *degreewise complete* if each $\mathbb{A}(d)$ is complete.

Now consider the subspace

$$\mathbb{A}(d) \cap (\mathbb{A} \cdot F_{-n-1}\mathbb{A}) = \sum_{k \leq -n-1} \mathbb{A}(d-k) \cdot \mathbb{A}(k) \qquad (1.2.3)$$

and let $I_n(\mathbb{A}(d))$ be its closure in $\mathbb{A}(d)$. We assume that $\{I_n(\mathbb{A}(d))\}$ forms a fundamental system of open neighborhoods of zero in each $\mathbb{A}(d)$.

Definition 1.2.1 A *compatible degreewise topological algebra* is a graded algebra $\mathbb{A}$ endowed with a degreewise topology such that all the conditions mentioned above are satisfied. A *compatible degreewise complete algebra* is a compatible degreewise topological algebra $\mathbb{A}$ such that it is degreewise complete.

Let $\mathbb{A}$ be a compatible degreewise topological algebra. Since the multiplication maps $\mathbb{A}(d) \times \mathbb{A}(e) \to \mathbb{A}(d+e)$ are continuous, we have

$$\mathbb{A}(d) \cdot I_n(\mathbb{A}(e)) \subset I_n(\mathbb{A}(d+e)), \quad I_n(\mathbb{A}(d)) \cdot \mathbb{A}(e) \subset I_{n-e}(\mathbb{A}(d+e)).$$
$$(1.2.4)$$

Therefore, for $a \in \mathbb{A}(d)$ and $b \in \mathbb{A}(e)$, we have

$$(a + I_{n+e}(\mathbb{A}(d))) \cdot (b + I_n(\mathbb{A}(e))) \subset a \cdot b + I_n(\mathbb{A}(d+e)). \qquad (1.2.5)$$

Instead of $I_n(\mathbb{A}(d))$, we may consider the closure $I_n^{\vee}(\mathbb{A}(d))$ of $\mathbb{A}(d) \cap (F_{-n-1}^{\vee}\mathbb{A} \cdot \mathbb{A})$. Since we have

$$I_n^{\vee}(\mathbb{A}(d)) = I_{n-d}(\mathbb{A}(d)), \qquad (1.2.6)$$

the topology defined by $I_n^{\vee}(\mathbb{A}(d))$ agrees with the one defined by $I_n(\mathbb{A}(d))$.

In the sequel, we will use the following terminologies: a graded subspace is said to be *degreewise dense* if each homogenesous subspace is dense; the sum of the closures of the homogeneous subspaces of a graded subspace U is called the *degreewise closure* of U.

Note 1.2.2 Let $\mathbb{A}$ be a graded algebra endowed with a degreewise topology. In general, there is no canonical way of extending the topologies on the homogeneous subspaces to the whole space $\mathbb{A}$ which makes $\mathbb{A}$ into a topological algebra.

1.3. Degreewise completion

Let $\mathbb{A}$ be a compatible degreewise topological algebra. Set $\hat{\mathbb{A}} = \bigoplus_d \hat{\mathbb{A}}(d)$ where $\hat{\mathbb{A}}(d)$ is the completion of the space $\mathbb{A}(d)$. We call $\hat{\mathbb{A}}$ the *degreewise completion* of $\mathbb{A}$. Since the multiplication maps $\mathbb{A}(d) \times \mathbb{A}(e) \to \mathbb{A}(d+e)$ are continuous, they induce continuous bilinear maps $\hat{\mathbb{A}}(d) \times \hat{\mathbb{A}}(e) \rightarrow \hat{\mathbb{A}}(d + e)$ which make $\hat{\mathbb{A}}$ into an algebra endowed with a degreewise topology.

Proposition 1.3.1   *The degreewise completion $\hat{\mathbb{A}}$ is a compatible degreewise complete algebra.*

For instance, let $\mathbb{A} = \bigoplus_d \mathbb{A}(d)$ be any graded algebra. We endow the space $\mathbb{A}(d)$ with the linear topology defined by $\mathbb{A}(d) \cap (\mathbb{A} \cdot F_{-n-1}\mathbb{A})$. Then $\mathbb{A}$ becomes a compatible degreewise topological algebra and the degreewise completion $\hat{\mathbb{A}}$ is degreewise complete. We call this degreewise topology on $\mathbb{A}$ the *standard degreewise topology* and the algebra $\hat{\mathbb{A}}$ the *standard degreewise completion*.

Note 1.3.2   Giving topologies on the homogeneous subspaces of a graded algebra in the way described above was considered by some authors, see e.g. [FeF], [Kac], [FrZ], [Mal], [LiW].

1.4. Associated filtered topological algebra

A topological algebra is said to be left linear if the topology is a linear topology defined by a sequence of left ideals. Right linearity is defined similarly.

Let $\mathbb{A}$ be a compatible degreewise topological algebra. Let us endow the algebra $\mathbb{A}$ with the associated filtration F. Consider the left ideal $I_n(\mathbb{A})$ of $\mathbb{A}$ defined by

$$I_n(\mathbb{A}) = \bigoplus_d I_n(\mathbb{A}(d)). \tag{1.4.1}$$

By the compatibility of $\mathbb{A}$, the space $I_n(\mathbb{A})$ is the degreewise closure of $\mathbb{A} \cdot F_{-n-1}\mathbb{A}$. Let us endow the space $\mathbb{A}$ with the topology defined by $I_n(\mathbb{A})$. Then the relative topology on $\mathbb{A}(d)$ induced from $\mathbb{A}$ agrees with the original topology on $\mathbb{A}(d)$. For any $a \in F_p\mathbb{A}$ and $b \in F_q\mathbb{A}$ we have $(a + I_{n+q}(\mathbb{A})) \cdot (b + I_n(\mathbb{A})) \subset a \cdot b + I_n(\mathbb{A})$. Hence the multiplication $\mathbb{A} \times \mathbb{A} \rightarrow \mathbb{A}$ is continuous. In particular, as $I_n(\mathbb{A})$ are left ideals, $\mathbb{A}$ is a left linear topological algebra. We simply call this topology *the left linear topology* of $\mathbb{A}$.

Now consider the opposite filtration $F^\vee$ and set

$$I_n^\vee(\mathbb{A}) = \bigoplus_d I_n^\vee(\mathbb{A}(d)). \tag{1.4.2}$$

The space $I_n^\vee(\mathbb{A})$ is the degreewise closure of $F_{-n-1}^\vee \mathbb{A} \cdot \mathbb{A}$. Then the linear topology on $\mathbb{A}$ defined by $I_n^\vee(\mathbb{A})$ now makes $\mathbb{A}$ into a right linear topological algebra. We call this topology *the right linear topology* of $\mathbb{A}$.

The left linear topology and the right linear topology on the same algebra $\mathbb{A}$ do *not* agree with each other in general although the restrictions to $\mathbb{A}(d)$ are the same.

1.5. Filterwise completion

Let $\mathbb{A}$ be a compatible degreewise topological algebra. Let us endow $\mathbb{A}$ with the left linear topology and let $F_p \mathscr{A}$ be the completion

$$F_p \mathscr{A} = \varprojlim_n F_p \mathbb{A}/F_p \mathbb{A} \cap I_n(\mathbb{A}) \tag{1.5.1}$$

with the projective limit topology. Then the multiplication $F_p \mathbb{A} \times F_q \mathbb{A} \to F_{p+q}\mathbb{A}$ induces a multiplication $F_p \mathscr{A} \times F_q \mathscr{A} \to F_{p+q}\mathscr{A}$. Consider the space

$$\mathscr{A} = \varinjlim_p F_p \mathscr{A} \tag{1.5.2}$$

and give it the linear topology defined by

$$I_n(\mathscr{A}) = \mathrm{Ker}\left(\mathscr{A} \to \mathbb{A}/I_n(\mathbb{A})\right). \tag{1.5.3}$$

Then $I_n(\mathscr{A})$ is a left ideal of $\mathscr{A}$ and, for any $a \in F_p \mathscr{A}$ and $b \in F_q \mathscr{A}$, we have $(a + I_{n+q}(\mathscr{A})) \cdot (b + I_n(\mathscr{A})) \subset a \cdot b + I_n(\mathscr{A})$. Thus the space $\mathscr{A}$ becomes a filtered left linear topological algebra such that the subspaces $F_p \mathscr{A}$ with the relative topologies are complete and that the image of $\mathbb{A}$ under the canonical map $\mathbb{A} \to \mathscr{A}$ is a dense subspace of $\mathscr{A}$. We call $\mathscr{A}$ the *left linear filterwise completion* of $\mathbb{A}$.

Now consider the right linear topology on $\mathbb{A}$ and the corresponding filterwise completion

$$\mathscr{A}^\vee = \varinjlim_p F_p^\vee \mathscr{A}^\vee, \quad F_p^\vee \mathscr{A}^\vee = \varprojlim_n F_p^\vee \mathbb{A}/F_p^\vee \mathbb{A} \cap I_n^\vee(\mathbb{A}). \tag{1.5.4}$$

We will call $\mathscr{A}^\vee$ the *right linear filterwise completion* of $\mathbb{A}$.

If $\mathbb{A}$ is degreewise complete then the canonical maps $\mathbb{A} \to \mathscr{A}$ and $\mathbb{A} \to \mathscr{A}^\vee$ are injective. We will then identify $\mathbb{A}$ with its images. Thus we have two inclusions

$$\mathscr{A} \leftarrow \mathbb{A} \to \mathscr{A}^\vee \tag{1.5.5}$$

such that the relative topologies on each $\mathbb{A}(d)$ induced from $\mathscr{A}$ and from $\mathscr{A}^\vee$ agree with the original topology of a compatible degreewise topological algebra. Note that the filterwise completions $\mathscr{A}$ and $\mathscr{A}^\vee$ are *not* complete in general.

1.6. Hamiltonian of a graded algebra

Let $\mathbb{A}$ be a graded algebra. An element $h \in \mathbb{A}$ is called a *Hamiltonian* of $\mathbb{A}$ if

$$\mathbb{A}(d) = \{ a \in \mathbb{A} \mid [h, a] = da \} \tag{1.6.1}$$

holds for any d, where $[h, a] = h \cdot a - a \cdot h$ denotes the commutator. If this is the case then any central element as well as h itself belongs to $\mathbb{A}(0)$ and an element h' is a Hamiltonian if and only if $h - h'$ is central.

An *algebra graded by Hamiltonian* is a pair $(\mathbb{A}, h)$ of a graded algebra $\mathbb{A}$ and a Hamiltonian $h \in \mathbb{A}(0)$. We will denote by H the subalgebra of $\mathbb{A}(0)$ generated by the Hamiltonian h.

Remark 1.6.1 Let $(\mathbb{A}, h)$ be a compatible degreewise complete algebra graded by Hamiltonian. Then the images of the canonical injections (1.5.5) are given respectively by $\sum_d \mathscr{A}(d)$ and $\sum_d \mathscr{A}^\vee(d)$, where

$$\mathscr{A}(d) = \{ a \in \mathscr{A} \mid [h, a] = da \}, \quad \mathscr{A}^\vee(d) = \{ a \in \mathscr{A}^\vee \mid [h, a] = da \}. \tag{1.6.2}$$

In the rest of Part I, we will be mainly concerned with a compatible degreewise complete algebra graded by Hamiltonian.

2. Quasi-finite algebras

In this section, we will formulate a finiteness condition on compatible degreewise complete algebras and describe its consequences. In particular, we will formulate an analogue of the regular bimodule as a degreewise dense subspace of the algebra.

2.1. Canonical quotient modules

Let $\mathbb{A}$ be a compatible degreewise topological algebra and recall the spaces $I_n(\mathbb{A})$. We set

$$Q_n(\mathbb{A}) = \mathbb{A}/I_n(\mathbb{A}). \tag{2.1.1}$$

Since $I_n(\mathbb{A})$ is a left ideal, the quotient $Q_n(\mathbb{A})$ is a left $\mathbb{A}$-module. We will call this module the *left canonical quotient module*.

By (1.2.4), we have $I_n(\mathbb{A}) \cdot F_0\mathbb{A} \subset I_n(\mathbb{A})$. Hence the multiplication $\mathbb{A} \times F_0\mathbb{A} \to \mathbb{A}$ induces a right action of $F_0\mathbb{A}$ on $Q_n(\mathbb{A})$ for any n. Thus the space $Q_n(\mathbb{A})$ is an $(\mathbb{A}, F_0\mathbb{A})$-bimodule. In particular, it is an $(\mathbb{A}(0), \mathbb{A}(0))$-bimodule.

Since $I_n(\mathbb{A})$ is a graded subspace, the grading of $\mathbb{A}$ induces a grading on the quotient by setting $Q_n(\mathbb{A})(d) = \mathbb{A}(d)/I_n(\mathbb{A}(d))$. If $d \leq -n - 1$ then since $\mathbb{A}(d) \subset I_n(\mathbb{A})$ we have $Q_n(\mathbb{A})(d) = 0$. Thus

$$Q_n(\mathbb{A}) = \bigoplus_{d=-n}^{\infty} Q_n(\mathbb{A})(d). \tag{2.1.2}$$

Lemma 2.1.1 *For any $v \in Q_n(\mathbb{A})$ there exists an m such that $I_m(\mathbb{A}) \cdot v = 0$.*

Proof. By (1.2.4) we have $I_{n+d}(\mathbb{A}) \cdot \mathbb{A}(d) \subset I_n(\mathbb{A})$ and hence $I_m(\mathbb{A}) \cdot Q_n(\mathbb{A})(d) = 0$ for $m = n + d$. $\qquad\square$

Note that the action $\mathbb{A} \times Q_n(\mathbb{A}) \to Q_n(\mathbb{A})$ is continuous when $\mathbb{A}$ is endowed with the left linear topology and $Q_n(\mathbb{A})$ with the discrete topology.

Lemma 2.1.2 *Let v be an element of $Q_n(\mathbb{A})$. Then $F_{-n-1}\mathbb{A} \cdot v = 0$ implies $I_n(\mathbb{A}) \cdot v = 0$.*

Proof. Suppose $F_{-n-1}\mathbb{A} \cdot v = 0$ and choose an m such that $I_m(\mathbb{A}) \cdot v = 0$. Since $I_n(\mathbb{A})$ is the closure of $\mathbb{A} \cdot F_{-n-1}\mathbb{A}$, we have $I_n(\mathbb{A}) \subset \mathbb{A} \cdot F_{-n-1}\mathbb{A} + I_m(\mathbb{A})$. Hence $I_n(\mathbb{A}) \cdot v \subset \mathbb{A} \cdot F_{-n-1}\mathbb{A} \cdot v + I_m(\mathbb{A}) \cdot v = 0$. $\qquad\square$

We may likewise consider the right canonical quotient module $Q_n^\vee(\mathbb{A}) = \mathbb{A}/I_n^\vee(\mathbb{A})$. We have analogous statements for the right canonical quotient modules as well. By the definitions of the filterwise completions $\mathscr{A}$ and $\mathscr{A}^\vee$, the spaces $Q_n(\mathbb{A})$ and $Q_n^\vee(\mathbb{A})$ are canonically isomorphic to the spaces $\mathscr{A}/I_n(\mathscr{A})$ and $\mathscr{A}^\vee/I_n^\vee(\mathscr{A}^\vee)$, respectively.

2.2. Quasi-finite algebra graded by Hamiltonian

Let $\mathbb{A}$ be a compatible degreewise topological algebra. Let us consider the subspace $\mathbb{A}(0)$ of degree zero, which is a subalgebra of $\mathbb{A}$. Then $I_n(\mathbb{A}(0))$ and $I_n^\vee(\mathbb{A}(0))$ agree with each other and they give a two-sided ideal of $\mathbb{A}(0)$. Therefore, the quotient $Q_n(\mathbb{A})(0) = \mathbb{A}(0)/I_n(\mathbb{A}(0))$ is an algebra.

Lemma 2.2.1   *The space $Q_n(\mathbb{A})(d)$ is a $(Q_{n+d}(\mathbb{A})(0), Q_n(\mathbb{A})(0))$-bimodule.*

Proof. By (1.2.4), we have $I_{n+d}(\mathbb{A}(0)) \cdot \mathbb{A}(d) \subset I_n(\mathbb{A}(d))$ and $\mathbb{A}(d) \cdot I_n(\mathbb{A}(0)) \subset I_n(\mathbb{A}(d))$. The result follows. $\square$

Let h be a Hamiltonian and let h_n be the image of h in the quotient $Q_n(\mathbb{A})(0)$. Let H_n be the subalgebra of $Q_n(\mathbb{A})(0)$ generated by h_n. By Lemma 2.2.1, the space $Q_n(\mathbb{A})(d)$ is in particular an (H_{n+d}, H_n)-bimodule.

Definition 2.2.2   A *weakly quasi-finite algebra* is a compatible degreewise complete algebra such that $Q_n(\mathbb{A})(0)$ are finite-dimensional for all n. A *weakly quasi-finite algebra graded by Hamiltonian* is a pair $(\mathbb{A}, h)$ of a weakly quasi-finite algebra $\mathbb{A}$ and a Hamiltonian h of $\mathbb{A}$.

Let $(\mathbb{A}, h)$ be a weakly quasi-finite algebra graded by Hamiltonian and let h_n and H_n be as above. Then H_n is a finite-dimensional commutative algebra for any nonnegative integer n. This last property is what we will need in practice in considering a weakly quasi-finite algebra graded by Hamiltonian.

Now let us consider the following conditions for each d:

(a) The spaces $Q_n(\mathbb{A})(d)$ are finite-dimensional for all n.
(b) The spaces $Q_n^\vee(\mathbb{A})(d)$ are finite-dimensional for all n.

Thanks to the relation (1.2.6), these conditions are actually equivalent.

Definition 2.2.3   A *quasi-finite algebra* is a compatible degreewise complete algebra such that the equivalent conditions (a) and (b) above are satisfied for all integers d. A *quasi-finite algebra graded by Hamiltonian* is a pair $(\mathbb{A}, h)$ of a quasi-finite algebra $\mathbb{A}$ and a Hamiltonian h of $\mathbb{A}$.

2.3. Spectrum of the Hamiltonian

Let $(\mathbb{A}, h)$ be a weakly quasi-finite algebra graded by Hamiltonian and recall the image h_n of h in $Q_n(\mathbb{A})(0)$. We let φ_n be the minimal polynomial of h_n and let Ω_n be the set of roots of φ_n. Then the set Ω_n agrees with the eigenvalues of the left action of h on $Q_n(\mathbb{A})(0)$.

Let us introduce a partial order on **k** by letting $\lambda \succcurlyeq \mu$ when $\lambda - \mu$ is a nonnegative integer and let Γ_0 be the set of minimal elements of Ω_0. We put

$$\Gamma_\infty = \left\{ \gamma + k \mid \gamma \in \Gamma_0 \text{ and } k = 0, 1, \ldots \right\}. \tag{2.3.1}$$

We will also use the following notation:

$$\Gamma_m = \left\{ \gamma + k \,\middle|\, \gamma \in \Gamma_0 \text{ and } k = 0, 1, \ldots, m \right\}. \tag{2.3.2}$$

Let g be the maximum of the integral differences among elements of Ω_0:

$$g = \max \left\{ \lambda - \mu \mid \lambda, \mu \in \Omega_0, \lambda \succcurlyeq \mu \right\}. \tag{2.3.3}$$

Then we have $\Gamma_0 \subset \Omega_0 \subset \Gamma_g$.

Recall the subalgebras $H \subset \mathbb{A}(0)$ and $H_n \subset Q_n(\mathbb{A})(0)$. For any left H-module W, we denote by $W[\lambda]$ the generalized eigenspace of the Hamiltonian h acting on W:

$$W[\lambda] = \left\{ v \in W \,\middle|\, (h - \lambda)^r \cdot v = 0 \text{ for some nonnegative integer } r \right\}. \tag{2.3.4}$$

If W is an H_n-module then $W[\lambda] \neq 0$ implies $\lambda \in \Omega_n$.

Now consider the space

$$K_m(Q_n(\mathbb{A})) = \left\{ v \in Q_n(\mathbb{A}) \,\middle|\, I_m(\mathbb{A}) \cdot v = 0 \right\}. \tag{2.3.5}$$

Then this is a left $Q_m(\mathbb{A})(0)$-module and hence a left H_m-module. Thus the set of the eigenvalues of the left action of h on $K_m(Q_n(\mathbb{A}))$ is contained in the set Ω_m. Note that $v \in K_m(Q_n(\mathbb{A}))$ if and only if $F_{-m-1}\mathbb{A} \cdot v = 0$ by Lemma 2.1.2.

Proposition 2.3.1 *The set Ω_n is a subset of Γ_{n+g}.*

Proof. Let λ be an element of Ω_n. Then there exists a generalized eigenvector v in $Q_n(\mathbb{A})(0) \subset K_n(Q_n(\mathbb{A}))$ with the eigenvalue λ. Since $v \neq 0$ and $F_{-n-1}\mathbb{A} \cdot v = 0$, there exists a nonnegative integer k with $0 \leq k \leq n$ for which $v \notin K_{k-1}(Q_n(\mathbb{A}))$ but $v \in K_k(Q_n(\mathbb{A}))$. Then $\mathbb{A}(-k) \cdot v$ is a nonzero subspace of $K_0(Q_n(\mathbb{A}))$. Since $\mathbb{A}(-k) \cdot v \subset Q_n(\mathbb{A})[\lambda - k]$, we have $\lambda - k \in \Omega_0$ and hence $\lambda \in \Gamma_{k+g} \subset \Gamma_{n+g}$. $\qquad\square$

Now Lemma 2.1.1 implies $Q_n(\mathbb{A}) = \bigcup_{m=0}^{\infty} K_m(Q_n(\mathbb{A}))$. Therefore, Proposition 2.3.1 implies that

$$Q_n(\mathbb{A}) = \bigoplus_{\lambda \in \Gamma_\infty} Q_n(\mathbb{A})[\lambda]. \tag{2.3.6}$$

Lemma 2.3.2 *The space $Q_n(\mathbb{A})[\lambda]$ with $\lambda \in \Gamma_m$ is a subspace of $K_m(Q_n(\mathbb{A}))$.*

Proof. Since $\lambda - m - 1 \notin \Gamma_\infty$, we have $F_{-m-1}\mathbb{A} \cdot Q_n(\mathbb{A})[\lambda] = 0$. Hence $Q_n(\mathbb{A})[\lambda] \subset K_m(Q_n(\mathbb{A}))$. $\qquad\square$

Proposition 2.3.3 *The multiplicities of the roots of the minimal polynomial φ_n do not exceed those of φ_g for any nonnegative integer n.*

Proof. Let ℓ be the maximum of the multiplicities of the roots of φ_g and let v be an element of $Q_n(\mathbb{A})$. We show that $(h - \lambda)^k \cdot v = 0$ for some k implies $(h - \lambda)^\ell \cdot v = 0$. Let λ be a minimal counterexample to this claim: there exists a nonzero vector v such that $(h - \lambda)^k \cdot v = 0$ for some k but $(h - \lambda)^\ell \cdot v \neq 0$. Then, by the minimality, we have $F_{-1}\mathbb{A} \cdot (h - \lambda)^\ell \cdot v = 0$ and hence $(h - \lambda)^\ell \cdot v \in K_0(Q_n(\mathbb{A}))$ by Lemma 2.1.2. Hence $\lambda \in \Omega_0 \subset \Gamma_g$ and so $v \in Q_n(\mathbb{A})[\lambda] \subset K_g(Q_n(\mathbb{A}))$ by Lemma 2.3.2. Hence $(h - \lambda)^\ell \cdot v = 0$, a contradiction. $\qquad\qquad\square$

2.4. The regular bimodule

Let $(\mathbb{A}, h)$ be a weakly quasi-finite algebra graded by Hamiltonian. We denote by $\mathbb{A}[\lambda, \mu]$ the simultaneous generalized eigenspace of the left and the right actions of the Hamiltonian h on $\mathbb{A}$:

$$\mathbb{A}[\lambda, \mu] = \left\{ a \in \mathbb{A} \mid (h - \lambda)^r \cdot a = a \cdot (h - \mu)^r = 0 \text{ for some } r \right\}. \quad (2.4.1)$$

We will call this λ the *left eigenvalue* and μ the *right eigenvalue* of h.

By the definition, $\mathbb{A}[\lambda, \mu] \neq 0$ implies that $d = \lambda - \mu$ is an integer and $\mathbb{A}[\lambda, \mu] \subset \mathbb{A}(d)$. We also note that

$$\mathbb{A}(d) \cdot \mathbb{A}[\lambda, \mu] \subset \mathbb{A}[\lambda + d, \mu], \quad \mathbb{A}[\lambda, \mu] \cdot \mathbb{A}(e) \subset \mathbb{A}[\lambda, \mu - e]. \quad (2.4.2)$$

It is easy to see that

$$\mathbb{A}[\kappa, \lambda] \cdot \mathbb{A}[\mu, \nu] = 0 \text{ if } \lambda \neq \mu. \quad (2.4.3)$$

Indeed, if $a \cdot (h - \lambda)^k = 0$ and $(h - \mu)^m \cdot b = 0$ with $\lambda \neq \mu$ then $a \cdot b = a \cdot 1 \cdot b = a \cdot (h - \lambda)^k f(h) \cdot b + a \cdot (h - \mu)^m g(h) \cdot b = 0$ for some polynomials $f(x)$ and $g(x)$. Also note that

$$\mathbb{A}[\kappa, \lambda] \cdot \mathbb{A}[\lambda, \mu] \subset \mathbb{A}[\kappa, \mu]. \quad (2.4.4)$$

We now set

$$B_\infty(\mathbb{A}) = \sum_d B_\infty(\mathbb{A})(d), \quad B_\infty(\mathbb{A})(d) = \sum_{\lambda - \mu = d} \mathbb{A}[\lambda, \mu]. \quad (2.4.5)$$

Then by (2.4.2) this is an $(\mathbb{A}, \mathbb{A})$-bimodule. We call $B_\infty(\mathbb{A})$ the *regular bimodule* of $\mathbb{A}$.

Remark 2.4.1 The structure of an $(\mathbb{A}, \mathbb{A})$-bimodule on $B_\infty(\mathbb{A})$ actually prolongs to a structure of an $(\mathscr{A}, \mathscr{A}^\vee)$-bimodule. (See Proposition 3.1.4.)

2.5. Denseness of the regular bimodule

To investigate the spaces $\mathbb{A}[\lambda, \mu]$ we consider the left canonical quotient module $Q_n(\mathbb{A})$, which is an $(\mathbb{A}, F_0\mathbb{A})$-bimodule. Consider the simultaneous generalized eigenspaces:

$$Q_n(\mathbb{A})[\lambda, \mu] = \left\{ v \in Q_n(\mathbb{A}) \mid (h - \lambda)^r \cdot v = v \cdot (h - \mu)^r = 0 \text{ for some } r \right\}. \tag{2.5.1}$$

Then $Q_n(\mathbb{A})[\lambda, \mu] \neq 0$ implies that $d = \lambda - \mu$ is an integer and that $Q_n(\mathbb{A})[\lambda, \mu] \subset Q_n(\mathbb{A})(d)$.

Recall that the space $Q_n(\mathbb{A})(d)$ is a $(Q_{n+d}(\mathbb{A})(0), Q_n(\mathbb{A})(0))$-bimodule. In particular, it is an (H_{n+d}, H_n)-bimodule. Hence we have

$$Q_n(\mathbb{A})(d) = \sum_{\mu \in \Omega_n} Q_n(\mathbb{A})[\mu + d, \mu]. \tag{2.5.2}$$

Note that $\mathbb{A}(d) \cdot Q_n(\mathbb{A})[\lambda, \mu] \subset Q_n(\mathbb{A})[\lambda+d, \mu]$ and that $Q_n(\mathbb{A})[\lambda, \mu] \cdot \mathbb{A}(e) \subset Q_n(\mathbb{A})[\lambda, \mu - e]$ whenever $e \leq 0$.

Lemma 2.5.1 *If* $\mathbb{A}[\lambda, \mu] \neq 0$ *then* $\lambda, \mu \in \Gamma_\infty$.

Proof. Let a be a nonzero element of $\mathbb{A}[\lambda, \mu]$ with $d = \lambda - \mu$. Since we have assumed that $\mathbb{A}(d)$ is complete, it is separated. Hence there exists an n such that the image v of a in $Q_n(\mathbb{A})(d)$ is nonzero. Then since $v \in Q_n(\mathbb{A})[\lambda, \mu]$, we have $\lambda \in \Gamma_\infty$ and $\mu \in \Gamma_n \subset \Gamma_\infty$. $\square$

Lemma 2.5.2 *If* $\mu \in \Gamma_m$ *and* $n \geq m$ *then the restriction* $Q_n(\mathbb{A})[\lambda, \mu] \to Q_m(\mathbb{A})[\lambda, \mu]$ *of the canonical surjection is an isomorphism for any* λ.

Proof. Set $d = \lambda - \mu$. Consider the kernel $I_m(\mathbb{A})/I_n(\mathbb{A})$ of the canonical surjection $Q_n(\mathbb{A}) \to Q_m(\mathbb{A})$. Consider the map $\mathbb{A} \times F_{-m-1}\mathbb{A} \to \mathbb{A}$ which induces $\mathbb{A} \times F_{-m-1}\mathbb{A} \to I_m(\mathbb{A})/I_n(\mathbb{A})$. Since $I_m(\mathbb{A})$ is the closure of $\mathbb{A} \cdot F_{-m-1}\mathbb{A}$, this map is surjective. Hence the induced map $Q_n(\mathbb{A}) \times F_{-m-1}\mathbb{A} \to I_m(\mathbb{A})/I_n(\mathbb{A})$ is also surjective. The right eigenvalues of h on $Q_n(\mathbb{A}) \cdot F_{-m-1}\mathbb{A}$ exceed those of $Q_n(\mathbb{A})$ by more than m. Hence it follows that the set of the right eigenvalues on the space $I_m(\mathbb{A})/I_n(\mathbb{A})$ does not intersect Γ_m. This implies the result. $\square$

By this lemma, we have the following result.

Proposition 2.5.3 *Let* $(\mathbb{A}, h)$ *be a weakly quasi-finite algebra graded by Hamiltonian and let m be a nonnegative integer. Then the canonical surjection* $\mathbb{A}[\lambda, \mu] \to Q_n(\mathbb{A})[\lambda, \mu]$ *is an isomorphism whenever* $\mu \in \Gamma_m$ *and* $n \geq m$.

Proof. By Lemma 2.5.2, we have a canonical splitting

$$Q_m(\mathbb{A})[\lambda, \mu] \to \varprojlim_n Q_n(\mathbb{A})[\lambda, \mu]. \tag{2.5.3}$$

Set $d = \lambda - \mu$. Since $\mathbb{A}(d)$ is complete, the projective limit is isomorphic to the subspace $\mathbb{A}[\lambda, \mu]$ of $\mathbb{A}(d)$. □

The following is the first main result of Part I.

Theorem 2.5.4 *Let $(\mathbb{A}, h)$ be a weakly quasi-finite algebra graded by Hamiltonian. Then the regular bimodule $B_\infty(\mathbb{A})$ is degreewise dense in $\mathbb{A}$.*

Proof. By Proposition 2.5.3, the map $B_\infty(\mathbb{A})(d) \to Q_n(\mathbb{A})(d) = \mathbb{A}(d)/I_n(\mathbb{A}(d))$ is surjective for any n. This means that $B_\infty(\mathbb{A})(d)$ is dense in $\mathbb{A}(d)$. □

Note that the whole space $B_\infty(\mathbb{A})$ is dense in $\mathbb{A}$ with respect to the left linear and the right linear topologies of $\mathbb{A}$.

2.6. The associated finite algebras

Let $(\mathbb{A}, h)$ be a weakly quasi-finite algebra graded by Hamiltonian. We set

$$\mathbb{A}_n = B_n(\mathbb{A}) = \sum_{\lambda, \mu \in \Gamma_n} \mathbb{A}[\lambda, \mu]. \tag{2.6.1}$$

Then it follows from (2.4.4) that the space $\mathbb{A}_n$ is closed under the multiplication of $\mathbb{A}$. Setting $\mathbb{A}_n(d) = \mathbb{A}(d) \cap \mathbb{A}_n$, we have $\mathbb{A}_n = \bigoplus_d \mathbb{A}_n(d)$.

By Proposition 2.5.3, we may identify $\mathbb{A}_n(0)$ with a subspace of $Q_n(\mathbb{A})(0)$. Recall the elements 1_n and h_n of $Q_n(\mathbb{A})(0)$, which are the images of 1 and h under the map $\mathbb{A}(0) \to Q_n(\mathbb{A})(0)$, respectively. Let

$$1_n = \sum_{\lambda \in \Omega_n} 1_n[\lambda, \lambda], \quad h_n = \sum_{\lambda \in \Omega_n} h_n[\lambda, \lambda] \tag{2.6.2}$$

be the decompositions to sums of simultaneous generalized eigenvectors. We set

$$1_{\mathbb{A}_n} = \sum_{\lambda \in \Gamma_n} 1_n[\lambda, \lambda], \quad h_{\mathbb{A}_n} = \sum_{\lambda \in \Gamma_n} h_n[\lambda, \lambda], \tag{2.6.3}$$

and regard them as elements of $\mathbb{A}_n$.

Proposition 2.6.1 *The graded algebra structure on $\mathbb{A}$ induces a graded algebra structure on $\mathbb{A}_n$ for which $1_{\mathbb{A}_n}$ is the unity and $h_{\mathbb{A}_n}$ is a Hamiltonian.*

We set

$$P_n(\mathbb{A}) = \sum_{\lambda \in \Gamma_\infty} \sum_{\mu \in \Gamma_n} \mathbb{A}[\lambda, \mu], \quad P_m^\vee(\mathbb{A}) = \sum_{\lambda \in \Gamma_m} \sum_{\mu \in \Gamma_\infty} \mathbb{A}[\lambda, \mu]. \qquad (2.6.4)$$

Then $P_n(\mathbb{A})$ is an $(\mathbb{A}, \mathbb{A}_n)$-bimodule and $P_m^\vee(\mathbb{A})$ is an $(\mathbb{A}_m, \mathbb{A})$-bimodule, which will play prominent roles in the next section.

Remark 2.6.2   The structure of an $(\mathbb{A}, \mathbb{A}_n)$-bimodule on $P_n(\mathbb{A})$ prolongs to a structure of an $(\mathscr{A}, \mathbb{A}_n)$-bimodule and the structure of an $(\mathbb{A}_m, \mathbb{A})$-bimodule on $P_m^\vee(\mathbb{A})$ to an $(\mathbb{A}_m, \mathscr{A}^\vee)$-bimodule. (See Remark 2.4.1 and Proposition 3.1.4.)

Let us consider the case when $\mathbb{A}$ is quasi-finite.

Proposition 2.6.3 *Let* $(\mathbb{A}, h)$ *be a quasi-finite algebra graded by Hamiltonian. Then* $\mathbb{A}_n$ *are finite-dimensional for all n.*

Proof. For any $\lambda, \mu \in \Gamma_n$, the space $\mathbb{A}[\lambda, \mu]$ is isomorphic to $Q_n(\mathbb{A})[\lambda, \mu]$ by Proposition 2.5.3. Since the range Γ_n of λ and μ is finite and $Q_n(\mathbb{A})[\lambda, \mu] \subset Q_n(\mathbb{A})(\lambda - \mu)$ is finite-dimensional, the space $\mathbb{A}_n$ is also finite-dimensional. $\qquad\qquad\square$

3. Exhaustive modules

We will define the notion of exhaustive modules and investigate the properties of the category of such modules over a weakly quasi-finite algebra $\mathbb{A}$ graded by Hamiltonian. In particular, we will show that the category of exhaustive $\mathbb{A}$-modules and the category of $\mathbb{A}_n$-modules are equivalent if $n \geq g$.

3.1. Exhaustive modules

Let $(\mathbb{A}, h)$ be a weakly quasi-finite algebra graded by Hamiltonian and endow it with the left linear topology.

Definition 3.1.1   A left $\mathbb{A}$-module M is *exhaustive* if for any $v \in M$ there exists an m such that $I_m(\mathbb{A}) \cdot v = 0$.

By this definition, it is easy to see that a left $\mathbb{A}$-module M is exhaustive if and only if the action $\mathbb{A} \times M \to M$ is continuous with respect to the left linear topology on $\mathbb{A}$ and the discrete topology on M. Note, however, that a topology on an exhaustive left $\mathbb{A}$-module M need *not* be the discrete topology in order for the action $\mathbb{A} \times M \to M$ to be continuous with respect to the left linear topology on $\mathbb{A}$.

Lemma 3.1.2 *Let M be an exhaustive left $\mathbb{A}$-module and let v be an element of M. Then $F_{-n-1}\mathbb{A} \cdot v = 0$ implies $I_n(\mathbb{A}) \cdot v = 0$.*

Proof. See the proof of Lemma 2.1.2.   □

Let us now consider the left linear filterwise completion $\mathscr{A}$. We may analogously define the notion of exhaustive left $\mathscr{A}$-modules as follows.

Definition 3.1.3   A left $\mathscr{A}$-module M is *exhaustive* if for any $v \in M$ there exists an m such that $I_m(\mathscr{A}) \cdot v = 0$.

It is fairly clear that results similar to those given above hold for exhaustive left $\mathscr{A}$-modules. The following proposition is a particular case of a general fact on topological algebras. (See e.g. [Bou].)

Proposition 3.1.4   *For any exhaustive left $\mathbb{A}$-module M, the action $\mathbb{A} \times M \to M$ induces an exhaustive left $\mathscr{A}$-module structure $\mathscr{A} \times M \to M$. Conversely, for any exhaustive left $\mathscr{A}$-module M, the action $\mathscr{A} \times M \to M$ restricts to an exhaustive left $\mathbb{A}$-module structure.*

Thus the notion of exhaustive left $\mathbb{A}$-modules and that of exhaustive left $\mathscr{A}$-modules have no essential differences.

Note 3.1.5   An exhaustive module is nothing else but a torsion module with respect to the left linear topology. (See [Gab].)

3.2. Generalized eigenspaces of exhaustive modules

Let $\mathbb{A}$ be a compatible degreewise topological algebra. For a left $\mathbb{A}$-module M and a nonnegative integer n, we set

$$K_n(M) = \left\{ v \in M \,\middle|\, I_n(\mathbb{A}) \cdot v = 0 \right\}. \tag{3.2.1}$$

Note that M is exhaustive if and only if $M = \bigcup_n K_n(M)$ and if this is the case then $K_n(M) = \left\{ v \in M \,\middle|\, F_{-n-1}\mathbb{A} \cdot v = 0 \right\}$ by Lemma 3.1.2.

Consider the case when $(\mathbb{A}, h)$ is a weakly quasi-finite algebra graded by Hamiltonian. Since the space $K_n(M)$ has a structure of a left $Q_n(\mathbb{A})(0)$-module, M decomposes into the sum of the generalized eigenspaces of the left action of h:

$$K_n(M) = \sum_{\lambda \in \Omega_n} K_n(M)[\lambda]. \tag{3.2.2}$$

Hence if M is exhaustive then the whole space M also decomposes into the sum of the generalized eigenspaces.

Let us set

$$E_n(M) = \sum_{\lambda \in \Gamma_n} M[\lambda]. \tag{3.2.3}$$

The following is one of the key observations in the present paper.

Proposition 3.2.1 *If M is exhaustive then $E_n(M) \subset K_n(M) \subset E_{n+g}(M)$.*

Proof. The containment $K_n(M) \subset E_{n+g}(M)$ follows from $\Omega_n \subset \Gamma_{n+g}$ by (3.2.1). The rest is similar to Lemma 2.3.2. Consider the space $M[\lambda]$ with $\lambda \in \Gamma_n$. Then we have $\mathbb{A}(d) \cdot M[\lambda] \subset M[\lambda + d]$. Therefore, since $\lambda - n - 1 \notin \Gamma_\infty$, we have $F_{-n-1}\mathbb{A} \cdot M[\lambda] = 0$. This implies that $E_n(M) \subset K_n(M)$. $\square$

Now consider the space $E_n^\perp(M) = \sum_{\lambda \in \Gamma_\infty \setminus \Gamma_n} M[\lambda]$ and set

$$R_n(M) = M/E_n^\perp(M). \tag{3.2.4}$$

Then the canonical map $E_n(M) \to R_n(M)$ is an isomorphism for each n.

3.3. Equivalence of categories

We will mean by the category of exhaustive left $\mathbb{A}$-modules the full sub-category of the category of left $\mathbb{A}$-modules consisting of exhaustive left $\mathbb{A}$-modules.

Let M be an exhaustive left $\mathbb{A}$-module. Then the space $E_n(M)$ is a left $\mathbb{A}_n$-module. A homomorphism $\phi : M' \to M''$ of left $\mathbb{A}$-modules induces a map

$$E_n(\phi) : E_n(M') \to E_n(M''). \tag{3.3.1}$$

Thus we have a functor $E_n(-)$ from the category of exhaustive left $\mathbb{A}$-modules to the category of left $\mathbb{A}_n$-modules.

Lemma 3.3.1 *If M is exhaustive then $K_0(M) = 0$ implies $M = 0$.*

Proof. Suppose $K_0(M) = 0$ and $M \neq 0$. Let v be a nonzero element of M. Since M is exhaustive, $I_n(\mathbb{A}) \cdot v = 0$ and hence $F_{-n-1}\mathbb{A} \cdot v = 0$ for sufficiently large n. Take the maximal integer n for which $F_{-n-1}\mathbb{A} \cdot v \neq 0$. Then by the maximality $F_{-1}\mathbb{A} \cdot (F_{-n-1}\mathbb{A} \cdot v) = 0$ and hence $I_0(\mathbb{A}) \cdot (F_{-n-1}\mathbb{A} \cdot v) = 0$ by Lemma 3.1.2. Hence $F_{-n-1}\mathbb{A} \cdot v \subset K_0(M) = 0$ which is a contradiction. $\square$

Lemma 3.3.2 *A homomorphism $\phi : M' \to M''$ of exhaustive left $\mathbb{A}$-modules is an isomorphism if and only if $E_g(\phi)$ is an isomorphism of vector spaces.*

Proof. Assume that $E_g(\phi)$ is an isomorphism of vector spaces. Then, by Lemma 3.3.1, we see $\mathrm{Ker}\,\phi = 0$ and $\mathrm{Coker}\,\phi = 0$ since $K_0(\mathrm{Ker}\,\phi) \subset E_g(\mathrm{Ker}\,\phi) = \mathrm{Ker}(E_g(\phi)) = 0$ and $K_0(\mathrm{Coker}\,\phi) \subset E_g(\mathrm{Coker}\,\phi) = \mathrm{Coker}(E_g(\phi)) = 0$. $\square$

Recall the $(\mathbb{A}, \mathbb{A}_n)$-bimodule $P_n(\mathbb{A})$ defined in Subsection 2.6, which is an exhaustive left $\mathbb{A}$-module. Therefore, for a left $\mathbb{A}_n$-module X, the space $P_n(\mathbb{A}) \otimes_{\mathbb{A}_n} X$ is an exhaustive left $\mathbb{A}$-module. We note that $\mathbb{A}_n = E_n(P_n(\mathbb{A}))$.

Lemma 3.3.3 *For a left $\mathbb{A}_n$-module X, the map $\mathbb{A}_n \otimes_{\mathbb{A}_n} X \to P_n(\mathbb{A}) \otimes_{\mathbb{A}_n} X$ induced by the inclusion $\mathbb{A}_n \to P_n(\mathbb{A})$ is injective.*

Proof. We set $E_n^\perp(P_n(\mathbb{A})) = \sum_{\lambda \in \Gamma_\infty \backslash \Gamma_n} P_n(\mathbb{A})[\lambda]$. Then the decomposition $P_n(\mathbb{A}) = E_n(P_n(\mathbb{A})) \oplus E_n^\perp(P_n(\mathbb{A})) = \mathbb{A}_n \oplus E_n^\perp(P_n(\mathbb{A}))$ is a direct sum decomposition of a right $\mathbb{A}_n$-module. Hence the map $E_n(P_n(\mathbb{A})) \otimes_{\mathbb{A}_n} X \to P_n(\mathbb{A}) \otimes_{\mathbb{A}_n} X$ is injective. $\square$

Now we come to the second main result of Part I. Recall the number g defined by (2.3.3).

Theorem 3.3.4 *Let $(\mathbb{A}, h)$ be a weakly quasi-finite algebra graded by Hamiltonian and let n be an integer such that $n \geq g$. Then the functors $E_n(-)$ and $P_n(\mathbb{A}) \otimes_{\mathbb{A}_n} -$ are mutually inverse equivalences of categories between the category of exhaustive left $\mathbb{A}$-modules and the category of left $\mathbb{A}_n$-modules.*

Proof. Let X be a left $\mathbb{A}_n$-module. By Lemma 3.3.3, the map $\mathbb{A}_n \otimes_{\mathbb{A}_n} X \to P_n(\mathbb{A}) \otimes_{\mathbb{A}_n} X$ is injective. Therefore

$$X \cong \mathbb{A}_n \otimes_{\mathbb{A}_n} X \cong E_n(P_n(\mathbb{A}) \otimes_{\mathbb{A}_n} X). \tag{3.3.2}$$

Let M be an exhaustive left $\mathbb{A}$-module. Then, by letting $X = \mathrm{E}_n(M)$ in (3.3.2), we have $\mathrm{E}_n(M) \cong \mathrm{E}_n(\mathrm{P}_n(\mathbb{A}) \otimes_{\mathbb{A}_n} \mathrm{E}_n(M))$. Since $n \geq g$, we have $\mathrm{P}_n(\mathbb{A}) \otimes_{\mathbb{A}_n} \mathrm{E}_n(M) \cong M$ by Lemma 3.3.2. $\qquad\square$

Corollary 3.3.5 *Let* $(\mathbb{A}, h)$ *be a weakly quasi-finite algebra graded by Hamiltonian and let n be an integer such that $n \geq g$. Then the module $\mathrm{P}_n(\mathbb{A})$ is a progenerator of the category of exhaustive left $\mathbb{A}$-modules.*

Let us consider the case when $\mathbb{A}$ is quasi-finite. Then the algebra $\mathbb{A}_n$ is finite-dimensional for all n by Proposition 2.6.3. Thus we have the following corollary.

Corollary 3.3.6 *If $(\mathbb{A}, h)$ is a quasi-finite algebra graded by Hamiltonian then the category of exhaustive left $\mathbb{A}$-modules is equivalent to the category of left modules over a finite-dimensional algebra.*

We may likewise define the notion of exhaustive right $\mathbb{A}$-modules and exhaustive right $\mathscr{A}^{\vee}$-modules by using the spaces $\mathrm{I}_n^{\vee}(\mathbb{A})$ and $\mathrm{I}_n^{\vee}(\mathscr{A}^{\vee})$, respectively. We have analogous results for the right modules as well.

Note 3.3.7 Theorem 3.3.4 may be viewed as a particular case of a topological variant of Morita equivalences. See [Gab] for general theory of equivalences between abelian categories and module categories and [Gre] and [Mez] for results closely related to Theorem 3.3.4.

4. Coexhaustive modules

We will now consider the notion of coexhaustive modules, which is *dual* to that of exhaustive modules. In this section, we will give the definition of coexhaustive modules and describe their topologies by means of the generalized eigenspaces. The precise duality will be considered in the next section under necessary finiteness assumptions.

4.1. Coexhaustive modules

Let $(\mathbb{A}, h)$ be a weakly quasi-finite algebra graded by Hamiltonian and let $\mathscr{A}^{\vee}$ be the right filterwise completion.

Let $\mathcal{M}$ be a left $\mathscr{A}^{\vee}$-module endowed with a linear topology such that the action $\mathscr{A}^{\vee} \times \mathcal{M} \to \mathcal{M}$ is continuous. Let us denote by $\mathrm{I}_n^{\vee}(\mathcal{M})$ the closure of the space $F^\vee_{-n-1}\mathscr{A}^\vee \cdot \mathcal{M}$. We assume that $\{I^\vee_n(\mathcal{M})\}$ forms a fundamental system of open neighborhoods of zero.

Definition 4.1.1 A *compatible topological left $\mathscr{A}^\vee$-module* is a left $\mathscr{A}^\vee$-module endowed with a linear topology which satisfies the conditions mentioned above.

Let $\mathcal{M}$ be a compatible topological left $\mathscr{A}^\vee$-module. Then we have

$$F^\vee_p\mathscr{A}^\vee \cdot I^\vee_n(\mathcal{M}) \subset I^\vee_{n-p}(\mathcal{M}), \quad I^\vee_n(\mathscr{A}^\vee) \cdot \mathcal{M} \subset I^\vee_n(\mathcal{M}) \tag{4.1.1}$$

because the action $\mathscr{A}^\vee \times \mathcal{M} \to \mathcal{M}$ is continuous.

Definition 4.1.2 A compatible topological left $\mathscr{A}^\vee$-module is *coexhaustive* if it is complete as a topological vector space.

Obviously, the completion of a compatible topological left $\mathscr{A}^\vee$-module has a canonical structure of a coexhaustive left $\mathscr{A}^\vee$-module.

4.2. Generalized eigenspaces of coexhaustive modules

Let $\mathcal{M}$ be a coexhaustive left $\mathscr{A}^\vee$-module. Consider the space

$$Q^\vee_n(\mathcal{M}) = \mathcal{M}/I^\vee_n(\mathcal{M}). \tag{4.2.1}$$

Then this is a left $Q^\vee_n(\mathbb{A})(0)$-module and hence decomposes into the sum of the generalized eigenspaces as follows:

$$Q^\vee_n(\mathcal{M}) = \sum_{\mu \in \Omega_n} Q^\vee_n(\mathcal{M})[\mu]. \tag{4.2.2}$$

Lemma 4.2.1 *If $\lambda \in \Gamma_m$ and $n \geq m$ then the restriction $Q^\vee_n(\mathcal{M})[\lambda] \to Q^\vee_m(\mathcal{M})[\lambda]$ of the canonical surjection is an isomorphism.*

Proof. See the proof of Lemma 2.5.2. □

This lemma implies that the subspace $\sum_{\lambda \in \Gamma_\infty} \mathcal{M}[\lambda]$ is dense in $\mathcal{M}$. It also implies that the space

$$E_n(\mathcal{M}) = \sum_{\lambda \in \Gamma_n} \mathcal{M}[\lambda] \tag{4.2.3}$$

is a discrete subspace for each n. Consider the space $E_n^\perp(\mathcal{M}) = \sum_{\lambda \in \Gamma_\infty \backslash \Gamma_n} \mathcal{M}[\lambda]$ and let $\widehat{E_n^\perp(\mathcal{M})}$ be its closure in $\mathcal{M}$. Then the quotient

$$R_n(\mathcal{M}) = \mathcal{M}/\widehat{E_n^\perp(\mathcal{M})} \qquad (4.2.4)$$

is canonically isomorphic to $E_n(\mathcal{M})$.

Lemma 4.2.2 *If $\mathcal{M}$ is a coexhaustive left $\mathscr{A}^\vee$-module then the canonical map $E_m(\mathcal{M}) \to Q_m^\vee(\mathcal{M})$ is injective and the canonical map $Q_m^\vee(\mathcal{M}) \leftarrow E_{m+g}(\mathcal{M})$ is surjective.*

Proof. By Lemma 4.2.1, we know that the map $\mathcal{M}[\lambda] \to Q_m^\vee(\mathcal{M})[\lambda]$ is injective if $\lambda \in \Gamma_m$. Hence the map $E_m(\mathcal{M}) \to Q_m^\vee(\mathcal{M})$ is injective. Now recall that $Q_m^\vee(\mathcal{M}) = \sum_{\lambda \in \Gamma_{m+g}} Q_m^\vee(\mathcal{M})[\lambda]$. Hence the map $Q_m^\vee(\mathcal{M}) \leftarrow E_{m+g}(\mathcal{M})$ is surjective. □

4.3. Exhaustion and coexhaustion

We will mean by the category of coexhaustive left $\mathscr{A}^\vee$-modules the category for which the objects are the coexhaustive left $\mathscr{A}^\vee$-modules and the morphisms are the continuous homomorphisms of left $\mathscr{A}^\vee$-modules.

Let M be a left $\mathscr{A}$-module and regard it as a left $\mathbb{A}$-module by identifying $\mathbb{A}$ with a subalgebra of $\mathscr{A}$ via the canonical map $\mathbb{A} \to \mathscr{A}$. We set

$$I_n^\vee(M) = F_{-n-1}^\vee \mathbb{A} \cdot M. \qquad (4.3.1)$$

Lemma 4.3.1 *If M is exhaustive then $I_n^\vee(M) = I_n^\vee(\mathbb{A}) \cdot M$.*

Proof. Since M is exhaustive, $I_{m+d}^\vee(\mathbb{A}(d)) \cdot v = I_m(\mathbb{A}(d)) \cdot v = 0$ holds for sufficiently large m for each d and $v \in M$. Since $I_n^\vee(\mathbb{A})$ is the degreewise closure of $F_{-n-1}^\vee \mathbb{A} \cdot \mathbb{A}$, we have $I_n^\vee(\mathbb{A}) \cdot v \subset F_{-n-1}^\vee \mathbb{A} \cdot M$ for each $v \in M$. Thus $I_n^\vee(\mathbb{A}) \cdot M = I_n^\vee(M)$ and the conclusion follows. □

Let M be an exhaustive left $\mathscr{A}$-module. Then the lemma implies that the action $\mathbb{A} \times M \to M$ is continuous with respect to the right linear topology on $\mathbb{A}$ and the linear topology on M defined by $I_n^\vee(M)$. Let $\mathcal{Q}_\infty^\vee(M)$ be the completion of M with respect to the linear topology defined by $I_n^\vee(M)$:

$$\mathcal{Q}_\infty^\vee(M) = \varprojlim_n Q_n^\vee(M), \quad Q_n^\vee(M) = M/I_n^\vee(M). \qquad (4.3.2)$$

Then $\mathcal{Q}_\infty^\vee(M)$ is a coexhaustive left $\mathscr{A}^\vee$-module.

Let M' and M'' be exhaustive left $\mathscr{A}$-modules and let $\phi : M' \to M''$ be a homomorphism of left $\mathscr{A}$-modules. Then ϕ gives rise to a continuous homomorphism of left $\mathbb{A}$-modules with respect to the right linear topology on $\mathbb{A}$. Hence it induces a continuous homomorphism $Q_\infty^\vee(\phi) : Q_\infty^\vee(M') \to Q_\infty^\vee(M'')$ of coexhaustive left $\mathscr{A}^\vee$-modules. We call the functor $Q_\infty^\vee(-)$ the *coexhaustion functor*.

Conversely, let $\mathcal{M}$ be a coexhaustive left $\mathscr{A}^\vee$-module and regard it as a left $\mathbb{A}$-module via the canonical map $\mathbb{A} \to \mathscr{A}^\vee$. Consider the space

$$\mathrm{K}_\infty(\mathcal{M}) = \bigcup_n \mathrm{K}_n(\mathcal{M}), \quad \mathrm{K}_n(\mathcal{M}) = \{v \in \mathcal{M} \mid \mathrm{I}_n(\mathbb{A}) \cdot v = 0\}. \quad (4.3.3)$$

Then this is an exhaustive left $\mathbb{A}$-module and hence an exhaustive left $\mathscr{A}$-module.

Let $\varphi : \mathcal{M}' \to \mathcal{M}''$ be a continuous homomorphism of coexhaustive left $\mathscr{A}^\vee$-modules. We regard φ as a homomorphism of left $\mathbb{A}$-modules via the canonical map $\mathbb{A} \to \mathscr{A}^\vee$. If $\mathrm{I}_n(\mathbb{A}) \cdot v = 0$ then $\mathrm{I}_n(\mathbb{A}) \cdot \varphi(v) = \varphi(\mathrm{I}_n(\mathbb{A}) \cdot v) = \varphi(0) = 0$. Therefore, φ restricts to a homomorphism $\mathrm{K}_\infty(\varphi) : \mathrm{K}_\infty(\mathcal{M}') \to \mathrm{K}_\infty(\mathcal{M}'')$ of left $\mathbb{A}$-modules and hence a homomorphism of left $\mathscr{A}$-modules. We call the functor $\mathrm{K}_\infty(-)$ the *exhaustion functor*.

For an exhaustive left $\mathscr{A}$-module M, we set

$$\mathcal{R}_\infty(M) = \varprojlim_n \mathrm{R}_n(M) \qquad (4.3.4)$$

and, for a coexhaustive left $\mathscr{A}^\vee$-module $\mathcal{M}$,

$$\mathrm{E}_\infty(\mathcal{M}) = \bigcup_n \mathrm{E}_n(\mathcal{M}). \qquad (4.3.5)$$

Here the space $\mathrm{R}_n(M)$ is defined by (3.2.4).

Lemma 4.3.2 *For an exhaustive left $\mathscr{A}$-module M its coexhaustion $Q_\infty^\vee(M)$ is canonically isomorphic to $\mathcal{R}_\infty(M)$ as topological vector spaces. For a coexhaustive left $\mathscr{A}^\vee$-module $\mathcal{M}$ its exhaustion $\mathrm{K}_\infty(\mathcal{M})$ is canonically isomorphic to $\mathrm{E}_\infty(\mathcal{M})$ as vector spaces.*

Proof. Conclusions are clear by the arguments in Subsections 3.2 and 4.2. □

Now the following theorem is an immediate consequence of this lemma.

Theorem 4.3.3 *Let $(\mathbb{A}, h)$ be a weakly quasi-finite algebra graded by Hamiltonian and let $\mathscr{A}$ and $\mathscr{A}^\vee$ be the left and the right filterwise completions, respectively. Then the functors $Q_\infty^\vee(-)$ and $\mathrm{K}_\infty(-)$ are mutually inverse equivalences of categories between the category of exhaustive left $\mathscr{A}$-modules and the category of coexhaustive left $\mathscr{A}^\vee$-modules.*

In particular, the category of coexhaustive left $\mathscr{A}^\vee$-modules is an abelian category.

We may likewise define the notion of coexhaustive right $\mathscr{A}$-modules by using the spaces $I_n(\mathscr{A})$. We have analogous results for the right modules as well.

5. Duality for quasi-finite modules

We now consider finiteness conditions for exhaustive modules and for coexhaustive modules and discuss the duality between the categories of such modules.

5.1. Quasi-finiteness of exhaustive modules

Let $(\mathbb{A}, h)$ be a weakly quasi-finite algebra graded by Hamiltonian and let $\mathscr{A}$ and $\mathscr{A}^\vee$ be the left and the right filterwise completions, respectively.

Definition 5.1.1 A *quasi-finite left $\mathscr{A}$-module* is an exhaustive left $\mathscr{A}$-module M such that the spaces $K_n(M)$ are finite-dimensional for all n.

For an exhaustive right $\mathscr{A}^\vee$-module N, we define the space $K_n^\vee(N)$ in the same way as $K_n(M)$ for an exhaustive left $\mathscr{A}$-module M:

$$K_n^\vee(N) = \left\{ v \in N \mid v \cdot I_n^\vee(\mathscr{A}^\vee) = 0 \right\}. \tag{5.1.1}$$

Definition 5.1.2 A *quasi-finite right $\mathscr{A}^\vee$-module* is an exhaustive right $\mathscr{A}^\vee$-module N such that the spaces $K_n^\vee(N)$ are finite-dimensional for all n.

The following proposition characterizes the quasi-finiteness of an algebra by means of the quasi-finiteness of the canonical quotient modules.

Proposition 5.1.3 *Let $(\mathbb{A}, h)$ be a weakly quasi-finite algebra graded by Hamiltonian. Then the following conditions are equivalent:*

(a) $\mathbb{A}$ *is a quasi-finite algebra.*
(b) *The left canonical quotient modules $Q_n(\mathbb{A})$ are quasi-finite for all n.*
(c) *The right canonical quotient modules $Q_n^\vee(\mathbb{A})$ are quasi-finite for all n.*

Proof. We will show the equivalence of (a) and (b). Since $K_m(Q_n(\mathbb{A}))$ is an (H_m, H_n)-bimodule, the pair of the left and the right eigenvalues of h on $K_m(Q_n(\mathbb{A}))$ are in the finite set $\Omega_m \times \Omega_n$. Then if $K_m(Q_n(\mathbb{A}))(d) \neq 0$ then $d = \lambda - \mu$ for some $\lambda \in \Omega_m$ and $\mu \in \Omega_n$. Therefore, we have

$$K_m(Q_n(\mathbb{A})) \subset \bigoplus_{-n \leq d \leq m+g} Q_n(\mathbb{A})(d). \tag{5.1.2}$$

On the other hand, it is easy to see that $Q_n(\mathbb{A})(d) \subset K_{n+d}(Q_n(\mathbb{A}))$. Therefore we immediately see that $K_m(Q_n(\mathbb{A}))$ are finite-dimensional for all m and n if and only if $Q_n(\mathbb{A})(d)$ are finite-dimensional for all d and n. The proof for the equivalence of (a) and (c) is similar.   □

Now consider the case when $(\mathbb{A}, h)$ is a quasi-finite algebra graded by Hamiltonian. Since $E_n(M)$ is a left $\mathbb{A}_n$-module and $\mathbb{A}_n$ is finite-dimensional, $E_n(M)$ is finite-dimensional if and only if it is finitely generated as a left $\mathbb{A}_n$-module.

Proposition 5.1.4 *Let* $(\mathbb{A}, h)$ *be a quasi-finite algebra graded by Hamiltonian. Then the following conditions for an exhaustive left $\mathscr{A}$-module M are equivalent:*

(a) *M is a quasi-finite left $\mathscr{A}$-module.*
(b) *M is finitely generated as a left $\mathscr{A}$-module.*
(c) *$E_n(M)$ are finite-dimensional for all n.*
(d) *$E_g(M)$ is finite-dimensional.*

Proof. We will show (a)⇒(c)⇒(d)⇒(b)⇒(a). Assume that M is quasi-finite. Then $E_n(M)$ are finite-dimensional since $E_n(M) \subset K_n(M)$ by Proposition 3.2.1. In particular, $E_g(M)$ is finite-dimensional. Next assume that $E_g(M)$ is finite-dimensional and let M' be the left $\mathscr{A}$-submodule of M generated by $E_g(M)$. Then $K_0(M/M') \subset E_g(M/M') = 0$. By Lemma 3.3.1, we have $M = M'$ and hence M is finitely generated. Now assume that M is finitely generated and let $v_1, \ldots, v_k$ be a set of generators. Since M is exhaustive, there exists $n_1, \ldots, n_k$ such that $I_{n_i}(\mathscr{A}) \cdot v_i = 0$ for $i = 1, \ldots, k$. This implies the existence of a surjective homomorphism $Q_{n_1}(\mathbb{A}) \times \cdots \times Q_{n_k}(\mathbb{A}) \to M$ of left $\mathscr{A}$-modules. Since $\mathbb{A}$ is quasi-finite and hence the modules $Q_{n_i}(\mathbb{A})$ are quasi-finite, so is the image M.   □

5.2. Quasi-finiteness of coexhaustive modules

Let us now turn to the quasi-finiteness of coexhaustive modules.

Definition 5.2.1 A *quasi-finite left $\mathscr{A}^\vee$-module* is a coexhaustive left $\mathscr{A}^\vee$-module $\mathcal{M}$ such that the spaces $Q_n^\vee(\mathcal{M})$ are finite-dimensional for all n.

For a coexhaustive right $\mathscr{A}$-module $\mathcal{N}$, we set

$$Q_n(\mathcal{N}) = \mathcal{N}/I_n(\mathcal{N}). \tag{5.2.1}$$

Definition 5.2.2   A *quasi-finite right $\mathscr{A}$-module* is a coexhaustive right $\mathscr{A}$-module $\mathcal{N}$ such that the spaces $Q_n(\mathcal{N})$ are finite-dimensional for all n.

Let M be an exhaustive left $\mathscr{A}$-module and let $\mathcal{M}$ be its coexhaustion. Then the canonical map $E_n(M) \to R_n(\mathcal{M})$ is an isomorphism for each n. Therefore, the consideration in Subsection 4.3 implies the following result.

Proposition 5.2.3   *Let $(\mathbb{A}, h)$ be a quasi-finite algebra graded by Hamiltonian. Let M be an exhaustive left $\mathscr{A}$-module and let $\mathcal{M}$ be its coexhaustion. Then the exhaustive module M is quasi-finite if and only if the coexhaustive module $\mathcal{M}$ is quasi-finite.*

We have analogous results for quasi-finite right $\mathscr{A}^\vee$-modules and for quasi-finite right $\mathscr{A}$-modules.

5.3. Duality

Let $(\mathbb{A}, h)$ be a weakly quasi-finite algebra graded by Hamiltonian. Let M be an exhaustive left $\mathscr{A}$-module and consider its full dual space $\mathcal{N}$:

$$\mathcal{N} = M^* = \mathrm{Hom}_{\mathbf{k}}(M, \mathbf{k}). \tag{5.3.1}$$

Then this space becomes a right $\mathscr{A}$-module.

Since M is exhaustive, it decomposes into the sum of the generalized eigenspaces $M = \bigoplus_{\lambda \in \Gamma_\infty} M[\lambda]$, which gives rise to

$$\mathcal{N} = \prod_{\lambda \in \Gamma_\infty} M[\lambda]^* \tag{5.3.2}$$

where $M[\lambda]^*$ is the set of linear functions $f : M \to \mathbf{k}$ such that $f(M[\mu]) = 0$ holds for any $\mu \neq \lambda$. Therefore, by setting

$$J_n(\mathcal{N}) = \left\{ f \in \mathcal{N} \mid f(E_n(M)) = 0 \right\}, \tag{5.3.3}$$

we have

$$\mathcal{N}/J_n(\mathcal{N}) = \prod_{\lambda \in \Gamma_n} M[\lambda]^*. \tag{5.3.4}$$

We give $\mathcal{N}$ the linear topology defined by $J_n(\mathcal{N})$.

Lemma 5.3.1   *The right $\mathscr{A}$-module $\mathcal{N}$ endowed with the topology as above is a coexhaustive right $\mathscr{A}$-module.*

Proof. Completeness of $\mathcal{N}$ as a topological vector space is clear from the definition. Since $(f \cdot a)(E_n(M)) = f(a \cdot E_n(M)) = 0$ holds for any $f \in \mathcal{N}$ and any $a \in I_n(\mathscr{A})$, we have $\mathcal{N} \cdot I_n(\mathscr{A}) \subset J_n(\mathcal{N})$. Hence the action $\mathcal{N} \times \mathscr{A} \to \mathcal{N}$ is continuous. Let $I_n(\mathcal{N})$ be the closure of $\mathcal{N} \cdot I_n(\mathscr{A})$. It remains to show that $\{I_n(\mathcal{N})\}$ forms a fundamental system of open neighborhoods of zero. Since $J_n(\mathcal{N})$ is closed, we have $I_n(\mathcal{N}) \subset J_n(\mathcal{N})$. Consider the quotient $\mathcal{N}/I_n(\mathcal{N})$. Since we have $I_n(\mathcal{N}) \cdot I_n(\mathscr{A}) \subset I_n(\mathcal{N})$, the quotient space $\mathcal{N}/I_n(\mathcal{N})$ is a right $Q_n(\mathbb{A})(0)$-module. In particular, it is a right H_n-module. Hence it has the generalized eigenspace decomposition with respect to the right action of h with the eigenvalues belonging to $\Omega_n \subset \Gamma_{n+g}$. Since $J_{n+g}(\mathcal{N})$ is the closure of $E_{n+g}^{\perp}(\mathcal{N})$, we have $J_{n+g}(\mathcal{N}) \subset I_n(\mathcal{N})$. Hence $I_n(\mathcal{N})$ form a basis of open neighborhoods of zero.   □

Thus we have shown that for any exhaustive left $\mathscr{A}$-module, its full dual naturally becomes a coexhaustive right $\mathscr{A}$-module.

Conversely, let $\mathcal{N}$ be a coexhaustive right $\mathscr{A}$-module and consider the continuous dual space M:

$$M = \mathrm{Hom}_{\mathbf{k}}^{\mathrm{cont}}(\mathcal{N}, \mathbf{k}). \tag{5.3.5}$$

Recall that the base field $\mathbf{k}$ is given the discrete topology.

Take any element of M, which is given by a continuous map $f : \mathcal{N} \to \mathbf{k}$. Then, for any $a \in \mathscr{A}$, the map $a \cdot f : \mathcal{N} \to \mathbf{k}$ is also continuous since for $a \in F_p\mathscr{A}$ we have $(a \cdot f)(v + I_n(\mathcal{N})) \subset f(v \cdot a) + f(I_{n-p}(\mathcal{N})) = f(v \cdot a)$ for sufficiently large n. We then have $(I_n(\mathscr{A}) \cdot f)(v) = f(v \cdot I_n(\mathscr{A})) \subset f(I_n(\mathcal{N})) = 0$ for all v. Therefore, the module M is exhaustive.

Thus we have defined contravariant functors $\mathrm{Hom}_{\mathbf{k}}(-, \mathbf{k})$ and $\mathrm{Hom}_{\mathbf{k}}^{\mathrm{cont}}(-, \mathbf{k})$ between the category of exhaustive left $\mathscr{A}$-modules and the category of coexhaustive right $\mathscr{A}$-modules.

Proposition 5.3.2   *Let $(\mathbb{A}, h)$ be a weakly quasi-finite algebra graded by Hamiltonian. Then the functors $\mathrm{Hom}_{\mathbf{k}}(-, \mathbf{k})$ and $\mathrm{Hom}_{\mathbf{k}}^{\mathrm{cont}}(-, \mathbf{k})$ give rise to mutually inverse duality of categories between the category of quasi-finite left $\mathscr{A}$-modules and the category of quasi-finite right $\mathscr{A}$-modules.*

Note 5.3.3   A quasi-finite right $\mathscr{A}$-module is nothing else but a linearly compact right $\mathscr{A}$-module. We may understand the above-mentioned duality as a version of the Lefschetz duality [Lef] between discrete modules and linearly compact modules. See also [Mac].

5.4. Involution

Let $(\mathbb{A}, h)$ be a quasi-finite algebra graded by Hamiltonian. Suppose given a linear involution $\theta : \mathbb{A} \to \mathbb{A}$ satisfying the following conditions:

(i) $\theta(a \cdot b) = \theta(b) \cdot \theta(a)$ for any $a, b \in \mathbb{A}$ and $\theta(h) = h$.

(ii) $\theta : \mathbb{A}(d) \to \mathbb{A}(-d)$ is continuous.

Note that by (i) we have $[h, \theta(a)] = [\theta(h), \theta(a)] = -\theta([h, a])$. Hence $\theta(\mathbb{A}(d)) = \mathbb{A}(-d)$ and the condition (ii) makes sense. We then have that $\theta(I_n(\mathbb{A})) = I_n^{\vee}(\mathbb{A})$. Hence θ extends to anti-isomorphisms

$$\theta_{\infty} : \mathscr{A} \to \mathscr{A}^{\vee}, \quad \theta_{\infty}^{\vee} : \mathscr{A}^{\vee} \to \mathscr{A} \tag{5.4.1}$$

of filtered topological algebras such that $\theta_{\infty} \circ \theta_{\infty}^{\vee} = 1$ and $\theta_{\infty}^{\vee} \circ \theta_{\infty} = 1$. We will denote the maps θ_{∞} and $\theta_{\infty}^{\vee}$ by the same symbol θ by abuse of notation.

Let M be a left $\mathscr{A}$-module and let $\vartheta(M)$ be the same space M as a vector space. We give $\vartheta(M)$ a structure of a right $\mathscr{A}^{\vee}$-module by letting $v \cdot a = \theta(a) \cdot v$ for $a \in \mathscr{A}^{\vee}$ and $v \in M$. Similarly, for a right $\mathscr{A}^{\vee}$-module N, we define a left $\mathscr{A}$-module $\vartheta^{\vee}(N)$ in a similar way.

Proposition 5.4.1 *The functors ϑ and $\vartheta^{\vee}$ are mutually inverse equivalences of categories between the category of left $\mathscr{A}$-modules and the category of right $\mathscr{A}^{\vee}$-modules.*

Let us compose the involution, the duality and the exhaustion. Then we get an auto-duality

$$D : M \mapsto \vartheta(K_{\infty}(\mathrm{Hom}_{\mathbf{k}}(M, \mathbf{k}))) \tag{5.4.2}$$

of the category of quasi-finite left $\mathscr{A}$-modules. Note that $K_{\infty}(\mathrm{Hom}_{\mathbf{k}}(M, \mathbf{k}))$ is the restricted dual space:

$$K_{\infty}(\mathrm{Hom}_{\mathbf{k}}(M, \mathbf{k})) = \left\{ f \mid f(E_n^{\perp}(M)) = 0 \text{ for some } n \right\}. \tag{5.4.3}$$

5.5. Finiteness theorems

Let $(\mathbb{A}, h)$ be a quasi-finite algebra and consider the following categories.

L1. The category of quasi-finite left $\mathscr{A}$-modules.
L2. The category of quasi-finite left $\mathscr{A}^{\vee}$-modules.
L3. The category of finitely generated left $\mathbb{A}_n$-modules.

R1. The category of quasi-finite right $\mathscr{A}^{\vee}$-modules.
R2. The category of quasi-finite right $\mathscr{A}$-modules.

R3. The category of finitely generated right $\mathbb{A}_n$-modules.

Recall that quasi-finite left $\mathscr{A}$-modules and right $\mathscr{A}^\vee$-modules are exhaustive whereas quasi-finite left $\mathscr{A}^\vee$-modules and right $\mathscr{A}$-modules are coexhaustive.

The following theorem summarizes the results obtained so far regarding the quasi-finite modules over quasi-finite algebras.

Theorem 5.5.1 *Let* $(\mathbb{A}, h)$ *be a quasi-finite algebra graded by Hamiltonian and let n be an integer such that* $n \geq g$.

(1) *The categories* L1, L2 *and* L3 *are equivalent to each other.*
(2) *The categories* R1, R2 *and* R3 *are equivalent to each other.*
(3) *The categories* L1, L2 *and* L3 *and the categories* R1, R2 *and* R3 *are dual to each other.*
(4) *If* $\mathbb{A}$ *has an involution* θ *then the categories* L1, L2 *and* L3 *and the categories* R1, R2 *and* R3 *are equivalent to each other.*

Part II

Quasi-finiteness and Zhu's Finiteness Condition

6. Vertex operator algebras and current algebras

We now turn our attention to vertex operator algebras. In this section, we will describe in detail the construction and properties of the universal enveloping algebra associated with a vertex operator algebra, which we will simply call the *current algebra*, in order to give precise statements which seem to have been overlooked in the literature.

6.1. Vertex operator algebra

Recall that a vertex operator algebra is a graded vector space V with the grading being indexed by integers equipped with countably many bilinear maps indexed by integers and two distinguished elements, called the vacuum vector and the conformal vector (or the Virasoro element), satisfying a number of axioms ([Bor], [FLM], [FHL]), which we will describe briefly below. See [MaN] for an account.

Let us denote the homogeneous subspaces of the grading of V by V^k where k is an integer. It is assumed that there exists a nonnegative integer m such that $V^k = 0$ for $k < -m$. Therefore, the grading of V is of the following form

$$V = \bigoplus_{k=-m_0}^{\infty} V^k. \tag{6.1.1}$$

We will write $\Delta(u) = k$ when u belongs to the subspace V^k and call $\Delta(u)$ the *weight* of the element u.

Let us denote the countably many bilinear maps by

$$\mu_n : V \times V \to V, \quad (u, v) \mapsto u_{(n)}v. \tag{6.1.2}$$

It is assumed that they satisfy $V^j_{(n)} V^k \subset V^{j+k-n-1}$. In other words, for homogeneous u and v, we have

$$\Delta(u_{(n)}v) = \Delta(u) + \Delta(v) - n - 1. \tag{6.1.3}$$

313

Then the sums in the following expression are finite:

$$\sum_{i=0}^{\infty} \binom{p}{i} (u_{(r+i)}v)_{(p+q-i)}w$$

$$= \sum_{i=0}^{\infty} (-1)^i \binom{r}{i} \left(u_{(p+r-i)}(v_{(q+i)}w) - (-1)^r v_{(q+r-i)}(u_{(p+i)}w) \right).$$

(6.1.4)

The bilinear maps (6.1.2) are assumed to satisfy (6.1.4) for any $u, v, w \in V$ and any integers p, q, r.

The identity (6.1.4), or its equivalent generating-function form called the *Jacobi identity* or the *Cauchy-Jacobi identity*, is the main identity of vertex operator algebras, as discovered by Frenkel et al. in [FLM], although an equivalent set of particular cases of the coefficient form (6.1.4) had been discovered by Borcherds in [Bor].

The vacuum vector $\mathbf{1}$ is an element of weight 0. It enjoys, for any $u \in V$, the relations $u_{(-1)}\mathbf{1} = u$ and $u_{(n)}\mathbf{1} = 0$ for $n \geq 0$. We set

$$Tu = u_{(-2)}\mathbf{1}.$$

(6.1.5)

Then the axioms imply the relations $\mathbf{1}_{(-1)}u = 0$ and $\mathbf{1}_{(n)}u = 0$ for $n \neq -1$ and that $T : V \to V$ is a derivation with respect to the operations $_{(n)}$ for every n.

The conformal vector ω is an element of weight 2. It satisfies

$$\omega_{(n)}\omega = 0, \ (n \geq 4 \text{ or } n = 2), \ \omega_{(1)}\omega = 2\omega, \ \omega_{(3)}\omega \in \mathbf{k1}.$$

(6.1.6)

Then the axioms imply that the operators $L_n : V \to V$ defined by $L_n u = \omega_{(n+1)}u$ satisfy the Virasoro commutation relation with the central charge c_V given by $2\omega_{(3)}\omega = c_V \mathbf{1}$. Among the Virasoro operators L_n, the ones with $n = 0$ and $n = -1$ have special roles: the weights of V are given by the eigenvalues of L_0 and the derivation T agrees with L_{-1}. In other words,

$$V^k = \left\{ v \in V \mid L_0 v = kv \right\}, \quad Tu = L_{-1}u.$$

(6.1.7)

The weight subspaces are usually assumed to be finite-dimensional.

6.2. The current Lie algebras

Let us consider the space

$$V[t, t^{-1}] = V \otimes_{\mathbf{k}} \mathbf{k}[t, t^{-1}].$$

(6.2.1)

We define a bilinear map $V[t, t^{-1}] \times V[t, t^{-1}] \to V[t, t^{-1}]$ by setting

$$[u \otimes t^m, v \otimes t^n] = \sum_{i=0}^{\infty} \binom{m}{i} (u_{(i)}v) \otimes t^{m+n-i}. \tag{6.2.2}$$

Consider the quotient space

$$\mathfrak{g} = V[t, t^{-1}] / \partial V[t, t^{-1}] \tag{6.2.3}$$

where $\partial : V[t, t^{-1}] \to V[t, t^{-1}]$ is defined by

$$\partial(u \otimes t^n) = Tu \otimes t^n + nu \otimes t^{n-1}. \tag{6.2.4}$$

Then the bracket operation on $V[t, t^{-1}]$ induces a bilinear operation on $\mathfrak{g}$ denoted by the same symbol which gives a structure of a Lie algebra on $\mathfrak{g}$ ([Bor]). We will call the Lie algebra $\mathfrak{g}$ the associated *current Lie algebra*. We will denote the image of an element of $V[t, t^{-1}]$ in $\mathfrak{g}$ by the same symbol.

Since $n \mathbf{1} \otimes t^{n-1} = \partial(\mathbf{1} \otimes t^n)$, we know that $\mathbf{1} \otimes t^n = 0$ in $\mathfrak{g}$ unless $n = -1$, when $\mathbf{1} \otimes t^{-1}$ is central.

It will be useful to introduce the following notation:

$$J_n(u) = u \otimes t^{n+\Delta(u)-1} \tag{6.2.5}$$

for a homogeneous u and extend it linearly. We denote its image in $\mathfrak{g}$ by the same symbol and assign the degree $-n$ to $J_n(u)$. Then the associated current Lie algebra is graded by the degree:

$$\mathfrak{g} = \bigoplus_d \mathfrak{g}(d). \tag{6.2.6}$$

Note the relation

$$[L_0, J_n(u)] = -n J_n(u) \tag{6.2.7}$$

for the element $L_0 = J_0(\omega)$, which follows from the axioms for vertex operator algebras.

Let $\mathbf{U}$ be the quotient algebra of the universal enveloping algebra of the Lie algebra $\mathfrak{g}$ by the two sided ideal generated by $J_0(\mathbf{1}) - 1$ and let us denote the image of $J_n(u)$ by the same symbol. We give the degree $d_1 + \cdots + d_k$ to the element of the form $J_{-d_1}(u_1) \cdots J_{-d_k}(u_k)$ with $u_1, \ldots, u_k \in V$. Let $\mathbf{U}(d)$ be the span of these vectors of degree d. Then we have

$$\mathbf{U} = \bigoplus_d \mathbf{U}(d). \tag{6.2.8}$$

by which the algebra $\mathbf{U}$ becomes a graded algebra. Note that the relation (6.2.7) says that the image of L_0 is a Hamiltonian of $\mathbf{U}$.

Consider the standard degreewise topology on $\mathbf{U}$ and let $\hat{\mathbf{U}}$ denote the degreewise completion. (See Subsection 1.3.)

6.3. The current algebras

For $u, v \in V$ and for integers m, n, r, consider the following expression in $\hat{\mathbf{U}}$:

$$
B_{m,n,r}(u, v) = \sum_{i=0}^{\infty} \binom{m + \Delta(u) - 1}{i} J_{m+n+r}(u_{(r+i)}v)
$$

$$
- \sum_{i=0}^{\infty} (-1)^i \binom{r}{i} J_{m+r-i}(u) \cdot J_{n+i}(v)
$$

$$
+ (-1)^r \sum_{i=0}^{\infty} (-1)^i \binom{r}{i} J_{n+r-i}(v) \cdot J_{m+i}(u). \qquad (6.3.1)
$$

Then the first sum in the right-hand side is actually a finite sum whereas the second and the last are infinite sums which converge in the linear topology of $\hat{\mathbf{U}}(-m - n - r)$. The relation $B_{m,n,r}(u, v) = 0$ turns our to be the counterpart of the identity (6.1.4) in the action of V on a module, where the infinite sums become finite when they act on each element of the module. (See the next subsection for the definition of modules.)

Let $\mathbf{B}$ be the ideal of $\hat{\mathbf{U}}$ generated by the elements of the form $B_{m,n,r}(u, v)$ with $u, v \in V$ and integers m, n, r, and let $\hat{\mathbf{B}}$ be the degreewise closure of $\mathbf{B}$. Then $\hat{\mathbf{B}}$ is also an ideal of $\hat{\mathbf{U}}$.

Remark 6.3.1 The ideal $\mathbf{B}$ is in fact generated by the elements of the form $B_{m,n,r}(u, v)$ with $m = -\Delta(u) + 1$. Alternatively, it is also generated by the elements of the form $B_{m,n,r}(u, v)$ with $r < 0$.

We now define the *current algebra* $\mathbb{U}$ associated with V to be the quotient algebra of $\hat{\mathbf{U}}$ by the ideal $\hat{\mathbf{B}}$:

$$
\mathbb{U} = \hat{\mathbf{U}} / \hat{\mathbf{B}}. \qquad (6.3.2)
$$

Then $\mathbb{U}$ is a graded algebra, since $\hat{\mathbf{B}}$ is a graded ideal, and the image of L_0 is a Hamiltonian.

Proposition 6.3.2 *The pair* $(\mathbb{U}, L_0)$ *is a compatible degreewise complete algebra graded by Hamiltonian.*

Note 6.3.3 The construction of $\mathbb{U}$ is essentially due to Frenkel and Zhu [FrZ]. The left linear filterwise completion of $\mathbb{U}$ as in Subsection 1.5 is isomorphic to the current algebra $\mathcal{U}(V)$ considered in [NaT].

6.4. Denseness of the current Lie algebra

Let $\mathbb{U}$ be the current algebra associated with a vertex operator algebra V. Let us regard the current Lie algebra $\mathfrak{g}$ as a subspace of $\mathbf{U}$ and let ϕ denote the composition of the canonical maps $\mathbf{U} \to \hat{\mathbf{U}} \to \mathbb{U}$. By construction, $\mathbf{U}(d)$ is a dense subspace of $\hat{\mathbf{U}}(d)$.

The following observation is insightful.

Proposition 6.4.1 *The image $\phi(\mathfrak{g}(d))$ is a dense subspace of $\mathbb{U}(d)$ for each d.*

Proof. It suffices to show that $\phi(\mathfrak{g})$ is dense in $\mathbb{U}$ with respect to the left linear topology on $\mathbb{U}$. Let us denote by $\phi_n : \mathbf{U} \to Q_n(\mathbb{U})$ the composite of ϕ with the canonical surjection $\mathbb{U} \to Q_n(\mathbb{U})$. By the relation (6.3.1), we have

$$J_s(u) \cdot J_t(v) \cdot 1_n$$

$$= \sum_{m=0}^{n} \sum_{j=0}^{\Delta(u)+n} (-1)^m \binom{n-s+m}{n-s} \binom{\Delta(u)+n}{j} J_{s+t}(u_{(s+\Delta(u)-m-j-1)}v) \cdot 1_n$$

$$\tag{6.4.1}$$

in the quotient $Q_n(\mathbb{U})$ for any integers s and t provided $s \leq n$. Hence by induction we have $\phi_n(\mathbf{U}) = \phi_n(\mathfrak{g})$ for any nonnegative integer n. Therefore, since $\phi(\mathbf{U}(d))$ is dense, $\phi(\mathfrak{g}(d))$ is also dense. $\qquad\square$

6.5. Exhaustive V-modules

Let M be a vector space and suppose given a series of bilinear maps $V \times M \to M$ indexed by integers which we denote by $(u, v) \mapsto \pi_n^M(u)v$. Such an M is said to be a *weak V-module* if it satisfies the conditions listed below.
  Set $J_n^M(u) = \pi_{n+\Delta(u)-1}^M(u)$ and let $B_{m,n,r}^M(u, v)$ be the expression (6.3.1) with J_n being replaced by J_n^M. Then the conditions are as follows:

  (i) For any $u \in V$ and $v \in M$ there exists an m such that $J_n^M(u)v = 0$ for all $n \geq m$.
  (ii) The operator $J_n^M(\mathbf{1})$ is the identity if $n = 0$ and is zero otherwise.
  (iii) The identity $B_{m,n,r}^M(u, v) = 0$ holds for any integers m, n, r and $u, v \in V$.

We will say that a weak V-module is an *exhaustive V-module* if instead of the condition (i) the following stronger condition is satisfied:

(i)$'$ For any $v \in M$ there exists an m such that $J_{n_1}^M(u_1) \cdots J_{n_k}^M(u_k)v = 0$ for all $u_1, \ldots, u_k \in V$ whenever $n_1 + \cdots + n_k \geq m$.

Remark 6.5.1   Thanks to the condition (iii), the condition (i)$'$ follows from the apparently weaker condition that for any $v \in M$ there exists an m such that $J_n^M(u)v = 0$ for all $u \in V$ and $n \geq m$ by successive use of the relation (6.4.1).

Let us consider the map $J_n : V \to \mathbb{U}$ which sends $u \in V$ to the image of $J_n(u) = u \otimes t^{n+\Delta(u)-1}$ in $\mathbb{U}$. Then any exhaustive left $\mathbb{U}$-module M becomes an exhaustive V-module by letting $J_n^M(u)$ be the action of $J_n(u)$ on M. We will call this V-module structure on M the *associated V-module structure*.

Proposition 6.5.2   *Let M be an exhaustive V-module. Then there exists a unique structure of an exhaustive $\mathbb{U}$-module on M such that the associated V-module structure agrees with the given V-module structure on M.*

Proof. Let $J_n^M : V \times M \to M$ be the given V-module structure on M. Then they induce a map $\mathfrak{g} \times M \to M$ which gives a $\mathfrak{g}$-module structure on M by the relation $B_{m,n,r}^M(u, v) = 0$ with $r \geq 0$. By the universal property of the universal enveloping algebra of $\mathfrak{g}$, this lifts to a $\mathbf{U}$-module structure on M because of the axiom (ii). Since M is an exhaustive V-module, the map $\mathbf{U}(d) \times M \to M$ is continuous for each d when M is endowed with the discrete topology. Hence this map prolongs to the action of the degreewise completion $\hat{\mathbf{U}}$. Now the axiom (iii) is nothing else but the defining relations of the algebra $\mathbb{U}$. Hence the $\hat{\mathbf{U}}$-module structure induces a $\mathbb{U}$-module structure on M, which is exhaustive by Lemma 3.1.2. The uniqueness is clear on each step. $\square$

Thus we have obtained the following result.

Theorem 6.5.3   *The category of exhaustive V-modules is canonically equivalent to the category of exhaustive $\mathbb{U}$-modules.*

Note 6.5.4   Recall from [DLM1] that a weak V-module M is said to be *admissible* if it is given a grading $M = \bigoplus_{d=0}^{\infty} M_d$ so that $J_n(u) \cdot M_d \subset M_{d-n}$ for any integer n and $u \in V$. Then it is easy to see that any admissible V-module is exhaustive. The converse is true if $\mathbb{U}$ is weakly quasi-finite. Indeed, if $\mathbb{U}$ is weakly quasi-finite and M is exhaustive then it decomposes into the sum of the generalized eigenspaces of the action of L_0 by the argument of

Subsection 3.2. Then setting $M_d = \sum_{\lambda \in \Gamma_d} M[\lambda]$ we have $M = \bigoplus_{d=0}^{\infty} M_d$ with $J_n(u) \cdot M_d \subset M_{d-n}$.

7. Associated Poisson algebras

We will consider the associated graded algebra with respect to a filtration on $\mathbb{U}$ and show that it has a structure of a degreewise complete Poisson algebra.

7.1. Zhu's Poisson algebra

Recall that a (commutative) *Poisson algebra* is a vector space $\mathfrak{p}$ equipped with two bilinear maps $\cdot : \mathfrak{p} \times \mathfrak{p} \to \mathfrak{p}$ and $\{ , \} : \mathfrak{p} \times \mathfrak{p} \to \mathfrak{p}$ called the multiplication and the Poisson bracket, respectively, such that $\mathfrak{p}$ is a commutative associative algebra with unity with respect to the multiplication, $\mathfrak{p}$ is a Lie algebra with respect to the Poisson bracket and the Leibniz identity holds:

$$\{x \cdot y, z\} = x \cdot \{y, z\} + y \cdot \{x, z\}. \tag{7.1.1}$$

We denote the unity of $\mathfrak{p}$ by $1_{\mathfrak{p}}$.

Let V be a vertex operator algebra. We let $C_2(V)$ be the span of the elements of the form $u_{(n)}v$ with $u, v \in V$ and $n \leq -2$. We set

$$u \cdot v = u_{(-1)}v \quad \text{and} \quad \{u, v\} = u_{(0)}v. \tag{7.1.2}$$

The following result is obtained in [Zhu].

Proposition 7.1.1 (Zhu) *The operations $\cdot$ and $\{ , \}$ induces a Poisson algebra structure on $V/C_2(V)$.*

Let us call this Poisson algebra *Zhu's Poisson algebra*.

7.2. Poisson filtrations and the associated graded algebras

Let V be a vertex operator algebra and let $\mathfrak{g}$, $\mathbf{U}$, $\hat{\mathbf{U}}$ and $\mathbb{U}$ as in the preceding section.

Let $G_p \mathfrak{g}$ be the image of $\bigoplus_{k \leq p} V^k \otimes_{\mathbf{k}} \mathbf{k}[t, t^{-1}]$ in $\mathfrak{g}$ and let $G_p \mathbf{U}$ be the sum of subspaces $G_{p_1}\mathfrak{g} \cdots G_{p_k}\mathfrak{g}$ with $k = 0, 1, \ldots$ and $p_1 + \cdots + p_k = p$ in $\mathbf{U}$. Then G is a separated filtration on $\mathbf{U}$ satisfying

$$G_p \mathbf{U} \cdot G_q \mathbf{U} \subset G_{p+q} \mathbf{U} \tag{7.2.1}$$

for any integers p and q.

Let $G_p\hat{U}(d)$ be the closure of the image of $G_pU(d) = G_pU \cap U(d)$ in $\hat{U}(d)$. Then the associated graded algebra is given by

$$\text{gr}^G\hat{U} = \bigoplus_d \bigoplus_p \text{gr}_p^G\hat{U}(d), \quad \text{gr}_p^G\hat{U}(d) = G_p\hat{U}(d) / G_{p-1}\hat{U}(d). \quad (7.2.2)$$

Considering the quotients by $\hat{B}$, we obtain the induced filtration G_pU of the current algebra U and the associated graded algebra:

$$\text{gr}^GU = \bigoplus_{d=-\infty}^{\infty} \text{gr}^GU(d), \quad \text{gr}^GU(d) = \bigoplus_p \text{gr}_p^GU(d). \quad (7.2.3)$$

Let us give the space gr^GU the induced degreewise topology and let $\widetilde{\text{gr}}^GU$ be the degreewise completion of the algebra gr^GU:

$$\widetilde{\text{gr}}^GU = \bigoplus_d \widetilde{\text{gr}}^GU(d), \quad \widetilde{\text{gr}}^GU(d) = \varprojlim_n \text{gr}^GU(d) / I_n(\text{gr}^GU(d)). \quad (7.2.4)$$

Then this is a compatible degreewise complete algebra. We will denote the image of $J_n(u)$ in $\widetilde{\text{gr}}^GU$ by $\psi_n(u)$.

Proposition 7.2.1 *The algebra U is quasi-finite if and only if $\widetilde{\text{gr}}^GU$ is so.*

Proof. By the construction, we have

$$Q_n(\widetilde{\text{gr}}^GU)(d) = \text{gr}^GQ_n(U)(d), \quad Q_n(U)(d) = \bigcup_p G_pQ_n(U)(d), \quad (7.2.5)$$

where $G_pQ_n(U)(d)$ denotes the induced filtration. Hence $Q_n(U)(d)$ is finite-dimensional if and only if $Q_n(\widetilde{\text{gr}}^GU)(d)$ is so. $\qquad\square$

7.3. Associated Poisson structure

Consider the operation of taking commutator of elements of U:

$$U \times U \to U, \quad (a, b) \mapsto [a, b] = a \cdot b - b \cdot a. \quad (7.3.1)$$

Then by the relation (6.3.1) we have

$$[J_m(u), J_n(v)] = \sum_{i=0}^{\infty} \binom{m + \Delta(u) - 1}{i} J_{m+n}(u_{(i)}v). \quad (7.3.2)$$

Lemma 7.3.1 $[G_pU, G_qU] \subset G_{p+q-1}U.$

Proof. The left-hand side of (7.3.2) belongs to $G_{\Delta(u)+\Delta(v)}U$ whereas the element $J_{m+n}(u_{(i)}v)$ in the right-hand side belongs to $G_{\Delta(u)+\Delta(v)-i-1}U$ for

$i \geq 0$. Hence an element of $[G_p\mathbb{U}, G_q\mathbb{U}]$ is written as a sum of elements of $G_{p+q-1}\mathbb{U}$. $\qquad\square$

Thanks to this lemma, the multiplication of $\mathrm{gr}^G\mathbb{U}$ is commutative and the operations $[\,,\,] : G_p\mathbb{U} \times G_q\mathbb{U} \to G_{p+q-1}\mathbb{U}$ induce operations

$$\mathrm{gr}^G_p\mathbb{U} \times \mathrm{gr}^G_q\mathbb{U} \to \mathrm{gr}^G_{p+q-1}\mathbb{U}, \quad (\alpha, \beta) \mapsto \{\alpha, \beta\} \qquad (7.3.3)$$

by letting $\{\alpha, \beta\}$ be the image of $[a, b]$ in $\mathrm{gr}^G_{p+q-1}\mathbb{U}$, where a and b are representatives of α and β, respectively.

In general, we will call a compatible degreewise topological algebra with a Poisson algebra structure for which the Poisson bracket is continuous a *compatible degreewise topological Poisson algebra*. In case the degreewise topology is complete then we will say that the Poisson algebra is a *compatible degreewise complete Poisson algebra*.

Proposition 7.3.2 *The multiplication and the bracket operation defined as above endow the space $\widetilde{\mathrm{gr}}^G\mathbb{U}$ with a structure of a compatible degreewise complete Poisson algebra.*

7.4. Relation to Zhu's Poisson algebra

Let us look more carefully at the relations $B_{m,n,r}(u, v) = 0$. Let p be an integer and let $m = -\Delta(u) + 1$. Then in case $m + n + r = p$ we have

$$J_p(u_{(r)}v) = \sum_{i=0}^{\infty}(-1)^i \binom{r}{i} J_{m+r-i}(u) \cdot J_{n+i}(v)$$

$$- (-1)^r \sum_{i=0}^{\infty}(-1)^i \binom{r}{i} J_{n+r-i}(v) \cdot J_{m+i}(u). \qquad (7.4.1)$$

Then the left-hand side belongs to $G_{\Delta(u)+\Delta(v)-r-1}\mathbb{U}$ whereas the right-hand side to $G_{\Delta(u)+\Delta(v)}\mathbb{U}$. Therefore,

$$J_p(u_{(r)}v) \equiv 0 \quad \text{if } r \leq -2. \qquad (7.4.2)$$

This implies that the map $J_p : V \to \mathrm{gr}^G\mathbb{U}$ factors a map

$$\psi_p : V/C_2(V) \to \mathrm{gr}^G\mathbb{U}. \qquad (7.4.3)$$

Now substitute $r = -1$ in (7.4.1) and replace m by $m + 1$. Then $m + n = p$ and

$$J_p(u_{(-1)}v) = \sum_{i=0}^{\infty} J_{m-i}(u) \cdot J_{n+i}(v) + \sum_{i=0}^{\infty} J_{n-i-1}(v) \cdot J_{m+i+1}(u). \quad (7.4.4)$$

Projecting (7.4.4) to the associated graded algebra, we have

$$\psi_p(u_{(-1)}v) = \sum_{i=0}^{\infty} \psi_{m-i}(u) \cdot \psi_{n+i}(v) + \sum_{i=0}^{\infty} \psi_{n-i-1}(v) \cdot \psi_{m+i+1}(u). \quad (7.4.5)$$

Note 7.4.1   The results in this section and the next are reformulations of the arguments in Subsection 3.2 of [NaT].

8. Poisson current algebras

In this section, we will construct a universal Poisson algebra satisfying the relations (7.4.5), which we will call a Poisson current algebra, and will investigate its properties.

8.1. Symmetric algebras on the loop Lie algebras

Let $\mathfrak{p}$ be a Poisson algebra. A *Poisson ideal* of $\mathfrak{p}$ means a subspace $\mathfrak{a}$ of $\mathfrak{p}$ such that both $\mathfrak{p} \cdot \mathfrak{a} \subset \mathfrak{a}$ and $\{\mathfrak{p}, \mathfrak{a}\} \subset \mathfrak{a}$ hold.

Consider the case when $\mathfrak{p}$ is given a grading $\mathfrak{p} = \bigoplus_k \mathfrak{p}^k$ indexed by integers satisfying

$$\{\mathfrak{p}^j, \mathfrak{p}^k\} \subset \mathfrak{p}^{j+k-1}, \quad \mathfrak{p}^j \cdot \mathfrak{p}^k \subset \mathfrak{p}^{j+k}. \quad (8.1.1)$$

Then the unity $1_{\mathfrak{p}}$ must belong to $\mathfrak{p}^0$. We will call a Poisson algebra endowed with such a grading a *graded Poisson algebra*. We denote $\Delta(x) = r$ when $x \in \mathfrak{p}^r$.

Recall the well-known fact that the symmetric algebra on a Lie algebra has a canonical structure of a Poisson algebra induced from the Lie bracket operation on the Lie algebra.

Let $\mathfrak{p} = \bigoplus_k \mathfrak{p}^k$ be a graded Poisson algebra and let $\mathfrak{p}[t, t^{-1}]$ be the loop Lie algebra

$$\mathfrak{p}[t, t^{-1}] = \mathfrak{p} \otimes_{\mathbf{k}} \mathbf{k}[t, t^{-1}] \quad (8.1.2)$$

with the Lie bracket defined by $[x \otimes t^m, y \otimes t^n] = \{x, y\} \otimes t^{m+n}$.

For each homogeneous x, put

$$\Psi_n(x) = x \otimes t^{n+\Delta(x)-1} \tag{8.1.3}$$

and extend it linearly to all x. Then the Lie bracket operation takes the following form: $[\Psi_m(x), \Psi_n(y)] = \Psi_{m+n}(\{x, y\})$.

Let $\mathfrak{q}$ be the quotient of $\mathfrak{p}[t, t^{-1}]$ by the span of the elements $\Psi_n(1_\mathfrak{p})$ with $n \neq 0$. We will denote the image of $\Psi_n(x)$ in $\mathfrak{q}$ by the same symbol. Since $[\Psi_m(1_\mathfrak{p}), \Psi_n(x)] = 0$, the space $\mathfrak{q}$ becomes a Lie algebra.

Now let $\mathbf{S}$ be the quotient algebra of the symmetric algebra on $\mathfrak{q}$ by the ideal generated by $\Psi_0(1_\mathfrak{p}) - 1$ and let us denote the image of $\Psi_n(x)$ in $\mathbf{S}$ by the same symbol. We give the degree $d_1 + \cdots + d_k$ to the element of the form $\Psi_{-d_1}(x_1) \cdots \Psi_{-d_k}(x_k)$ with $x_1, \ldots, x_k \in \mathfrak{p}$. Let $\mathbf{S}(d)$ be the span of these vectors of degree d. Then we have $\mathbf{S} = \bigoplus_d \mathbf{S}(d)$. The algebra $\mathbf{S}$ becomes a Poisson algebra for which we have

$$\mathbf{S}(d) \cdot \mathbf{S}(e) \subset \mathbf{S}(d + e), \quad \{\mathbf{S}(d), \mathbf{S}(e)\} \subset \mathbf{S}(d + e). \tag{8.1.4}$$

Recall the standard degreewise topology on $\mathbf{S}$ defined by

$$I_n(\mathbf{S}(d)) = \sum_{k \leq -n-1} \mathbf{S}(d - k) \cdot \mathbf{S}(k), \tag{8.1.5}$$

which is separated. Then the multiplication maps $\mathbf{S}(d) \times \mathbf{S}(e) \to \mathbf{S}(d + e)$ are continuous. Moreover we have the following.

Lemma 8.1.1 *The Poisson bracket operation* $\mathbf{S}(d) \times \mathbf{S}(e) \to \mathbf{S}(d + e)$ *is continuous with respect to the standard degreewise topology.*

Proof. Let i be any integer with $i \leq -n - 1$. Then we have $\{\mathbf{S}(d), \mathbf{S}(e - i) \cdot \mathbf{S}(i)\} \subset \{\mathbf{S}(d), \mathbf{S}(e-i)\} \cdot \mathbf{S}(i) + \{\mathbf{S}(d), \mathbf{S}(i)\} \cdot \mathbf{S}(e-i) \subset \mathbf{S}(d+e-i) \cdot \mathbf{S}(i) + \mathbf{S}(e-i) \cdot \mathbf{S}(d + i)$. Therefore, $\{u + I_k(\mathbf{S}(d)), v + I_m(\mathbf{S}(e))\} \subset \{u, v\} + I_n(\mathbf{S}(d + e))$ if k and m satisfy $k, k - e, m, m - d \geq n$. $\qquad\square$

Let $\hat{\mathbf{S}}$ be the degreewise completion of $\mathbf{S}$ with respect to the standard degreewise topology. Then by Lemma 8.1.1 the Poisson algebra structure on $\mathbf{S}$ extends to $\hat{\mathbf{S}}$. Let us denote the image of $\Psi_n(x)$ under the canonical map $\mathbf{S} \to \hat{\mathbf{S}}$ again by the same symbol.

The algebra $\hat{\mathbf{S}}$ is a compatible degreewise complete Poisson algebra.

8.2. Poisson current algebras

Let x, y be elements of $\mathfrak{p}$ and let p be an integer. Motivated by the relations (7.4.4) and (7.4.5), choose integers m, n with $m + n = p$ and set

$$N_p(x, y) = \Psi_{m+n}(x \cdot y) - \sum_{i=0}^{\infty} \left(\Psi_{m-i}(x) \cdot \Psi_{n+i}(y) + \Psi_{n-i-1}(y) \cdot \Psi_{m+i+1}(x) \right).$$
$$(8.2.1)$$

By the commutativity of $\hat{\mathbf{S}}$, this does not depend on the choice of m, n.

Let $\hat{\mathbf{N}}$ denote the degreewise closure of the ideal of $\hat{\mathbf{S}}$ generated by the elements of the form $N_p(x, y)$ with $x, y \in \mathfrak{p}$ and p an integer.

Lemma 8.2.1 *The ideal $\hat{\mathbf{N}}$ is a Poisson ideal of $\hat{\mathbf{S}}$.*

We let $\mathbb{S} = \mathbb{S}(\mathfrak{p})$ be the quotient of $\hat{\mathbf{S}}$ by the Poisson ideal $\hat{\mathbf{N}}$:

$$\mathbb{S} = \hat{\mathbf{S}} / \hat{\mathbf{N}}. \qquad (8.2.2)$$

We will call the Poisson algebra $\mathbb{S}$ the *Poisson current algebra* associated with the Poisson algebra $\mathfrak{p}$.

Proposition 8.2.2 *The algebra $\mathbb{S}$ is a compatible degreewise complete Poisson algebra.*

8.3. Quasi-finiteness

Now assume that our Poisson algebra $\mathfrak{p}$ is finite-dimensional and let $x_1, \ldots, x_r$ be a basis of a linear complement of $\mathbf{k}1_{\mathfrak{p}}$ in $\mathfrak{p}$. Then the algebra $\mathbf{S}$ is spanned by the elements of the form $\Psi_{-d_1}(x_{i_1}) \cdots \Psi_{-d_k}(x_{i_k})$ with $k \geq 0$ and $d_1 \geq \cdots \geq d_k$.

Consider the left canonical quotient module $Q_n(\mathbb{S}) = \mathbb{S}/I_n(\mathbb{S})$. Then $Q_n(\mathbb{S})$ is generated as a left $\mathbb{S}$-module by the image 1_n of the unit of $\mathbb{S}$.

For each $k \geq 0$ and $d \geq -n$, consider the set

$$\Pi_k(d) = \left\{ (d_1, \ldots, d_k) \mid d_1 \geq \cdots \geq d_k \geq -n \text{ and } d_1 + \cdots + d_k = d \right\}.$$
$$(8.3.1)$$

We will call a vector of the form $\Psi_{-d_1}(x_{i_1}) \cdots \Psi_{-d_k}(x_{i_k}) \cdot 1_n$ a *vector with index* $(d_1, \ldots, d_k)$. We put

$$\overset{\circ}{\Pi}_k(d) = \left\{ (d_1, \ldots, d_k) \mid d_1 > \cdots > d_k \geq -n \text{ and } d_1 + \cdots + d_k = d \right\}$$
$$(8.3.2)$$

and

$$\Pi(d) = \bigcup_{k=0}^{\infty} \Pi_k(d), \quad \overset{\circ}{\Pi}(d) = \bigcup_{k=0}^{\infty} \overset{\circ}{\Pi}_k(d). \tag{8.3.3}$$

Lemma 8.3.1 *The space* $Q_n(\mathbb{S})(d)$ *is spanned by vectors with indices in* $\overset{\circ}{\Pi}(d)$.

Proof. Since the module $Q_n(\mathbb{S})$ is exhaustive and the image of each $\mathbb{S}(d)$ is dense in $\mathbb{S}(d)$, the space $Q_n(\mathbb{S})(d)$ is spanned by vectors with indices in $\Pi(d)$. Let us show that any vector with index in $\Pi_k(d)$ is a linear combination of vectors with indices in $\overset{\circ}{\Pi}(d)$.

Introduce the lexicographic order on the set $\Pi_k(d)$: we define $(d_1, \ldots, d_k) < (e_1, \ldots, e_k)$ by $d_1 < e_1$ and in case $d_1 = e_1$ recursively by $(d_2, \ldots, d_k) < (e_2, \ldots, e_k)$. Then as this is a total order on a finite set, there exists a maximum element: that is $(d + (k-1)n, -n, \ldots, -n)$.

We now proceed by induction on the length k. The case $k = 1$ is trivial. Assume that the claim is true for any vector with index shorter than k and suppose given a vector

$$\Psi_{-d_1}(x_{i_1}) \cdots \Psi_{-d_k}(x_{i_k}) \cdot 1_n \tag{8.3.4}$$

with $d_1 \geq \cdots \geq d_k > -n - 1$. If $d_1 > \cdots > d_k > -n - 1$ then we have nothing to prove so we consider the case when $d_i = d_{i+1}$ at some position i. Recall the relations $N_p(x, y) = 0$, which imply

$$\Psi_{-d_i}(x_i) \cdot \Psi_{-d_i}(x_{i+1}) = \Psi_{-2d_i}(x_i \cdot x_{i+1}) - \sum_{j=1}^{\infty} \Big(\Psi_{d_i-j}(x_i) \cdot \Psi_{d_i+j}(x_{i+1})$$
$$+ \Psi_{d_i-j}(x_{i+1}) \cdot \Psi_{d_i+j}(x_i) \Big) \tag{8.3.5}$$

Hence the vector (8.3.4) is rewritten as the sum of a shorter vector and a finite number of vectors greater in the lexicographic order. By the inductive hypothesis, the shorter vector is written by the vectors with indices in $\overset{\circ}{\Pi}(d)$. Then apply the same argument to the vectors with greater indices in the rest of the sum. The recursion stops within a finite number of steps, at most at the maximum. □

Note 8.3.2 The argument described above is a refined variation of the proof of Theorem 3.2.7 in [NaT]. The idea of utilizing the relation (8.3.5) goes back to [GaN]. See [Buh] and [Li2] for related results.

We will say that a compatible degreewise complete Poisson algebra is *quasi-finite* if the conditions of quasi-finiteness for a compatible degreewise complete algebras are satisfied except the existence of a Hamiltonian. Now the following result is an immediate consequence of Lemma 8.3.1.

Theorem 8.3.3 *If $\mathfrak{p}$ is finite-dimensional then the algebra $\mathbb{S}$ is quasi-finite.*

Proof. By Lemma 8.3.1, the space $Q_n(\mathbb{S})(d)$ is spanned by the vectors with indices in $\overset{\circ}{\Pi}(d) = \bigcup_{k=0}^{\infty} \overset{\circ}{\Pi}_k(d)$, which is a finite set. Since $\mathfrak{p}$ is finite-dimensional, the number of the vectors with a fixed index is finite. Hence $Q_n(\mathbb{S})(d)$ is finite-dimensional. $\qquad\square$

9. Current algebras and Poisson current algebras

In this section, we will show that Zhu's finiteness condition on a vertex operator algebra implies the quasi-finiteness of the associated current algebra. This will be done by relating the results of the preceding section to the Poisson algebra $\widetilde{\mathrm{gr}}^G \mathbb{U}$ associated with $\mathbb{U}$.

9.1. Relation to the current algebras

A *homomorphism of degreewise topological Poisson algebras* is a map from a degreewise topological Poisson algebra to another such that it is a homomorphism of Poisson algebras that preserves the gradings for which the restriction to each homogeneous subspace is continuous.

Let V be a vertex operator algebra and let $\mathfrak{p}$ be Zhu's Poisson algebra $V/C_2(V)$. Recall the notations in the previous sections of Part II.

By the definition of Zhu's Poisson algebra, (7.4.2) implies that the maps $V \to \mathrm{gr}^G \mathbb{U}$ which sends $u \in V^k$ to the image $\psi_p(u)$ of $J_p(u)$ in $\mathrm{gr}_k^G \mathbb{U}$ factors a map

$$\psi_p : \mathfrak{p} \to \mathrm{gr}^G \mathbb{U}. \qquad (9.1.1)$$

Then the set of the maps $\psi_p : \mathfrak{p} \to \mathrm{gr}^G \mathbb{U}$ gives rise to a single map

$$\psi : \mathfrak{q} \to \mathrm{gr}^G \mathbb{U} \qquad (9.1.2)$$

which sends $\Psi_n(x)$ to $\psi_n(u)$, where $x = \bar{u}$ is the class of $u \in V$ in $\mathfrak{p}$.

Now the relations (7.3.2) and (7.4.2) imply that the map ψ is a homomorphism of Lie algebras. Since $\mathrm{gr}^G \mathbb{U}$ is a Poisson algebra, this map induces a unique homomorphism $\mathbf{S} \to \mathrm{gr}^G \mathbb{U}$ of Poisson algebras by the universal property of the symmetric algebra. We denote this map by the same symbol ψ.

Lemma 9.1.1   *The map ψ prolongs to a surjective homomorphism $\hat{\mathbb{S}} \to \widetilde{gr}^G \mathbb{U}$ of degreewise topological Poisson algebras.*

Proof. The assertion follows immediately from the construction by noting the relation $\psi(I_n(\mathbb{S})) = I_n(gr^G \mathbb{U})$, where the latter space is the one induced from $I_n(\mathbb{U})$. □

Now let $\mathbb{S}$ be the Poisson current algebra of $\mathfrak{p}$ as defined in Subsection 8.2. The relation (7.4.5) implies that the ideal $\hat{\mathbb{N}}$ is mapped to the closure of $gr^G \hat{\mathbb{B}}$. Therefore, the map $\hat{\mathbb{S}} \to \widetilde{gr}^G \mathbb{U}$ induces a homomorphism $\mathbb{S} \to \widetilde{gr}^G \mathbb{U}$ of degreewise topological Poisson algebras.

Thus we have verified the following result.

Theorem 9.1.2   *Let V be a vertex operator algebra and let $\mathbb{U}$ be the associated current algebra. Let $\widetilde{gr}^G \mathbb{U}$ be the degreewise completion of $gr^G \mathbb{U}$ and let $\mathbb{S}$ be the Poisson current algebra associated with Zhu's Poisson algebra $V/C_2(V)$. Then there exists a surjective homomorphism $\mathbb{S} \to \widetilde{gr}^G \mathbb{U}$ of degreewise topological Poisson algebras.*

9.2. Consequences of Zhu's finiteness condition

A vertex operator algebra V is said to satisfy *Zhu's finiteness condition* or said to be *C_2-finite* if Zhu's Poisson algebra $\mathfrak{p} = V/C_2(V)$ is finite-dimensional.

By combining Proposition 7.2.1, Theorem 9.1.2 and Theorem 8.3.3, we immediately see that Zhu's finiteness condition implies quasi-finiteness. Namely, we have the following theorem which is the main result of Part II.

Theorem 9.2.1   *If a vertex operator algebra V satisfies Zhu's finiteness condition then the associated current algebra $\mathbb{U}$ is quasi-finite.*

Let V be a vertex operator algebra satisfying Zhu's finiteness condition and let $\mathbb{U}$ be the associated current algebra. Let $\mathbb{U}_n$ be the finite-dimensional algebra associated with $\mathbb{U}$ as defined in Subsection 2.6 and let g be the number defined by (2.3.3). Then Theorem 9.2.1 allows us to apply the results of Part I to $(\mathbb{U}, L_0)$. For instance, we have the following.

Corollary 9.2.2   *Let V be a C_2-finite vertex operator algebra and let n be an integer such that $n \geq g$. Then the category of exhaustive V-modules is canonically equivalent to the category of left $\mathbb{U}_n$-modules.*

Moreover, the finiteness theorems in Subsection 5.5 hold for the various categories of $\mathbb{U}$-modules.

References

[Bor] R. E. Borcherds: Vertex algebras, Kac-Moody algebras, and the Monster. Proc. Nat. Acad. Sci. U.S.A. **83**, (1986), no. 10, 3068–3071.

[Bou] N. Bourbaki: Éléments de mathématique. Topologie générale. Chapitres 1 à 4. Hermann, Paris, 1971.

[Buh] G. Buhl: A spanning set for VOA modules. J. Algebra **254**, (2002), no. 1, 125–151.

[DLM1] C.-Y. Dong, H.-S. Li and G. Mason: Regularity of rational vertex operator algebras. Adv. Math. **132**, (1997), no. 1, 148–166.

[DLM2] C.-Y. Dong, H.-S. Li and G. Mason: Twisted representations of vertex operator algebras and associative algebras. Internat. Math. Res. Notices 1998, (1998), no. 8, 389–397.

[DLM3] C.-Y. Dong, H.-S. Li and G. Mason: Vertex operator algebras and associative algebras. J. Algebra **206**, (1998), no. 1, 67–96.

[DLi] C.-Y. Dong and Z.-Z. Lin: Induced modules for vertex operator algebras. Comm. Math. Phys. **179**, (1996), no. 1, 157–183.

[EGA] A. Grothendieck and J. Dieudonné: Eléments de géométrie algébrique I, Springer Verlag, Heidelberg, New York, 1971.

[FeF] B. I. Feigin and D. B. Fuks: Casimir operators in modules over the Virasoro algebra. Dokl. Akad. Nauk SSSR **269**, (1983), no. 5, 1057–1060.

[FHL] I. B. Frenkel, Y.-Z. Huang and J. Lepowsky: On axiomatic approaches to vertex operator algebras and modules. Mem. Amer. Math. Soc. **104**, (1993), no. 494,

[FLM] I. B. Frenkel, J. Lepowsky and A. Meurman: Vertex operator algebras and the Monster. Pure and Appl. Math. 134, Academic Press, Boston, 1989.

[FrZ] I. B. Frenkel and Y.-C. Zhu: Vertex operator algebras associated to representations of affine and Virasoro algebras. Duke Math. J. **66**, (1992), no. 1, 123–168.

[GaN] M. R. Gaberdiel and A. Neitzke: Rationality, quasirationality and finite W-algebras. Comm. Math. Phys. **238**, (2003), no. 1-2, 305–331.

[Gab] P. Gabriel: Des catégories abéliennes. Bull. Soc. Math. France **90**, (1962), 323–448.

[Gre] E. Gregorio: On a class of linearly compact rings. Comm. Algebra **19**, (1991), no. 4, 1313–1325.

[Kac] V. G. Kac: Laplace operators of infinite-dimensional Lie algebras and theta functions. Proc. Nat. Acad. Sci. U.S.A. **81**, (1984), no. 2, Phys. Sci., 645–647.

[Lef] S. Lefschetz: Algebraic Topology. American Mathematical Society Colloquium Publications, v. 27. American Mathematical Society, New York, 1942.

[Li1] H.-S. Li: Some finiteness properties of regular vertex operator algebras. J. Algebra **212**, (1999), no. 2, 495–514.

[Li2] H.-S. Li: Abelianizing vertex algebras. Preprint math.QA/0409140.

[LiW] H.-S. Li and S.-Q. Wang: On **Z**-graded associative algebras and their **N**-graded modules. Recent developments in quantum affine algebras and related topics (Raleigh, NC,1998), 341–357, Contemp. Math., 248, Amer. Math. Soc., Providence, RI, 1999.

[Mac] I. G. Macdonald: Duality over complete local rings. Topology **1**, (1962), 213–235.

[Mal] F. Malikov: On a duality for **Z**-graded algebras and modules. Unconventional Lie algebras, 103–114, Adv. Soviet Math., 17, Amer. Math. Soc., Providence, RI, 1993.

[MaN] A. Matsuo and K. Nagatomo: Axioms for a vertex algebra and the locality of quantum fields. MSJ Memoirs, 4. Mathematical Society of Japan, Tokyo, 1999.

[Men] C. Menini and A. Orsatti: Topologically left Artinian rings. J. Algebra **93**, (1985), 475–508.

[Mez] G. Mezzetti: Topological Morita equivalences induced by ideals generated by dense idempotents. J. Algebra **201**, (1998), 167–188.

[Mor] K. Morita: Duality for modules and its applications to the theory of rings with minimum condition. Sci. Rep. Tokyo Kyoiku Daigaku Sect. A **6**, (1958), 83–142.

[NaT] K. Nagatomo and A. Tsuchiya: Conformal field theories associated to regular chiral vertex operator algebras I: theories over the projective line. Duke Math. J. 128 (2005), no. 3, 393–471.

[Zhu] Y.-C. Zhu: Modular invariance of characters of vertex operator algebras. J. Amer. Math. Soc. **9**, (1996), no. 1, 237–302.

On Certain Automorphic Forms Associated to Rational Vertex Operator Algebras

Antun Milas

Department of Mathematics, University at Albany,
SUNY Albany, NY 12222
E-mail address: amilas@math.albany.edu

Abstract

To every rational vertex operator algebra V we associate an automorphic form on $\Gamma'(1)$ that we call the *Wronskian* of V. We have previously shown [M2], [M3] that in the case of Virasoro minimal models it is possible to give qualitative arguments about the Wronskian by using the representation theoretic methods. Here we apply the theory of automorphic forms and extend our previous work to a larger class of vertex operator algebras. We also give a detailed analysis of two-dimensional modular invariant spaces that arise from affine Kac-Moody Lie algebras.

As a main byproduct of our analysis we provide new proofs of certain Dyson-Macdonald's identities for powers of the Dedekind η–function for C_l, BC_l and D_l series, and related identities (e.g., Jacobi's Four Square Theorem).

0. Introduction and notation

The existence of a fusion ring and modular invariance of graded dimensions, or characters, are the most interesting features of every rational conformal field theory [MS]. When it comes to vertex operator algebra theory, proving modular invariance [Zh] (cf. [DLM1]) and ultimately the Verlinde formula [Hu1] is a formidable task. A key ingredient in proving modular invariance is played by the so-called C_2–cofiniteness [Zh] which, in particular, guarantees the convergence of all one-point functions on the torus. The C_2–cofiniteness plays an important role in the proof of the Verlinde conjecture as well [Hu1]. Even though the vector space spanned by irreducible characters is a $PSL(2, \mathbb{Z})$–module (i.e., a modular invariant space), one cannot state the Verlinde formula without having an action of $SL(2, \mathbb{Z})$ (the operator S^2 does not act as the identity in general–*charge conjugation*). In fact, irreducible characters are

330

sometimes linearly dependent. A proper algebraic framework in which *all* irreducible characters are accounted is the one of *modular data* and *modular tensor categories* [Hu1], [Hu3]. Modular data were studied intensively by several authors [Ga], [CG], [Ban], etc. There are classification results for modular data with a small number of irreducible modules (or primaries) so one hopes that modular data approach will help in classification of rational conformal field theories. Individual irreducible characters have also interesting properties. The most important result in this direction was obtained in [Ban], where it was proven that every irreducible character is in fact a (meromorphic) function on an appropriate modular curve $X(N)$. One should say that this result does not apply verbatim to rational vertex operator algebras. In fact, Bantay's construction [Ban] uses some facts that are still conjectural in the setting of rational vertex operator algebras.

In this paper we focus on certain number theoretic properties of modular invariant spaces spanned by irreducible characters, viewed as $PSL(2, \mathbb{Z})$-modules (so we will ignore the Verlinde formula at this point). The main motivation is our previous work [M2] [M3] where we have proved various q-series identities by using representation theoretic methods. Here, by using classical theory of automorphic forms, we extend our results to a larger class of rational vertex operator algebras.

Let us outline the content of the paper. To every rational vertex operator algebra V, or more precisely, to a modular invariant space spanned by the irreducible characters of V, we associate an automorphic form $\mathcal{W}_V(\tau)$ that we call the Wronskian of V. Since our space is a $PSL(2, \mathbb{Z})$-module, we show that $\mathcal{W}_V(\tau)$ is a meromorphic modular form on $\Gamma'(1)$ (see Theorem 1.4). Then we utilize this fact to show that, under certain conditions, $\mathcal{W}_V(\tau)$ is a power of the Dedekind η-function (see Theorem 2.2 and Proposition 2.4). Even though this phenomena has been observed in a variety of important models, there are cases where the Wronskian fails to be an η-power (see Section 6). It is an open question to describe $\mathcal{W}_V(\tau)$ as an infinite product in terms of some easily calculated data. Such a formula would have to capture arithmetic properties of the zeros of $\mathcal{W}_V(\tau)$ (see Proposition 2.3).

As an application of our results to familiar rational vertex operator algebras we have (cf. Theorem 3.1 and Theorem 4.3):

Theorem 0.1. *Let V be one of the following:*

(i) *Vertex operator algebra $L_{A_1}(k-1, 0)$ associated to the affine Kac-Moody Lie algebra of type $A_1^{(1)}$, $k \in \mathbb{N}$,*

(ii) *Virasoro vertex operator algebra $L(c_{p,p'}, 0)$ associated to $\mathcal{M}(p, p')$ Virasoro minimal models, $k = \frac{(p-1)(p'-1)}{2}$,*

(iii) *Rank one lattice vertex operator algebra V_L, where $L = \mathbb{Z}\alpha$ and $\langle \alpha, \alpha \rangle = 2N$, $k = N + 1$.*

Then

$$\mathcal{W}_V(\tau) = \eta(\tau)^{2k(k-1)}.$$

In particular, for appropriate choices of k, p and p', the above formula give a series of Dyson-Macdonald's identities for C_l, BC_l and D_l series of affine root systems, respectively.

The part (ii) in the theorem was proven earlier in [M3] by using representation theoretic methods.

We analyze further the case of two-dimensional modular invariant spaces that arise from rational vertex operator algebras. In particular, we compute $\mathcal{W}_V(\tau)$ for a class of rational vertex operator algebras associated to affine Kac-Moody Lie algebras that have exactly two inequivalent modules (see Proposition 5.2). These models are essentially powers of $\eta(\tau)$, so by using explicit formulas for the characters we can prove several classical q-series identities. Here we give two examples that are due to Ramanujan [M2], and Jacobi, respectively (see Corollary 4.2):

Corollary 0.2.

$$1 - 5\sum_{n=1}^{\infty} \left(\frac{n}{5}\right) \frac{nq^n}{1 - q^n} = \frac{\eta(\tau)^5}{\eta(5\tau)}, \tag{0.1}$$

$$1 + 8\left(\sum_{n=1}^{\infty} \frac{2nq^{2n}}{1 + q^{2n}} - \frac{(2n-1)q^{2n-1}}{1 + q^{2n-1}}\right) = \left(\sum_{n\in\mathbb{Z}}(-q)^{n^2}\right)^4. \tag{0.2}$$

It is worth saying here that considering Wronskians in this context is motivated by a more or less classical situation in number theory and the theory of Riemann surfaces where Weierstrass points on modular curves can be studied via holomorphic differentials (i.e., cusp forms of weight 2) and Wronskians [Mir], [Ro]. We hope that $\mathcal{W}_V(\tau)$ and their zeros play a similar role and may unravel some hidden arithmetic properties of rational vertex operator algebras. A word of caution here. In the vertex operator algebra setting irreducible characters give rise to modular invariant spaces (of weight 0), not of weight 2. Also, in the vertex operator algebra setting we are dealing with meromorphic modular forms (there will be poles at cusps in general). A possible resolution is to consider the Wronskian of *derivatives* of irreducible modules which are of weight two. We explored this direction further in [MMO].

Notation: Throughout the text we will be using the following notation: $\mathbb{N}$; the set of non-negative integers, $\mathbb{H}$; the upper half-plane (which does not include the cusp at ∞) and $q = e^{2\pi i \tau}$. To simplify the exposition we will often use "characters" instead of "modified graded dimensions". As usual, for $N \geq 1$ we let

$$\Gamma(N) = \left\{ \begin{bmatrix} a & b \\ c & d \end{bmatrix} \in SL(2, \mathbb{Z}) : a \equiv d \equiv 1 \bmod N, b \equiv c \equiv 0 \bmod N \right\},$$

$$\Gamma_0(N) = \left\{ \begin{bmatrix} a & b \\ c & d \end{bmatrix} \in SL(2, \mathbb{Z}) : c \equiv 0 \bmod N \right\}.$$

The Eisenstein series are given by

$$G_{2k}(\tau) = -\frac{B_{2k}}{(2k)!} + \frac{2}{(2k-1)!} \sum_{n=1}^{\infty} \frac{n^{2k-1} q^n}{1 - q^n}, \ k \geq 1,$$

$$E_{2k}(\tau) = \frac{-(2k)!}{B_{2k}} G_{2k}(\tau).$$

Acknowledgment: Some portions of this paper were presented at the conference *Moonshine-the First Quarter Century and Beyond*, Edinburgh, July 2004. I would like to thank the organizers for the invitation. My thanks also go to Sunil Mukhi for informing me of his old work on RCFTs and modular forms.

1. Wronskians and rational vertex operator algebras

In this part to every rational vertex operator algebra V we associate a canonical automorphic form $\mathcal{W}_V(\tau)$ that we call the *Wronskian* of V. We shall see that this automorphic form coincides with a certain graded trace computed on a tensor product of several copies of V.

Let us recall first the main parts in the definition of vertex operator algebra (for more details see [LL], for example). Vertex operator algebra $(V, Y, \mathbf{1}, \omega)$ consists of a $\mathbb{Z}$–graded vector space

$$V = \coprod_{n \in \mathbb{Z}} V_n, \ \dim(V_n) < +\infty,$$

with two distinguished vectors $\mathbf{1} \in V_0$ (the *vacuum vector*) and $\omega \in V_2$ (the *conformal vector*), equipped with the vertex operator map

$$Y : V \longrightarrow \mathrm{End}(V)[[x, x^{-1}]],$$

subject to certain grading conditions, the truncation condition, the creation property and most importantly the Jacobi identity. In addition, the conformal vector ω defines a representation of the Virasoro algebra (i.e., essentially unique central extension of the Lie algebra of vector fields on the circle), so that

$$Y(\omega, x) = \sum_{n \in \mathbb{Z}} L(n) x^{-n-2},$$

$$[L(m), L(n)] = (m-n)L(m+n) + \frac{m^3 - m}{12} \delta_{m+n,0} c, \qquad (1.1)$$

$$Y(L(-1)v, x) = \frac{d}{dx} Y(v, x),$$

where $c \in \mathbb{C}$ (the number c is called the central charge of V). Moreover, the operator $L(0)$ is compatible with the grading of V; $L(0) \cdot v = nv$, for every $v \in V_n$. Finally, $Y(\mathbf{1}, x)$ is the identity operator on V.

Naturally, we can define the notion of a module for a vertex operator algebra, with an important difference in a relaxation of the grading condition for a module M, now being a $\mathbb{C}$–graded vector space

$$M = \coprod_{r \in \mathbb{C}} M_r, \ \dim(M_r) < +\infty,$$

where, again, M admits an action of the Virasoro algebra (with the same central charge) and the grading stems from the action of $L(0)$.

We say that a vertex operator algebra V satisfies C_2–cofiniteness condition if $\dim(V/C_2(V)) < +\infty$, where

$$C_2(V) = \{u_{-2}v \ : \ u, v \in V\}.$$

The following definition is from [DLM1] [DLM2]:

Definition 1.1. *We say that a vertex operator algebra V is rational if V is C_2–cofinite and every admissible V–module is completely reducible.*

For various reasons it is important to study graded dimensions, or simply *characters*, of irreducible V–modules, where the character of a V–module M is defined as

$$\mathrm{tr}|_M q^{L(0) - \frac{c}{24}} = q^{-c/24} \sum_{r \in \mathbb{C}} \dim(M_r) q^r$$

where c is as in (1.1). Of course, for rational vertex operator algebras it suffices to study irreducible characters, i.e., characters associated to irreducible representations. Then an important result proven by Zhu [Zh], and improved slightly by Dong, Li and Mason [DLM1], states that for every rational vertex operator algebra V the vector space spanned by irreducible characters is modular invariant, i.e., an $SL(2, \mathbb{Z})$–module. Moreover, it is known [DLM1] (see also [AM]) that in the rational case the irreducible characters take the form

$$\mathrm{tr}|_M q^{L(0) - c/24} = q^r \sum_{n=0}^{\infty} a_n q^n,$$

where r is a rational number.

Suppose that V is a rational vertex operator algebra. Denote by $\mathcal{M}_V$ the finite-dimensional vector space spanned by irreducible characters of V. Let

$$d_V = \dim(\mathcal{M}_V).$$

If a vertex operator algebra is fixed we will often omit writing the subscript V and write d instead. It is known that in general d_V *does not* equal the number of mutually inequivalent irreducible modules of V (see Section 3.). Now, pick a basis $\{f_1, ..., f_{d_V}\}$ for $\mathcal{M}_V$. Consider the Wronskian

$$W_{\left(q\frac{d}{dq}\right)}(f_1, ..., f_{d_v}) = \begin{vmatrix} f_1 & f_2 & \cdot & \cdot & f_{d_v} \\ f_1' & f_2' & \cdot & \cdot & f_{d_v}' \\ \cdot & \cdot & \cdot & \cdot & \cdot \\ \cdot & \cdot & \cdot & \cdot & \cdot \\ f_1^{(d_v-1)} & f_2^{(d_v-1)} & \cdot & \cdot & f_{d_v}^{(d_v-1)} \end{vmatrix},$$

where

$$f^{(j)} = \left(q\frac{d}{dq}\right)^j f(q), \ j \geq 1.$$

Clearly, $W_{\left(q\frac{d}{dq}\right)}(f_1, ..., f_{d_V})(\tau)$ is holomorphic and has a q-expansion given by expanding the determinant. Keep in mind that $\left(q\frac{d}{dq}\right) = -i\frac{1}{2\pi}\frac{d}{d\tau}$.

Definition 1.2. Let V be a rational vertex operator algebra. Then the *Wronskian* $\mathcal{W}_V(\tau)$ of V is defined as a (non-zero) multiple of $W_{\left(q\frac{d}{dq}\right)}(f_1, ..., f_{d_V})$ with the property that the leading coefficient in the q-expansion is 1.

The previous definition clearly does not depend on the choice of a basis for $\mathcal{M}_V$. In what follows we will need a few basic properties of the Wronskian determinant.

Lemma 1.3.

(a) *Let A be a linear operator on $\mathcal{M}_V$ and f and h holomorphic functions in $\mathbb{H}$, then*

$$W_{\left(q\frac{d}{dq}\right)}(A \cdot f_1, ..., A \cdot f_d) = \det(A) \, W_{\left(q\frac{d}{dq}\right)}(f_1, ..., f_d).$$

(b)

$$W_{\left(q\frac{d}{dq}\right)}(f \cdot f_1, ..., f \cdot f_d) = f^d W_{\left(q\frac{d}{dq}\right)}(f_1, ..., f_d)$$

(c)

$$W_{\left(h(\tau)q\frac{d}{dq}\right)}(f_1, ..., f_d) = h(\tau)^{d(d-1)} W_{\left(q\frac{d}{dq}\right)}(f_1, ..., f_d).$$

Proof: The first formula is clear. Part (b) follows from the Leibnitz formula

$$(fg)^{(n)} = \sum_{k=0}^{n} \binom{n}{k} f^{(k)} g^{(n-k)}$$

and properties of the determinant (apply row operations!). Similarly with (c).

∎

Let us recall a few basic definitions in the theory of automorphic forms and functions (for details consult [Miy] or [Mil]). Let $\Gamma \subset \Gamma(1)$ be a congruence subgroup (e.g., $\Gamma_0(N)$ or $\Gamma(N)$). We say that a holomorphic function f in $\mathbb{H}$ is a modular form of weight k with respect to Γ, if it has a q-expansion $f(\tau) = \sum_{n \geq 0} a_n q^{n/r}$, where $r \in \mathbb{N}$, and it satisfies $f(\gamma \cdot \tau) = j(\gamma, \tau)^k f(\tau)$ where $\gamma \in \Gamma$, $\gamma = \begin{bmatrix} a & b \\ c & d \end{bmatrix}$ and $j(\gamma, \tau) = (c\tau + d)$. The number r is also denoted as $\mathrm{ord}_{i\infty}(f)$. Cusps of $\Gamma \subset \Gamma(1)$ are defined as equivalence classes of $\mathbb{Q} \cup \{i\infty\}$ under the action of $SL(2, \mathbb{Z})$. If $\Gamma = \Gamma(1)$ then there is only one orbit and the corresponding cusp is $i\infty$. A function f is said to be holomorphic at $i\infty$ if $r \geq 0$. Similarly, f is said to be holomorphic at the cusp $r \in \mathbb{Q} \cup \{i\infty\}$ if $f(v \cdot \tau)$ is holomorphic at $i\infty$, where $v \cdot i\infty = r$. For a function $f(\tau)$ defined on $\mathbb{H}$ we define an action of $\gamma \in \Gamma(1)$ by $(f|_k \gamma)(\tau) = j(\gamma, \tau)^{-k} f(\gamma \cdot \tau)$. We say that f is an automorphic form on $\Gamma(1)$ of weight k, with the multiplier system χ, if $(f|_k)(\gamma \cdot \tau) = \chi(\gamma) f(\tau)$, where χ is a character of $\Gamma(1)$.

In the following theorem we do not use much of the theory of vertex operator algebras so it applies for an arbitrary modular invariant vector space $\mathcal{M}$.

Theorem 1.4. *Let V be as above, then the Wronskian $\mathcal{W}_V(\tau)$ is a modular form of weight $k(k - 1)$ for $\Gamma'(1) := \Gamma(1)/[\Gamma(1), \Gamma(1)]$. Moreover, the sixth power of $\mathcal{W}_V(\tau)$ is a holomorphic modular form for $SL(2, \mathbb{Z})$, with a possible pole at the infinity.*

Proof: For simplicity let $d = d_V$. Let $f_1, ..., f_d$ be a basis for $\mathcal{M}_V$. As usual let

$$S = \begin{bmatrix} 0 & -1 \\ 1 & 0 \end{bmatrix}, \quad T = \begin{bmatrix} 1 & 1 \\ 0 & 1 \end{bmatrix}.$$

Let us introduce constants S_a^b via

$$f_a(S \cdot \tau) = \sum_{b=1}^{d} S_a^b f_b(a), \quad a = 1, ..., d.$$

The vector space $\mathcal{M}_V$ is in fact a $PSL(2, \mathbb{Z})$–module, because S^2 acts as the identity operator. Also, $(ST)^3 = I$ and $S^4 = I$, where I is the identity matrix.

From these formulas it follows that $\det(S) = \pm 1$ and $(\det(T))^6 = 1$. We have to compute

$$W_{\left(q\frac{d}{dq}\right)}(f_1, ..., f_d)(S \cdot \tau).$$

For that purpose notice that

$$f_a^{(i)}(S \cdot \tau) = \left(\tau^2 \frac{d}{d(2\pi i\tau)}\right)^i (f_a(S \cdot \tau)) = \sum_{a=1}^{d} S_a^b \left(\tau^2 \frac{d}{d(2\pi i\tau)}\right)^i f_b(\tau).$$

The previous formula, combined with Lemma 1.3 (a),(c), yields

$$W_{\left(\tau^2 \frac{d}{2\pi i d\tau}\right)}(f_1(S \cdot \tau), ..., f_d(S \cdot \tau)) = \det(S)\tau^{d(d-1)} W_{\left(q\frac{d}{dq}\right)}(f_1(\tau), ..., f_d(\tau)).$$

Thus

$$\mathcal{W}_V\left(\frac{-1}{\tau}\right) = \pm \tau^{d(d-1)} \mathcal{W}_V(\tau).$$

For the T action a similar computation yields

$$\mathcal{W}_V(\tau + 1) = \zeta_6 \mathcal{W}_V(\tau),$$

where ζ_6 is a sixth root of unity. Let us recall again [DLM1] that every irreducible character of a rational vertex operator algebras has rational q-powers truncated from below. Thus, the Wronskian $\mathcal{W}(\tau)$ has q-expansion

$$\mathcal{W}_V(\tau) = q^{\frac{r}{6}} \sum_{n=0}^{\infty} a_n q^n, \quad r \in \mathbb{Z}.$$

This is a cusp form if $r > 0$. To prove that $\mathcal{W}_V(\tau)$ is a modular form on the commutator subgroup $\Gamma'(1)$ it is enough to recall that, by the definition, $\Gamma'(1)$ is generated by elements of the form $aba^{-1}b^{-1}$, which act trivially on any one-dimensional representation of $SL(2, \mathbb{Z})$. ∎

Remark 1. The group $\Gamma'(1)$ is closely related to a congruence subgroup $\Gamma(6) \triangleleft SL(2, \mathbb{Z})$, In fact, the group $\Gamma(6)$ is the smallest principal congruence subgroup under which the multiplier system for $\mathcal{W}_V$ trivializes. If we use a presentation for $SL(2, \mathbb{Z}/6)$ given in [CG] it easily follows that the multiplier system factors through the quotient $SL(2, \mathbb{Z}/6)$, thus $\mathcal{W}_V(\tau)$ is a modular form for $\Gamma(6)$.

Remark 2. Let $L(c_{p,p'}, 0)$ be the Virasoro vertex operator algebra associated to $\mathcal{M}(p, p')$–minimal models [LL], [FZ], [M2], [M3]. Then

$$\mathcal{W}_{L(c_{p,p'},0)}(\tau) = \eta(\tau)^{2k(k-1)},$$

where $k = \frac{(p-1)(p'-1)}{2}$ is the number of irreducible modules for $L(c_{p,p'}, 0)$. In the case of $\mathcal{M}(2, 5)$ minimal models [M2] the Wronskian is $\eta(\tau)^4$. This fact is useful for proving the Ramanujan's formula (0.2).

Remark 3. The previous theorem does not apply for rational vertex operator superalgebras. The vector space spanned by irreducible characters in the vertex operator superalgebra setting is no longer modular invariant. However, if we add appropriate twisted modules and super graded traces (i.e., supercharcaters) we can construct three Wronskians that (conjecturally) get permuted under the action of $SL(2, \mathbb{Z})$. For an explicit constructions of these automorphic forms in the case of $N = 1$ superconformal minimal models we refer the reader to [M4].

The next result gives another interpretation of $\mathcal{W}_V$. In fact, the following construction works for vertex operator (super)algebras as well. Let us introduce some notation first. For every homogeneous $v \in V_m$, $Y(v, x) = \sum_{n \in \mathbb{Z}} v_n x^{-n-1}$, acting on a V–module M, let $o(v) = v_{m-1}$; a grading preserving operator on M. Thus we have a well–defined trace map (which for a moment we consider only as a formal q–series):

$$v \mapsto \mathrm{tr}|_M o(v) q^{L(0)-c/24}, \qquad (1.2)$$

and its multi-linear extension

$$v_1 \otimes \cdots \otimes v_k \mapsto \mathrm{tr}|_M o(v_1) \otimes \cdots \otimes o(v_k) q^{L(0)-c/24}.$$

The antisymmetrization map

$$\mathrm{Alt} : \Lambda^k V \longrightarrow V^{\otimes k}$$

is defined as usual. Consider now

$$\mathrm{Alt}\left\{\mathbf{1} \wedge L[-2]\mathbf{1} \wedge L[-2]^2\mathbf{1} \wedge \cdots \wedge L[-2]^{k-1}\mathbf{1}\right\} \in V^{\otimes k}.$$

Proposition 1.5. *Suppose that V is rational and M a $V^{\otimes d_V}$–module. Then*

$$\mathrm{tr}|_M o\left(\mathrm{Alt}\left\{\mathbf{1} \wedge L[-2]\mathbf{1} \wedge \cdots \wedge L[-2]^{k-1}\mathbf{1}\right\}\right) q^{L(0)-c/24} = \lambda \mathcal{W}_V(\tau),$$

where $\lambda \in \mathbb{Q}$ (possibly zero).

Proof: Firstly, we may assume that M is an irreducible $V^{\otimes d}$–module. Thus, $M \cong M_1 \otimes \cdots \otimes M_d$, where M_i are irreducible V–modules [FHL]. Now,

$$\text{tr}|_{M_1 \otimes M_2 \otimes \cdots \otimes M_d} o \left(\text{Alt} \left\{ \mathbf{1} \wedge L[-2]\mathbf{1} \wedge \cdots \wedge L[-2]^{d-1}\mathbf{1} \right\} \right) q^{L(0)-c/24} =$$

$$\begin{vmatrix} \text{tr}|_{M_1} q^{L(0)-c/24} & \text{tr}|_{M_2} q^{L(0)-c/24} \\ \text{tr}|_{M_1} o(L[-2]\mathbf{1}) q^{L(0)-c/24} & \text{tr}|_{M_2} o(L[-2]\mathbf{1}) q^{L(0)-c/24} \\ \cdot & \cdot \\ \cdot & \cdot \\ \cdot & \cdot \\ \text{tr}|_{M_1} o(L[-2]^{d-1}\mathbf{1}) q^{L(0)-c/24} & \text{tr}|_{M_2} o(L[-2]^{d-1}\mathbf{1}) q^{L(0)-c/24} \\ \end{vmatrix}$$

$$\begin{matrix} \cdot \cdot & \text{tr}|_{M_d} q^{L(0)-c/24} \\ \cdot \cdot & \text{tr}|_{M_d} o(L[-2]\mathbf{1}) q^{L(0)-c/24} \\ \cdot \cdot & \cdot \\ \cdot \cdot & \cdot \\ \cdot \cdot & \text{tr}|_{M_d} o(L[-2]^{d-1}\mathbf{1}) q^{L(0)-c/24} \end{matrix}$$

If the characters of $M_1, \ldots, M_d$ are linearly dependent then

$$\text{tr}|_{M_1 \otimes \cdots \otimes M_d} \left(\text{Alt} \left\{ \mathbf{1} \wedge L[-2]\mathbf{1} \wedge \cdots \wedge L[-2]^{k-1}\mathbf{1} \right\} \right) q^{L(0)-c/24} = 0,$$

because the determinant is equal to zero. Let us compute entries of the determinant. It is known [Zh] that

$$\text{tr}|_M o(L[-2]^i \mathbf{1}) q^{L(0)-c/24}$$

$$= \sum_{j=0}^{i} P_{i,j}(G_2, G_4, G_6) \left(q \frac{d}{dq} \right)^j \text{tr}|_M q^{L(0)-c/24},$$

where $P_{i,j}(G_2, G_4, G_6)$ are certain polynomials in Eisenstein series G_2, G_4 and G_6, and more importantly $P_{i,i} = 1$, for every i. Now, from properties of the determinant it follows that

$$\text{tr}|_{M_1 \otimes M_2 \otimes \cdots \otimes M_d} o \left(\text{Alt} \left\{ \mathbf{1} \wedge L[-2]\mathbf{1} \wedge \cdots \wedge L[-2]^{d-1}\mathbf{1} \right\} \right) q^{L(0)-c/24}$$

is a multiple of the Wronskian $\mathcal{W}_V(\tau)$.

Now, the statement obviously works for an arbitrary $V^{\otimes d}$–module simply because every $V^{\otimes d}$–module is a sum of irreducible modules (cf. [LL]). ∎

2. Differential equations and $\mathcal{W}_V(\tau)$

The Wronskian determinant is a very useful gadget for studying homogeneous ordinary differential equations. We review this well-known relationship applied in the special case where solution spaces are modular invariant subspaces. We should say here that some results from this session have been known for a while by physicists and number theorists (e.g., for a related discussion in the setup of rational conformal field theory setting see [AM], [Mu1], [Mu2] and [MMS]).

Denote by $\Omega(\mathbb{H})$ the space of holomorphic functions in the upper half-plane that have q-expansion of the form

$$\sum_{i=1}^{s} q^{r_i} \sum_{m=0}^{\infty} a_i^{(m)} q^m.$$

Let $Gr_{(k)}$ be the Grassmannian of k-dimensional subspaces of $\Omega(\mathbb{H})$ and $\mathcal{D}_{(k)}$ the affine space of k-th order differential operators

$$\left(q\frac{d}{dq}\right)^k + \sum_{i=0}^{k-1} P_i(q) \left(q\frac{d}{dq}\right)^i,$$

where $P_i(q)$ are meromorphic functions in $\mathbb{H}$. Then there is an injective map Ψ from $Gr_{(k)}$ to $\mathcal{D}_{(k)}$ given by

$$(f_1, ..., f_k) \mapsto (-1)^k \frac{W_{(q\frac{d}{dq})}(f, f_1, ..., f_k)}{W_{(q\frac{d}{dq})}(f_1, ..., f_k)}, \tag{2.1}$$

where $\{f_1, ..., f_k\}$ is any basis of $(f_1, ..., f_k) \in Gr_{(k)}$. The preimage of Ψ is obtained by solving the differential equation $(-1)^k \frac{W_{(q\frac{d}{dq})}(f, f_1, ..., f_k)}{W_{(q\frac{d}{dq})}(f_1, ..., f_k)} = 0$. The coefficients $P_i(q)$ in

$$(-1)^k \frac{W_{(q\frac{d}{dq})}(f, f_1, ..., f_k)}{W_{(q\frac{d}{dq})}(f_1, ..., f_k)} = \left(q\frac{d}{dq}\right)^n + \sum_{i=0}^{k-1} P_i(q) \left(q\frac{d}{dq}\right)^i,$$

are meromorphic functions in general, but if the Wronskian $W_{(q\frac{d}{dq})}(f_1, ..., f_k)$ is non-vanishing in $\mathbb{H}$, then $P_i(q)$ will be holomorphic.

Instead of $\left(q\frac{d}{dq}\right)^i$, it is more convenient to work with a slightly modified differential operators. As in [Ro] we let

$$\Theta_\tau = q\frac{d}{dq},$$

$$\Theta_\tau^2 = (q\frac{d}{dq} + 2G_2)\Theta_\tau,$$

$$\Theta_\tau^k = (q\frac{d}{dq} + 2kG_2)\Theta_\tau^{k-1}, \ k \geq 3.$$

From

$$G_2(-1/\tau) = \tau^2 G_2(\tau) - \frac{\tau}{2\pi i},$$
$$G_2(\tau + 1) = G_2(\tau),$$

it follows

$$\Theta^k_{\tau+1} = \Theta^k_\tau,$$
$$\Theta^k_{-1/\tau} = \tau^{2k}\Theta^k_\tau. \tag{2.2}$$

Now, we introduce generalized Wronskian determinants

$$W^{i_1,\dots,i_k}_{\Theta_\tau}(f_1, \dots, f_k) = \det[\Theta^{i_j}_\tau f_l]_{1\le j\le k, 1\le l\le k},$$
$$0 \le i_1 < i_2 < \cdots < i_{k-1} < i_k. \tag{2.3}$$

In particular $W^{1,2,\dots,k}_{\Theta_\tau} = W_{\Theta_\tau}$. It is not hard to see (see [Ro] for instance) that

$$W_{\Theta_\tau}(f_1, \dots, f_k) = \lambda_k W_{(q\frac{d}{dq})}(f_1, \dots, f_k), \tag{2.4}$$

for some nonzero constant λ_k. From the above formulas we have:

Lemma 2.1. *Let* $\{f_1, \dots, f_k\}$ *be a basis of* $\mathcal{M}_V$. *Then for every* $i_1, \dots i_k$ *as in (2.3)*

$$\frac{W^{i_1,\dots,i_k}_{\Theta_\tau}(f_1, \dots, f_k)}{W_{\Theta_\tau}(f_1, \dots, f_k)}$$

is a (meromorphic) modular form of weight $2i_i + 2i_2 + \cdots + 2i_k$.

Now, we consider again the map Ψ where instead of $\left(q\frac{d}{dq}\right)$-derivative we work with the Θ-derivative. This time to an element $(f_1, \dots, f_k) \in \mathrm{Gr}_{(k)}$ we associate

$$(-1)^k \frac{W_{\Theta_\tau}(f, f_1, \dots, f_k)}{W_{\Theta_\tau}(f_1, \dots, f_k)}. \tag{2.5}$$

Theorem 2.2. *Let* $S = \{f_1, \dots, f_k\}$ *be a basis of a modular invariant space* $\mathcal{M}_V$ *and suppose further that* $\mathcal{W}_V(\tau)$ *is non-vanishing in the upper half-plane. Then there is a unique* k^{th} *order differential equation with coefficients being quasimodular forms, with a possible pole at the infinity, with a basis of solution being S. Moreover,*

$$\mathcal{W}_V(\tau) = \eta(\tau)^{2k(k-1)}.$$

In particular, if

$$f_i = q^{h_i-c/24} \sum_{n=0}^{\infty} a_n^{(i)} q^n,$$

and the exponents satisfy

$$h_i \neq h_j, \text{ for } i \neq j,$$

then the coefficients $P_i(q)$ are holomorphic at the infinity and

$$24 \sum_{i=1}^{k} h_i = k(2k - 2 + c). \qquad (2.6)$$

Proof: The first part follows immediately from Lemma 2.1 and (2.5) (the uniqueness is obvious). Suppose now that the Wronskian $\mathcal{W}_V(\tau)$ is non-vanishing in $\mathbb{H}$. Then

$$(-1)^k \frac{W_{\Theta_\tau}(f, f_1, ..., f_k)}{W_{\Theta_\tau}(f_1, ..., f_k)} = \Theta^k f + \sum_{i=0}^{k-1} \tilde{P}_i(\tau) \Theta^i f = 0, \qquad (2.7)$$

where $\tilde{P}_i(q)$ are quotients of certain generalized Wronskians appearing in Lemma 2.1. From Lemma 2.1 we know that $\tilde{P}_i(q)$ is a holomorphic modular form of the weight $2(k-i)$, with at most polynomial growth at the infinity. Because of (2.4) it follows we see that $\tilde{P}_i$ will be holomorphic in $\mathbb{H}$. Thus, $\tilde{P}_i \in \mathbb{C}[G_4, G_6, j]$. In particular, $\tilde{P}_{k-1}(q) = 0$ (i.e., there is no holomorphic modular form of weight 2 with a polynomial growth at the infinity). Let us recall that the ring of holomorphic quasimodular forms is invariant under the action of $(q\frac{d}{dq})$, i.e., for $i = 1, 2, 3$, $\left(q\frac{d}{dq}\right) G_{2i}(\tau)$ is expressible in terms of the basic Eisenstein series $G_2(\tau)$, $G_4(\tau)$ and $G_6(\tau)$. Thus, in

$$\Theta^k f + \sum_{i=0}^{k-1} \tilde{P}_i(\tau) \Theta^i f = \left(q\frac{d}{dq}\right)^k f + \sum_{i=0}^{k-1} P_i(\tau) \left(q\frac{d}{dq}\right)^i f,$$

P_i are quasimodular forms. It is easy to see that $P_{k-1}(\tau) = k(k-1)G_2(\tau)$. To show

$$\mathcal{W}_V(\tau) = \eta(\tau)^{2k(k-1)},$$

we just recall the Abel's lemma for ODEs and proceed as in Theorem 8.1 [M2] or as in [M3]. The equality (2.6) is obtained by comparing the leading powers in the q-expansion of $\eta(\tau)^{2k(k-1)}$ and $\mathcal{W}_V(\tau)$. ∎

If $\mathcal{W}_V(\tau)$ has a zero in $\mathbb{H}$, then the previous theorem fails and $P_{k-1}(q)$ has poles in general. For instance, consider

$$y' + \frac{E_6}{E_4} y = 0,$$

where $\mathcal{W}_V(\tau) = j(\tau)$. Modular forms such as $j(\tau)$ admit beautiful product formulas (after Gross, Zagier and Borcherds) so it is tempting to seek for an infinite-product formula for $\mathcal{W}_V$ in general. One idea is to apply a result of Bruinier, Kohnen and Ono [BKO]:

Proposition 2.3.

$$P_{k-1}(\tau) = \frac{k(k-1)E_2(\tau)}{12} - \frac{E_4^2(\tau)E_6(\tau)}{\Delta(\tau)} \sum_{z \in \mathbb{H}/\Gamma(1)} \frac{e_z \operatorname{ord}_z(\mathcal{W}_V)}{j(\tau) - j(z)},$$

where

$$e_z := \begin{cases} \frac{1}{2} & \text{if } z = i, \\ \frac{1}{3} & \text{if } z = e^{\pi i/3}, \\ 1 & \text{otherwise.} \end{cases}$$

Proof: Firstly, we recall Theorem 1 from [BKO]: Let $F(\tau) = q^h \sum_{n=0}^{\infty} a_n q^n$, be a meromorphic weight k modular form for $\Gamma(1)$, then

$$\frac{(q\frac{d}{dq})F(\tau)}{F(\tau)} = \frac{kE_2(\tau)}{12} - \frac{E_4(\tau)^2 E_6(\tau)}{\Delta(\tau)} \sum_{z \in \mathbb{H}/\Gamma(1)} \frac{e_z \operatorname{ord}_z(F)}{j(\tau) - j(z)}.$$

Because of $P_{k-1}(\tau) = \frac{\mathcal{W}_V'}{\mathcal{W}_V}$, we just have to show that Theorem 1 in [BKO] also applies for modular forms with a character and, in particular, for $\mathcal{W}_V(\tau)$. According to Theorem 1.4, $\mathcal{W}_V(\tau)^6$ is a modular form for $SL(2, \mathbb{Z})$, which together with

$$\frac{(\mathcal{W}_V(\tau)^6)'}{\mathcal{W}_V(\tau)^6} = 6\frac{\mathcal{W}_V'(\tau)}{\mathcal{W}_V(\tau)} = 6P_{k-1}(\tau)$$

now proves the claim. ∎

From the previous proposition one can in practice obtain an infinite-product expression for $\mathcal{W}_V(\tau)$ (cf. [BKO]).

There is an elegant way of checking whether a modular form is non-vanishing in $\mathbb{H}$ (the same result was obtained in [MMS] as well):

Proposition 2.4. *Suppose that $\mathcal{W}_V(\tau)$ is of weight $k(k-1)$, $k \in \mathbb{N}$ and satisfies*

$$\operatorname{ord}_{i\infty}(\mathcal{W}_V) = \frac{k(k-1)}{12},$$

then $\mathcal{W}_V(\tau)$ is non-vanishing in $\mathbb{H}$.

Proof: Let us recall that $\mathcal{W}_V(\tau)$ is actually an automorphic form with a character, so that the sixth power of $\mathcal{W}_V(\tau)$ is a modular form for $\Gamma(1)$.

An application of the Riemann-Roch formula in the case of a meromorphic modular form f on the genus zero surface $X = X(1)$, gives (see [Mil], [MMS]):

$$\text{ord}_{i\infty}(f) + \frac{\text{ord}_i(f)}{2} + \frac{\text{ord}_\rho(f)}{3} + \sum_Q \text{ord}_Q(f) = \frac{k}{6},$$

where $\rho = e^{\pi i/3}$ and the summation is over the remaining points in $\mathbb{H}/\Gamma(1)$. Suppose now that $f = \mathcal{W}_V(\tau)^6$. Then, $2k = 6m(m-1)$ and $\text{ord}_{i\infty}(f) = \frac{m(m-1)}{2}$. Because f is holomorphic, it follows that $\mathcal{W}_V(\tau)^6$ has no zeros in the upper half plane. Clearly, the same holds for $\mathcal{W}_V(\tau)$. ∎

3. Lattice vertex operator algebras of rank one

In this part we apply the results from the previous section in the case of rank one lattice vertex operator algebras.

Let $L = \mathbb{Z}\alpha$ be a rank one even lattice such that

$$\langle \alpha, \alpha \rangle = 2N,$$

where $N \in \mathbb{N}$. We denote by V_L the corresponding lattice vertex operator algebra [D] [LL]. It was proven in [D] and [DLM2] (cf. [LL]) that V_L is rational for every N and for a set of representatives of equivalence classes of irreducible V_L-modules we can take

$$V_{L+\lambda}, \ \lambda \in L^\circ/L,$$

where L° is the dual lattice of L (cf. [LL]). For a set of representatives of cosets of L°/L we choose $i\frac{\alpha}{2N}, i = 0, 1, \ldots 2N - 1$. Clearly,

$$|L^\circ/L| = 2N.$$

The central charge of V_L is one and

$$\begin{aligned}
\text{ch}_{2N,i}(q) &= \text{tr}|_{V_{L+i\frac{\alpha}{2N}}} q^{L(0)-1/24} \\
&= \frac{\sum_{v \in L+\frac{i\alpha}{2N}} q^{\frac{\langle v,v \rangle}{2}}}{\eta(\tau)} = \frac{\sum_{m \in \mathbb{Z}} q^{N(m+i/2N)^2}}{\eta(\tau)}.
\end{aligned}$$

It is easy to see that for every i we have

$$\text{ch}_{2N,i}(q) = \text{ch}_{2N,2N-i}(q). \tag{3.1}$$

If we analyze the leading terms in the q-expansion of $\text{ch}_{2N,i}(q)$, it follows that (3.1) are essentially the only linear relations between characters. Thus, for a basis of the vector space $\mathcal{M}_{V_L}$ we may choose

$$\text{ch}_{2N,i}(q), \ i = 0, \ldots, N.$$

The lowest weights of $ch_{2N,i}$ are given by

$$h_{2N,i} = \frac{i^2}{4N}, \quad i = 0, ..., N$$

Therefore

$$\sum_{i=0}^{N} \left(h_{2N,i} - \frac{1}{24} \right) = \frac{N(N+1)}{12}.$$

Now from Proposition 2.4 and Theorem 2.2 we have:

Theorem 3.1. *Let $L = \mathbb{Z}\alpha$, $\langle \alpha, \alpha \rangle = 2N$, $N \geq 1$, and V_L as the above, then*

$$d_V = N + 1,$$

and

$$\mathcal{W}_{V_L}(\tau) = \eta(\tau)^{2N(N+1)}.$$

If we let $l := N + 1$ in the previous formula and factor $\frac{1}{\eta(\tau)}$ from each of $ch_{2N,i}(q)$ in the Wronskian $\mathcal{W}_V(\tau)$, we get the following series of Dyson-Macdonald's identities for the affine root system of type D_l (see [Mac], p.138):

Proposition 3.2.

$$\eta(\tau)^{2l^2-l} = \nu_l \sum_{\mathbf{v}} \chi_D(\mathbf{v}) q^{\frac{||\mathbf{v}||}{4(l-1)}},$$

where in the summation $\mathbf{v} = (v_1, ..., v_l) \in \mathbb{Z}^l$, such that $v_i \equiv i-1 \pmod{2l-2}$, $||\mathbf{v}|| = \sum_{i=1}^{l} v_i^2$, $\chi_D(\mathbf{v}) = \prod_{i<j} (v_i^2 - v_j^2)$ and $\nu_l \in \mathbb{Q}$.

Proof: Follows along the lines of the proof of Proposition 4.4 given below. ∎

4. Vertex operator algebra $L_{A_1}(k, 0)$

In this section we study the Wronskians $\mathcal{W}_V(\tau)$, where $V = L_{A_1}(k, 0)$ is the level k vertex operator algebra associated to $A_1^{(1)}$ [LL]. One would hope that for these vertex operator algebras we can apply methods from [M2], [M3], to obtain the differential equation satisfied by irreducible characters directly from the C_2-cofiniteness. Even though we had some success in the $k = 1$ case, the C_2-cofiniteness did not provide us with a suitable ODE for $k \geq 2$. Thus, let us first focus on the vertex operator algebra $L_{A_1}(1, 0)$ with exactly two inequivalent irreducible modules. Recall the well-known formulas for irreducible characters [FL] (cf. [K]):

$$\text{tr}|_{L_{A_1}(1,0)}q^{L(0)-c/24} = \frac{\sum_{n\in\mathbb{Z}}q^{n^2}}{\eta(\tau)}$$

and

$$\text{tr}|_{L_{A_1}(1,1)}q^{L(0)-c/24} = \frac{\sum_{n\in\mathbb{Z}}q^{(n+\frac{1}{2})^2}}{\eta(\tau)}.$$

Theorem 4.1. *Irreducible characters* $\text{tr}|_{L_{A_1}(1,i)}q^{L(0)-1/24}$, $i = 0, 1$ *form a basis for the solutions of*

$$\left(q\frac{d}{dq}\right)^2 y(\tau) + 2G_2(\tau)\left(q\frac{d}{dq}\right)y(\tau) - \frac{25}{4}G_4(\tau)y(\tau) = 0. \qquad (4.1)$$

Proof: Let $V = L_{A_1}(1,0)$. Firstly, recall that the vertex operator algebra V is C_2–cofinite and $\dim(V/C_2(V)) \le 8$ (cf. [DLM2]). Moreover, we have

$$\coprod_{i\ge 4} V_i \subseteq C_2(V).$$

Thus,

$$L^2[-2]\mathbf{1} \in C_2(V).$$

Let M be a $L_{A_1}(1, 0)$-module. Then, according to Zhu [Zh] there exists $v_1 \in V_2$ such that

$$\begin{aligned}\text{tr}|_M o(L^2[-2]\mathbf{1})q^{L(0)-1/24} &= \lambda G_2(\tau)\text{tr}|_M o(v_1)q^{L(0)-1/24}\\ &+ (\nu_1 G_2^2(\tau) + \nu_2 G_4(\tau))\text{tr}|_M q^{L(0)-1/24},\end{aligned} \qquad (4.2)$$

where $\lambda, \nu_1, \nu_2 \in \mathbb{C}$. The vector space V_2 is four-dimensional, and only non-zero multiples of $h_\alpha[-1]^2\mathbf{1}$, where $\{x_\alpha, x_{-\alpha}, h_\alpha\}$ is the standard basis of $\mathbf{sl}_2$, contribute with a non-zero trace in (4.2). Therefore we may assume $v_1 = L[-2]\mathbf{1}$. After some computation, in parallel to [M2], Theorem 6.3, we obtain

$$\left(q\frac{d}{dq}\right)^2 y(\tau)+(2-\lambda)G_2(\tau)\left(q\frac{d}{dq}\right)y(\tau)+(\nu_1 G_2(\tau)^2+\nu_2 G_4(\tau))y(\tau) = 0,$$

where $y = \text{tr}|_M q^{L(0)-1/24}$.

From the asymptotic behavior of characters it follows that $\frac{2-\lambda}{12} = \frac{1}{4} - \frac{1}{12}$, thus $\lambda = 0$. Similarly, from the q-expansion we see that $\nu_1 = 0$ and $\nu_2 = \frac{-25}{4}$. ∎

Remark 4. Of course, it would have been much easier to use modular invariance directly and apply results from Section 2, to prove (4.1). In some sense our proof is representation theoretic since it relies only on the C_2-cofiniteness.

The previous Theorem 4.1 and Theorem 6.2 in [M2] give the following classical identity.

Corollary 4.2. *(Jacobi's Four Square Theorem)*

$$1 - 8\sum_{n=1}^{\infty}\left(\frac{(2n-1)q^{2n-1}}{1+q^{2n-1}} - \frac{(2n)q^{2n}}{1+q^{2n}}\right) = \left(\sum_{n\in\mathbb{Z}}(-1)^n q^{n^2}\right)^4. \qquad (4.3)$$

Proof: Apply Jacobi Triple Product Identity to write two θ constants

$$\theta_2(\tau) = \sum_{n\in\mathbb{Z}}q^{\left(n+\frac{1}{2}\right)^2}$$

and

$$\theta_3(\tau) = \sum_{n\in\mathbb{Z}}q^{n^2}$$

as infinite products:

$$\theta_2(\tau) = 2q^{1/4}\prod_{n=1}^{\infty}(1-q^{2n})(1+q^{2n})^2,$$

$$\theta_3(\tau) = \prod_{n=1}^{\infty}(1-q^{2n})(1+q^{2n-1})^2.$$

Now,

$$W_{\left(q\frac{d}{dq}\right)}(\mathrm{ch}_{L(1,0)}(\tau), \mathrm{ch}_{L(1,1)}(\tau))$$

$$= 2q^{1/4}\left(\frac{1}{4} - 2\sum_{n=1}^{\infty}\frac{nq^{2n}}{1+q^{2n}} + 2\sum_{n=1}^{\infty}\frac{(2n-1)q^{2n-1}}{1+q^{2n-1}}\right)\prod_{n=1}^{\infty}(1+q^n)^4 (4.4)$$

Formulas

$$\theta_4(\tau) = \sum_{n\in\mathbb{Z}}(-1)^n q^{n^2} = \prod_{n=1}^{\infty}\frac{1-q^n}{1+q^n},$$

$$\mathcal{W}_{L_{A_1}(1,0)}(\tau) = \eta(\tau)^4$$

now imply (4.3). ∎

As in the previous section we have the following description of $\mathcal{W}_{L_{A_1}(k,0)}(\tau)$.

Theorem 4.3. *For every $k \geq 1$*

$$\mathcal{W}_{L_{A_1}(k,0)}(\tau) = \eta(\tau)^{2k(k+1)}.$$

Proof: Let us denote by $h_{i,k}$ the lowest weight of $L_{A_1}(k, i-1)$ and by c_k the central charge. It is known (cf. [LL]) that

$$h_{i,k} = \frac{i^2 - 1}{4(k+2)},$$

$$c_k = \frac{3k}{k+2}.$$

Then

$$\sum_{i=1}^{k+1} \left(h_{i,k} - \frac{c_k}{24} \right) = \frac{k(k+1)}{12}.$$

The proof now follows directly from Proposition 2.4 and Theorem 2.2.   ∎

As a consequence we have the following result which is a series of Dyson-Macdonald's identity for C_l–series (cf. p. 136 in [Mac]):

Proposition 4.4. *For $l \geq 2$,*

$$\eta(\tau)^{2l^2+l} = \mu_l \sum_{\mathbf{n}} \chi_C(\mathbf{n}) q^{\frac{||\mathbf{v}||}{4(l+1)}},$$

where in the summation $\mathbf{n} = (n_1, ..., n_l) \in \mathbb{Z}^l$, such that $n_i \equiv i \bmod 2(l+1)$, $\chi_C(\mathbf{n}) = (\prod_{i=1}^l n_i) \prod_{i<j}(n_i^2 - n_j^2)$, $||\mathbf{n}|| = \sum_{i=1}^l n_i$, and $\mu_l \in \mathbb{Q}$.

Proof: Let us recall the formula for specialized characters of irreducible $L_{A_1}(k, 0)$–modules (see [K]):

$$\mathrm{ch}|_{L_{A_1}(k,i-1)}(\tau) = \frac{1}{2} \frac{\sum_{\substack{n \in \mathbb{Z} \\ n \equiv i \bmod 2(k+2)}} nq^{n^2/4(k+2)} - \sum_{\substack{n \in \mathbb{Z} \\ n \equiv -i \bmod 2(k+2)}} nq^{n^2/4(k+2)}}{\eta(\tau)^3}$$

$$= \frac{\sum_{\substack{n \in \mathbb{Z} \\ n \equiv i \bmod 2(k+2)}} nq^{n^2/4(k+2)}}{\eta(\tau)^3}, \tag{4.5}$$

where $i = 1, ..., k+1$. By Theorem 4.3 we have

$$\eta(\tau)^{3k+3} \mathcal{W}_V(\tau) = \eta(\tau)^{2(k+1)^2 + (k+1)}. \tag{4.6}$$

On the other, $\mathcal{W}_V(\tau)$ can be computed by expanding the determinant, so by using the formula (4.5) we have

$$\eta(\tau)^{3k+3} W_{q\frac{d}{dq}}(\mathrm{ch}|_{L_{A_1}(k,0)}, ..., \mathrm{ch}|_{L_{A_1}(k,k)})$$

$$= \sum_{\substack{(n_1, ..., n_{k+1}) \in \mathbb{Z}^{k+1} \\ n_i \equiv i \bmod 2(k+2)}} (\prod_{i=1}^{k+1} n_i) V(\frac{n_1^2}{4(k+2)}, ..., \frac{n_{k+1}^2}{4(k+2)}) q^{\sum_{i=1}^{k+1} \frac{n_i^2}{4(k+2)}},$$

where $V(z_1, ..., z_k)$ is the Vandermonde determinant, where $\tilde{\mu}_k$ is a rational number. Now, take $k = l - 1$. ∎

5. Modular invariant spaces with $d_V = 2$ and the Schwarzian

In this section we study the $d_V = 2$ case in more details. Let V be a rational vertex operator algebra with exactly two inequivalent irreducible modules that have linearly independent characters. We are actually not aware of examples of simple vertex operator algebras with exactly two inequivalent irreducible modules whose characters are proportional. On the contrary, there are several known examples of rational vertex operator algebras for which $d_V = 2$ and the number of irreducible inequivalent modules is > 2. In the $d_V = 2$ case the Wronskian is of weight 2 for $\Gamma'(1)$ with a non-trivial multiplier system (otherwise there would be a holomorphic weight two modular form on $SL(2, \mathbb{Z})$ with at most polynomial growth at the infinity). Let us recall that the modular curve $X'(1) = \overline{\mathbb{H}/\Gamma'(1)}$ is of genus one, so the space of holomorphic differential is one-dimensional [Miy]. More precisely, if we denote by $S_k(\Gamma)$ the vector space of cusp forms of weight k for Γ, then

Lemma 5.1.

$$\dim_{\mathbb{C}} \mathcal{S}_2(\Gamma'(1)) = 1,$$

$$\dim_{\mathbb{C}} \mathcal{S}_2(\Gamma(6)) = 1,$$

where both vector spaces of cusp forms are spanned by $\eta(\tau)^4$.

In fact $X(6) = \overline{\mathbb{H}/\Gamma(6)}$ is also a genus one modular curve with a holomorphic differential $\eta(\tau)^4 d\tau$.

Let us recall

$$j^{1/3}(\tau) = \frac{E_4(\tau)}{\eta(\tau)^8},$$

which is the third root of Klein's absolute modular invariant $j(\tau) = q^{-1} + 744 + \cdots$. Here, $E_4(\tau)$ is the normalized Eisenstein series of weight 4 (see the introduction).

Even though there are no classification results of rational vertex operator algebras with $d_V = 2$, it is possible to classify modular data with exactly two inequivalent irreducible modules. By solving the constraints from the Verlinde formula and the 2-dimensional fusion algebra one obtains 12 inequivalent modular data [Ga], which all can be constructed via affine Kac-Moody Lie algebras [Ga]. In what follows we compute Wronskians for these twelve examples. This is by no means the classification result for $\mathcal{W}_V(\tau)$ with $d_V = 2$.

Proposition 5.2. *Let $L_{X_r}(m, 0)$ stand for the level m vertex operator algebra associated to the affine Kac–Moody Lie algebra of type $X_r^{(1)}$ (cf. [LL]), and let c_V be the central charge of V.*

(i) *We have the following table:*

$X_r^{(1)}$	V	c_V	T_V	$\mathcal{W}_V(\tau)$
$A_1^{(1)}$	$L_{A_1}(1,0)$	1	$\frac{1}{6}$	$\eta(\tau)^4$
$E_7^{(1)}$	$L_{E_7}(1,0)$	7	$\frac{1}{6}$	$\eta(\tau)^4$
$F_4^{(1)}$	$L_{F_4}(1,0)$	$\frac{26}{5}$	$\frac{1}{6}$	$\eta(\tau)^4$
$G_2^{(1)}$	$L_{G_2}(1,0)$	$\frac{14}{5}$	$\frac{1}{6}$	$\eta(\tau)^4$

where

$$T_V = \left(h_0 - \frac{c_V}{24}\right) + \left(h_1 - \frac{c_V}{24}\right),$$

and h_0 and h_1 are the lowest weights of two irreducible V–modules.

(ii) *Let*

$$V = L_{X_r}(1, 0) \otimes L_{E_8}(1, 0),$$

where $L_{X_r}(1, 0)$ is a vertex operator algebra from the table. Then

$$\mathcal{W}_V(\tau) = j(\tau)^{2/3}\eta(\tau)^4,$$

with $T_V = -\frac{1}{2}$,

(iii) *Let*

$$V = L_{X_r}(1, 0) \otimes L_{E_8}(1, 0) \otimes L_{E_8}(1, 0),$$

where $L_{X_r}(1, 0)$ is a vertex operator algebra from the table. Then

$$\mathcal{W}_V(\tau) = j(\tau)^{4/3}\eta(\tau)^4$$

and $T_V = -\frac{7}{6}$.

Proof: The third and the fourth column in the table can be computed by using the theory of affine Lie algebra and the Sugawara construction [K], [KW]. In particular, $h_0 \neq h_1$ in all cases. Thus, for every V from the table

$$\mathcal{W}_V(\tau) = q^{1/6} + \cdots .$$

Therefore $\mathcal{W}_V(\tau)$ is a cusp form on $\Gamma'(1)$. Now, Lemma 5.1 implies that $\mathcal{W}_V(\tau)$ is $\eta(\tau)^4$, which proves Part (a).

Recall that $L_{E_8}(1, 0)$ is a holomorphic[1] vertex operator algebra. Also, it is known (cf. [K]) that $\mathrm{tr}|_{L_{E_8}(1,0)} q^{L(0)-1/3} = j^{1/3}(\tau)$. Thus, we have a one-to-one correspondence between the equivalence classes of irreducible V-modules and the equivalence classes of $V \otimes L_{E_8}(1, 0)$-modules [FHL]. The same fact applies for the vertex operator algebra $V \otimes L_{E_8}(1, 0)^{\otimes^2}$. Parts (b) and (c) now follow from (a) and the second formula in Lemma 1.3. ∎

Remark 5. For completeness let us recall here that the vector space of modular functions for $\Gamma'(1)$ with a polynomial growth at infinity is given by $\mathbb{C}[j^{1/3}, (j - 1728)^{1/2}]$ (cf. [KZ]). Thus, the space of modular forms of weight two for $\Gamma'(1)$, with the same behavior at the cusps, is given explicitly by $\eta(\tau)^4 \mathbb{C}[j^{1/3}, (j - 1728)^{1/2}]$.

Let f_1, f_2 be a basis for $\mathcal{M}_V$. Then the action of $\Gamma(1)$ on $\mathbb{H}$ induces an action of $\Gamma(1)$ on the $\frac{f_1}{f_2}$-plane in the natural way. Explicitly,

$$\frac{f_1(\gamma \cdot \tau)}{f_2(\gamma \cdot \tau)} = \frac{a_{11} \frac{f_1(\tau)}{f_2(\tau)} + a_{12}}{a_{21} \frac{f_1(\tau)}{f_2(\tau)} + a_{22}},$$

where

$$\begin{bmatrix} a_{11} & a_{12} \\ a_{21} & a_{22} \end{bmatrix} \in GL(2, \mathbb{C}).$$

Let

$$g(\tau) = \frac{f_1(\tau)}{f_2(\tau)}.$$

We shall denote by $\{g; \tau\}$ the Schwarzian derivative of g. Explicitly,

$$\{g; \tau\} = 2 \left(\frac{g''}{g'} \right)' - \left(\frac{g''}{g'} \right)^2 .$$

Let us also recall the chain rule for the Schwarzian

$$\{g \circ h; \tau\} = (h')^2 \{g; h(\tau)\} + \{h; \tau\}.$$

---

1 A vertex operator algebra V is called holomorphic if, up to equivalence, the only irreducible V–module is V itself.

If we take now

$$h(\tau) = \frac{a\tau + b}{c\tau + d},$$

such that $ad - bc = 1, a, b, c, d \in \mathbb{Z}$, then by using

$$\{h; \ \tau\} = 0,$$

we have

$$\{g; \ h(\tau)\} = (c\tau + d)^4 \{g \circ h; \ \tau\}$$
$$= (c\tau + d)^4 \left\{ \frac{a_{11}g + a_{12}}{a_{21}g + a_{22}}; \ \tau \right\} = (c\tau + d)^4 \{g; \ \tau\}.$$

Thus

$$\{g; \ \tau\}|_4(\gamma \cdot \tau) = \{g; \ \tau\},$$

for every $\gamma \in \Gamma(1)$.

The Schwarzian derivative also appear in the context of second order ordinary differential equations. Let us recall this basic but important fact. Every second order ODE of the form

$$\left(q\frac{d}{dq}\right)^2 y(\tau) + P_1(\tau)\left(q\frac{d}{dq}\right) y(\tau) + P_2(\tau)y(\tau) = 0, \qquad (5.1)$$

can be brought to its *projective normal* form

$$\left(q\frac{d}{dq}\right)^2 \tilde{y}(\tau) + \left(P_2(\tau) - \frac{1}{2}\left(q\frac{d}{dq}\right)P_1(\tau) - \frac{P_1^2(\tau)}{4}\right)\tilde{y}(\tau) = 0, \quad (5.2)$$

where $\tilde{y}(\tau) = y(\tau)/\sqrt{W}$ and W is the Wronskian of a fundamental system. Let y_1, y_2 be a pair of linear independent solutions of (5.1) or (5.2). Then the ratio $f(\tau) = \frac{y_1(\tau)}{y_2(\tau)}$ satisfies

$$\{f; \ \tau\} = 4P_2(\tau) - 2\left(q\frac{d}{dq}\right)P_1(\tau) - P_1^2(\tau). \qquad (5.3)$$

Proposition 5.3. *Let f_1 and f_2 form a basis for $\mathcal{M}_V$, then*

$$\left\{ \frac{f_1}{f_2}; \ \tau \right\} \qquad (5.4)$$

is a meromorphic modular form of weight 4. In particular, if $\mathcal{W}_V$ is non-vanishing then (5.4) is holomorphic, including at the infinity. Thus, it is a multiple of $E_4(\tau)$.

Proof: We already proved the first part. Because of (5.3) it follows that (5.4) can have a pole only at the infinity. However, because of $P_1(q) = 2G_2(q) = a_0 + a_1 q + \cdots$,

$$\text{ord}_{i\infty}(y'' + P_1(q)y') \geq \text{ord}_{i\infty} y.$$

On the other hand, having a pole at the infinity would imply

$$\text{ord}_{i\infty}(y'' + P_1(q)y') = \text{ord}_{i\infty} P_2(q)y < \text{ord}_{i\infty}(y),$$

which gives the contradiction. ∎

Expressions of the form $\{f; \tau\}$, where f is a *Hauptmodul* for certain genus zero congruence subgroups such as $\Gamma_0(n)$ or $\Gamma_0^+(n)$ were studied in [McS]. In particular, it was shown that the Schwarzian derivative of the Hauptmodul for the torsion free subgroups is always a multiple of $E_4(\tau)$. Some of our examples can also be used to prove related formulas. For instance, in the case of $L_{A_1}(1, 0)$ we have

Corollary 5.4. *Let* $\theta_2(\tau) = \sum_{n \in \mathbb{Z}} q^{\left(n + \frac{1}{2}\right)^2}$ *and* $\theta_3(\tau) = \sum_{n \in \mathbb{Z}} q^{n^2}$. *Then*

$$\left\{ \frac{\text{ch}_{L_{A_1}(1,1)}(\tau)}{\text{ch}_{L_{A_1}(1,0)}(\tau)}; \tau \right\} = \left\{ \frac{\theta_2}{\theta_3}; \tau \right\} = -45G_4(\tau).$$

Also, for $V = L(-22/5, 0)$ studied in [M2], $d_V = 2$ and we have

Corollary 5.5. *Let* $\chi_5(n) = \left(\frac{n}{5}\right)$, *(Legendre symbol). Then*

$$j_{X(5)}(\tau) = \frac{\text{ch}_{L(-22/5,-1/5)}(\tau)}{\text{ch}_{L(-22/5,0)}(\tau)} = q^{-1/5} \prod_{n=1}^{\infty} (1 - q^n)^{-\chi_5(n)n}$$

and

$$\left\{ j_{X(5)}; \tau \right\} = -\frac{144}{5} G_4(\tau).$$

(The function $j_{X(5)}$ *is the Hauptmodul for the genus zero curve* $X(5) = \mathbb{H}/\Gamma(5)$.)

6. Conclusion and an outlook

In this note we examined various number theoretic aspects of the Wronskian determinant associated to a rational vertex operator algebra. Even though our methods are those of automorphic forms rather than vertex operator algebra, it is interesting to see that many classical modular q–series identities

arise from considerations of graded dimensions of modules. Having in mind
that the coefficients of irreducible characters are non-negative integers, often
with interesting combinatorial interpretation, it is an open problem to find a
link between divisibility properties of partitions and vertex operator algebra
theory.

There are several closely related directions that we have already explored
related to our present work:

- In [M4], in parallel with [M2] and [M3], we derived several infinite series
  of q-series identities from consideration of $N = 1$ super minimal models.
  As an example, in the case of $N = 1$ minimal models of type $(2, 8)$, we
  obtained a classical Carlitz's q-series identity :

$$1 - 2\sum_{n=1}^{\infty} \chi_8(n) \frac{nq^n}{1 - q^n} = \frac{\eta^3(4z)\eta(2\tau)\eta^2(\tau)}{\eta^2(8\tau)},$$

  where $\chi_8(\cdot)$ is the Kronecker symbol mod 8.

- It would be interesting to find a proof of Theorem 3.1 and Theorem 4.3
  directly from the C_2–cofiniteness, in parallel with [M2] and [M3].

- It is not hard to construct examples of Wronskians $\mathcal{W}_V(\tau)$ with zeros in
  the upper half plane. We already indicated that one can just tensor a VOA
  with a holomorphic vertex operator algebra such as $L_{E_8}(1, 0)$. Here are
  two additional more interesting examples:
  Let $V = L(c_{2,7}, 0) \otimes L(c_{2,7}, 0)$ where $L(c_{2,7}, 0)$ be the VOA associated
  to $\mathcal{M}(2, 7)$–minimal models [M2]. Then $d_V = 6$ (symmetric square) and

$$\mathcal{W}_V(\tau) = \eta(\tau)^{48} E_6(\tau),$$

  with the zero at $\tau = \sqrt{-1}$.
  The zeros of $\mathcal{W}_V(\tau)$ can be more complicated (transcendental?). Let $V =
  L(c_{2,9}, 0) \otimes L(c_{2,9}, 0)$, for which $d_V = 10$. Then

$$\mathcal{W}_V(\tau) = \eta(\tau)^{168} E_6(\tau)(j(\tau) + \frac{12202}{143}).$$

- In [MMO] we investigated the following modular form for the full
  modular group:

$$\frac{W_{(q\frac{d}{dq})}(\text{ch}_1', ..., \text{ch}_k')}{W_{(q\frac{d}{dq})}(\text{ch}_1, ..., \text{ch}_k)}, \tag{6.2}$$

where $ch_1,...,ch_k$ are irreducible characters of $\mathcal{M}(2, 2k+1)$ Virasoro minimal models. We proved that an appropriate normalization of (6.2) give rise to *supersingular j-invariants*.

References

[AM] G. Anderson and G. Moore, Rationality in conformal field theory. *Comm. Math. Phys.* **117** (1988), 441–450.

[Ban] P. Bantay, The kernel of the modular representation and the Galois action in RCFT, *Comm. Math. Phys.* **233** (2003), 423–438.

[BKO] J. Bruinier, W. Kohnen and K. Ono, The arithmetic of the values of modular functions and the divisors of modular forms, *Compos. Math.* **140** (2004),552–566.

[CG] A. Coste and T. Gannon, Congruence subgroups and rational conformal field theory, math.QA/9909080.

[D] C. Dong, Vertex algebras associated with even lattices, *J. of Algebra*, **161** (1993), 245–265.

[DLM1] C. Dong, H. Li and G. Mason, Modular–invariance of trace functions in orbifold theory and generalized Moonshine, *Comm. Math. Phys.* **214** (2000), 1–56.

[DLM2] C. Dong, H. Li and G. Mason, Regularity of rational vertex operator algebras, *Adv. Math.* **132** (1997), 148–166.

[DM] C. Dong and G. Mason, Monstrous Moonshine of higher weight, *Acta Math.* **185** (2000), no. 1, 101–121.

[FL] A. Feingold and J. Lepowsky, The Weyl-Kac character formula and power series identities, *Advances in Math.* **29** (1978), 271–309.

[FHL] I. B. Frenkel, Y.-Z. Huang and J. Lepowsky, On axiomatic approaches to vertex operator algebras and modules, *Mem. Amer. Math. Soc.* **494**, 1993.

[FZ] I. Frenkel and Y.Zhu, Vertex operator algebras associated to representations of affine and Virasoro algebras, *Duke Math J.* **66** (1992), 123–168.

[Ga] T. Gannon, Modular data: the algebraic combinatorics of conformal field theory, math.QA/0103044.

[Hu1] Y.-Z. Huang, Vertex operator algebras and the Verlinde conjecture, math.QA/0406291.

[Hu2] Y.-Z. Huang, Differential equations, duality and modular invariance, *Comm. Contem. Math.*, **7** (2005), 375–400.

[Hu3] Y.-Z. Huang, Vertex operator algebras, the Verlinde conjecture and modular tensor categories, *Proc Natl Acad Sci U S A.* ; **102** (15) (2005) 5352Ð5356.

[K] V. Kac, *Infinite-dimensional Lie algebras*, Third edition, Cambridge University Press, Cambridge, 1990.

[KW] V. Kac and M. Wakimoto, Modular and conformal invariance constraints in representation theory of affine algebras, *Adv. in Math.* **70** (1988), 156–236.

[KZ] M. Kaneko and D. Zagier, Supersingular j-invariants, hypergeometric series, and Atkin's orthogonal polynomials, AMS/IP Series in Advanced Mathematics, **7** (1998), 97–126.

[LL] J. Lepowsky and H. Li, *Introduction to vertex operator algebras and their representations*, Progress in Mathematics, 227, Birkhäuser Boston, Inc., Boston, MA, 2004

[Mac] I. Macdonald, Affine root system and Dedekind's η–function, *Invent. Math.* **15** (1972), 91–143.

[MMS] S. Mathur, S. Mukhi and A. Sen, Reconstruction of conformal field theory from modular geometry on the torus, *Nucl. Phys.* B318 (1989), 483–540.

[McS] J. McKay and A. Sebbar, Fuchsian groups, automorphic functions and Schwarzians, *Math. Ann.* **318** (2000), 255–275.

[M1] A. Milas, Formal differential operators, vertex operator algebras and zeta–values, I, II, *J. Pure Appl. Alg.* **183** (2003), 129-190, 191–244.

[M2] A. Milas, Ramanujan's "Lost Notebook" and the Virasoro algebra, *Comm. Math. Phys.* **251** (2004), 657–678.

[M3] A. Milas, Virasoro algebra, Dedekind eta-function and Specialized Macdonald's identities, *Transf. Groups*, **9** (2004), 273–288.

[M4] A. Milas, Characters, Supercharacters and Weber modular functions, *Crelle's Journal*, **608**,(2007), 35–64.

[MMO] A. Milas, E. Mortenson and K. Ono, Number theoretic properties of Wronskians of Andrews-Gordon series, to appear in *International Journal of Number Theory*.

[Mil] J. Milne, *Modular Functions and Modular Forms*, available from http://www.jmilne.org/math

[Mir] R. Miranda, *Algebraic curves and Riemann surfaces*, Graduate Studies in Mathematics, American Mathematical Society, Providence, RI, 1995.

[Miy] T. Miyake, *Modular Forms*, Springer Verlag, 1989.

[MS] G. Moore and N. Seiberg, Classical and quantum conformal field theory, *Comm. Math. Phys.* **123** (1989), 177–254.

[Mu1] S. Mukhi, Feigin-Fuchs integrals and Rogers-Ramanujam identities in rational conformal field theory, *Recent developments in conformal field theories* (Trieste, 1989), 70–80, World Sci. Publishing, 1990.

[Mu2] S. Mukhi, Modular geometry and classification of rational conformal field theory, *Mathematical Physics* 1989, 252–282, World. Sci. Publishing, 1990.

[Ro] D. Rohrlich, Weierstrass points and modular forms, *Illinois J. Math.* **29** (1985), 134–141.

[Zh] Y. Zhu, Modular invariance of characters of vertex operator algebras, *J. Amer. Math. Soc.* **9** (1996), 237–307.

Moonshine and Group Cohomology

C. B. Thomas

Introduction

If $\mathbb{M}$ is the Monster simple group, then to each rational conjugacy class $\langle g \rangle$ there is associated a formal q-expansion

$$j_g(q) = q^{-1} + \sum_{n=1}^{\infty} a_g(n)q^n$$

with integral coefficients, such that

(M1) For each $n \geqslant 1$ the function $g \mapsto a_g(n)$ is the character of a representation space H_n.

For each $\langle g \rangle$ there is an integer h dividing the greatest common divisor $(24, |g|)$, and a discrete Γ sandwiched between the congruence subgroup $\Gamma_0(N)$ and its normaliser in $SL_2(\mathbb{R})$ ($N = h|g|$) and commensurable with $SL_2(\mathbb{Z})$. If $\bar{Z}$ denotes the upper half-plane compactified by the cusps of Γ, then

(M2) The character $j_g(q)$ generates the field of meromorphic functions defined on the Riemmann surface $\Gamma \backslash \bar{Z}$ of genus zero.

We shall refer to an infinite representation space of type (M1) as a McKay-Thompson series, and to (M2) as the 'genus zero' property. Similar Moonshine modules can be constructed for other groups, notably for centralisers $C_{\mathbb{M}}(g)$ of elements in $\mathbb{M}$. Simon Norton proposed a rule for relating these (see Section 5

2000 *Mathematics Subject Classification*. Primary 20J06, 20D08; Secondary 55N20, 55N34.
Key words and phrases. sporadic simple group, group cohomology, generalised cohomology theory.
After submitting this paper, sadly Charles Thomas died suddenly in December 2005. He had pursued Moonshine-related mathematics for well over a decade and did much to alert algebraic topologists to its connections with elliptic cohomology through his participation in many conferences and coordinating an EU network. He is greatly missed by those who knew him and enjoyed his company, his love of life and his mathematics.
The final version of this paper was produced by A. Baker in consultation with the Editors.

below) and we shall regard McKay-Thompson series as being doubly indexed in this sense, once by the degree n and once by a family $\{g\}$ of group elements representing conjugacy classes. Quite independently of the work of Conway and Norton topologists were becoming interested in a 2-variable generalisation of K-theory called 'elliptic cohomology', and it has since become clear that, at least for some finite groups G, there is a close relation between doubly-indexed McKay-Thompson series and the elliptic cohomology of the classifying space BG. This explains our use of the term 'elliptic object', for which however the 'genus zero' condition remains a mystery. One motivation for writing this survey is to encourage the calculation of $Ell^*(BG)$ for subgroups G of $\mathbb{M}$ in the hope of identifying a cohomological condition equivalent to the more complex-analytic conditions (M2). That this is not an impossible undertaking is shown by the fact that, depending on which variant of elliptic cohomology one uses, one can neglect the 2 or 3-Sylow structure of G.

For any cohomology theory h^* the calculation of $h^*(BG)$ begins with that of $H^*(BG, \mathbb{Z})$ or $H^*(BG, \mathbb{F}_p)$. In part because $K^0(BG)$ is given by a certain completion of the representation ring $R(G)$, the subring of ordinary cohomology generated by characteristic classes of representations is particularly important. Indeed for some groups and some theories, such as Morava K-theory $K(n)^*$, such classes serve to generate the cohomology as a module over the coefficients. For example this holds for the elementary non-abelian group p_+^{1+2} of order p^3; $K(n)^* = K(n)^{\text{even}}$ is generated by Chern classes, but $H^{\text{odd}}(p_+^{1+2}, \mathbb{Z})$, although small, is non-trivial.

The contents of the paper are as follows: the first two sections are devoted to ordinary cohomology and include what is known about $H^*(B\mathbb{M}, \mathbb{Z})$. Section 3 is concerned with a single example, the eventual aim being to calculate the cohomology of Co_1 at the prime 5. In Section 4 we introduce increasingly complicated coefficients $h^*(\text{point})$, starting with variants of K-theory, first to build up to the universal theory (cobordism localised at a single prime p), and then drop down again to v_2-periodic theories. The last section is devoted to Moonshine modules and elliptic objects.

Some readers are warned that this is not an expanded version of my Edinburgh lecture, but may be a version of the lecture which I should have given. The original lecture, devoted mainly to calculations in ordinary cohomology, will be published elsewhere.

1. Cohomology of finite groups

We start with a topological definition. If G is a finite (or more generally a discrete) group a classifying space $BG = K(G, 1)$ is a space X such that

$\pi_1(X) \cong G$ and the universal covering space $\widetilde{X}$ is contractible. Such spaces can be shown to exist, either by inductively adding cells to a 2-complex realising G geometrically (1-cells corresponding to generators and 2-cells to relations), or by allowing G to act on an infinite-dimensional Stiefel manifold via a unitary representation. The latter construction has the advantage of showing that BG may be approximated by smooth finite-dimensional manifolds. If A is an abelian coefficient group, we then define

$$H^*(G, A) = H^*(BG, A).$$

This definition is independent of the particular model BG, since this is unique up to homotopy type. Those expecting a more algebraic definition should note that we assume a trivial G-structure on A, which will usually be the integers $\mathbb{Z}$ or the finite field $\mathbb{F}_p$. With attention paid to a G-action we must define our cochains on the equivariant chains of $\widetilde{X}$. If A is a ring, $H^*(G, A)$ can be given the structure of a graded ring; we shall be particularly interested in the commutative subring

$$H^{\text{even}}(G, A) = \bigoplus_{k \geqslant 0} H^{2k}(G, A).$$

Let H be a subgroup of G, so that BH is a covering space of BG with covering map Bi (say). We write the induced map as $i^*: H^*(G, A) \to H^*(H, A)$ (restriction), and if $j: G \to G/H$ (H normal in G) there is a similarly defined map $j^*: H^*(G/H, A) \to H^*(G, A)$ (inflation). Conjugation by an element $g, x \mapsto gxg^{-1}$, induces $c_g: H^*(H, A) \to H^*(H^g, A)$, with $H^g = gHg^{-1}$, and we can also define a covariant map $i_*: H^*(H, A) \to H^*(G, A)$ (corestriction/transfer), which is an $H^*(H, A)$-module rather than a ring homomorphism (Frobenius reciprocity). One way to define this at the level of cochains is to mimic the construction of an induced representation. Thus, let G_1 be contained in G_2 with index s and let F be some subgroup of the symmetric group S_s. The wreath product $G_1 \wr F$ is the semi-direct product of s copies of G_1 with F, the latter group acting by permutation on the former. Embed G_2 in the product by first choosing left coset representatives $\{1 = g_1, \ldots, g_s\}$, noting that $yg_j = g_{\sigma(j)}x_j$ for each $y \in G_2$, some $x \in G_1$, and some permutation $\sigma \in S_s$, and mapping y to $(x_1, \ldots, x_s, \sigma)$. At the cochain level first extend f to a map defined on the chains of $(G_1 \times \cdots \times G_1) \rtimes F$ and then restrict down to the subgroup G_2.

With these relations with and between subgroups we can prove the basic proposition that the cohomology of G is detected by the cohomology of a representative family of Sylow p-subgroups G_p. Restricting attention to p-torsion we have that

$$H^*(G, A)_{(p)} \cong \text{stable elements in } H^*(G_p, A).$$

Here an element x is said to be *stable* if, for all $g \in G$, the restriction of x to $H^*(G_p \cap gG_pg^{-1})$ is unchanged if we first conjugate by g. From the definition one sees that the image of $H^*(G, A)_{(p)}$ in $H^*(G_p, A)$ certainly consists of stable elements; the converse depends on properties of the composition i^*i_*. Note that the stable elements are contained in the set of normaliser-invariant elements $H^*(G_p, A)^N$, where $N = N_G(G_p)$, which is usually easier to calculate. Hence the usefulness of

Proposition 1.1. *If G_p is abelian $H^*(G, A)_{(p)} \cong H^*(G_p, A)^N$.*

The main point of the proof of Proposition 1.1 is that, if G_p is abelian, the centraliser Z of $G_p \cap G_p^g$ contains both G_p and G_p^g as (conjugate) Sylow subgroups. It is also important to note that the argument definitely requires trivial G-action on the coefficients A.

Much of what we will prove in later sections depends on the relation between cohomology and the representation ring $R(G)$. One can show that, given a complex representation ρ of degree n, there exist (Chern) classes $c_k(\rho) \in H^{2k}(G, \mathbb{Z})$, $1 \leqslant k \leqslant n$ with the following properties:

(C1) If $\varphi: G_1 \to G_2$ is a homomorphism and $\rho: G_2 \to U_n$ a representation, then $\varphi^*(c_k(\rho)) = c_k(\varphi^*\rho)$, where $\varphi^*\rho$ denotes the representation $\rho\varphi: G_1 \to U_n$.

(C2) Denote the total Chern class by $c_\bullet(\rho) = 1 + c_1(\rho) + \cdots + c_n(\rho)$. Then $c_\bullet(\rho_1 + \rho_2) = c_\bullet(\rho_1)c_\bullet(\rho_2)$.

(C3) The first Chern class c_1 defines an isomorphism between $H^2(G, \mathbb{Z})$ and the 1-dimensional representations $\text{Hom}(G, \mathbb{C}^*)$.

One possible proof of existence and uniqueness goes as follows. Use (C3) as a definition of the class c_1. Because as an additive group $\mathbb{C}$ is divisible the coboundary map $H^1(G, \mathbb{C}^*) \to H^2(G, \mathbb{Z})$ associated with the short exact sequence of coefficients

$$\mathbb{Z} \longrightarrow \mathbb{C} \xrightarrow{\text{exp}} \mathbb{C}^*$$

is an isomorphism. Property (C2) provides an immediate extension to sums of 1-dimensional representations. From the previous discussion it suffices to consider groups of prime power order, which are monomial, i.e., such that an irreducible representation of degree greater than one is induced up from a 1-dimensional representation of some proper subgroup. In principle the Chern classes of an induced representation can be calculated in the same way that we have defined corestriction, i.e., via $H^*(H \wr S_n, \mathbb{Z})$, and an intermediate determination of the classes for the permutation representations of the symmetric groups S_m ($m \leqslant n$). See the original paper of L. Evens [7], together with that by V. Snaith [20], on explicit induction.

Definition 1.2. The Chern ring $\text{Ch}(G) \leqslant H^{\text{even}}(G, \mathbb{Z})$ is the subring generated by the Chern classes of the irreducible representations of G.

(It is sometimes useful to consider the larger subring $\text{Tre}(G)$ generated by transferred Euler classes of real representations.)

Proposition 1.3 (Evens-Venkov). *$H^*(G, \mathbb{Z})$ is finitely generated as a module over the Chern subring $\text{Ch}(G)$.*

For a proof see [23].

This proposition allows us to regard $\text{Ch}(G)$ as a calculable approximation to $H^*(G, \mathbb{Z})$ for many groups G. Using property (C2) we can also define the total Chern class of a virtual representation $[\rho_1] - [\rho_2]$ as $c_\bullet(\rho_1)/c_\bullet(\rho_2)$, and one way of extending the theory to modular representations is to apply $c_\bullet$ to the Brauer lift of a representation in characteristic ℓ. More elegant constructions than the one outlined above apply both to finite and zero characteristic, see [11].

First examples

(1) Cyclic group $p = C_p$ generated by A.

Let $\widehat{\alpha}$ be the one-dimensional representation mapping A to $\zeta = e^{2\pi i/p}$, a primitive pth root of unity. Then $H^*(C_p, \mathbb{Z}) = \mathbb{F}_p[\alpha]$ where $\alpha = c_1(\widehat{\alpha}) \in H^2$.

(2) Elementary abelian group of rank 2 $(p, p) = C_p^A \times C_p^B$. With the obvious notation $H^{\text{even}}(C_p \times C_p, \mathbb{Z}) = \mathbb{F}_p[\alpha, \beta]$, and there is a 3-dimensional exterior algebra generator μ.

Warning: with increasing rank the $\mathbb{Z}$-cohomology of an elementary abelian group becomes complicated. Thus $H^4(3C_p, \mathbb{Z})$ contains an element outside the Chern subring. With $\mathbb{F}_p$-coefficients the situation is much simpler (use the Künneth formula). One obtains $H^{\text{even}}(rC_p, \mathbb{Z})$ from the kernel of a 'coboundary derivation' $\delta \colon H^k(rC_p, \mathbb{F}_p) \to H^{k+1}(rC_p, \mathbb{F}_p)$ associated with the sequence

$$\mathbb{Z} \xrightarrow{\times p} \mathbb{Z} \longrightarrow \mathbb{F}_p.$$

(3) Elementary non-abelian or order p^3, p_+^{1+2} with presentation

$$\langle A, B, D : A = [B, D], A = \text{central} \rangle.$$

The abelianisation of p_+^{1+2} is isomorphic to $C_p^{\overline{B}} \times C_p^{\overline{D}}$, providing generators β and δ in H^2 as in Example 2 above. For the remaining generators of H^{even} let $\widehat{\alpha}$ be the 1-dimensional representation of $\langle A, B \rangle$ which maps A to ζ and B to 1; then the induced representation $i_*\widehat{\alpha}$ is a typical representative of the family of p-dimensional irreducible representations of p_+^{1+2}. A formula of Riemann-Roch type shows that $s_k(i_*\widehat{\alpha}) = \text{cor}(\alpha^k)$ for $2 \leqslant k \leqslant p - 1$, where s_k is the k-th Newton polynomial in the Chern classes c_j.

$\text{Ch}(p_+^{1+2}) = H^{\text{even}}(p_+^{1+2}, \mathbb{Z})$ is generated by β, δ, $\text{cor}(\alpha^k)$ ($2 \leqslant k \leqslant p - 1$) and $c_p(i_*\widehat{\alpha})$. Perhaps surprisingly, given that the group has exponent p, the last generator has order p^2 rather than p. There are two exterior algebra generators μ and γ in dimension 3. We omit the numerous relations in H^{even}.

For an expanded version of the results in this section the reader is referred to [23] and the references therein.

2. An elementary non-abelian group $E(p)$ of order p^4

We quote an elementary proposition from representation theory, see [13, pp295–297].

- Let G be a non-abelian p-group which contains an abelian subgroup H of index p. Then there exists a normal subgroup K of G of order p such that K is contained in $H \cap [G, G] \cap Z(G)$.

- If $H, K \leqslant G$ are as above then every irreducible character of G is given by either (i) the inflation of an irreducible character of G/K or (ii) the transfer of some 1-dimensional character ψ of H, which satisfies $K \not\subseteq \text{Ker}\,\psi$.

- If $|G| = p^n$ and $|G/[G, G]| = p^m$, then G has p^m 1-dimensional characters and $p^{n-2} - p^{m-2}$ irreducible characters of p-dimensional representations.

By way of an example consider G with presentation

$$G = \langle A, B, C, D : A = \text{central}, [B, C] = 1, [C, D] = B, [B, D] = A \rangle.$$

Write

$$H = G_1 = \langle A, B, C \rangle \cong (p, p, p), \quad [G, G] = \langle A, B \rangle, \quad K = \langle A \rangle,$$
$$G/[G, G] = C_p^{\overline{C}} \times C_p^{\overline{D}}.$$

We have a diagram of subgroups and quotients:

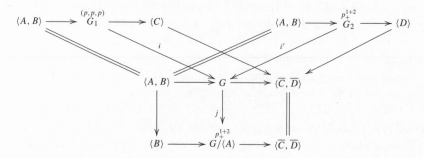

Here the class of B is central in $G/\langle A \rangle$. The maps i and i' are inclusions, and j is a quotient map.

The quoted proposition shows that the irreducible characters of degree greater than 1 are either inflated from the quotient $\langle \overline{B}, \overline{C}, \overline{D} \rangle \cong p_+^{1+2}$ or transferred from the subgroup $\langle A, B, C \rangle \cong (p, p, p)$. Call these types (a) and (b) respectively. Taking both together we have $p^2 - 1$ irreducible characters of degree p, obtained from representations of G_1 mapping either A or B to a root of unity and the remaining two generators to 1. Arguing as for p_+^{1+2} in Example 3, and with self-explanatory notation, we see that the Chern subring of $H^{\text{even}}(G, \mathbb{Z})$ is generated by classes α, δ (of degree 2), $\text{cor}(\alpha^r)\text{cor}(\beta^s)$ ($2 \leqslant r$, $s \leqslant p - 1$) and the top-dimensional classes $c_p(\rho_1)$, $c_p(\rho_2)$ for representative representations ρ_1, and ρ_2 of types (a) and (b) respectively.

The calculation of the Chern subring suggests how to obtain the complete cohomology via a triple comparison of spectral sequences for the restrictions i^* and i'^* and the inflation j^* above. (The spectral sequence of a group extension is explained and used in [23, Chapter 4].) The problem is to understand the spectral sequence for the extension G. That for G_1 (abelian) is easy to understand, that for G_2 is described briefly by G. Lewis in [15] (for total degree $\leqslant 2p$ it collapses after page 3 (Corollary 6.34)), while that for $G/\langle A \rangle$ is the main object of study in the same paper (collapse after page 4 (Lemma 6.18)).

In general one complicating factor is the fact already referred to in Example 2 of Section 1, namely that $\text{Ch}(3C_p)$ is *properly* contained in $H^{\text{even}}(3C_p, \mathbb{Z})$. This does not occur for $\mathbb{F}_p$-coefficients, or for some among the generalised cohomology theories to be considered in Section 4, and is the one of the reasons why the analogous calculation seems to be much easier.

For $p = 5$ we will return to the inclusion $i': G_2 \to G$ when discussing the Conway groups. Again, with special attention paid to the prime 5, $H^*(G, \mathbb{F}_p)$ has been studied by D. Green using the programme described in [10] and briefly considered in the recent survey [5] by J. Carlson.

3. The sporadic simple groups M_{24}, Co_1 and $\mathbb{M}$

In this section we put together the result that $H^*(G, \mathbb{Z})_{(p)}$ is detected by the stable elements in $H^*(G_p, \mathbb{Z})$ with the calculations for p-groups of low rank. Hence our permitted Sylow subgroups are cyclic (p), elementary abelian of rank 2 (p, p), p_+^{1+2} and the group $E(p)$ of exponent p and order p^4. With little in the way of additional proof we will list the p-torsion in $\mathbb{Z}$-cohomology for

$$M_{24} \ (p \geqslant 3), Co_1 \ (p \geqslant 5, \text{Chern ring only for } p = 5), \mathbb{M} \ (11 \leqslant p \leqslant 31).$$

For primes $p \geqslant 41$ dividing the order of the Monster calculation is easy and well-known. Another reason for their omission is a desire to understand the various relations between $\mathbb{M}$ and the exceptional Lie group E_8. We therefore concentrate on those primes dividing the order of $E_8(\mathbb{F}_q)$ for small values of q.

An exception is the prime $p = 23$, which from the cohomological point of view shows itself to be slightly anomalous.

The first result combines an easy calculation for cyclic groups with a 3-primary calculation in [9]. It does however appear to be typical, and illustrates the close relation between cohomology and modular characters. For M_{24} the 2-modular character τ (in honour of J. A. Todd) takes the following values on regular conjugacy classes of prime order

1^{24}	$1^6 3^6$	$1^4 5^4$	$1^3 7^3$	$1^3 7^3$	$1^2 11^2$	$1\,23$	$1\,23$	3^8
11	2	1	$\dfrac{1-i\sqrt{7}}{2}$	$\dfrac{1+i\sqrt{7}}{2}$	0	$\dfrac{-1+i\sqrt{23}}{2}$	$\dfrac{-1-i\sqrt{23}}{2}$	-1

Proposition 3.1. $H^{\mathrm{even}}(M_{24}, \mathbb{Z})_{\mathrm{odd}}$ *is generated by the Chern classes of the representation* τ. $H^{\mathrm{odd}}(M_{24}, \mathbb{Z})_{(p \geqslant 5)} = 0$ *and* $H^{\mathrm{odd}}(M_{24}, \mathbb{Z})_{(3)}$ *has an exterior algebra generator in degree* 11.

Conway's group Co_1 of order $2^{21}\, 3^9\, 5^4\, 7^2\, 11\, 13\, 23$

$p \geqslant 11$: $H^*(Co_1, \mathbb{Z})_{(p)} \subseteq \mathbb{F}_p[x]$ with generator a power of x determined by the index of a Sylow p-centraliser in its normaliser. With the notation of the Atlas, if ρ_2 has degree 276 and ρ_{17} has degree 673500, the three generators concerned may be taken to be $c_{10}(\rho_2)$, $c_{12}(\rho_2)$ and $c_{11}(\rho_{17})$.

$p = 7$: a Sylow 7-subgroup is abelian, so the stable elements coincide with those invariant under the normaliser-mod-centraliser (Weyl group) action. For $Co_{1,7}$ the group acting is isomorphic to $3 \times T^*$ ($T^* = $ binary tetrahedral group), and the invariant elements (ξ_i) are generated in degrees 12, 36, 48 and 35, subject to the relations

$$\xi_{35}^2 = 0, \quad \xi_{36}^2 + \xi_{12}^6 + 3\xi_{12}^2\xi_{48} = 0.$$

The first of these restricts to a generator of $H^*(Co_2, \mathbb{Z})_{(17)}$, which can be identified with $c_6(\rho_2)$, where ρ_2 is the natural representation of the group in the Leech lattice. Since there are two conjugacy classes of elements of order 7 in Co_1 at least one other representation contributes to the Chern subring. Inspection of the character table suggests that ρ_{12} (see above) may suffice.

$p = 5$: the partial calculation in Section 2 points to the classes $c_i(\rho_j)$ ($j = 2, 17$) again exhausting the (proper) Chern subring of $H^*(Co_1, \mathbb{Z})_{(5)}$. As at the prime 7 it is instructive to restrict to the subgroup Co_2, for which $Co_{2,5}$ is isomorphic to the subgroup G_2 of $E(5)$ of order 5^4. It is known (see for example [2, Theorem 5.6]) that $\mathrm{Ch}(Co_2)_{(5)} = H^{\mathrm{even}}(Co_2, \mathbb{Z})_{(5)}$ is generated by the Chern classes of ρ_2 of degree 23. Since, working either in characteristic 2 or with the 276-dimensional representation of Co_1 in characteristic zero, $\mathrm{Ch}(Co_1)_{(5)}$ maps onto $\mathrm{Ch}(Co_2)_{(5)}$, we also have a surjection for H^{even}. At least one other representation is needed for the kernel; compare the roles of the inflated $(i^*\widehat{\beta})$ and transferred $(i^*\widehat{\alpha})$ representations of $E(5)$ itself.

The reader may find the character values below helpful in following the discussion.

	1	5C	5A	5B	7X	7Y
Co_1	276	21	6	1	10	3
	673750	0	0	0	0	0
Co_2	23	–	3	-2	–	2
	275	–	5	0	–	2

The Monster (order divisible by 7^6 11^2 13^3 17 19 23 29 31)

$17 \leqslant p \leqslant 31$: Arguing as for the Conway group ($p \geqslant 1$) and with irreducible representations numbered as in the Atlas we have generators $c_{16}(\rho_2)$, $c_8(\rho_2)$, $c_{11}(\rho_{16})$, $c_{28}(\rho_2)$ and $c_{15}(\rho_{26})$.

$p = 13$: the cohomology has been calculated by M. Tezuka and M. Yagita [11, Theorem 6.6].

The Chern subring is properly contained in $H^{\mathrm{even}}(\mathbb{M}, \mathbb{Z})_{(13)}$ which is properly contained in the normaliser-invariant subring. The existence of two conjugacy classes again suggests that more than one representation (ρ_2) is needed to obtain the Chern subring. This is confirmed by the presence of elements constructed both from $\mathrm{cor}(\alpha^{12})$ and $\mathrm{cor}(\alpha^{12}) - \beta^{12} - \delta^{12}$ (notation as in Section 1, Example 3) in the stable cohomology.

$p = 11$: this is much easier than for $p = 13$. There are generators in dimensions 40, 60 (two), 80 (two), 160 and 240. Since there is only one conjugacy class of elements of order 11, a single representation provides generators of the Chern ring, and inspection of the character table shows that these are $c_{110}(\rho_2)$ and $c_{120}(\rho_2)$. They may be identified with the so-called Dickson invariants for the action of the Weyl group $5 \times I^*$ ($I^* = $ binary icosahedral group) on $\mathbb{F}_{11}[x, y]$.

Remarks for $p = 7$. This is an extremely interesting question, which is probably just within the bounds of human calculation. A Sylow subgroup $\mathbb{M}_7$ is isomorphic to an extension of C_7 by an extra-special group of order 7^5, contained in an extension of $C_3 \times S_7^*$ by 7_+^{1+4}. (Here the $*$ again denotes 'binary'.) A related, possibly easier question is the determination of $H^*(G_2(p), \mathbb{Z})_{(p)}$ for small values of p. Besides helping with the Monster ($p = 7$) this would also throw light on the Lyons group ($p = 5$).

4. Generalised cohomology theories

Our aim is to interpret a module of Moonshine type as an element of $h^*(BG)$, where h^* is a cohomology theory admitting complex orientations. This means that $h^* = \{h^n : n \in \mathbb{Z}\}$ is a family of functors from a suitable class of cell complexes to abelian groups, satisfying the same (Eilenberg-Steenrod) axioms as $H^* = \{H^n : n \geqslant 0\}$, except for the so-called dimension axiom. Thus we allow $h^*(\text{point})$ to be more complicated than $\{H^0(\text{point}) \cong \mathbb{Z}, H^n(\text{point}) = 0 \ (n \geqslant 0)\}$, and we shall also assume that like $H^*(X)$, $h^*(X)$ is a graded commutative ring. The theory h^* is said to be *complex oriented* if an Euler class $e^h(L)$ is defined for any complex line bundle L over X, such that

- $e^h(L)$ is natural for bundle maps,
- $e^h(L) \in h^2(X)$, and
- $h^*(\mathbb{C}P^n)$ is the truncated polynomial algebra over $h^*(\text{point})$ generated by x, the Euler class of the canonical line bundle over the projective space, with $x^{n+1} = 0$.

Such a theory admits Chern and Euler classes as in Section 1. The relation

$$e^h(L_1 \otimes L_2) = \sum_{i,j} a_{ij} \, e^h(L_1)^i e^h(L_2)^j,$$

with $a_{ij} \in h^{2(1-i-j)}(\text{point})$, determines a formal group law for h^*,

$$F_h(x, y) = \sum_{i,j} a_{ij} \, x^i y^j.$$

Furthermore, at least if 2 is invertible in h^0(point), the theory h^* is very nearly determined by the pair $(h^*(\text{point}), F_h)$, and hence there is a universal cohomology theory corresponding to M. Lazard's universal formal group. In case this seems too abstract here are some examples:

(i) H^*(point, $\mathbb{Z}$), $F_h = x + y$,

(ii) K^*(point) $= \mathbb{Z}[u, u^{-1}]$, $F_k = x + y - uxy$, complex K-theory ($u \in K^{-2}$(point) is the Bott periodic element). For 'real K-theory' we shall use $KO^*(X) \otimes \mathbb{Z}[\frac{1}{2}]$ with coefficients $\mathbb{Z}[\frac{1}{2}][v, v^{-1}]$ (degree $v = -4$). Increasing the complexity of the coefficients we have

(iii) $Ell_J^*(X) = Ell^*(X)$ with Ell^*(point) $= \mathbb{Z}[\frac{1}{2}][\delta, \varepsilon, \varepsilon^{-1}]$ (degree $\delta = -4$, degree $\varepsilon = -8$) and formal group

$$F(x, y) = x\sqrt{R(y)} + y\sqrt{R(x)}/1 - \varepsilon x^2 y^2,$$

$R(x) = 1 - 2\delta x^2 + \varepsilon x^4$. Here J stands for Jacobi; there is a related theory $Ell_W^*(X)$ (here W stands for Weierstrass) with coefficients $\mathbb{Z}[\frac{1}{6}][g_1, g_2, \Delta^{-1}]$.

As parent 'universal' theory we shall take oriented cobordism, modulo 2-torsion. This has coefficients

$$\Omega_{SO}^*(\text{point}) = \mathbb{Z}[\frac{1}{2}][x_4, x_8, x_{12}, \dots],$$

where in the dual homology theory the generator x_{4j} can be represented by the complex projective space $\mathbb{C}P^{2j}$. For those with a non-topological background we should point out that for 'bundle-based' theories such as K^* and KO^* it is the cohomology which has a nice geometric interpretation, while for 'manifold-based' theories such as Ω_{SO}^* it is the dual homology. It is possible (author's unpublished manuscript from the 1960s) to describe $\Omega^n(X)$ in terms of maps into X from manifolds modelled on an n-codimensional subspace of a separable Hilbert space.

The universality of cobordism is illustrated by results of the kind

$$\Omega_U^*(X) \otimes_{\Omega_U^*} \mathbb{Z} \cong K^*(X),$$

both sides being regarded as $\mathbb{Z}/2$-graded. Variants of this Conner-Floyd isomorphism hold both for KO^* and elliptic cohomology.

In Section 5 we will show that a good bundle-theoretic approximation can be found for $Ell^*(BG)$, provided that G is a finite group for which $Ell^{\text{odd}}(BG) = 0$. What is the significance of this condition?

The problem with the variants of the universal cohomology theory is that the coefficients are polynomial rings on infinitely many generators, and that some way must be found to cut these down to a more usable size. The first step is to localise at a single prime p. As a building block for the local theory we obtain BP^* with coefficients $\mathbb{Z}_{(p)}[v_1, v_2, \ldots]$ with $\deg(v_j) = 2(p^j - 1)$, where $\mathbb{Z}_{(p)}$ denotes the ring of integers localised at p, and BP^* inherits a complex orientation from Ω^*. Concentrating on a single degree, and passing from $\mathbb{Z}_{(p)}$ to $\mathbb{F}_p$ we obtain the Morava K-theory $K(n)^*$ with coefficients $\mathbb{F}_p[v_n, v_n^{-1}]$, and an inherited complex orientation. In case this seems too abstract it is worth noting that $K(1)^*$ turns out to be the mod p reduction of a certain (J.F. Adams') summand of p-local K-theory, that $K(2)^*$ has a similar interpretation (A. Baker's) in elliptic cohomology, and that by mapping v_n to zero in the non-periodic theory with coefficients $\mathbb{F}_p[v_n]$ we recover non-surprisingly $H^*(\ , \mathbb{F}_p)$.

The advantage of Morava K-theory is that it has good applicable properties, and that, by considering all values of n, some results can be pulled back to localised cobordism, and then pushed forward to geometrically interesting theories such as elliptic cohomology. For $p \geqslant 3$, $K(n)^*(X)$ is a commutative graded ring, the coefficients form a graded field in the sense that all graded modules are free, and hence there is a Künneth isomorphism

$$K(n)^*(X \times Y) \cong K(n)^*(X) \otimes_{K(n)^*} K(n)^*(Y).$$

Let $X = BG$, an infinite cell complex having only finitely many cells in each dimension.

Definition 4.1. The finite group G is *good* in the sense of Hopkins, Kuhn and Ravenel [12] if $K(n)^*BG$ is additively generated as a $K(n)^*$-module by transferred Euler classes of complex representations of subgroups of G.

If G is good, then $K(n)^{\mathrm{odd}}(BG) = 0$. This is immediate from the definition.

Proposition 4.2 (I. Kriz). *If H is a normal subgroup of G and $G/H \cong C_p$, the goodness of H implies that of G if and only if a certain spectral sequence with $E_2^{r,s} = H^r(C_p, K(n)^*(BH))$ collapses.*

Now consider a split extension of the form

$$\text{(elementary abelian } p\text{-group)} \lhd G \twoheadrightarrow C_p^t.$$

Proposition 4.3 ([14, Thm 3.1]). *An extension of the form above is good.*

This can be generalised, for example by allowing the normal subgroup to be elementary non-abelian. For variants see papers by M. Tezuka and N. Yagita, in particular [21, §2]. Kriz's result is stronger than the one stated, since, besides proving the vanishing of $K(n)^{\mathrm{odd}}$, he also obtains the structure of $K(n)^{\mathrm{even}}(BG)$.

Corollary 4.4. *The groups* C_p, $C_p \times C_p$, p_+^{1+2} *and* $E(p)$ *(of order* p^4*) are all good.*

Proof. C_p is good, hence so is $C_p \times \cdots \times C_p$, and the remaining groups are extensions of the kind allowed in Proposition 4.2. Arguing more directly Tezuka and Yagita (op. cit. supra [21, Thm. 4.10]) show that

$$\mathrm{Ch}_{BP}(p_+^{1+2}) = BP^*(BG). \qquad \square$$

Comparison of Proposition 4.2 with Section 2 shows that K-theoretic calculations for finite groups can be easier than for ordinary cohomology. Case by case analysis shows that $(p \geqslant 5)$ all groups of order p^4 are good. Examples of groups of order p^6 are known to be non-good, and I suspect that the same holds for p^5.

Theorem 4.5. *If G is a finite group such that for each odd prime p dividing the order of a Sylow subgroup G_p is good, then* $Ell^{\mathrm{odd}}(BG) = 0$.

Proof. For an arbitrary cohomology theory h^*, such that $1/2 \in h^*(\text{point})$, $h^*(BG)$ is detected by $\bigoplus_{\substack{p \| |G| \\ p=\mathrm{odd}}} h^*(BG_p)$. A theorem of D. Ravenel, S. Wilson and N. Yagita [18] implies that $BP^*(BG)$ is concentrated in even dimensions and that $BP^{\mathrm{even}}(BG)$ is flat over the coefficients, provided that G_p is good. Hence delocalising, the same holds for $\Omega^*_{SO}(BG)$, and for $Ell^*(BG)$ as a quotient of this universal theory. $\qquad \square$

Examples 4.6.

$$Ell^{\mathrm{odd}}_J(BM_{24}) = Ell^{\mathrm{odd}}_W(BCo_1) = 0.$$

If we restrict attention to extensions G such that the normal subgroup is of type $(p, \cdots, p)$ and the quotient of type (p^t), then $BP^*(BG)$ can be obtained more directly from a spectral sequence with $E_2^{r,s} = H^r(C_{p^t}, BP^s(BH))$. $BP^s(BH)$ vanishes for s odd, and is p-torsion free for s even. Hence $E_2^{r,s} = 0$ unless r and s are both even, the spectral sequence collapses and $E_2^{r,s} = E_\infty^{r,s}$. In particular $BP^{\mathrm{odd}}(BG) = 0$.

Algebraically the results of this section suggest that $Ell^*(BM_{24})$ is generated as an Ell^*-module by the Chern classes of the Todd representation τ. In the next section we consider this more geometrically.

5. Elliptic Objects

Let $\varepsilon\colon R(G) \to \mathbb{Z}$ map an element of the complex representation ring of the finite group G to its virtual dimension, and let $R(G)\widehat{}$ be the completion with respect to powers of $I = \operatorname{Ker}\varepsilon$. Then, if we regard K^* as a $\mathbb{Z}/2$-graded cohomology theory

$$R(G)\widehat{} \cong K^0(BG), \qquad\qquad\qquad\text{(i)}$$

$$K^1(BG) = 0. \qquad\qquad\qquad\text{(ii)}$$

At least for a restricted class of groups G we would like to set up a similar isomorphism in elliptic cohomology, and result (ii) above explains the emphasis on groups which are HKR-good in the previous section. Arguing in terms of characters localised at the prime p one can prove

Theorem 5.1 ([12], [19], [24]). *For any finite group G, with $G_p^{(2)}$ equal to conjugacy classes of commuting p-power pairs, and $R = Ell^*(\text{point})$, let $\overline{R}$ be the algebraic closure of the quotient field of the p-adic completion. Then, if $Ell^*(BG)_p$ is the p-adic completion of $Ell^*(BG)$, there is an isomorphism*

$$Ell^*(BG)_p \otimes_R \overline{R} \to \operatorname{Map}(G_p^{(2)}, \overline{R}).$$

This result motivates the still-conjectural definition of an 'elliptic representation ring' given in [2]. We start with a bundle construction due to J. L. Brylinski [4], who restricts his attention to the loop space LM of a finite-dimensional, simply-connected, C^∞-manifold M. LM is a Fréchet manifold modelled on the space of unbased C^∞-maps from S^1 into M, and admits an action by $A = \operatorname{Diff}^+(S^1)$, whose fixed point set is the subspace of constant loops diffeomorphic to M. Let E be an infinite dimensional complex vector bundle over LM admitting an A-action covering that on M. Infinitesimally we can consider the action of A on local sections of the bundle E to be an action of the Virasoro Lie algebra

$$\mathfrak{a} = \widetilde{\operatorname{Vect}}(S^1) \cong \operatorname{Vect}(S^1) \oplus \mathbb{R},$$

and this restricts to a fibrewise action of $\mathfrak{a}$ on the restriction of E to the fixed point set M.

Definition 5.2. The restricted bundle $E|_M$ is *admissible* if, for some fixed integer m, it admits a decomposition

$$E|_M = \widehat{\bigoplus_{n \in \frac{1}{m}\mathbb{Z}}} E_n$$

under the action of the infinitesimal generator L_0 of the rotation subgroup of $\text{Diff}^+(S^1)$, such that

(i) E_n has finite fibre dimension, and

(ii) $E_n = 0$ for $n \leqslant n_0 \in \mathbb{Z}$.

Denote by $K_V(LM)$ the Grothendieck group of admissible $\mathfrak{a}$-equivariant vector bundles over LM, and note that there is a forgetful map

$$K_V(LM) \to K(M)[[q^{1/m}]]$$

$$[E] \mapsto \sum_{n \in \frac{1}{m} \cdot \mathbb{Z}} [E_n]q^n \quad \text{(Laurent series)}.$$

Now replace M by the non-simply-connected space BG, which we have already noted can be approximated by C^∞-manifolds. Let

$$G^{(2)} = \{(g_1, g_2) \in G \times G : g_1 g_2 = g_2 g_1\},$$

$$\Gamma_0(2) = \left\{ \begin{pmatrix} a & b \\ c & d \end{pmatrix} \in SL_2(\mathbb{Z}) : c = 0 \mod 2 \right\},$$

and

$$Z = \{\tau \in \mathbb{C} : \text{im}(\tau) > 0\}.$$

As usual $q = e^{2\pi i \tau}$. Then $\Gamma_0(2) \times G$ acts on $G^{(2)} \times Z$ via

$$(A, g) \cdot (g_1, g_2, \tau) = \left(g g_1^d g_2^{-b} g^{-1}, g g_1^{-c} g_2^a g^{-1}, \frac{a\tau + b}{c\tau + d} \right).$$

This is a sightly modified version of S. Norton's rule, introduced in the Appendix to [17], quite independently of the development of elliptic cohomology.

Denote by $BC_G(g)$ the classifying space for the centraliser of g in G, as g runs through a family of representatives for the conjugacy classes in G.

Definition 5.3. The *graded elliptic representation ring* $REll^*(G) = \{REll^{2k}(G) : k \geqslant 0\}$ consists, up to equivalence of finite-dimensional representations on subspaces, of the following data:

- For each centraliser $C_G(g)$ a representation space

$$\rho_g = \widehat{\bigoplus_{n \in \frac{1}{m} \cdot \mathbb{Z}}} H_{g,n},$$

such that $H_{n,g} \in R(C_G(g))$ and $H_{g,n} = 0$ for $n \leqslant n_0(g)$.

- If $\operatorname{char}_q \rho = \sum_n \chi_{C_G(g)}(H_{g,n})q^n$, so that $\operatorname{char}_q \rho : G^{(2)} \times Z \to \mathbb{C}$, we require that for each pair $(g_1, g_2) \in G^{(2)}$, $\operatorname{char}_q \rho_{g_1}(g_2)$ is a modular form of weight $2k$ for some congruence subgroup $\Gamma \subseteq \Gamma_0(2)$.
- As (g_1, g_2) varies in $G^{(2)}$ the class function $F = \operatorname{char}_q \rho$ satisfies the equivariant modularity condition

$$F(A \cdot (g_1, g_2, \tau)) = (c\tau + d)^{-2k} F(g_1, g_2, \tau).$$

Up to finite index $REll^*(G)$ coincides with the coefficient ring $\mathcal{Ell}^*_G$ for Devoto's equivariant cohomology theory. If completion is with respect to the kernel of the augmentation map $\varepsilon : \mathcal{Ell}^*_G \to \mathcal{Ell}^*_1$, induced by the inclusion of the identity in G, then

$$\mathcal{Ell}^*(BG)[1/|G|] \cong \widehat{\mathcal{Ell}}^*_G,$$

see [6, 24].

For a good group G $Ell^{\mathrm{odd}}(BG) = 0$ and elements of $Ell^{\mathrm{even}}(BG)$ are detected by generalised characters. These two statements are analogues of (i) and (ii) for K-theory.

Our definition has been chosen to fit in with Brylinski's, each representation $H_{n,g}$ being associated with a *flat* bundle over $BC_G(g)$. The next lemma, due to R. Steiner, shows how these combine to give an admissible bundle over LBG.

Lemma 5.4. *The free loop space LBG satisfies*

$$LBG \simeq \coprod_g BC_G(g),$$

where g runs through a set of representatives for the conjugacy classes of G.

Norton's formula thus serves to tie together the graded flat bundles over the disjoint components of LBG. Our definition also fits in with that of an elliptic object proposed by G. B. Segal in [19]. His bundles come equipped with a connection, which we can neglect in the case of BG, because of the built-in flatness. The action of the Virasoro algebra is reflected in the modularity conditions. Without being too precise we can refer to an element in $REll^*(G)$ as being defined by a compatible family of McKay-Thompson series. Restricting to the component $BG = BC_G(1)$ we obtain an element in $R(G)[[q^{1/m}]]$, having some of the properties of a Moonshine module.

For each of the groups $\mathbb{M}$, Co_1, and M_{24} the candidates for a Monster or almost-Monster module have characters which, when restricted the conjugacy class of g, are modular functions rather than forms, i.e., the class concerned belongs to $Ell^0(BG)$ (restricted to $K^0(BG)[[q]]$). What of the 'genus zero' condition, that is, how can one give a cohomological condition for the character to generate the function field of $\Gamma\backslash\overline{Z}$ for some subgroup Γ sandwiched between a congruence subgroup $\Gamma_0(N) \subseteq SL_2(\mathbb{Z})$ and its normaliser in $SL_2(\mathbb{R})$? So far as I know this remains an open problem, although a clue to its solution may lie with the *replication formulae* establishing relations between the finite dimensional summands of the Monster module for $\mathbb{M}$. If $H_n = H_{n,1}$ is the n-th summand in the McKay-Thompson series, then J. Conway and S. Norton claimed (and later R. Borcherds [3] proved it for the Moonshine module $V^\natural$ of I. Frenkel, J. Lepowsky and A. Meurman [8]) that H_n can be expressed in terms of H_1, H_2, H_3 and H_5 under the action of Hecke operators $\{T_m\}$. In the first instance these are defined for modular forms, can be extended to McKay-Thompson series by the modularity of the characters, and more systematically to the cohomology theory $Ell^*(X)$, see various papers of A. Baker starting with [1].

In the final section of [2] an attempt is made to apply this general programme to an 'almost Moonshine module' associated to the group Co_0, the quotient of which by a central subgroup of order 2 gives Co_1. Thus let L be an even, unimodular $24d$-dimensional lattice admitting the finite group G as a group of automorphisms. Write $f: L \to 2\mathbb{Z}$ for the associated quadratic form, V for the representation space $L \otimes_{\mathbb{Z}} \mathbb{Q}$ and

$$\theta_L(q) = \sum_{n=0}^{\infty} a_n q^n,$$

where

$$(a_n = |\{x \in L : f(x) = 2n\}|).$$

Then θ_L is the character of an infinite dimensional representation space Θ_G, which can be more explicitly described as the complex group algebra $\mathbb{C}[L]$ of the lattice. For each conjugacy class $\langle g \rangle$ the character is a modular form of specific weight $2k$, level N and twisting factor ε. We obtain meromorphic modular forms with the same labels for a second representation space Ω_G as follows. If $\rho: G \to SL_{24d}(\mathbb{Q})$ describes the group action on V, let $\rho(g)$ have characteristic polynomial equal to

$$\prod_{i \geqslant 1} \eta(iq)^{a_i},$$

where

$$\eta(\tau) = q^{1/24} \prod_{n} (1 - q^n).$$

The label N actually equals $h|g|$ with h dividing the g.c.d. $(24, |g|)$, the same level appearing in the original Moonshine data. Taking characters of both sides we see that

$$\Omega_G = q^d \Lambda(-(Vq + Vq^2 + \cdots)),$$

where Λ denotes the exterior algebra. As a candidate for a Moonshine module write

$$\Theta / \Omega_G = q^{-d}(\mathbb{C}[L] \otimes \mathrm{Sym}(Vq + Vq^2 + \cdots)),$$

where Sym denotes the symmetric algebra. For the group Co_0, $d = 1$ and L is the Leech lattice. For technical reasons to do with low-dimensional cohomology (satisfied for example by the Mathieu group $M_{23} \subseteq Co_1$) the construction given in [2] is not quite as general as one might hope. But with more care these should be removable, the basic idea being transparent and elegant.

In the case of Co_0 it turns out that for some conjugacy classes $\langle g \rangle$ the character of Θ / Ω_G does not satisfy the 'genus zero' condition. There may be some connection between this and the fact that 'classical' elliptic cohomology is defined away only from the prime 2.

There are other examples which illustrate the significance of the Hecke operators as cohomology operations. In [16], G. Mason studies the McKay-Thompson series Ω_G for $G = M_{24}$. If g represents a conjugacy class in M_{24} regarded as a subgroup of the symmetric group S_{24}, and g decomposes as a product of a_i cycles on length i, where $1 \leqslant i \leqslant r$ and $a_1 + 2a_2 + \cdots + ra_r = 24$, then the value of the character of Ω_G at g equals $\eta(q)$ above.

Note that with $g = 1$,

$$\eta_1(q) = q \prod_{n \geqslant 1} (1 - q^n)^{24} = \Delta(q),$$

the usual discriminant function. It turns out that the form η_g are eigenforms (cusp forms invariant under the Hecke operators) since the space of forms (level N, weight $2k$ and twisting function ε) taking the value zero at each cusp, containing η_g, has dimension one. What makes this example particularly interesting is that, on the one hand the coefficients of $\Delta(q)$ are the Ramanujan numbers $\tau(n)$ and on the other they equal the dimensions of certain (virtual) representations of M_{24}. The first few terms of the sequence of dimensions are

$$1 \quad -24 \quad 252 \quad -14722 \quad 4830 \quad -6048 \quad -16744 \quad 84480 \quad \cdots,$$

and, given the definition of Ω_{24} in terms of the exterior powers of a virtual representation built from copies of the 24-dimensional representation

V, it should be possible to find replication formulae for the corresponding summands.

By way of conclusion here are two further problems:

(1) Identify those sporadic simple groups G for which, at least locally, $Ell^*(BG)$ is generated over Ell^*(point) by transferred Euler classes. In particular, for which groups is $Ell^{odd}(BG) = 0$? Study the action of the Hecke algebra.

(2) Relate the elliptic representation ring of M_{24} to $Ell^*(BM_{24})$ as a module generated by the Chern classes of the Todd representation. Start locally with Devoto's description of $Ell^*(BC_p)$, and extend this to the non-abelian group p_+^{1+2}. Can one throw any additional light on Moonshine restricted to the Mathieu groups?

References

[1] A. Baker, Hecke operators as operations in elliptic cohomology, J. Pure and App. Alg. **63** (1990), 1–11.

[2] A. Baker, C. Thomas, Classifying spaces, Virasoro equivariant bundles, elliptic cohomology and Moonshine, http://www.maths.gla. ac.uk/~ajb/dvi-ps/ab-cbt.pdf.

[3] R. E. Borcherds, Monstrous moonshine and monstrous Lie superalgebras, Invent. Math. **109** (1992), 405–444.

[4] J. L. Brylinski, Representations of loop groups, Dirac operators on loop space, and modular forms, Topology **29** (1990), 461–480.

[5] J. F. Carlson, Cohomology, computations and commutative algebra, Not. Amer. Math. Soc. **52** (2005), 426–434.

[6] J. Devoto, Equivariant elliptic cohomology and finite groups, Mich. Math. J. **43** (1996), 3–32.

[7] L. Evens, On the Chern classes of representations of finite groups, Trans. Amer. Math. Soc. **115** (1965), 180–193.

[8] I. B. Frenkel, J. Lepowsky & A. Meurman, *Vertex Operator Algebras and the Monster*, Academic Press (1988).

[9] D. J. Green, The 3-local cohomology of the Mathieu group M_{24}, Glas. Math. J. **38** (1996), 69–75.

[10] ——, Gröbner bases and the computation of group cohomology, Lect. Notes in Math. **1827** (2003).

[11] A. Grothendieck, *Classes de Chern et représentations linéaires des groupes discrets. (10 exposés sur la cohomologie étale des schémas)*, North Holland (1968) 215–305.

[12] M. Hopkins, N. Kuhn, D. Ravenel, Morava K-theories of classifying spaces and generalised characters for finite groups, Lect. Notes in Math. **1509** (1992), 186–209.

[13] G. James, M. Liebeck, *Representations and characters of groups*, Cambridge University Press (1993).

[14] I. Kriz, Morava K-theory of classifying spaces – some calculations, Topology **36** (1997), 1247–73.

[15] G. Lewis, Integral cohomology rings of groups of order p^3, Trans. Amer. Math. Soc. **132** (1968), 501–529.

[16] G. Mason, M_{24} and certain automorphic forms, Contemp. Math. **45** (1985), 223–244.

[17] —— (with an appendix by S.P. Norton), Finite groups and modular functions, Proc. of Symp. in Pure Math. **47** (1987), 181–210.

[18] D.C. Ravenel, W.S. Wilson, N. Yagita, Brown-Peterson cohomology from Morava K-theory, K-Theory **15** (1998), 147–199.

[19] G.B. Segal, Elliptic cohomology, Astérisque **161–162** (1988), 187–201.

[20] V. Snaith, Explicit Brauer induction, Invent. Math. **94** (1988), 455–478.

[21] M. Tezuka, N. Yagita, Cohomology of finite groups and Brown-Peterson cohomology II, Lect. Notes in Math. **1418** (1990), 57–69.

[22] ——, On odd prime components of cohomologies of sporadic simple groups and the rings of universal stable elements, J. of Alg. **183** (1996), 483–516.

[23] C.B. Thomas, *Characteristic classes and the cohomology of finite groups*, Cambridge University Press (1986).

[24] ——, *Elliptic cohomology*, Kluwer Academic/Plenum publishers (1999).

Monstrous and Generalized Moonshine and Permutation Orbifolds

Michael P. Tuite

Department of Mathematical Physics,
National University of Ireland,
Galway, Ireland.

Abstract

We consider the application of permutation orbifold constructions towards a new possible understanding of the genus zero property in Monstrous and Generalized Moonshine. We describe a theory of twisted Hecke operators in this setting and conjecture on the form of Generalized Moonshine replication formulas.

1. Introduction

The Conway and Norton Monstrous Moonshine Conjectures [CN], the construction of the Moonshine Module [FLM] as an orbifold vertex operator algebra and the completion of the proof of Monstrous Moonshine by Borcherds [Bo2] provided much of the motivation for the development of Vertex Operator Algebras (VOAs) e.g. [Bo1], [FLM], [Ka], [MN]. Another highlight of VOA theory is Zhu's study of the modular properties of the partition function (and n-point functions) for generic classes of VOAs [Z]. Zhu's ideas were generalized to include orbifold VOAs [DLM] whose relevance to Monstrous Moonshine is emphasized in refs. [T1], [T2]. Norton's Generalized Moonshine Conjectures [N2] concerning centralizers of the Monster group has yet to be generally proven using either Borcherds' approach or orbifold partition function methods although some progress has recently been made in refs. [H] and [T3], [IT1], [IT2] respectively.

In this note, we sketch a possible new approach to these areas based on permutation orbifold VOA constructions [DMVV], [BDM]. In particular, we introduce a theory of twisted Hecke operators generalizing classical Hecke operators in number theory e.g. [Se]. We then discuss permutation orbifold constructions where the classical Hecke operators naturally appear and finite

378

group and permutation orbifold constructions where the twisted Hecke operators appear. Using these ideas we formulate a conjecture on the nature of Generalized Moonshine replication formulas generalizing replication formulas for the classical J function and McKay-Thompson series in Monstrous Moonshine. Detailed proofs will appear elsewhere [T4].

2. Replication Formula for the J Function

We begin with a brief review of Faber polynomials, Hecke algebras and the replication formula for the classical J function. Consider

$$t(q) = q^{-1} + 0 + \sum_{k \geq 1} a(k) q^k,$$

the formal series in q. Define the Faber polynomial $P_n(x)$ for $t(q)$ to be the unique n^{th} order polynomial with coefficients in $\mathbb{Z}[a(1), \ldots, a(n-1)]$ such that

$$P_n(t(q)) = q^{-n} + O(q). \tag{1}$$

Thus $P_1(x) = x$, $P_2(x) = x^2 - 2a(1)$, $P_3(x) = x^3 - 3a(1)x - 3a(2)$ etc. The Faber polynomials for t satisfy the following generating relation e.g. [Cu], [N1]

$$\exp\left(-\sum_{n \geq 1} \frac{p^n}{n} P_n(x)\right) = p(t(p) - x), \tag{2}$$

for formal parameter p.

Let $f(\tau)$ be a meromorphic function of $\tau \in \mathbb{H}$, the upper half complex plane. Then for integer $k \geq 1$ define a right modular group action on f for $\gamma = \begin{pmatrix} a & b \\ c & d \end{pmatrix} \in \Gamma = SL(2, \mathbb{Z})$ as follows

$$(f|_k \gamma)(\tau) = (c\tau + d)^{-k} f(\gamma \tau), \tag{3}$$

with $\gamma \tau = \frac{a\tau + b}{c\tau + d}$. Then f is a modular form of (necessarily even) weight k if f is holomorphic in τ and

$$f|_k \gamma = f.$$

Define the standard Hecke operators $T(n)$ for $n \geq 1$ with the following action on a modular form f of weight k [Se]

$$T(n)f(\tau) = \frac{1}{n} \sum_{a \geq 1, ad=n} a^k \sum_{0 \leq b < d} f\left(\frac{a\tau + b}{d}\right). \tag{4}$$

These satisfy the Hecke algebra relations

$$T(mn) = T(m)T(n), \quad (m, n) = 1,$$
$$T(p)T\left(p^m\right) = T\left(p^{m+1}\right) + p^{k-1}T\left(p^{m-1}\right), \quad m \geq 1, \text{ prime } p. \quad (5)$$

One finds

$$(T(n)f)|_k\gamma = T(n)f, \quad (6)$$

i.e. $T(n)f$ is also a modular form of weight k.

The classical example is the Eisenstein series G_k of even weight $k \geq 4$ (with $G_k(\tau) = 0$ for odd k) [op.cit.]

$$G_k(\tau) = \sum_{\substack{m,n\in\mathbb{Z} \\ (m,n)\neq(0,0)}} \frac{1}{(m\tau + n)^k}$$

$$= 2\zeta(k) + 2\frac{(2\pi i)^k}{(k-1)!} \sum_{n\geq 1} \sigma_{k-1}(n)q^n, \quad (7)$$

where $q = \exp(2\pi i\tau)$ and $\sigma_k(n) = \sum_{d|n} d^k$. Furthermore, G_k is an eigenfunction of $T(n)$ with eigenvalue determined by the coefficient of q^n normalized to the coefficient of q i.e.

$$T(n)G_k = \sigma_{k-1}(n)G_k. \quad (8)$$

Thus it follows from (5) that for k odd

$$\sigma_k(mn) = \sigma_k(m)\sigma_k(n), \quad (m, n) = 1,$$
$$\sigma_k(p)\sigma_k(p^m) = \sigma_k\left(p^{m+1}\right) + p^k\sigma_k\left(p^{m-1}\right), \quad m \geq 1, \text{ prime } p. \quad (9)$$

In fact, it is easy to check directly that (9) holds for all $k \in \mathbb{C}$.

The classical modular invariant function of weight 0 is given by

$$J(\tau) = 1728\frac{G_4^3}{G_4^3 - G_6^2} - 744$$

$$= \sum_{k\in\mathbb{Z}} c(k)q^k = q^{-1} + 0 + 196884q + 21493760q^2 + \ldots$$

with standard normalization $c(-1) = 1$ and $c(0) = 0$. J is a hauptmodul for the genus zero group Γ and is thus a generator for the field of modular invariants. Thus

$$T(n)J(\tau) = \sum_{a\geq 1, a|n} \frac{1}{a} \sum_{s\in\mathbb{Z}} c\left(\frac{ns}{a}\right) q^{as} \tag{10}$$

$$= \frac{1}{n}q^{-n} + O(q). \tag{11}$$

is a polynomial in J which from (1) and (11) must be

$$T(n)J(\tau) = \frac{1}{n}P_n(J(\tau)), \tag{12}$$

where P_n is the Faber polynomial corresponding to J. Eqn. (12) is called the replication formula for J.

Eqn. (10) also implies that

$$\sum_{n\geq 1} p^n T(n)J(\tau) = \sum_{r\geq 1, s\in\mathbb{Z}} c(rs) \sum_{a\geq 1} \frac{1}{a}p^{ar}q^{as} = -\sum_{r\geq 1, s\in\mathbb{Z}} c(rs)\log(1-p^r q^s).$$

Then (2) implies the famous J function denominator formula [N1], [Bo2]

$$\exp\left(-\sum_{n\geq 1} p^n T(n)J(\tau)\right) = \prod_{r\geq 1, s\in\mathbb{Z}} (1 - p^r q^s)^{c(rs)}$$

$$= p(J(p) - J(q)). \tag{13}$$

This formula is one of the cornerstones of Borcherds' celebrated proof of the genus zero Moonshine property where (13) is a denominator formula for a particular generalized Kac-Moody algebra constructed from the Moonshine Module $V^\natural$ [Bo2].

3. Twisted Hecke Operators and Eisenstein Series

The definitions of modular functions and Hecke operators above can be generalized to "twisted" versions as follows. We define a twisted modular form of integer weight k to be a holomorphic (in τ) function $f = f((\theta, \phi), \tau)$ for $(\theta, \phi) \in U(1) \times U(1)$ such that

$$f|_k\gamma = f,$$

where

$$(f|_k\gamma)((\theta, \phi), \tau) = (c\tau + d)^{-k} f(\gamma(\theta, \phi), \gamma\tau),$$

with left group action

$$\gamma(\theta, \phi) = (\theta^a\phi^b, \theta^c\phi^d). \tag{14}$$

Clearly the case $(\theta, \phi) = (1, 1)$ defines a standard modular form of weight k.

We can extend the definition of the Hecke operator $T(n)$ to twisted modular forms as follows:

$$T(n)f((\theta, \phi), \tau) = \frac{1}{n} \sum_{a \geq 1, ad=n} a^k \sum_{0 \leq b < d} f\left((\theta^a \phi^b, \phi^d), \frac{a\tau + b}{d}\right), \quad (15)$$

which includes the standard definition in the case $(\theta, \phi) = (1, 1)$. For $\phi = 1$ and $\theta^m = 1$ for integer m, this Hecke operator is essentially that which appears in Borcherds' proof [Bo2] and is discussed at length in ref. [F].

We also define a homothety operator[1]

$$R(n)f((\theta, \phi), \tau) = f((\theta^n, \phi^n), \tau). \quad (16)$$

These operators satisfy the Hecke algebra [T4]

$$R(mn) = R(m)R(n)$$
$$R(m)T(n) = T(n)R(m)$$
$$T(mn) = T(m)T(n), \quad (m, n) = 1,$$
$$T(p)T(p^m) = T(p^{m+1}) + p^{k-1}T(p^{m-1})R(p), \quad m \geq 1, \text{ prime } p, \quad (17)$$

and one again finds

$$(T(n)f)|_k \gamma = T(n)f, \quad (18)$$

i.e. $T(n)f$ is also a twisted modular form of weight k.

A twisted Eisenstein series $G_k((\theta, \phi), \tau))$ of weight $k \geq 1$ can also be defined [DLM], [MTZ]. In particular, for $k \geq 4$ we define[2]

$$G_k((\theta, \phi), \tau) = \sum_{\substack{m,n \in \mathbb{Z} \\ (m,n) \neq (0,0)}} \frac{\theta^m \phi^n}{(m\tau + n)^k},$$

for $\theta, \phi \in U(1)$ with $G_k((1, 1), \tau)) = G_k(\tau)$ [MTZ]. $G_k((\theta, \phi), \tau)$ is not an eigenfunction of $T(n)$ in general. However, for prime p

$$T(p)G_k((\theta, \phi), \tau) = p^{k-1}G_k((\theta, \phi), \tau) + R(p)G_k((\theta, \phi), \tau). \quad (19)$$

Hence if $(\theta, \phi) = (\theta^p, \phi^p)$ then $T(p)G_k((\theta, \phi), \tau) = \sigma_{k-1}(p)G_k((\theta, \phi), \tau)$.

---

1 A similar operator is defined in the standard case [Se].
2 The notation used here differs from that of op.cit.

4. The Permutation Orbifold of a $C = 24$ Holomorphic Vertex Operator Algebra

4.1. The Orbifold of a Holomorphic VOA

We now consider a Vertex Operator Algebra V (VOA) of central charge 24 e.g. [FLM], [Ka], [MN]. We assume that V is a Holomorphic VOA (HVOA) so that V is the unique irreducible module for itself with modular invariant meromorphic partition function [Sch], [DM]

$$Z_V(\tau) = Tr_V\left(q^{L(0)-1}\right) = \sum_{k \geq -1} a(k)q^k$$

$$= J(\tau) + a(0).$$

For example, the Moonshine Module $V^\natural$ is a HVOA with $Z_{V^\natural}(\tau) = J(\tau)$ whereas $Z_{V_L}(\tau) = J(\tau) + 24$ for the Leech lattice HVOA V_L [FLM].

Let G be a finite subgroup of the automorphism group of V. Then for $g \in G$ define the orbifold trace function

$$Z_V((g, 1), \tau) = Tr_V\left(gq^{L(0)-1}\right),$$

(so that $Z_V((1, 1), \tau) = Z_V(\tau)$). For the Moonshine Module $V^\natural$

$$T_g(\tau) = Z_{V^\natural}((g, 1), \tau), \qquad (20)$$

is the McKay-Thompson series for $g \in \mathbb{M}$, the Monster group of automorphisms of $V^\natural$.

Since V is holomorphic, there is a unique twisted module M_h for each $h \in G$ [DLM]. For $g \in C(h)$, the h centralizer, g induces a class of linear maps $\phi(g)$ on M_h so that we may define a twisted orbifold trace function[3]

$$Z((g, h), q) = Z((g, h), \tau) = Tr_{M_h}\left(\phi(g)q^{L(0)-1}\right), \qquad (21)$$

a meromorphic function for $\tau \in \mathbb{H}$ [op.cit.]. We define a right action of the modular group for $\gamma \in \Gamma$ as follows[4]

$$(Z|_0\gamma)((g, h), \tau) = Z(\gamma(g, h), \gamma\tau), \qquad (22)$$

with

$$\gamma(g, h) = \left(g^a h^b, g^c h^d\right).$$

The trace function enjoys the modular invariance property

$$(Z|_0\gamma)((g, h), \tau) = \epsilon_\gamma(g, h)Z((g, h), \tau). \qquad (23)$$

---

3 denoted by $Z(h, g^{-1}, \tau)$ in ref. [DLM],
4 The 0 subscript denotes the modular weight of $Z((g, h), \tau)$.

for cocycle $\epsilon_\gamma(g, h) \in \mathbb{C}^*$ [op.cit.] generalizing earlier ideas of Zhu concerning trace functions [Z]. Specializing to the McKay-Thompson series (20), these results imply that $T_g(\tau)$ is a meromorphic function on $\mathbb{H}$ satisfying the modular invariance property (23) for $h = 1$.

Let us consider orbifolds without a global phase anomaly i.e. where each $\phi(g)$ acting on M_h can be chosen such that $\epsilon_\gamma(g, h) = 1$ so that [Va]

$$(Z|_0\gamma)((g, h), \tau) = Z((g, h), \tau). \tag{24}$$

In particular, this condition implies that for h of order m

$$Z((1, h), \tau) = \sum_{k \in \mathbb{Z}} a\left((1, h), \frac{k}{m}\right) q^{k/m}, \tag{25}$$

for some integers $a((1, h), \frac{k}{m}) \geq 0$. We next define the G−orbifold partition function by

$$Z^{G-\mathrm{orb}}(\tau) = \frac{1}{|G|} \sum_{\substack{g, h \in G \\ gh=hg}} Z((g, h), \tau)$$

$$= \sum_{[h] \in G} \frac{1}{|C(h)|} \sum_{g \in C_G(h)} Z((g, h), \tau), \tag{26}$$

for centralizer $C(h) = \{hg = gh | g \in G\}$ and where $[h]$ denotes a conjugacy class of G. Clearly $Z^{G-\mathrm{orb}}(\tau)$ is also modular invariant. The most well-known example is the original construction for the Moonshine Module $V^\natural$ as a $\mathbb{Z}_2$ orbifold of the Leech lattice VOA [FLM].

4.2. Permutation Orbifolds

Let $V^{\otimes n} = V \otimes V \otimes \ldots V$ denote the nth tensor product VOA with partition function $Z_{V^{\otimes n}}(\tau) = Z(\tau)^n$ and central charge $24n$. The symmetric group S_n naturally acts on $V^{\otimes n}$ as an automorphism group. For each $\beta \in S_n$ there is a unique β-twisted $V^{\otimes n}$ module M_β which can be explicitly constructed from the original HVOA V [DMVV], [BDM], [Ba]. Furthermore, we may explicitly compute the permutation orbifold partition function $Z^{S_n-\mathrm{orb}}(\tau)$.

We illustrate this in the first non-trivial case for $V \otimes V$. Then $S_2 = \langle \sigma \rangle$ for 2−cycle σ where $\sigma : u \otimes v \to v \otimes u$ for all $u \otimes v \in V \otimes V$. We thus find

$$Z_{V \otimes V}((\sigma, 1), \tau) = Z(2\tau).$$

The σ−twisted module M_σ has partition function

$$Z_{V \otimes V}((1, \sigma), \tau) = Tr_{M_\sigma}\left(q^{L(0)-1}\right) = Z\left(\frac{\tau}{2}\right),$$

with

$$Z_{V \otimes V}((\sigma, \sigma), \tau) = Z\left(\frac{\tau + 1}{2}\right),$$

following (24). Thus we obtain

$$Z^{S_2-\text{orb}}(\tau) = \frac{1}{2}Z(\tau)^2 + T(2)Z(\tau), \tag{27}$$

for Hecke operator $T(2)$ of (4) for weight zero.

In general, for $\beta \in S_n$, consider the cycle decomposition

$$\beta = \sigma_1^{m_1} \sigma_2^{m_2} \ldots \sigma_n^{m_n}, \tag{28}$$

where σ_k denotes a k−cycle. The conjugacy classes of S_n are enumerated by the set of partitions of $n = \sum_{1 \le k \le n} km_k$. The centralizer is then

$$C(\beta) = S_{m_1} \times (S_{m_2} \rtimes C_2^{m_2}) \times \ldots \times (S_{m_n} \rtimes C_n^{m_n}), \tag{29}$$

of order $\prod_{1 \le k \le n} k^{m_k} m_k!$ with cyclic group $C_k = \langle \sigma_k \rangle$ and S_{m_k} the permutation group on the m_k cycles σ_k. We may construct $M_\beta = \otimes_k M_{\sigma_k}^{\otimes m_k}$ which has partition function [BDM]

$$Z((1, \beta), \tau) = Tr_{M_\beta}\left(q^{L(0)-1}\right) = \prod_{1 \le k \le n} Z\left(\frac{\tau}{k}\right)^{m_k}.$$

One eventually finds that the S_n permutation orbifold partition function is [DMVV]

$$Z^{S_n-\text{orb}}(\tau) = \sum_{[\beta] \in S_n} \frac{1}{|C(\beta)|} \sum_{\alpha \in C(\beta)} Z((\alpha, \beta), \tau)$$

$$= \sum_{\substack{m_1, \ldots m_n \\ \sum km_k = n}} \prod_{1 \le k \le n} \frac{1}{m_k!}(T(k)Z(\tau))^{m_k}, \tag{30}$$

for the classical Hecke operator $T(k)$ of (4).

It is natural to define a permutation orbifold generating function by

$$Z^{\text{perm}}(p, q) = 1 + \sum_{n \ge 1} p^n Z^{S_n-\text{orb}}(\tau). \tag{31}$$

for a formal parameter p. Thus we obtain [op.cit.]

$$Z^{\text{perm}}(p, q) = \exp\left(\sum_{n\geq 1} p^n T(n) Z(\tau)\right) = \prod_{r\geq 1, s\in\mathbb{Z}} \frac{1}{(1 - p^r q^s)^{a(rs)}}, \quad (32)$$

where $Z(\tau) = \sum_{k\geq -1} a(k)q^k$. This is clearly of the form of the **inverse** of the LHS of denominator formula (13). Thus such expressions canonically arise in the context of permutation orbifolds for $C = 24$ HVOAs.

Specializing to the case of the Moonshine module $V^\natural$ where $Z(\tau) = J(\tau)$ we obtain

$$Z_{V^\natural}^{\text{perm}}(p, q) = \exp\left(\sum_{n\geq 1} p^n T(n) J(\tau)\right) = \frac{1}{pJ(p) - pJ(q)}$$

$$= 1 + pJ(q) + \left(J(q)^2 - c(1)\right) p^2 + \left(J(q)^3 - 2J(q)c(1) - c(2)\right) p^3 +$$

$$\left(J(q)^4 - 3c(1)J(q)^2 - 2J(q)c(2) - c(3) + c(1)^2\right) p^4 + \dots$$

This formula and the infinite product formula of (32) very strongly suggest that $Z_{V^\natural}^{\text{perm}}(p, q)$ is the partition function for a doubly graded *symmetric* bosonic module with Monster characters which is, algebraically speaking, the inverse of the alternating homological structure constructed by Borcherds [Bo2]. Furthermore, infinite product formulas such as that of (32) have been given the interesting interpretation as a "second quantized" string partition function in the physics literature [DMVV]. A rigorous VOA construction for such a structure would be of obvious interest.

5. Finite Group and Permutation Orbifolds

Let us now consider orbifolding $V^{\otimes n}$ with respect to $G \times S_n$ where G acts diagonally on $V^{\otimes n}$ and S_n is the permutation group for $V^{\otimes n}$. We consider again a $C = 24$ holomorphic VOA, with modular invariant partition function and where (24) is holds. We may construct unique twisted sectors for each $(h, \beta) \in G \times S_n$ [BDM] to find [T4]

$$Z^{S_n-\text{orb}}((g, h), \tau) = \sum_{\substack{m_1 \dots m_n \\ \sum km_k = n}} \prod_{1\leq k\leq n} \frac{1}{m_k!} (T(k)Z((g, h), \tau))^{m_k}, \quad (33)$$

where here $T(k)$ is the twisted Hecke Operator of (15). Then (18) implies

$$(T(n)Z((g, h), \tau))|_0\gamma = T(n)Z((g, h), \tau). \quad (34)$$

We may define a permutation orbifold generating function generalizing (31) as follows:

$$Z^{\text{perm}}((g,h),p,q) = \sum_{n\geq 1} p^{\frac{n}{m}} Z^{S_n-\text{orb}}((g,h),\tau)$$

$$= \exp\left(\sum_{n\geq 1} p^{\frac{n}{m}} T(n) Z((g,h),\tau)\right). \tag{35}$$

where h is of order m and using (33). For $g = 1$, this expression reduces to an infinite product formula generalizing (32) to find

$$Z^{\text{perm}}((1,h),p,q) = \prod_{r\geq 1, s\in\mathbb{Z}} \left(1 - p^{\frac{r}{m}} q^{\frac{s}{m}}\right)^{-a\left((1,h^r),\frac{rs}{m}\right)}, \tag{36}$$

where $Z((1,h^r),\tau)) = \sum_{k\in\mathbb{Z}} a\left((1,h^r),\frac{k}{m}\right) q^{\frac{k}{m}}$.

6. Monstrous and Generalized Moonshine - the Genus Zero Property

We now consider the FLM Moonshine Module VOA $V^\natural$ [FLM] and its relationship to Moonshine. The original Monstrous Moonshine paper of Conway and Norton described evidence for an unexpected relationship between properties of the Monster finite group and the theory of modular forms [CN]. Many of these relationships are now understood to be generic to orbifold constructions in conformal field theory/VOA theory e.g. [FLM], [T1], [T2], [DLM]. However, the special feature that sets the Moonshine Module $V^\natural$ apart from other VOAs is the Genus Zero Property [CN]. This states that for each $g \in \mathbb{M}$, the McKay-Thompson series $T_g(\tau)$ of (20) is a hauptmodul for some genus zero modular group Γ_g. Thus for g of prime order $o(g) = p$, one finds (excluding one class of order 3) that either $\Gamma_g = \Gamma_0(p)$ with $g = p-$ (in the notation of [CN]) or else with $g = p+$ with $\Gamma_g = \Gamma_0(p)+ = \langle\Gamma_0(p), W_p\rangle$ where $\Gamma_0(p) = \left\{\begin{pmatrix} a & b \\ c & d \end{pmatrix} \middle| c = 0 \bmod p\right\}$ and $W_p : \tau \to -1/p\tau$ is a Fricke involution. In general, we say that $g \in \mathbb{M}$ is Fricke if T_g is invariant under a Fricke involution $W_N : \tau \to -1/N\tau$ where $N = ko(g)$ and $k|24$ and is otherwise non-Fricke. $k = 1$ in the global phase anomaly free cases where (24) holds.

The distinction between Fricke and non-Fricke classes is particularly important in the orbifold interpretation of Moonshine [T1]. There is very significant evidence for the general conjecture that the genus zero property for a McKay-Thompson series is equivalent to the statement that for any global phase anomaly free element g, orbifolding $V^\natural$ with respect to $\langle g\rangle$ for g Fricke results in $V^\natural$ again whereas orbifolding $V^\natural$ with respect to $\langle g \rangle$ for g non-Fricke results in the Leech lattice VOA [T2].

Generalized Moonshine refers to the still generally unproven conjecture of Norton [N2] that for each commuting pair $g, h \in \mathbb{M}$, then $Z_{V^\natural}((g, h), \tau)$ is either a hauptmodul for a genus zero modular group or is a constant. It is easy to show using (23) that (1) $Z_{V^\natural}((g, h), \tau)$ is constant iff $g^c h^d$ is non-Fricke for all $(c, d) = 1$ [N2], [T3] and (2) if $g, h \in \langle k \rangle$ for some $k \in \mathbb{M}$ then $Z_{V^\natural}((g, h), \tau)$ is a hauptmodul (since it can then be modular transformed to a McKay-Thompson series [T3], [IT1], [DLM]). In the remaining "non-trivial" cases, we may use (23) again to transform $Z_{V^\natural}((g, h), \tau)$ to a trace over a Fricke twisted module so that the genus zero property reduces to an analysis of h Fricke cases alone. The case of $h = 2+$, with centralizer $2.B$ for the Baby Monster B, has now been proved by Hoehn [H]. The relationship between orbifoldings of the Moonshine module and the genus zero property of Generalized Moonshine for $h = p+$ is discussed at length in [T3], [IT1], [IT2], [I].

The approach taken in Borcherds' proof of the Monstrous Moonshine genus zero property is to firstly prove a twisted denominator identity generalizing (13) [N1], [Bo2]

$$\exp\left(-\sum_{n \geq 1} p^n T(n) T_g(\tau)\right) = p\left(T_g(p) - T_g(q)\right). \tag{37}$$

This is the defining formula for completely replicable functions [N1], [FMN] from which it follows that the leading coefficients of $T_{g^i}(\tau)$ for $i = 1, 2, \ldots$ determine $T_g(\tau)$. Koike showed that the list of hauptmoduln appearing in the Moonshine Conjectures are completely replicable [Ko]. Based on this result and an analysis of the leading coefficients (using the explicit form for $T_g(\tau)$ found by FLM for $2-$ centralizers in the Monster [FLM]) Borcherds then demonstrated that indeed $T_g(\tau)$ obeys the genus zero property. This part of the proof was improved upon in [CG] where meromorphicity, modularity and the genus zero property are shown to generally follow from the infinitely many replication formulas that follow from (1) and (37), namely

$$T(n) T_g(q) = \frac{1}{n} F_n\left(T_g(q)\right), \tag{38}$$

where F_n is the Faber polynomial for $T_g(q)$.

However, as already noted, $T_g(q)$ is known to be meromorphic on $\mathbb{H}$ from [DLM]. Furthermore, by combining (38) with the general modular transformation property (23) (or (24) in the absence of global anomalies) one can also expect to obtain the genus zero property for T_g in a more direct fashion along the lines of the methods described in refs. [T1], [T2].

On the other hand, in the permutation orbifold construction based on $V^\natural$, (37) reads

$$Z_{V^\natural}^{\text{perm}}((g, 1), p, q) = \exp\left(\sum_{n \geq 1} p^n T(n) Z_{V^\natural}((g, 1), q)\right)$$

$$= \frac{1}{p(Z_{V^\natural}((g, 1), p) - Z_{V^\natural}((g, 1), q))}.$$

Recall our previous remarks concerning the reduction of Generalized Moonshine to Fricke classes. Consider h a global phase anomaly-free Fricke element of order $o(h) = m$ so that $Z_{V^\natural}((1, h), q^m) = Z_{V^\natural}((h, 1), q)$. It follows that

$$\exp\left(\sum_{n \geq 1} p^n T(n) Z\left((1, h), q^m\right)\right) = \frac{1}{p\left(Z_{V^\natural}((1, h), p^m) - Z_{V^\natural}((1, h), q^m)\right)}.$$

Hence from (36) we find for such Fricke h that

$$Z_{V^\natural}^{\text{perm}}((1, h), p, q) = \frac{1}{p^{\frac{1}{m}}\left(Z_{V^\natural}((1, h), p) - Z_{V^\natural}((1, h), q)\right)}. \tag{39}$$

This together with the general result (36) again suggests the existence of a symmetric bosonic construction forming a doubly-graded module for the centralizer of $C(h)$ for each such Fricke element h. It is thus natural to conjecture that for all order m global phase anomaly-free Fricke elements h the following holds

$$Z_{V^\natural}^{\text{perm}}((g, h), p, q) = \exp\left(\sum_{n \geq 1} p^{\frac{n}{m}} T(n) Z_{V^\natural}((g, h), q)\right)$$

$$= \frac{1}{p^{\frac{1}{m}}\left(Z_{V^\natural}((g, h), p) - Z_{V^\natural}((g, h), q)\right)}. \tag{40}$$

From (1), this is equivalent to the following Generalized Moonshine replication formula for global phase anomaly-free Fricke elements h

$$T(n) Z_{V^\natural}((g, h), q) = \frac{1}{n} F_n(Z_{V^\natural}((g, h), q)), \tag{41}$$

where F_n is the Faber polynomial for $Z_{V^\natural}((g, h), q^m))$. The equivalence of this replication formula to the genus zero property for Generalized Moonshine would therefore require a suitable generalization of the various Monstrous Moonshine arguments.

We conclude with the example of $n = 2$. Then (41) implies

$$Z_{V^\natural}\left(\left(g^2, h\right), 2\tau\right) + Z_{V^\natural}\left(\left(g, h^2\right), \frac{\tau}{2}\right) + Z_{V^\natural}\left(\left(gh, h^2\right), \frac{\tau+1}{2}\right)$$
$$= Z_{V^\natural}((g, h), \tau)^2 - 2a\left((g, h), \frac{1}{m}\right),$$

where $Z_{V^\natural}((g, h), \tau) = q^{-\frac{1}{m}} + 0 + a((g, h), \frac{1}{m}) q^{\frac{1}{m}} + \dots$ which corresponds to an example quoted in ref. [N2]. In particular, for $h = 2+$ and g of order 2 then using Fricke invariance we find

$$T_h(\tau) + T_g\left(\frac{\tau}{2}\right) + T_{gh}\left(\frac{\tau+1}{2}\right) = Z_{V^\natural}((g, h), \tau)^2 - 2a\left((g, h), \frac{1}{2}\right),$$

which can be easily verified in each case.

References

[Ba] Bantay, P.: Permutation orbifolds, Nucl.Phys. **B633**, 365–378 (2002).

[BDM] Barron, K., Dong, C. and Mason, G.: Twisted sectors for tensor product vertex operator algebras associated to permutation orbifolds. Commun.Math.Phys. **227**, 349–384 (2002).

[Bo1] Borcherds, R.: Vertex algebras, Kac-Moody algebras and the Monster. Proc.Natl.Acad.Sci.U.S.A. **83**, 3068–3071 (1986).

[Bo2] Borcherds, R.: Monstrous moonshine and monstrous Lie superalgebras. Invent.Math. **109**, 405–444 (1992).

[CG] Cummins, C.J. and Gannon, T.: Modular equations and the genus zero property of moonshine functions, Invent.Math. **129,** 413–443 (1997).

[Cu] Curtiss, J. H.: Faber Polynomials and the Faber Series, Am. Math. Mon. **78** 577–596 (1971).

[CN] Conway, J.H. and Norton, S.P.: Monstrous moonshine. Bull. London Math. Soc. **11**, 308–339 (1979).

[DLM] Dong, C., Li, H. and Mason, G.: Modular-invariance of trace functions in orbifold theory and generalized moonshine. Commun.Math.Phys. **214**, 1–56 (2000).

[DM] Dong, C. and Mason, G., Holomorphic vertex operator algebras of small central charge, *Pac.J.Math.*, 2004, **213** 253–266.

[DMVV] Dijkgraaf, R., Moore, G., Verlinde, E. and Verlinde, H.: Elliptic genera of symmetric products and the second quantized string, Commun.Math.Phys. **185**, 197–209 (1997).

[F] Ferenbaugh, C.: Lattices and generalized Hecke operators, in *Groups, difference sets and the Monster*, Walter de Gruyter, (1996).

[FMN] Ford, D., McKay, J. and Norton, S.P.: More on replicable functions. Commun.Algebra **22**, 5175–5193 (1994).

[FLM] Frenkel, I., Lepowsky, J. and Meurman, A.: *Vertex operator algebras and the Monster.* New York:Academic Press, 1988.

[H] Hoehn, G.: Generalized moonshine for the baby monster, Talk presented at this conference.

[I] Ivanov, R.I.: *Generalised Moonshine from Abelian Orbifolds of the Moonshine Module.* Ph.D. Thesis, National University of Ireland, Galway, 2002.

[IT1] Ivanov, R.I. and Tuite, M.P.: Rational generalised moonshine and abelian orbifolds. Nucl.Phys. **B635** (2002), 435–472.

[IT2] Ivanov, R.I. and Tuite, M.P.: Some irrational generalised moonshine from orbifolds. Nucl.Phys. **B635** (2002), 473–491.

[Ka] Kac, V.: *Vertex Operator Algebras for Beginners.* University Lecture Series, Vol. 10, Boston:AMS 1998.

[Ko] Koike, M.: On replication formula and Hecke operators, Nagoya University preprint.

[MN] Matsuo, A. and Nagatomo, K.: Axioms for a vertex algebra and the locality of quantum fields, Math.Soc.Japan Memoirs. **4**, (1999).

[MTZ] Mason, G., Tuite, M.P. and Zuevsky, A.: Torus n-Point Functions for $\mathbb{R}$-graded Vertex Operator Superalgebras and Continuous Fermion Orbifolds, Commun.Math.Phys. **283**, 305–342 (2008).

[N1] Norton, S. P.: More on moonshine, in *Computational Group Theory*, Academic Press, 185–193 (1984).

[N2] Norton, S.P.: Generalised moonshine. Proc.Symp.PureMath. **47**, 208–210 (1987).

[Sch] Schellekens, A.N.: Meromorphic c = 24 conformal field theories, Commun.Math.Phys. **153**, 159–186 (1993).

[Se] Serre, J-P.: *A course in arithmetic*, Springer-Verlag (Berlin 1978).

[T1] Tuite, M. P.: Monstrous moonshine from orbifolds. Commun.Math.Phys. **146** 277–309 (1992).

[T2] Tuite, M. P.: On the relationship between monstrous moonshine and the uniqueness of the moonshine module. Commun.Math.Phys. **166**, 495–532 (1995).

[T3] Tuite, M. P.: Generalised moonshine and abelian orbifold constructions. Contemp.Math. **193**, 353–368 (1996).

[T4] Tuite, M. P.: To appear.

[Va] Vafa, C.: Modular invariance and discrete torsion on orbifolds. Nucl.Phys. **B273**, 592–606 (1986).

[Z] Zhu, Y.: Modular invariance of characters of vertex operator algebras. J.Amer.Math.Soc. **9**, 237–302 (1996).

New computations in the Monster

Robert A. Wilson

School of Mathematical Sciences
Queen Mary, University of London
Mile End Road,
London E1 4NS

Abstract

We survey recent computational results concerning the Monster sporadic simple group. The main results are: progress towards a complete classification of the maximal subgroups, including showing that $L_2(27)$ is not a subgroup; showing that the 196882-dimensional module over $GF(2)$ supports a quadratic form; a complete set of explicit conjugacy class representatives; small representations of most of the maximal subgroups; and a partial classification of the 'nets' (in the sense of Norton).

1. Introduction

Our aim in this paper is to update the survey [27] by describing the various explicit computations which have been performed in the Monster group, and the new information about the Monster which has resulted from these calculations. We begin by summarising [27] for the benefit of readers who do not have that paper to hand.

The smallest matrix representations of the Monster have dimension 196882 in characteristics 2 and 3, and dimension 196883 in all other characteristics. Three of these representations (over the fields of orders 2, 3, and 7) are now available explicitly [9, 15, 25]. It is hoped that the data and programs to manipulate them will be made available in the next release of MAGMA [16]. The generating matrices are stored in a compact way, and never written out in full. The basic operation of the system is to calculate the action of a generator on a vector of the underlying module.

Our first construction [15] was carried out over the field $GF(2)$ of two elements in the interests of speed, and proceeded by amalgamating various 3-local subgroups. Unfortunately, these 3-local subgroups are too small to contain many useful subgroups, so we embarked on a second construction [9] over $GF(3)$, in order to utilise the much larger 2-local subgroups. In [27] we described how Beth Holmes used this construction to find four new maximal subgroups, and obtain a complete classification of subgroups of the Monster isomorphic to one of 11 listed simple groups (out of 22 still unclassified). The third construction [25] was over $GF(7)$, again using the 3-local subgroups, and the same generators as in the $GF(2)$ case. Thereby one can calculate character values modulo 14, and obtain good conjugacy class invariants.

2. The 2-local construction

The 2-local construction, although not the first, is easier to describe than the 3-local constructions, and is closely related to the Griess construction [6]. We shall not describe the construction itself, merely the outcome, and refer the reader to [9] for details. The idea is first to construct the involution centralizer $2^{1+24} \cdot Co_1$, in such a way that we can both calculate in this subgroup, and calculate its action on the module of dimension 196882 over $GF(3)$. Then we make a special 'triality' element which normalizes a subgroup $2^2.2^{11}.2^{22}.M_{24}$, the centralizer of a 4-group.

Now the 3-modular irreducible representation of degree 196882 for the Monster restricts to the subgroup $2^{1+24} \cdot Co_1$ as the direct sum of three constituents, of degrees 98304, 98280 and 298. The constituent of degree 98280 is monomial, and that of degree 98304 is a tensor product of representations of the double cover, of degrees 24 and 4096. Any element of this subgroup can therefore be specified by three matrices (over $GF(3)$, or more generally, any field of characteristic not 2), of sizes 24, 4096, and 298, and a monomial permutation on 98280 points. (Note however that this representation is not unique: negating the matrices of size 24 and 4096 gives a second representation of the same element.)

By careful choice of basis we can ensure that the triality element can be written as a monomial permutation on 147456 points, followed by 759 identical 64×64 matrices, and an 850×850 matrix. In particular its action on a vector can be quickly computed.

It is important to realise that the only elements of the Monster which are stored in one of these two compact formats are the elements of $2^{1+24} \cdot Co_1$ and the triality element (or rather, eight triality elements, being the elements of order 3 in the A_4 generated by the normal 2^2 and a triality element). Every other element of the Monster is stored as a word in these generators. (Some improvements on this are possible, but seem not to be worth the extra effort. For example, it would be possible to devise a compact format for most, if not all, of the subgroup $2^2.2^{11}.2^{22}.(M_{24} \times S_3)$.)

3. The 3-local constructions

When we first seriously considered a computer construction of the Monster some ten years ago, we decided to produce matrices over $GF(2)$, since calculation with such matrices is much faster than with matrices over any other field. The disadvantage, however, is that the maximal 2-local subgroups are no longer available as ingredients of the construction. Thus we decided to use maximal 3-local subgroups instead. Here again we give only a sketch of the construction, and refer to [15] for details.

The role of the involution centralizer is now taken by a maximal subgroup of shape $3^{1+12} \cdot 2 \cdot Suz{:}2$. The restriction of the representation to this subgroup consists again of a 'tensor product' part, of dimension 131220, a 'monomial' part, of dimension 65520, and a 'small' part. The small part has dimension 142 over $GF(2)$, or dimension 143 in any characteristic bigger than 3. The 'monomial' part is in reality induced from a 2-dimensional representation of a subgroup of index 32760. The 'tensor product' part is again not exactly a tensor product: if we restrict to the subgroup of index 2, it is the direct sum of two (dual) tensor products over $GF(4)$, each tensor being the product of one 90-dimensional and one 729-dimensional representation.

To generate the Monster, we adjoined a 'duality' element normalizing a certain subgroup of shape $3^2.3^5.3^{10}.(M_{11} \times 2^2)$. Again, by careful choice of basis we were able to write this extra element as a combination of a 'monomial' permutation on 87480 subspaces of dimension 2, two 324×324 matrices (repeated 11 and 55 times respectively), and a 538×538 matrix.

In fact these calculations are considerably simplified if there is a cube root of unity in the field. For this reason, we repeated the calculations over the field of order 7, and obtained the same set of generators for the Monster in this different representation [25].

4. Basic calculations

There are just two basic operations available to us in any of the constructions we have described. The first is to multiply together elements in our chosen maximal subgroup to create new generators in this subgroup. The second is to act on a vector by one of these generators, or by the extra 'triality' or 'duality' element.

An element of the Monster is stored as a word $x_1 t_1 x_2 t_2 \ldots$, where the x_i are in our maximal subgroup, and the t_i are equal to the extra generator (or possibly its inverse, in the 2-local version). If we take a 'random' vector v in the underlying module, the chances are extremely good that it lies in a regular orbit under the Monster. Thus the order of an element x is, with probability very close to 1, equal to the smallest positive integer n such that $vx^n = v$.

In [15, 27] we described how to improve this probability to exactly 1 at the expense of taking two (carefully chosen) vectors instead of one.

The first serious calculations we attempted used the $GF(2)$ construction to try to improve estimates for the symmetric genus of the Monster. By character calculations alone, Thompson had shown that the Monster was a quotient of the triangle group $\Delta(2, 3, 29) = \langle x, y, z \mid x^2 = y^3 = z^{29} = xyz = 1 \rangle$, and the challenge was to find the minimal value of n such that the Monster is a quotient of $\Delta(2, 3, n)$. From Norton's work on maximal subgroups [18] it seemed very likely that this minimal value was 7. However, the probability that a random pair of elements of orders 2 and 3 has product of order 7 is around 10^{-8}, so we would need to look at something like 100 million pairs to have a reasonable chance of finding $(2, 3, 7)$-generators for the Monster. This took some 10 years of processor time. See [24] for more details.

5. The quadratic form

The 196882-dimensional representation of the Monster over the field of two elements is self-dual, so the Monster preserves a symplectic form on the module, and embeds in the symplectic group $Sp_{196882}(2)$. The question as to whether the Monster also preserves a quadratic form seems difficult to answer from a theoretical perspective. Beth Holmes and Steve Linton (and independently Jon Thackray) calculated explicitly a quadratic form which is invariant. They did not determine whether this form is of $+$ or $-$ type.

6. Traces and conjugacy classes

The trace of a matrix is easy to calculate, but it is less obvious how to calculate the trace of a linear transformation given in the form of a computer program. Ultimately it seems to be necessary to calculate the corresponding matrix, and extract the diagonal entries. This is obviously rather time-consuming compared to the tracing of individual vectors we have been doing up till now.

Now if p is any prime, the trace modulo p can only distinguish between different p'-parts of elements, since modulo p we have $Tr(x^p) = Tr(x)$. Thus in order to distinguish conjugacy classes, it is necessary to calculate traces modulo two distinct primes. Since we used exactly the same generators in the representations over $GF(2)$ and over $GF(7)$, we can calculate the trace mod 2 and the trace mod 7 for the same element of the group, thus obtaining the value of the degree 196883 character modulo 14. Combining this invariant with the order of the element and the traces of its powers, we are able to identify the conjugacy class of any element, up to a few ambiguities.

With this apparatus Richard Barraclough has produced a list of conjugacy class representatives [3]. To do this, he first improved the efficiency of our programs so that a trace modulo 7 now takes only a few hours to calculate. Then he conducted a wide search through words of length 1 and 2. Most classes turned up in this way, and the few that did not had representatives in the subgroup $3^{1+12} \cdot 2 \cdot Suz{:}2$. Thus a more targeted search was conducted in this subgroup. For example, this subgroup contains representatives of both classes $27A$ and $27B$, lying above class $9A$ in Suz. By finding elements of this type, and explicitly calculating their centralizers, it was possible to find representatives of classes $27A$ and $27B$, since they have different centralizer orders in the Monster.

7. Shortening words

As is well-known, the main difficulty in computing with a group whose elements are given as words is in preventing the words getting too long. We were able to find two tricks which in combination overcome this obstacle in most cases. The first trick takes two commuting $2B$-involutions, and produces a short word conjugating one to the other. The second trick is a method of rewriting a word known to be in the involution centralizer $2^{1+24} \cdot Co_1$, as a word of length 1.

To take the second part first, note that if we find a word in the generators, representing an element which commutes with the original $2B$-element, then it belongs to the original subgroup $2^{1+24} \cdot Co_1$. Therefore it can be written in 'standard' form (in two ways) as a combination of a 24×24 matrix, a 4096×4096 matrix, a monomial permutation on 98280 points, and a 298×298 matrix. This standard form can be determined by calculating just 36 rows of the full 196882×196882 matrix for this element, so can be obtained fairly quickly. Moreover, if necessary we can even express this standard form as a word in the original generators for the subgroup.

The first trick relies on the fact that all $2B$-elements in $2^{1+24} \cdot Co_1$ can be obtained from the central involution by a subset of the operations: (1) conjugate by the triality element to take it to a non-central involution of 2^{1+24}, (2) conjugate by a random element of $2^{1+24} \cdot Co_1$, (3) conjugate by the triality element again to move it outside 2^{1+24}, and (4) conjugate again by a random element of $2^{1+24} \cdot Co_1$. Thus to conjugate an arbitrary $2B$-element in this group to the central involution, it suffices to conduct two random searches to find the correct conjugating elements to reverse the above operation.

Combining these tricks with Ryba's method for conjugating an involution in a group to an involution in a known subgroup [14], we can in principle shorten any word to one of length less than about 20. Specifically, given an arbitrary element g which powers to a $2B$-element x, there is a good chance that xz will power to a $2B$-element y, where z is our original $2B$-element. Since x and z both centralize y, we can use the first trick to conjugate y to z, say $y^{w_1} = z$ where w_1 has length at most 4. Using the trick again, we can conjugate x^{w_1} to z, say $x^{w_1 w_2} = z$ where $w_1 w_2$ has length at most 8. We then use the second trick to write $g^{w_1 w_2}$ as a word of length 1, and thus obtain a word of length at most 17 for g. More generally, if h is an arbitrary word, we can multiply it by a random word of short length (preferably length 1) until we find an element g satisfying the above hypotheses. This is likely to produce a word of length at most 18 for h.

8. Maximal subgroups

A great deal of theoretical work on classifying the maximal subgroups of the Monster has been done in [17, 18, 19, 23], which reduced the problem to classifying conjugacy classes of simple subgroups of just 22 isomorphism types, subject to a variety of other conditions. In her PhD thesis [7] Beth Holmes dealt with 11 of the 22 isomorphism types, namely $L_2(q)$ for $q = 9, 11, 19, 23, 29, 31, 59, 71$ and $L_3(4)$, $U_4(2)$ and M_{11}. Since then she has completed the cases $L_2(q)$ for $q = 7, 8, 16, 17, 27$, and $L_3(3)$, $U_3(3)$, and $U_3(4)$. This leaves just the cases $L_2(13)$, $U_3(8)$ and $Sz(8)$.

The only really effective method of classifying such simple subgroups in a computational setting is to choose an abstract amalgam generating the desired isomorphism type of subgroup, and to classify all embeddings of that amalgam in the Monster. We then look at each embedding to decide whether it indeed generates a subgroup of the required isomorphism type.

The most successful calculation of this type has been the classification of subgroups generated by two copies of A_5 intersecting in D_{10} (see [7]). This amalgam can generate $L_2(q)$, for any $q \equiv \pm 1 \pmod 5$, as well as $L_3(4)$, so this deals with eight of the required cases. In particular, we found four new maximal subgroups by this method, including subgroups isomorphic to $L_2(59)$ and $L_2(71)$, thus answering a long-standing question. In addition, we found new maximal subgroups $L_2(29){:}2$ and $L_2(19){:}2$. (In fact, the $L_2(29)$ case was done by a different method, but with hindsight it would have been easier to use this method.)

Four more of these cases, namely $L_2(7)$, $L_2(17)$, $L_3(3)$ and $U_3(3)$, were dealt with by an amalgam of two copies of S_4, intersecting in D_8 (see [8]). The case $U_3(4)$ used a subgroup $5 \times A_5$, extending a diagonal C_5 (there are two classes, so both need to be considered) to D_{10}. In the case $L_2(8)$ we can assume the 7-element is in class $7B$, so from the 2-local analysis [17] we know the $2^3{:}7$ centralizes a $2B$-element, and most of the calculation can then be done inside the corresponding subgroup $2^{1+24}{\cdot}Co_1$.

The case $L_2(27)$ relies on an amalgam of $3^3{:}13$ and D_{26} intersecting in 13, and the fact that there are just two classes of $3^3{:}13$ in the Monster (this follows fairly easily from the results of [23]). In one case a simple counting argument shows that there is no such $L_2(27)$, while in the other case we needed to check a handful of cases computationally. In particular, there is no subgroup isomorphic to $L_2(27)$ in the Monster, which answers another long-standing question.

Regarding the three outstanding cases, $L_2(13)$, $U_3(8)$ and $Sz(8)$, our computers are currently working through the cases for $L_2(13)$. After that, the case of $U_3(8)$ should present no serious problems. Our strategy in this case is to take a subgroup $3 \times L_2(8)$, and extend one of the diagonal elements of order 9 to a D_{18}.

The final case, $Sz(8)$, is proving more tricky. The only approach we can think of is to start with a group $2^3{:}7$ and extend a 7-element to D_{14}. We can use the fact that $Sz(8)$ contains $2^{3+3}{:}7$ to reduce the number of possibilities for the $2^3{:}7$. Nevertheless, it is not easy to classify these subgroups. We know that the involutions are in class $2B$. Now there are three classes of $2B$-pure subgroups of order 4, whose normalizers involve composition factors M_{24}, M_{12} and A_8 respectively. A fairly easy counting argument shows that the first of these cannot occur in a putative subgroup $Sz(8)$.

In the second case, the normalizer of the 4-group has the shape $(2^2 \times 2^{1+20}){\cdot}(S_3 \times M_{12}{:}2)$ inside $2^{1+24}{\cdot}3{\cdot}Suz{:}2$ inside $2^{1+24}{\cdot}Co_1$. Now in $Sz(8)$ we have $2^{3+3}/2^2 \cong 4 \circ Q_8$, which embeds uniquely (up to conjugacy) in $M_{12}{:}2$. In this embedding the central involution is of M_{12}-class $2B$. Thus the 2^3 we are looking for is either entirely inside 2^{1+24}, or maps to a $2B$-element in M_{12}. In the former case, the whole of $2^{3+3}{:}7$ must lie inside $2^{1+24}{\cdot}Co_1$, and it is straightforward to show that this does not happen. In the latter case it turns out that the $2^3{:}7$ lies in the maximal subgroup $2^3.2^6.2^{12}.2^{18}.(L_3(2) \times 3S_6)$, with the 2^3 lying in the normal $2^3.2^6.2^{12}$ but not in the $2^3.2^6$. It can be shown that it is unique up to conjugacy. At this stage it seems to be necessary to resort to computer calculations.

A similar analysis of the third type of $2B^2$ is in progress.

9. Explicit representations of subgroups

The Monster contains many interesting subgroups, which it may be useful to study independently. To facilitate such study, we have tried to construct small representations of these groups, whenever such representations exist [4]. These representations are available from the Monster page of [26]. In many cases one of these subgroups may be described as a certain non-split extension of a group acting (not necessarily faithfully) on a module. While previous constructions have concentrated on representing p-local subgroups irreducibly in characteristic different from p, the smallest faithful (reducible) representations are usually to be found in characteristic p. John Bray has developed effective methods of constructing such non-split extensions explicitly by gluing together indecomposable (but reducible) modules for the quotient group. Various techniques are then employed to ensure that the group constructed is indeed isomorphic to the desired subgroup of the Monster.

In two of the larger cases, namely the 3-local subgroups $3^2.3^5.3^{10}.(M_{11} \times 2S_4)$ and $3^3.3^2.3^6.3^6.(L_3(3) \times SD_{16})$, we felt that the only reliable method of ensuring that we obtained a group of the right isomorphism type was to find it explicitly as a subgroup of the Monster. We then employed ad hoc techniques to try to find some smaller representations—in this case permutation representations.

To date we have representations of all the maximal subgroups except some of the 2-local subgroups. The latter do not appear to have faithful permutation representations of reasonable degree, and new methods will be required for these cases.

10. Character tables

Richard Barraclough has calculated the character table of the group $3^{1+12}.2.Suz:2$ used in some of our constructions of the Monster, along with various closely related groups [2]. There are many subtleties which make this calculation difficult, not the least of which is the fact that there are two non-isomorphic groups of this shape, whose character tables look very similar.

It would be interesting to have the character tables of other maximal subgroups. From the representations described in the previous section, it should be possible to calculate some of these character tables without difficulty. However, the larger subgroups still present a formidable challenge.

11. Nets and their classification

Norton has generalised the ideas of Moonshine to commuting pairs of elements of the Monster, introducing functions F which are invariant under the action of the modular group via $F(g, h) = F(g^\alpha h^\beta, g^\gamma h^\delta)$ when $\alpha\delta - \beta\gamma = 1$. This even makes sense for non-commuting elements g and h, in the case when

$g = ab$ and $h = bc$, and a, b, c are involutions. In this case, the action of the modular group corresponds to an action of the three-string braid group on triples of involutions.

In the case when a, b, c are in class $2A$, there are about 1.4×10^6 conjugacy classes of triples (a, b, c), which fall into about 14,000 orbits under the action of the braid group. These orbits are (roughly speaking) what Norton calls 'nets': they have a combinatorial structure of a polyhedron of genus 0 or 1. A complete classification of these nets would be of great interest in clarifying and developing the ideas of generalised moonshine.

There are various ways of dividing up the set of nets into more manageable subsets, for example according to the product abc, or the group generated by a, b, c, or the centralizer of a, b, c. So far, Richard Barraclough has a complete classification of the nets which are centralized by any element of prime order bigger than 3, and is working on the ones centralized by an element of order 3 [1].

The classification of nets with trivial centralizer will be difficult, however. Ultimately it requires calculating the orbits of certain groups on the nearly 10^{20} involutions in class $2A$. This is a major challenge for the future.

12. A presentation for the Monster, and a new existence proof?

Norton has shown how to produce a presentation for the Monster on generators closely related to the 2-local subgroups we used in one of our constructions. The proof of this presentation, however, requires deep arguments. We hope to be able to verify that certain elements in our group satisfy the relations of this presentation. It may then be possible to provide for the first time a computational proof of existence of the Monster, independent of Griess's proof.

References

[1] R. W. Barraclough, Some calculations related to the Monster group, Ph. D. thesis, The University of Birmingham, UK, 2005.

[2] R. W. Barraclough and R. A. Wilson, The character table of a maximal subgroup of the Monster, *LMS J. Comput. Math.* **10** (2007), 161–175.

[3] R. W. Barraclough and R. A. Wilson, Conjugacy class representatives in the Monster group, *LMS J. Comput. Math.* **8** (2005), 205–216.

[4] J. N. Bray and R. A. Wilson, Explicit construction of maximal subgroups of the Monster, *J. Algebra* **300** (2006), 835–857.

[5] J. H. Conway, R. T. Curtis, S. P. Norton, R. A. Parker and R. A. Wilson, *An ATLAS of Finite Groups*, Oxford University Press, 1985.

[6] R. L. Griess, Jr., The friendly giant, *Invent. Math.* **69** (1982), 1–102.

[7] P. E. Holmes, Computing in the Monster, Ph. D. thesis, The University of Birmingham, UK, 2001.

[8] P. E. Holmes, A classification of subgroups of the Monster isomorphic to S_4 and an application, *J. Algebra* **319** (2008), 3089–3099.

[9] P. E. Holmes and R. A. Wilson, A new computer construction of the Monster using 2-local subgroups, *J. London Math. Soc.* **67** (2003), 349–364.

[10] P. E. Holmes and R. A. Wilson, A new maximal subgroup of the Monster, *J. Algebra* **251** (2002), 435–447.

[11] P. E. Holmes and R. A. Wilson, $PSL_2(59)$ is a maximal subgroup of the Monster, *J. London Math. Soc.* **69** (2004), 141–152.

[12] P. E. Holmes and R. A. Wilson, On subgroups of the Monster containing A_5's, *J. Algebra* **319** (2008), 2653–2667.

[13] P. E. Holmes and R. A. Wilson, The maximal subgroups of the Monster, in preparation.

[14] P. E. Holmes, S. A. Linton, E. A. O'Brien, A. J. E. Ryba and R. A. Wilson, Constructive membership testing in black-box groups, *J. Group Theory*, to appear.

[15] S. A. Linton, R. A. Parker, P. G. Walsh and R. A. Wilson, Computer construction of the Monster, *J. Group Theory* **1** (1998), 307–337.

[16] Computational Algebra Group, School of Mathematics and Statistics, University of Sydney, *The Magma Computational Algebra System for Algebra, Number Theory and Geometry*, 2005. (http://magma.maths.usyd.edu.au/magma/).

[17] U. Meierfrankenfeld and S. V. Shpektorov, The maximal 2-local subgroups of the Monster and Baby Monster, Preprint.

[18] S. P. Norton, Anatomy of the Monster, I, in *The Atlas of Finite Groups Ten Years On (ed. R. T. Curtis and R. A. Wilson)*, 198–214. Cambridge University Press, 1998.

[19] S. P. Norton and R. A. Wilson, Anatomy of the Monster, II, *Proc. London Math. Soc.* **84** (2002), 581–598.

[20] R. A. Parker, The computer calculation of modular characters (The 'Meat-axe'), in *Computational Group Theory (ed. M. D. Atkinson)*, Academic Press, 1984, pp. 267–274.

[21] R. A. Parker and R. A. Wilson, Computer construction of matrix representations of finite groups over finite fields, *J. Symbolic Comput.* **9** (1990), 583–590.

[22] M. Ringe, *The C Meat-axe 2.3, documentation*, RWTH Aachen, 1995.

[23] R. A. Wilson, The odd-local subgroups of the Monster, *J. Austral. Math. Soc. (A)* **44** (1988), 1–16.

[24] R. A. Wilson, The Monster is a Hurwitz group, *J. Group Theory* **4** (2001), 367–374.

[25] R. A. Wilson, Construction of the Monster over $GF(7)$, and an application. Preprint 2000/22, School of Mathematics and Statistics, The University of Birmingham.

[26] R. A. Wilson, S. J. Nickerson and J. N. Bray, A world-wide-web Atlas of Group Representations, http://brauer.maths.qmul.ac.uk/Atlas/

[27] R. A. Wilson, Computing in the Monster, in *Groups, Combinatorics and Geometry (ed. A. A. Ivanov, M. W. Liebeck and J. Saxl)*, 327–335. World Scientific, 2003.